Sedimentary Rocks

Melissa Stewart

www.heinemann.co.uk/library
Visit our website to find out more information about Heinemann Library books.

To order:
 Phone 44 (0) 1865 888066
 Send a fax to 44 (0) 1865 314091
Visit the Heinemann Bookshop at www.heinemann.co.uk/library to browse our catalogue and order online.

First published in Great Britain by Heinemann Library, Halley Court, Jordan Hill, Oxford OX2 8EJ a division of Reed Educational and Professional Publishing Ltd. Heinemann is a registered trademark of Reed Educational and Professional Publishing Ltd.

OXFORD MELBOURNE AUCKLAND JOHANNESBURG BLANTYRE
GABORONE IBADAN PORTSMOUTH (NH) USA CHICAGO

Produced for Heinemann Library by Editorial Directions
Designed by Ox and Company
Originated by Ambassador Litho Ltd
Printed in Hong Kong

ISBN 0 431 14375 7
06 05 04 03 02
10 9 8 7 6 5 4 3 2 1

British Library Cataloguing in Publication Data
Stewart, Melissa
Sedimentary rocks. – (Rocks and minerals)
1. Rocks, Sedimentary – Juvenile literature
I. Title
552.5

Acknowledgements
The Publishers would like to thank the following for permission to reproduce photographs:

Photographs ©: Cover background, EyeWire; cover foreground, David Johnson/Reed Consumer Books, Ltd.; p. 4 top, Jeffrey Meyers/Frozen Images/The Image Works; p. 4 bottom, Ric Ergenbright/Corbis; p. 5, Ann & Rob Simpson/Simpson's Nature Photography; p. 8 top, Ross Frid/Visuals Unlimited, Inc; p. 8 bottom, Sharon Gerig/Tom Stack & Associates; p. 9, Doug Sokell/Tom Stack & Associates; p. 10, Sven Martson/The Image Works; p. 11, NASA/TSADO/Tom Stack & Associates; p. 13, A.J. Copley/Visuals Unlimited, Inc.; p. 14, Tom Bean; p. 15, Grace Davies Photography; p. 16, Adam Woolfitt/Corbis; p. 17, Charles & Judy Walker/Liaison International/Hulton Archive; p. 18, James P. Rowan; p. 19, Grace Davies Photography; p. 21, Judyth Platt/Ecoscene/Corbis; p. 22, Gianni Dagli Orti/Corbis; p. 23, Mike Okoniewski/Gamma Liaison/Hulton Archive; p. 24, James P. Rowan; pp. 25, 26, Tom Bean; p. 27, Jonathan Blair/Corbis; pp. 28, 29, B. Daemmrich/The Image Works.

Our thanks to Alan Timms and Martin Lawrence of the Natural History Museum, London for their assistance in the preparation of this edition.

Contents

Any words appearing in the text in bold, **like this**, are explained in the Glossary.

What is a rock?

It is quite surprising to find that blackboard chalk is made of rock. Chalk is made of limestone, a sedimentary rock formed by the shells and skeletons of sea animals.

Take a look around your classroom at school. How many kinds of rock do you see? Blackboard chalk is made from rock, and so is the concrete in the walls and the glass in the windows. The paper you write on is made from trees, but it also contains a little bit of ground-up rock.

Chalk, concrete, glass and paper all contain sedimentary rock. Sedimentary rock is one of three groups of rocks found in the world. The others are **igneous rocks** and **metamorphic rocks**. Each kind of rock forms in a different way, but all rocks are made of **minerals**.

Rocks come in all shapes and sizes. The chalky White Cliffs of Dover (right) along the southeastern coast of England are formed out of sedimentary rock made of minerals from sealife.

SCIENCE IN ACTION

Petrologists – scientists who study rocks – can identify a rock by knowing where it came from and by looking at the **properties** of its minerals. For example, petrologists examine the colour, the shininess and hardness of the minerals in a rock. They also study the size, shape and arrangement of the crystals.

Minerals

A mineral is a natural solid material. No matter where you find it, a mineral always has the same chemical makeup and the same structure. In other words, the **atoms** that mix together to form a mineral always arrange themselves in the same way. Most minerals have a **crystal** structure. Crystals usually have a regular shape and smooth, flat sides called **faces**.

These limestone cliffs in a river valley in the US state of Virginia are mostly calcite. Pure calcite is clear or white, but other minerals in limestone give it colour.

Limestone

Limestone is a sedimentary rock that usually contains one or more of the minerals calcite, dolomite and aragonite. The crystal structure of calcite is made up of calcium, carbon and oxygen atoms that are always arranged in the same way. Calcit and other minerals can sometimes be seen as 'fur' or 'scale' in kettles, boilers and pipes. The fur is deposited when water containing the minerals is boiled.

Land on the move

Sedimentary rock is the most common kind of rock on the Earth's surface. But just a few miles underground, the Earth's makeup changes. Below the Earth's solid **crust** is the **mantle** – a layer of hot, soft rock in its liquid form, known as **magma**. The mantle surrounds the Earth's central **core**. The outer core is made of melted metals, but the inner core is solid. The Earth's central core is sizzling hot – more than 5000°C. As heat from the solid metallic core moves first into the liquid outer core, and then into the mantle, the hottest magma moves upward toward the surface. Meanwhile, cooler magma at the top of the mantle moves down to take its place. Over millions of years, magma slowly circles through the Earth's mantle.

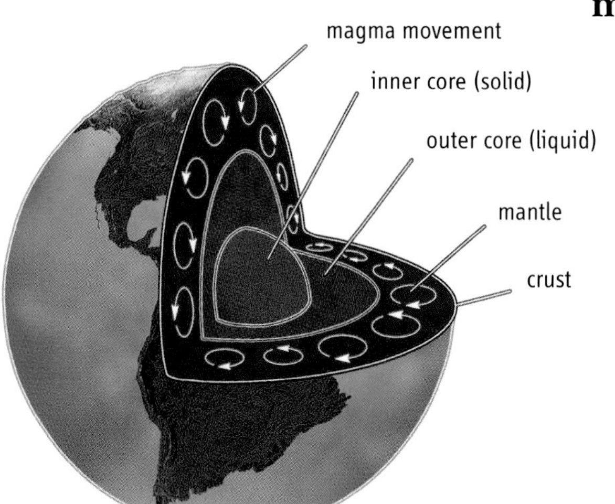

The Earth is made up of layers. The thin outer layer of the Earth is the crust. The next layer, the mantle, is made of magma that is constantly moving. The core is made of an outer liquid layer and an inner solid core.

Tectonic plates

The Earth's crust is made up of giant slabs of rock called **tectonic plates**. These plates fit together like the pieces of a jigsaw puzzle. As magma swirls within the mantle, the plates move, too. Over time, this process reshapes the land and moves it from place to place.

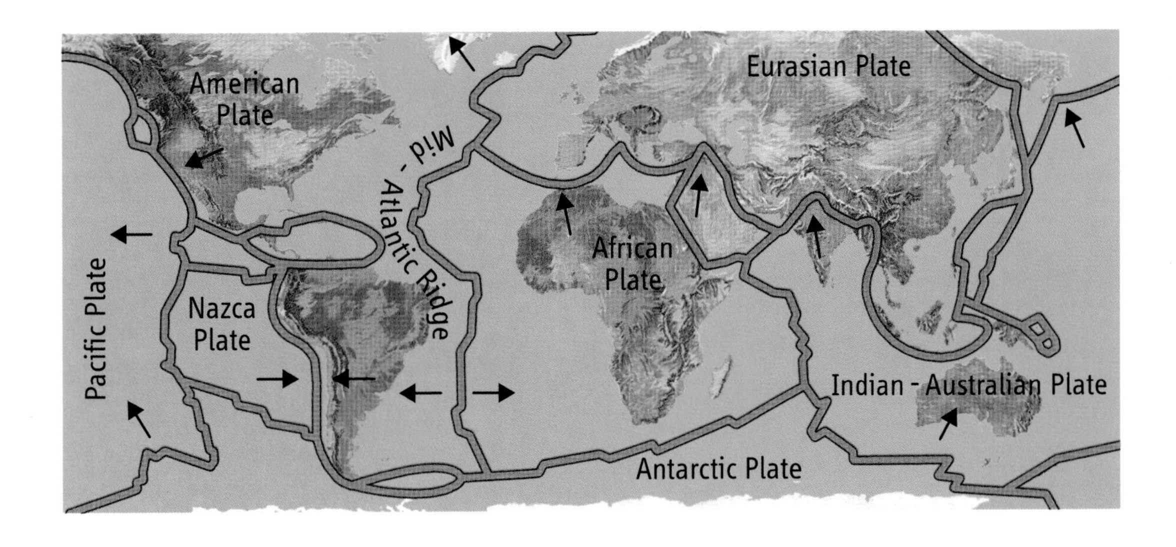

Earth's surface is broken into many tectonic plates. The major plates are shown on this map. The plates are moving constantly, though very slowly, in the direction of the arrows. The Mid-Atlantic Ridge is a gap, called a rift, formed by two plates moving apart.

Continents on the move

Scientists have evidence that the Earth formed about 4.6 billion years ago, but it has looked the way it does today for only about 65 million years. Before that, the continents were in different places. About 1.1 billion years ago, the Earth had one large continent known as Rodinia. The centre of Rodinia was close to the South Pole.

Then, about 600 million years ago, Rodinia broke into three large pieces, and most of the land drifted north. Nearly 180 million years ago, all of the Earth's land formed a second giant continent known as Pangaea. As time passed, the plates continued to travel across the Earth's surface to form the continents we know today – and they are still on the move. Perhaps, one day, all of the Earth's land will come together again and form a third giant continent.

Three groups of rocks

The Earth has three major groups of rocks – sedimentary, **igneous** and **metamorphic**. Each kind of rock forms in a different way. Sedimentary rocks are made of layers of mud, clay and sand that have been compressed and stuck together over time. Limestone, shale, graywacke, siltstone, sandstone, breccia, chert, flint and rock salt are all examples of sedimentary rock.

Shale (above) is a sedimentary rock formed by layers of clay and mud. It feels smooth and greasy and is soft enough to scratch. Basalt (below) is a kind of igneous rock. It forms when lava cools and hardens.

Igneous rocks form when **magma** from the Earth's **mantle** or lower crust cools and forms crystals. Sometimes the magma forces its way up to the Earth's surface and spills onto the land through a **volcano**. This kind of igneous rock cools quickly and has very small **crystals**. In other cases, pools of magma are trapped at some depth underground and cool slowly over thousands of years.

Some of the largest and most beautiful crystals in the world were formed in this way. Granite, gabbro, basalt and obsidian are examples of igneous rock.

DID YOU KNOW?

The word 'sedimentary' means 'related to or containing **sediments**'. The word 'sediment' comes from a Latin term meaning 'settling' or 'sinking'.

Heat and pressure

Metamorphic rocks form when heat or pressure changes the **minerals** within sedimentary rock, igneous rock or another metamorphic rock. This often happens when the Earth's **tectonic plates** collide and push up tall mountain ranges. Metamorphic rock also forms when a stream of magma bursts into the **crust** and cooks the surrounding rock. Marble, slate and gneiss are examples of metamorphic rock.

Gneiss is a metamorphic rock. The beautiful bands that run through it form as granite or schist is heated, twisted and squeezed deep under the ground.

WHAT A DISCOVERY!

In 1785, a Scottish doctor named James Hutton noticed layers of sedimentary rock in a riverbank near his home. At that time, most scientists thought the Earth was about 6000 years old. But Hutton thought it must have taken much longer for these layers of rock to build up. He suggested that the Earth might be more than a million years old. Many scientists laughed at Hutton's theory, but today we know that the Earth is about 4.6 billion years old.

Wear and tear

As this tree grows around a large rock, the rock slowly undergoes weathering. On the ground, you can see pieces that have already broken off.

Many rocks are so hard that they may seem indestructible. But over time, a number of different forces can cause them to wear away or break apart.

Erosion

The force of crashing ocean waves, fast-flowing streams, whipping winds and gigantic glaciers can slowly wear away, or **erode**, even the hardest of rocks.

Weathering

Rocks can also be broken down by a process called **weathering**. In places with hot days and cool nights, rock expands and shrinks again and again.

DID YOU KNOW?

Sedimentary rocks erode and weather more quickly than other types because they are softer and are arranged in layers that break apart easily.

Eventually, the rock weakens and crumbles. In addition, when water flows into a rock's nooks and crannies and then freezes, the rock may shatter. Plant roots can grow into cracks in rocks and slowly split them apart. **Acid rain** and snow are also tough enough to destroy some rocks. So are chemicals released by some tiny creatures that live in the soil.

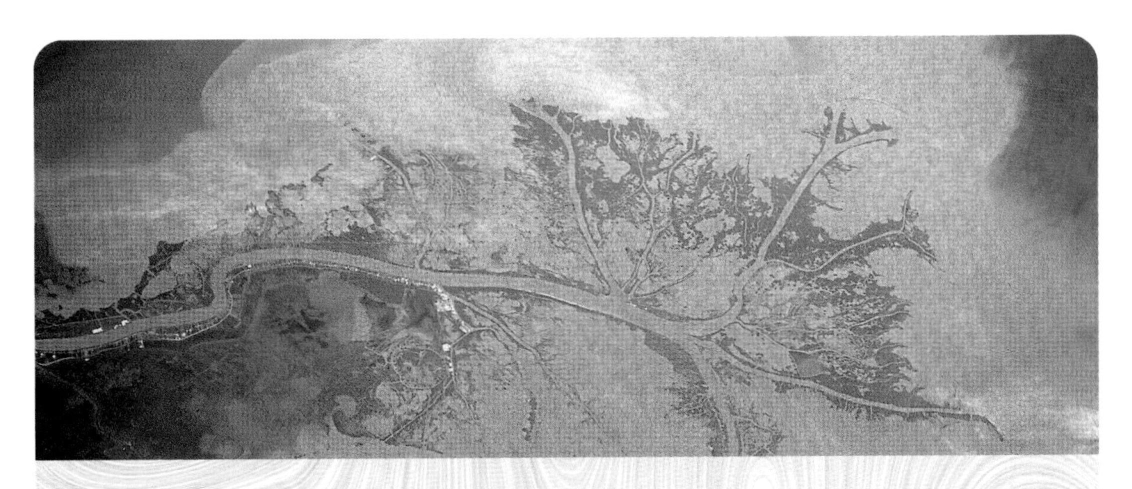

THE POWER OF WATER

It is quite surprising that water has the power to break down most of the world's rocks. Actually, water that is still doesn't do much damage, but as soon as water starts to move, everything changes. To feel this for yourself, turn on your bathroom tap and place your hand underneath! As **gravity** tugs on the water in mountain streams, the water picks up speed. The faster the water moves, the more force it has. Also, the faster a stream moves, the more **sediments** it can hold. In fact, if a stream's speed doubles, it can carry four times more material.

Eventually, the stream empties into a larger river. When a river's raging waters empty into the sea, they suddenly encounter strong ocean currents moving in different directions. The river water slows down and all its sediments drop down to the ocean floor. The Mississippi River dumps about 220 million tonnes of new sediments into the Gulf of Mexico every year (above). If you look at a photo of the Mississippi River taken from space, you can see all the sediments. You can also see how the build up of sediment blocks the river, causing the water to fork out in many directions and form a **delta**.

How does sedimentary rock form?

Each year, rivers dump tonnes of sand, mud and tiny pebbles into the Earth's oceans. As microscopic sea creatures die, their tiny skeletons and shells also fall to the ocean floor. The action of the waves sorts all these **sediments** into horizontal layers, or beds. The largest, heaviest materials fall through the water more quickly, so they form the bottom layers. Layers of lighter materials form on top of them.

Sedimentary rock forms slowly over millions of years as sediments slowly build up, compact and harden. Then **erosion,** earthquakes or other natural forces may lift or shift the layers so they rise to the surface.

Layers under pressure

Over time, these materials build up. The weight of the sediments at the top presses down on the lower layers. All that pressure squeezes out the water and cements the sediments together to form rock. The sediments are glued together by some of the **minerals** from the rocks. If you look closely at a sedimentary rock, you may be able to see its layers.

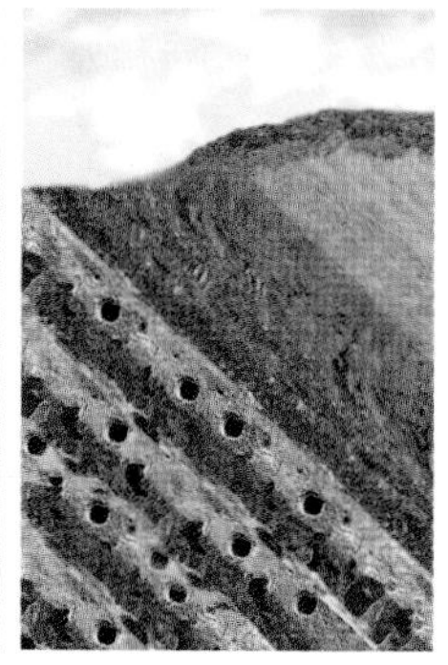

Different kinds of sediments form different kinds of sedimentary rock. For example, conglomerate and breccia contain a variety of different-sized materials. Sandstone is made mostly of sandy

SEDIMENTS IN SEDIMENTARY ROCK

SEDIMENTARY ROCK	SEDIMENTS INSIDE
Breccia	A variety of stones and pebbles with sharp edges
Conglomerate	A variety of stones and pebbles with rounded edges
Diatomite	Shells of tiny marine creatures called **diatoms**
Limestone	Shells of marine creatures, especially **foraminifers**
Rock salt	Salt from the ocean
Sandstone	Sand
Shale	Mud or clay

sediments. Smaller particles are the main ingredients of siltstone and shale. Rock salt is made up of salt from the ocean. Most of the shell sediments that make up limestone and diatomite came from animals that are smaller than the head of a pin.

Conglomerate (above) is formed when pebbles of sedimentary rock with rounded edges become cemented together. Some conglomerates are used in the road-making process.

SEE FOR YOURSELF

To find out how wave action sorts rocky material by size and weight, add a few spoonfuls of pebbles, sand and soil to a large glass jar. Then fill the jar with water, screw on the lid tightly, grasp the jar with both hands and shake it up. Place the jar on a flat surface and let the sediments settle overnight. The next morning, you should see that the sediments have arranged themselves in layers, or beds. The largest and heaviest materials should be on the bottom of the jar, while the smaller, most lightweight materials should make up the top layer.

Sedimentary stories

Sedimentary rock tells a story about how and when it formed, and what weather conditions were like at the time. Ripples in sandstone show which way the wind was blowing or the water was moving when the materials were deposited. For example, rock formations in Zion National Park, Utah, USA, show how drifting sand dunes shifted over time as winds changed.

In the US, the Grand Canyon formed over millions of years as the Colorado River **eroded** layer after layer of sedimentary rock – more than 2000 metres of rock in all. Over twenty different layers of sandstone and other kinds of sedimentary rock can be seen in the walls of the Grand Canyon. By studying the canyon's colourful walls, scientists can find clues about how the temperature and climate of the area has changed over time.

The Grand Canyon in the USA is an amazing example of how rock layers form over time. The Colorado River began carving the canyon about six million years ago, but scientists think that the rock layers at the bottom may be a billion years old!

Many sedimentary rocks contain **fossils**. Fossils are the remains of plants, animals and other living things. Ancient animal tracks and trails, eggs and nests have also formed fossils.

Fossils

At one time, this fossil of a fish was buried deep underground. As the land changed over time, it eventually came to the surface.

Plants and animals die all the time, but most do not become fossils. Fossils are rare because conditions must be just right for them to form. Body parts can become fossils only if a dead creature falls into the water or is buried quickly by **sediments**. Otherwise, the corpse will be eaten by animals or rot because it is exposed to air.

Most sedimentary rock forms in shallow seas near the continents. That is why fossils of sea creatures are quite common. Sediments may also build up on lake beds, in deserts, along river valleys or at the foot of large mountain ranges.

Each fossil find can help scientists learn about an ancient creature and its habitat. By studying many fossils, palaeontologists – scientists who study prehistoric life by examining fossils – can piece together how life on the Earth has **evolved** and how the Earth has changed over time.

DID YOU KNOW?

Long ago, people had no idea what fossils were or how they formed. Some people claimed mammoth tusks were horns from unicorns. Others believed fossils of ancient sea creatures called ammonites were coiled snakes that had been turned to stone by a powerful enchantress. A few even thought the fossils of oyster shells were the devil's toenails.

Where on Earth is sedimentary rock?

No matter where in the world you go, it is easy to find sedimentary rock. It is found on land on the surface of the Earth because, over time, ancient seas have disappeared and forces inside the Earth have lifted the land.

The Yorkshire Dales in England feature many limestone cliffs and caves. The sedimentary layers were laid down about 300 million years ago, and were exposed as glaciers moved over the land during the Ice Ages.

The Rock of Gibraltar, located near the south of Spain, is made of limestone. So are the Yorkshire Dales in northern England and an area in China called the Stone Forest. Uluru (Ayer's Rock) may be the most famous natural sedimentary structure in the world. Located in the Australian outback, this red sandstone formation rises 348 metres above the desert floor and is about 9.5 kilometres wide.

DID YOU KNOW?

About 130 million years ago, an ancient sea covered much of what is now the South Downs in the southeast United Kingdom. These chalky downlands built up under this sea from sediments made of shells and other fragments. The chalk is very pure, being 98 per cent calcium carbonate.

Scientists believe that at least two-thirds of the giant rock is still underground. The aboriginal peoples of Australia believe Uluru is a sacred place, and over many years have painted animals and other scenes on the walls of **caves** within the formation.

Travertine

Beautiful **travertine** terraces drape the land in an area of Turkey known as Pamukkale, which means 'cotton castle'. In Yellowstone National Park, a travertine structure known as Minerva Terrace looks like a frozen waterfall. This rocky structure grows and changes almost as quickly as a real waterfall. A layer of travertine about 30 centimetres thick is added to Minerva Terrace each year.

Water trickles over Pamukkale in Turkey. The 'falls' are made of travertine, a **mineral** deposited by water from hot and cold springs.

At Petrified Forest National Park in Arizona, USA, large tepee-shaped hills of sandstone and siltstone show signs of thousands of years of erosion. The nearby Painted Desert has more than a dozen colourful layers of limestone, sandstone and shale.

A ROCKY LIFE

The creatures that make up limestone and diatomite are still common on the Earth today, but they are so small that you need a microscope to see them. **Foraminifers** and **diatoms** are neither plants nor animals. They belong to a group of living things called **protists**.

A look at caves

A cave is a natural opening in the ground that is large enough to hold a person. Most caves form when water slowly dissolves sedimentary rock, such as limestone. As water flows through the ground, it picks up bits of carbon dioxide and forms a weak acid. When the water seeps through cracks between layers of limestone, the acid slowly dissolves the rock. Over time, the cracks grow wider and deeper. Eventually, large caverns and passages form. Years later, the land is lifted up, most of the water drains out, and the passages fill with air.

Deep inside most limestone caves, **mineral**-rich waters drip off the ceilings and flow along the floors. When that water **evaporates**, a kind of limestone called **travertine** is left behind.

Caves form slowly as water dissolves soft sedimentary rock (centre of picture). After creating sinkholes at the surface, the water trickles downward and wears away the weakest layers (at right of picture).

DID YOU KNOW?

The deepest cave in the world is Voronja Cave in the Georgian Republic. It plunges more than 1670 metres below the Earth's surface. Mammoth Cave in Kentucky, USA is the longest cave in the world. Its maze of passages is more than 550 kilometres (340 miles) long.

Stalactites and stalagmites

Giant travertine structures form in some caves. Thin, icicle-shaped stalactites grow down from the ceiling, while stubby stalagmites grow up from the floor. Sometimes the two meet and join up, forming thick columns.

Cave colour

Other stone structures may cover the cave walls. Some are shaped like flowers, bubbles or pearls, while others look similar to pieces of coral or twisted strands of hair. These delicate features are usually white or grey, but they may be pink, orange, yellow or even blue. Their colour depends on the minerals that make up the sedimentary rock they are formed from.

Ingleborough Cave in Yorkshire, England, features dozens of beautiful stalactites and stalagmites. The colours are created by impurities in the minerals that formed the structures.

How do people use sedimentary rock?

Thousands of years ago, some early peoples used a sedimentary rock called flint to make tools. Flint is easy to break, and it has sharp edges. It was perfect for making axes to cut down trees, knives to skin animals and spades to make dugout canoes. When ancient peoples rubbed two pieces of flint together, the resulting spark could be used to start fires. Much later, the world's first firearms depended on flint for the sparks that ignited gunpowder.

These flint tools were used by people who lived on Earth thousands of years ago. You can see where the rock was chipped away to make sharp edges.

DID YOU KNOW?

People living near the spot where the Niagara River crosses the United States-Canadian border can enjoy the spectacular beauty of Niagara Falls any time. But they also depend on the falls for electricity and for money from the tourist trade.

Niagara Falls consists of two waterfalls – the Horseshoe Falls and the American Falls. Both formed where the river wears away at a 24-metre thick layer of hard rock.

Today we make window glass, crockery and spectacle lenses from ground sandstone and limestone. Ground limestone is sometimes added to paper and toothpaste. The ground rock gives these products their white colour.

Rock salt is a sedimentary rock that is mined from large underground beds that formed when ancient seas evaporated. We use ground-up rock salt to preserve and add flavour to the food we eat.

Rock salt

You probably can't imagine eating rocks, but one sedimentary rock can make your food taste better. Table salt is ground-up rock salt, and rock salt is a sedimentary rock. Rock salt comes from mines all over the world. It formed when ancient seas **evaporated**, leaving all their salt behind. As time passed, other **sediments** buried the salty sediments deep under the ground.

THE VALUE OF FLINT

Christina Rossetti (1830–1894) was a well-known British poet. She wrote the following poem about flint.

An emerald is green as grass;
A ruby red as blood;
A sapphire shines blue as heaven;
A flint lies in the mud.

A diamond is a brilliant stone,
To catch the world's desire.
An opal holds a fiery spark;
But only flint holds fire.

Sedimentary structures

The Pyramid of Kukulkan in Chichen Itza, Mexico, is made of limestone. It was built by the Maya about 1000 years ago.

Limestone has been a popular building material for thousands of years. It is attractive, readily available and easy to work with. The ancient Egyptians used it to build their great pyramids and the ancient Maya also constructed impressive pyramids from limestone. One of these structures, the Pyramid of Kukulkan, still stands today in Chichen Itza on the Yucatán Peninsula of Mexico.

St Paul's Cathedral in London is also made of limestone, as is Nôtre Dame, a famous cathedral in Paris, France. Today, builders use limestone blocks to construct many public buildings, such as schools and libraries.

Limestone is not the only popular sedimentary building material. Polished breccia was used to build some parts of the Opera House in Paris and the Pantheon in ancient Rome (in modern Italy).

Sandstone

The ancient Egyptians used sandstone, another sedimentary rock, to create the Great Sphinx more than 4500 years ago. This huge statue, with the head of a man and the body of a lion, may have been modelled on a ruler buried in a nearby pyramid. Some public buildings in modern towns and cities, such as courthouses and town halls, have also been built with sandstone blocks.

DID YOU KNOW?

Because sedimentary rock erodes more easily than igneous and metamorphic rock, parts of the Great Sphinx are missing. Its nose wore away long ago. The destructive forces of wind and water also severely damaged the Sphinx's beard, so it was removed. It is now on display in the British Museum in London.

ANCIENT ARCHITECTS

Cement is made of limestone that has been ground up and heated. When sand, gravel and water are added to cement, we call it concrete. Concrete was invented by the ancient Egyptians, but it was greatly improved by the Romans. The ancient Romans were expert architects and builders. They constructed great arenas, such as the Colosseum (above), impressive bridges, such as the Pont du Gard in France, and extraordinary temples, such as the Pantheon. All of these structures are made of a mixture of concrete and brick that has been covered with more attractive rock.

The rock cycle

Over millions of years, waves and ocean currents along the coast of Australia have carved these structures, called the Twelve Apostles. The giant pillars are made of limestone.

Rocks are always changing. Forces deep underground and on the Earth's surface destroy some rocks and, at the same time, create new ones. But the process takes place so slowly that we often don't even notice it.

Erosion and **weathering** slowly break down even the hardest rocks on the Earth. The tiny rocky bits are picked up by streams and rivers. Each day, rivers dump tons of **sediments** into the Earth's oceans. At the same time, ocean currents and waves also wear down rocky coastlines. Over thousands of years, sediments build up. As the weight of the upper layers presses down on materials below, the rock compacts and sticks together to form giant beds of sedimentary rock.

Movement and change

As the Earth's **tectonic plates** move, some of the sedimentary rock is twisted and folded. When **minerals** in the rock break down or change, **metamorphic rock** forms. Shifting plates pull other regions of sedimentary rock into the **mantle**, where the rock melts into **magma**.

Eventually, some of the magma spills onto the land and cools to form **igneous rock**. Pools of magma also become trapped where the **crust** and mantle meet. As it cools, it too becomes igneous rock. As even more time passes, the new igneous and metamorphic rocks also break down and wear away. Eventually, they will also be carried back to the ocean and form new layers of sedimentary rock. The result is a never-ending process – all the minerals that make up the Earth's rock are continually recycled as the rocks themselves are created and destroyed.

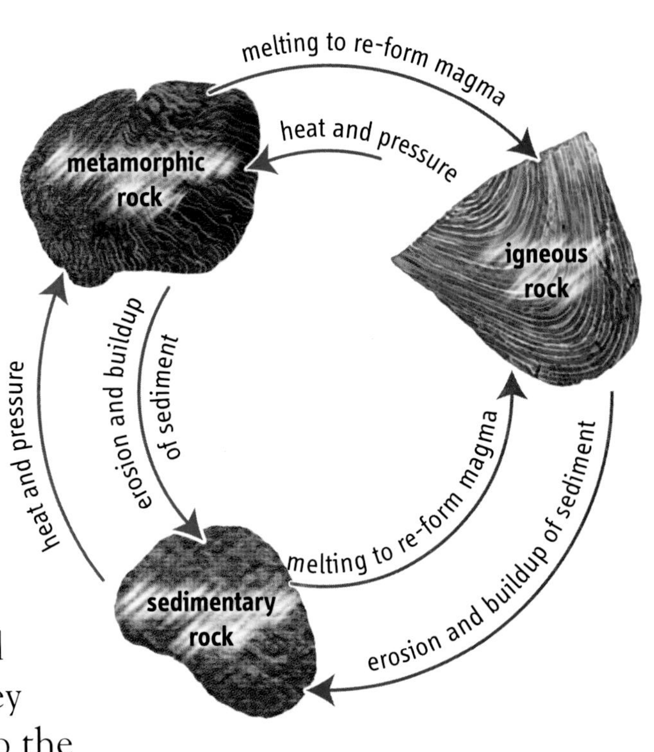

The rock we see on the Earth today has not always been here. Rock forms and is broken down in a continuous cycle.

DID YOU KNOW?

Wear and tear from the wind and waves has formed natural bridges, rock arches and **blow holes** in many parts of the world. At Arches National Park in Utah, USA, shown here, impressive sandstone formations crisscross the land. One well-known blow hole is found on one of the Galápagos Islands off the shore of Ecuador. It is called Nature's Toilet because the water flowing in and out looks similar to a flushing toilet.

Identifying sedimentary rocks

Colourful rock layers in the Painted Desert in Arizona, USA suggest that they are made of sedimentary rock. The rock was lifted to the surface millions of years ago and has been eroded by wind and water ever since.

Geologists can identify a rock by studying its **minerals** closely. Are the minerals hard or soft? Are they shiny or dull? Are their **crystals** large or small? Luckily, you do not need to know as much as a geologist does to decide whether a rock is sedimentary, **igneous** or **metamorphic**. All you need is a little bit of information about the area where the rock was found. For example, if your town was once covered by an ancient sea or lake, you are likely to find plenty of sedimentary rocks in your local area.

Asking questions

Next, think about how sedimentary rock forms, and ask yourself some questions about your rock sample. Does it have colourful layers? Can you see fragments of pebbles or shells in the rock? Are there ripple marks? Does the rock contain any **fossils**?

If the answer to one or more of these questions is 'yes', you probably have a sedimentary rock. In addition, many sedimentary rocks are soft. If you can break a rock with your hands, there's a good chance it's a sedimentary rock.

A TELLTALE TEST

If you think you've found limestone, try a simple test. Pour some warm vinegar over the rock. If it fizzes up, it's limestone. The chemicals in the vinegar react with the minerals that make up limestone. This chemical reaction breaks down the limestone, so you shouldn't try it on a limestone building.

Minerals

Also, take a quick look at the minerals. The minerals that make up most sedimentary rock are soft and dull with small crystals. Of course, one of the best ways to identify a rock is to study a guide to rocks and minerals. These books show pictures of rocks and give detailed descriptions.

IN SEARCH OF SHALE

Shale can be a valuable sedimentary rock. This soft, gray rock often contains oil deposits. Oil is the remains of tiny ocean plants and animals called **plankton**. When plankton die, their bodies sink to the bottom of the ocean and are buried by **sediments**. As time passes, the plankton turn into a thick, black liquid, and most of the sediments turn into shale.

Rock collecting

These young rock collectors are finding rock samples during a class trip. Do you think they are examining a sedimentary rock?

Now that you know how to spot sedimentary rocks, would you like to see or collect some? You can buy all kinds of beautiful and interesting sedimentary rock samples. You can also see them at a natural history museum, but it might be more fun to hunt for your own rocks.

Equipment

Before you begin planning your first rock hunting trip, you will need to gather together a few pieces of basic equipment. You will also need to learn a few rules.

Once you have identified the rocks, you may want to create a system for labelling, organizing and storing them. Then you will always be able to find a specific sample when you need it.

DID YOU KNOW?

A geologist is a scientist who studies rocks and rock formations to learn how the Earth formed and how it has changed over time. Sir Henry Thomas De la Beche (1796–1855) was the first director of the Geological Survey of England.

There are many ways to organize a rock collection. You can use a box with dividers or glue specimens to a piece of thick card.

You can arrange your specimens any way you like – by colour, by **crystal** shape, by collection site or even alphabetically. As your collection grows, being organized will become more important.

WHAT YOU NEED TO KNOW

- Never go rock hunting alone. Go with a group that includes a qualified adult.
- Know how to use a map and compass.
- Always get a landowner's permission before walking on private property. If you find interesting rocks, ask the owner if you can remove them.
- Before removing samples from public land, make sure collecting is allowed. Many natural rock formations are protected by law.
- Respect nature. Do not hammer out samples. Do not disturb living things and do not leave litter.

WHAT YOU NEED

- Strong boots or wellies
- A map and compass
- A small paintbrush to remove dirt and extra rock chips from samples
- A camera to take photographs of rock formations
- A hand lens to get an up-close look at **minerals**
- A notebook for recording when and where you find each rock
- A spotter's guide to rocks and minerals

Glossary

acid rain rain that is polluted with acid in the atmosphere and that damages the environment

atom smallest unit of an element that has all of the properties of that element

blow hole opening in a rock formation created by wind erosion

cave natural opening in the ground that is large enough to hold a person

core centre of the Earth. The inner core is solid, and the outer core is liquid.

crust outer layer of the Earth

crystal repeating structure within most minerals

delta formation of deposits that a river dumps at the spot where it empties into the ocean

diatom one-celled sea creature that may live alone or cluster with other diatoms. Some are round, others are long and thin.

erode to slowly wear away rock over time by the action of wind, water or glaciers

evaporate to change from a liquid to a gas

evolved changed slowly, adapted to changing environmental conditions

face flat side of a crystal

foraminifer one-celled sea creature with a simple shell

fossil remains or evidence of ancient life

gravity force that pulls objects toward the Earth's centre

igneous rock kind of rock that forms when magma from the Earth's mantle cools and hardens

magma hot, liquid rock that makes up the Earth's mantle. When magma spills out onto the surface, it is called lava.

mantle layer of the Earth between the crust and outer core. It is made of rock in its liquid form, known as magma.

metamorphic rock kind of rock that forms when heat or pressure changes the minerals within igneous rock, sedimentary rock or another metamorphic rock

mineral natural solid material with a specific chemical makeup and structure

plankton tiny creatures that live near the surface of the water. Plankton is an important source of food for fish and other sea animals.

property trait or characteristic that helps make identification possible

protist tiny creature. All plants, animals and fungi evolved from protists.

sediment mud, clay or bits of rock picked up by rivers and dumped in the ocean

tectonic plate one of the large slabs of rock that make up the Earth's crust

travertine mineral consisting of many layers of calcium carbonate

volcano opening in the Earth's surface that extends into the mantle

weathering breaking down of rock by plant roots or by repeated freezing and thawing

Further information

BOOKS

The Kingfisher book of planet Earth, Martin Redfern, Kingfisher, 1999

The pebble in my pocket, Meredith Hopper, Francis Lincoln, 1997

The best book of fossils, rocks and minerals, Chris Pellant, Kingfisher, 2000

Tourists rock, fossil and mineral map of Great Britain, British Geological Survey, 2000

ORGANIZATIONS

British Geological Survey
www.bgs.ac.uk
Kingsley Dunham Centre, Keyworth,
Nottingham NG12 5GG
UK

Rockwatch
www.geologist.demon.co.uk/rockwatch/
The Geologists' Association
Burlington House, Piccadilly
London W1V 9AG
UK

The Natural History Museum
www.nhm.ac.uk
Cromwell Road
London SW7 5BD
UK

The Geological Society of Australia
www.gsa.org.au/home
Suite 706
301 George Street
Sydney NSW 2000
Australia

Geological Survey of Canada
601 Booth Street
Ottawa, Ontario
KIA 0E8
Canada

US Geological Survey (USGS)
507 National Center
12201 Sunrise Valley Drive
Reston, Virginia 22092
USA

Index

Titles in the *Rocks and Minerals* series include:

Hardback 0 431 14370 6

Hardback 0 431 14371 4

Hardback 0 431 14372 2

Hardback 0 431 14373 0

Hardback 0 431 14374 9

Hardback 0 431 14375 7

Hardback 0 431 14376 5

Find out about the other titles in this series on our website www.heinemann.co.uk/library

TRANSPLANTATION

A Companion to Specialist Surgical Practice

Series Editors

O. James Garden
Simon Paterson-Brown

TRANSPLANTATION

FOURTH EDITION

Edited by

John L.R. Forsythe
MD FRCS(Ed) FRCS

Consultant Transplant Surgeon
Transplant Unit
Royal Infirmary of Edinburgh, UK

Edinburgh London New York Oxford Philadelphia St Louis Sydney Toronto 2009

SAUNDERS
ELSEVIER

First edition 1997
Second edition 2001
Third edition 2005
Fourth edition 2009
Reprinted 2009

ISBN 9780702030130

British Library Cataloguing in Publication Data
A catalogue record for this book is available from the British Library

Library of Congress Cataloging in Publication Data
A catalog record for this book is available from the Library of Congress

Notice
Knowledge and best practice in this field are constantly changing. As new research and experience broaden our knowledge, changes in practice, treatment and drug therapy may become necessary or appropriate. Readers are advised to check the most current information provided (i) on procedures featured or (ii) by the manufacturer of each product to be administered, to verify the recommended dose or formula, the method and duration of administration, and contraindications. It is the responsibility of the practitioner, relying on their own experience and knowledge of the patient, to make diagnoses, to determine dosages and the best treatment for each individual patient, and to take all appropriate safety precautions. To the fullest extent of the law, neither the Publisher nor the Editors assumes any liability for any injury and/or damage to persons or property arising out of or related to any use of the material contained in this book.

The Publisher

Printed in China

Commissioning Editor: Laurence Hunter
Development Editor: Elisabeth Lawrence
Project Manager: Andrew Palfreyman
Text Design: Charlotte Murray
Cover Design: Kirsteen Wright
Illustration Manager: Gillian Richards
Illustrators: Martin Woodward and Richard Prime

Contents

Contributors

Murat Akyol, MD, FRCS
Consultant Transplant Surgeon
Royal Infirmary of Edinburgh;
Honorary Clinical Senior Lecturer
University of Edinburgh
Edinburgh, UK

Simon T. Ball, MA, PhD, MRCP
Consultant Nephrologist
Department of Nephrology
University Hospital
Birmingham, UK

Daniel C. Brennan, MD
Professor of Medicine
Internal Medicine/Renal Division
Washington University
St Louis, MO, USA

Stephen C. Clark, BMedSci, BM, BS, DM, FRCS(CTh)
Consultant Cardiothoracic Surgeon
Cardiothoracic Unit
Freeman Hospital
Newcastle upon Tyne, UK

Juan L. Contreras, MD
Assistant Professor of Surgery
University of Alabama at Birmingham
Birmingham, AL, USA

Margaret J. Dallman, DPhil
Professor of Immunology
Department of Life Sciences
Imperial College of Science, Technology and Medicine
London, UK

Philip A. Dyer, PhD, FRCPath
Consultant Clinical Scientist
Transplantation Laboratory
Manchester Royal Infirmary;
Honorary Reader in Transplantation Science
University of Manchester
Manchester, UK

Devin E. Eckhoff, MD
Professor of Surgery
University of Alabama at Birmingham
Birmingham, AL, USA

Robert S. Gaston, MD
Professor of Medicine
Medical Director, Kidney and Pancreas Transplantation
University of Alabama at Birmingham
Birmingham, AL, USA

Keith P. Graetz, MA, MB, BChir, MRCS
Specialist Registrar
Nottingham City Hospital
Nottingham, UK

Rajesh Hanvesakul, MB, BS, PhD
Lecturer in Medicine
Department of Nephrology
University Hospital
Birmingham, UK

Asif Hasan, FRCS(CTh)
Consultant Cardiothoracic Surgeon
Freeman Hospital
Newcastle upon Tyne, UK

Benjamin E. Hippen, MD
Internist and Transplant Nephrologist
Carolinas Medical Center
Charlotte, North Carolina, USA

Nicholas Inston, FRCS(GS), PhD
Consultant Surgeon
Queen Elizabeth Hospital
Birmingham, UK

Alison Logan, BSc, MPhil
Senior Clinical Scientist
Transplantation Laboratory
Manchester Royal Infirmary
Manchester, UK

Derek M. Manas, BSc, MBBCh, Mmed, FCS, FRCS
Professor of Hepatobiliary and Transplantation Surgery
Department of Hepatobiliary and Transplant Surgery
The Freeman Hospital
Newcastle Upon Tyne, UK

Lorna P. Marson, MD, FRCS
Senior Lecturer in Transplant Surgery
Transplant Unit
Royal Infirmary of Edinburgh
Edinburgh, UK

Herwig-Ulf Meier-Kriesche, MD
Medical Director of Renal and
Pancreas Transplant Program
Professor of Medicine
University of Florida
Gainesville, FL, USA

Imran A. Memon, MD
Nephrologist and Paediatric Nephrologist,
Children's Place,
St. Louis, MO, USA

William D. Plant, BSc, MB, MRCPI, PRCP
Consultant Renal Physician
Cork University Hospital
Cork Ireland

Kay Poulton, BSc, PhD
Honorary Lecturer in Immunogenetics
Transplantation Laboratory
Manchester Royal Infirmary
Manchester, UK

Keith M. Rigg, MD, FRCS
Consultant Surgeon
Nottingham City Hospital
Nottingham, UK

Amy F. Rosenberg, PharmD, BCPS
Clinical Assistant Professor
University of Florida College of Pharmacy
Department of Pharmacy Services
Shands at the University of Florida
Gainesville, FL, USA

Christopher J. Rudge, BSc, MB, BS, FRCS
Managing and Transplant Director
UK Transplant
Bristol, UK

Steve A. White, MB, ChB, MD, FRCPS, FRCS
Consultant Laparoscopic, Hepatobiliary and Liver
Transplant Surgeon
Freeman Hospital;
Honorary Senior Lecturer
The Cellular Institute
The University of Newcastle
Newcastle Upon Tyne, UK

Christopher H. Wigfield, MD, FRCS
Consultant Cardiothoracic Surgeon
Cardiothoracic Unit
Freeman Hospital
Newcastle upon Tyne, UK

Stephen J. Wigmore,
BSc, MB, BS, MD, FRCS(Ed), FRCS
Professor of Transplantation Surgery
University of Edinburgh
Royal Infirmary of Edinburgh
Edinburgh, UK

Karen Wood, BSc MSc
Senior Clinical Scientist
Transplantation Laboratory
Manchester Royal Infirmary
Manchester, UK

Judith Worthington, BSc
Senior Clinical Scientist
Transplantation Laboratory
Manchester Royal Infirmary
Manchester, UK

Series preface

Since the publication of the first edition in 1997, the *Companion to Specialist Surgical Practice* series has aspired to meet the needs of surgeons in higher training and practising consultants who wish contemporary, evidence-based information on the subspecialist areas relevant to their general surgical practice. We have accepted that the series will not necessarily be as comprehensive as some of the larger reference surgical textbooks which, by their very size, may not always be completely up to date at the time of publication. This Fourth Edition aims to bring relevant state-of-the-art specialist information that we and the individual volume editors consider important for the practising subspecialist general surgeon. Where possible, all contributors have attempted to identify evidence-based references to support key recommendations within each chapter.

We remain grateful to the volume editors and all the contributors of this Fourth Edition. Their enthusiasm, commitment and hard work has ensured that a short turnover has been maintained between each of the editions, thereby ensuring as accurate and up-to-date content as possible. We remain grateful for the support and encouragement of Laurence Hunter and Elisabeth Lawrence at Elsevier Ltd. We trust that our aim of providing up-to-date and affordable surgical texts has been met and that all readers, whether in training or in consultant practice, will find this fourth edition an invaluable resource.

O. James Garden MB, ChB, MD, FRCS(Glas), FRCS(Ed), FRCP(Ed), FRACS(Hon), FRCSC(Hon)

Regius Professor of Clinical Surgery, Clinical and Surgical Sciences (Surgery), University of Edinburgh, and Honorary Consultant Surgeon, Royal Infirmary of Edinburgh

Simon Paterson-Brown MB, BS, MPhil, MS, FRCS(Ed), FRCS

Honorary Senior Lecturer, Clinical and Surgical Sciences (Surgery), University of Edinburgh, and Consultant General and Upper Gastrointestinal Surgeon, Royal Infirmary of Edinburgh

Editor's preface

Although part of a surgical series, this volume is not just for surgeons. It is for all those who play an important role in the transplant procedure. Thus it has been designed to interest nursing staff who care for organ failure patients, transplant coordinators, theatre staff, immunology laboratory staff, paramedical personnel, physicians and surgeons. Modern techniques in transplantation and new forms of immunosuppression, emphasised throughout this volume, have increased the complexity of clinical and ethical dilemmas which face the whole team caring for the transplant patient. Appropriate response to such dilemmas is required to ensure the continued success of transplantation medicine.

As in previous editions of this volume, this is less of a new edition and more a new book. For many chapters a new author has been commissioned, therefore giving a different view on a subject whilst bringing the factual information up to date. In other chapters the authors from the previous editions have been retained but the brief has been to write chapters in a slightly different way, emphasising newer techniques or dilemmas in transplantation.

As before, the book has been written for the whole multidisciplinary team looking after the prospective transplant patient. Transplant medicine is one of the best examples of multidisciplinary team care and it is this fact which makes it both challenging and rewarding. If this book has been successful in previous volumes, it is partly because the spirit of that team care is represented in the pages of the book.

John L.R. Forsythe
Edinburgh

Evidence-based practice in surgery

Critical appraisal for developing evidence-based practice can be obtained from a number of sources, the most reliable being randomised controlled clinical trials, systematic literature reviews, meta-analyses and observational studies. For practical purposes three grades of evidence can be used, analogous to the levels of 'proof' required in a court of law:

1. **Beyond all reasonable doubt.** Such evidence is likely to have arisen from high-quality randomised controlled trials, systematic reviews or high-quality synthesised evidence such as decision analysis, cost-effectiveness analysis or large observational datasets. The studies need to be directly applicable to the population of concern and have clear results. The grade is analogous to burden of proof within a criminal court and may be thought of as corresponding to the usual standard of 'proof' within the medical literature (i.e. $P < 0.05$).
2. **On the balance of probabilities.** In many cases a high-quality review of literature may fail to reach firm conclusions due to conflicting or inconclusive results, trials of poor methodological quality or the lack of evidence in the population to which the guidelines apply. In such cases it may still be possible to make a statement as to the best treatment on the 'balance of probabilities'. This is analogous to the decision in a civil court where all the available evidence will be weighed up and the verdict will depend upon the balance of probabilities.
3. **Not proven.** Insufficient evidence upon which to base a decision, or contradictory evidence.

Depending on the information available, three grades of recommendation can be used:

a. Strong recommendation, which should be followed unless there are compelling reasons to act otherwise.
b. A recommendation based on evidence of effectiveness, but where there may be other factors to take into account in decision-making, for example the user of the guidelines may be expected to take into account patient preferences, local facilities, local audit results or available resources.
c. A recommendation made where there is no adequate evidence as to the most effective practice, although there may be reasons for making a recommendation in order to minimise cost or reduce the chance of error through a locally agreed protocol.

Strong recommendation

Evidence where a conclusion can be reached **'beyond all reasonable doubt'** and therefore where a **strong recommendation** can be given.

This will normally be based on evidence levels:

- Ia. Meta-analysis of randomised controlled trials
- Ib. Evidence from at least one randomised controlled trial
- IIa. Evidence from at least one controlled study without randomisation
- IIb. Evidence from at least one other type of quasi-experimental study.

Expert opinion

Evidence where a conclusion might be reached **'on the balance of probabilities'** and where there may be other factors involved which influence the recommendation given. This will normally be based on less conclusive evidence than that represented by scalpel icons:

- III. Evidence from non-experimental descriptive studies, such as comparative studies and case–control studies
- IV. Evidence from expert committee reports or opinions or clinical experience of respected authorities, or both.

Evidence in each chapter of this volume which is associated with either a strong recommendation or expert opinion is annotated in the text by either a **scalpel** or **pen-nib** icon as shown above. References associated with **scalpel** evidence will be highlighted in the reference lists, along with a short summary of the paper's conclusions where applicable.

1

Controversies in the ethics of organ transplantation

Stephen J. Wigmore
William D. Plant

"Best practice in organ transplantation requires sensitivity to the ethical dimension within clinical situations, fluency in the language of ethical discussion and skill in the practical resolution of ethically complex scenarios."

W.D. Plant (1958–present)

"Philosophy, n. A route of many roads leading from nowhere to nothing."

Ambrose Bierce (1842–1914),
The Devil's Dictionary

Introduction

The history of transplantation is relatively brief at 40–50 years compared with the history of medicine, and the unique processes involved in this branch of medicine continue to surprise and challenge conventional medical and ethical doctrine. Many of the ethical dilemmas surrounding transplantation centre on the twin peaks of procurement and allocation. The questions of 'how do we get organs' and 'who should receive them' permeate all aspects of transplantation ethics and clinical practice. Indeed, ethics is an important part of medical education as acknowledged by the General Medical Council's publication *Tomorrow's Doctors*, which placed ethics squarely in the core curriculum for medical training so that doctors would be able 'to understand and analyse ethical problems so as to enable patients, their families, society, and the doctor to have proper regard to such problems in reaching decisions'.[1]

This chapter will outline the foundations of modern ethical principles and then discuss two controversial areas: conditional and directed donation; and payment for organ donation and transplant tourism.

Systems and principles in medical ethics and bioethics

The ethical basis of Western medicine is characterised by a number of traditions,[2] which find expression in codes of practice and in the culture of healthcare workers. Prominent amongst these are the deontological (duty-based) and the utilitarian (consequence-based) traditions.

The deontological tradition[3] stresses the duties of practitioners and the rights of patients. Many codes[3] of practice, such as the Hippocratic Oath, the Declaration of Geneva and the General Medical Council Statements[4] on 'Duties and Responsibilities of Doctors', are of this tradition. There is a strand of rational, universalising deliberation within deontology that may, in some circumstances, seem uncompromising and excessively rigid. Clinicians 'ought' to act in particular ways because this is 'right', often irrespective of the consequences. Any act should be capable of being expressed as a universal law. The great strength of the deontological tradition is its stress on the autonomy of the patient and on the primacy of doctor–patient relationships. It asserts that individuals should always be viewed as ends in themselves, never as **means** to an end. Its weakness is in application. 'Absolute' principles that are contradictory may co-present in particular circumstances (e.g. 'never cause pain', 'always preserve life' – what if life-saving interventions cause pain?).

On the other hand, the utilitarian tradition[3] aspires always to do that which leads to a 'good' or 'best' outcome. It is commonly paraphrased as 'Seek the greatest good for the greatest number'. If the 'right' action does not lead to the 'best' outcome then it should be reviewed or abandoned. The necessity for rationing of healthcare resources brings utilitarian analyses into particular prominence. On occasion, the individual may be the 'loser' in the greater scheme of things – a common criticism of this approach.

It is important to remember that utilitarian analysis can be applied to an individual case – what is in a patient's, as opposed to all patients', best interests (or will lead to the best outcomes)? Two strands may be identified – 'rule' utilitarianism and 'act' utilitarianism. 'Rule' utilitarianism weighs up the consequences of acting according to a general **moral rule** (e.g. that patients over 65 years with type II diabetes should never receive transplants as the long-term survival of this group is likely to be less than that of a younger non-diabetic group). 'Act' utilitarianism weighs up the consequences of a **particular act** (e.g. deciding not to proceed with living donation in a case where there is a high risk of recurrence of a primary disease in the recipient).

It is important to stress that these traditions are not opposite ways of viewing and acting upon ethical dilemmas. Most analyses will show that they tend to derive the same conclusions when 'working' moral issues are at stake.

> *"The medical profession has long subscribed to a body of ethical statements developed primarily for the benefit of the patient. As a member of this profession, a physician must recognise responsibility to patients first and foremost, as well as to society, to other health professionals, and to self."*
>
> Principles of Medical Ethics, American Medical Association

Both traditions have a strong flavour of universalisation. However, as all ethical analyses ultimately focus on **particular** cases, the importance of **context** may well have been understated in the past. Some current intellectual traditions, notably existentialism, situation ethics and postmodernism,[3] are less convinced of the existence of general laws/principles that can be applied to particular cases. Rather, they focus on how to solve specific problems as they arise and are open to the insights of other religious, racial, philosophical and cultural traditions. The principal criticism of these traditions is that moral relativism may easily drift into moral anarchy (although, of course, moral absolutism may similarly drift into moral fascism).

In real life, we recognise a spectrum between cases in which context and situation need to be the dominant consideration and those in which universally derived general principles can be applied. No one tradition or perspective is 'more correct' – they offer different perspectives to problem solving.

Principles

A number of prima facie principles (commonly distilled to four) are widely accepted as forming the bedrock of medical ethics.[3,4] As we shall see, in individual cases there is often conflict between the simultaneous adherence to all of these. Depending on the context (and on whether a deontological or utilitarian approach is favoured), a 'least unsatisfactory' trade-off between principles must be negotiated or achieved. Skill in identifying and achieving the best balance is that which characterises best practice in dealing with moral dilemmas.

1. Beneficence: doing good[3,4]

A central tenet of medical ethics is the obligation to strive at all times to do good for the patient. Deontologists view this as a universal moral duty, utilitarians as achieving the universally desired best outcome. It is instructive to reflect upon the secondary obligations that follow from acceptance of this principle. Beneficence demands competence. Accreditation and continuing medical education are central to this, as are elements such as professional development, clinical and basic research, and audit. Transplant programmes with clinical governance processes in place are expressing vigorous adherence to this principle. Communication skills are vital – weighing up possible outcomes is one thing, sharing them with the patient and negotiating choices quite another.

2. Non-maleficence: avoiding harm[3,4]

Since the days of Hippocrates, clinicians have endorsed the principle of *Primum non nocere* ('First, do no harm'). All interventions, however well intentioned, may cause harm. Making sure that the balance between benefit and harm is appropriate is an important clinical judgement. In the past, professional decisions (i.e. that the balance achieved by a particular course of action was acceptable) often paid scant attention to the patients' perspectives of the balance. This might be called **paternalism** – classically according predominant weight to clinicians' judgement on the balance between beneficence and non-maleficence, with little weight given to patient autonomy.

The other end of the spectrum is where overwhelming emphasis is placed on respect for patient autonomy, with little reflection on professional judgement of beneficence/non-maleficence trade-offs. This is equally undesirable and might be called **consumerism**.

3. Respect for autonomy[3–5]

Individuals should be treated as ends, not as means. Respect for the dignity, integrity and authenticity of the person is a basic human right. Deriving from this principle are the important issues of consent and confidentiality. Patients with capacity to understand relevant information (explained in broad terms and with simple language), to consider its implications in terms of their own values, and come to a communicable decision, are deemed to have decision-making capacity. (The legal default is that conscious adults are assumed to have capacity unless evidence to the contrary can be advanced.) Some patients, because of transient or irreversible cognitive impairment, may not have this capacity. It is important to note that even a decision that seems irrational to the physician **must** be respected if the patient has decision-making capacity. The issue is more complex regarding children under 16 years.

Informed consent is central to the doctor–patient relationship. Its definition is difficult – in countries such as Australia and Canada there is a **legal** duty to provide patients with information that a **prudent or reasonable patient**, in that patient's particular circumstances, would wish to know in order to arrive at a decision. In the UK, the amount of information that a **reasonable doctor** would provide, in that patient's particular circumstances, is required.[6] (However, a court might decide that, in unusual circumstances, failure to disclose certain risks of procedures might be negligent, **even if this is accepted practice** as attested by a responsible body of medical opinion.) Information about certain possible risks may be retained under so-called **therapeutic privilege**, if it is honestly felt that this would needlessly harm the patient psychologically (justified by an 'act' utilitarian attitude with heavy emphasis on non-maleficence). This privilege cannot be invoked when informed consent to participation in a research study is being obtained – for obvious reasons. No overt or covert pressure to consent should exist. Many guidelines[6] exist to help deal with difficulties in obtaining consent.

In the case considered in Box 1.1 it could be argued that transplantation in an individual who has expressed, by his actions, a wish to die is a breach of his autonomy. In general terms, interventions made when patients do not have the capacity to offer consent should follow certain principles. Any advance directives should be consulted; persons with knowledge of the patient's previous views should be consulted for their insight into the possible choices that the patient might have made were he able to participate. Family cannot consent to any procedure on his behalf, unless there is a legal guardianship in existence. The decision as to how best to proceed should be taken by the doctor with overall responsibility for the case. Choices should be limited to those that are in the patient's best interests and that least limit the patient's future choices should he regain the capacity to consent. In this situation the clinicians have used the principle of beneficence to justify the act of transplantation, in the hope that

Box 1.1 • Case history

A 23-year-old man has developed fulminant hepatic failure after taking an overdose of paracetamol following the breakdown of a long-term relationship. This is his first attempt at self-harm and after consultation with his family he is listed for urgent liver transplantation. A donor organ becomes available and the transplant is successful. During the recovery period the recipient expresses his dismay at having received a transplant. He states that he still wishes to die and refuses to take his immunosuppressive medication.

the attempt at self-harm was, at best, accidental or, at worst, not intended to result in death. A utilitarian perspective might argue against transplantation in this situation, giving priority in the allocation of organs to individuals who have unequivocally expressed a desire to live. Alternatively, it could be argued that the circumstances precipitating this act of self-harm are transient or that the normal mental state of the individual may have been temporarily disturbed. This could respond to a change in circumstances or to treatment with appropriate counselling or medications. In this unusual scenario, the reiteration by the patient that he wishes to die, and his refusal to continue with treatment necessary to keep the transplant functional, presents a complex matrix of interaction between differing principles. If he has regained the capacity to consent, then his autonomy must be respected. On the other hand, this should prompt a discussion as to the reasons why the original choice was made. Beneficence and non-maleficence may prompt the clinicians strongly to counsel him to continue with treatment. It should be pointed out (in the spirit of justice) that he has received a scarce resource and that a range of individuals have done their best to make the right choice for him. One might wish to use this as a pragmatic 'emotional lever' in the discussion – however, primacy must be given to his autonomy.

4. Justice: promoting fairness[3,4]

This is a very important principle in the ethics of transplantation, where demand far outstrips supply. In that context, the allocation of organs requires a rank-ordering system with some philosophical justification for the method chosen. There are many theories of justice, some deriving from the deontological and utilitarian traditions. If scarce resources are to be allocated, then how should this be done? A 'rule' utilitarian approach might suggest that we always seek to maximise the overall welfare of the group or minimise the waste of resources. Many transplant services have allocation systems loosely built upon the premise that selecting and organising allocation around a limited number of criteria predictive of best outcome (such as human leucocyte antigen (HLA) types) best serves this purpose.

This disadvantages some patients. Other systems operate very strict acceptance criteria for listing – again meeting some of the requirements for fairness, but not all. Discrimination on the basis of race, gender, age or 'social worth' obviously violates the principle of justice.

It is important to acknowledge that justice and fairness need to be applied broadly. These principles apply to the individual patient, but also to other patients whose circumstances may be influenced by events relating to that patient. Similarly, we need to be fair to other members of the transplant team and to the broader needs of society.

Implementation of the four principles: interactions

Although the four principles listed above are the central principles in medical ethics and bioethics, there are different interactions between different 'players' in their implementation. Issues relating to beneficence and non-maleficence lie very much in the domain of the clinician/doctor–patient relationship, with particular focus on individual patient events. Issues of respect for autonomy and justice interact much more widely, with greater roles for the law, social policy, politics and culture. In addition, the latter apply more readily to groups of patients rather than to individuals. Despite this, all four principles should be given due and equal consideration in case analyses.

A second trend is worthy of comment: autonomy and justice are themes enjoying considerably more attention now than in the past, when medical paternalism may have been a more culturally dominant phenomenon than in the present.

Conditional and directed donation

> *"I recently attracted the attention of friends and acquaintances by donating a kidney to the NHS, taking advantage of the change in legislation last year, which allows donations to be made anonymously. My motive for doing so can be summed up in the old rule of thumb 'Do as you would be done by' which may sound philosophically unsophisticated but has always been useful to me."*
>
> M. Harris, J Med Ethics 2008; 34:511–12

The following scenarios set out a series of events concerning organ donation which challenge ethical principles. These scenarios have been adapted from cases devised by Professor James Neuberger and Mr A. David Mayer of the Queen Elizabeth Hospital Birmingham and are based on actual cases.[7,8]

Example 1

A 45-year-old diabetic man is on the kidney transplant waiting list. He had a kidney transplant 7 years ago from a deceased, unrelated donor but the kidney was lost to chronic rejection 2 years before. His 21-year-old unmarried child is involved in a road traffic accident, suffers a severe head injury and is declared brainstem dead on an intensive care unit. The family are seen by the transplant coordinator. The father, as 'next of kin', will only agree to the donation of a single kidney to himself. No other organs can be taken. The transplant coordinator points out that, dependent upon the blood group, tissue type and crossmatch, the kidney may not be suitable for him. In this case the father states that the kidney must be returned to the child's body and not used for anyone else.

Should the proposed transplant go ahead?

When discussing the issues of conditional and directed donation it is important to understand the potential problems that these seemingly innocent terms may hide. Conditional donation can be defined as **donation which can only occur under certain proscribed conditions**. The term conditional in this context contains a threat that, should these conditions not be met, donation will not be permitted to continue. Cadaveric organ donation in the UK has always been considered as a gift freely given without conditions. This notion rests comfortably with the notions of equity of access to organs for transplant recipients. The stance of not allowing conditional donation in the UK is because of a potential breach of this principle of equity of access. In Example 1 we would consider this case to represent conditional donation: in that the father imposes conditions on the use of the organs, i.e. that they can only be used for him. This is perhaps a tame example but suppose the condition he imposed was that the organs should only be donated to people of a certain race or religion.

Is it reasonable that live-donor donation can be directed to a named individual whereas deceased-donor donation must be unconditional?

It may seem unreasonable to many that directed donation is legally permissible in the context of a living donor donating an organ to a specific recipient but that directed donation is not considered acceptable in the context of a deceased donor. Where does this apparent paradox come from? The difference between the living and the deceased donor is always raised in the context of conditional or directed donation. A living donor is a conscious being with autonomy who has the right to make choices which affect his or her being. In this context it seems appropriate that an individual should have the right to choose whether or not to donate an organ and similarly to whom it should be donated. When life is extinct autonomy ceases. How can a dead person make choices and how can any event affect their life since life has already ceased? In the UK recent changes in legislation give greater priority to the choice of individuals about what happens to their organs in the event of their death. This stance is more out of respect for the individuals' choices when alive rather than a granting of autonomy after death. Still, however, current practice does not permit conditional donation after death since this is considered to violate the more important principles of justice and equity of access in organ allocation.

Should patients who refuse to donate for religious or other reasons be entitled to receive a transplant?

The question of whether individuals whose religion prohibits organ donation should be allowed to receive a transplant is fairly straightforward. There are very few religions which do prohibit organ donation and religious leaders from all of the world's major religions have been consulted on this issue. Where conflicts arise it is frequently because of variance of interpretation of religious doctrine rather than a fundamental problem based on religious belief. For example, some of the practical difficulties presented by a requirement for rapid internment after death can be avoided by consultation with local religious leaders. To deal with the central question, however, it seems reasonable to assume that people do not become members of a religion with the sole purpose of avoiding the potential of them proceeding to organ donation and so it would seem churlish to refuse organ transplantation because of membership of a particular faith. Other people take a different view and believe that an unwillingness to donate an organ should preclude receiving one. On a practical level the reluctance of certain ethnic minorities to donate organs in the UK may prejudice the likelihood of a member of that ethnic group receiving a transplant because of conservation of rare HLA haplotypes in some minorities.

Example 2

A 30-year-old white male is involved in a road traffic accident. He suffers a severe head injury and is declared brainstem dead. He carries an organ donor card, agreeing to multiorgan donation. The family are seen by the transplant coordinator at 9 p.m. They state that he had expressed a wish that his organs should be given only to a white person.

Should the donation go ahead in accordance with the wishes of the donor?

This is perhaps the most extreme example of conditional donation. Such discrimination or bias undermines justice and equity and has the potential to divide societies. There is an understandable concern that allowing conditions to be attached to organ donation represents the thin end of the wedge and has the potential to open the door to all manner of unsavoury conditions which violate the principles of our society. Medicine in general and transplantation in particular should be above social and religious discrimination. On the other hand, the argument could be made that a number of patients could benefit from multiorgan donation in this man:

- two kidneys;
- one pancreas;
- one liver which can be split for two recipients;
- one heart;
- two lungs;
- two corneas;
- tissue donation.

In this context is it fair to the waiting list to not proceed with donation over a matter of principle?

The notion that, in the above example, not proceeding to donation would be 'cutting off one's nose to spite one's face' is understandable. Indeed, an argument could be made that by reducing the size of the waiting list, the donation will produce a net benefit for all patients on the list whatever their race. This short-term loss of organs is considered preferable to uphold the principles of justice and equity of access. As a separate but related issue, imagine how a black or Asian surgeon would feel if required to transplant a kidney from a donor who had specifically excluded black or Asian individuals as recipients?

Should the principle of unconditional donation be an overriding principle even if it results in patients dying on the waiting list?

UK Transplant check the donor's blood group and tissue type against those of patients at the top of the national transplant waiting list. They note that, notwithstanding the family's insistence on conditional donation, if the donation proceeds then this man's organs will be transplanted into Caucasian patients.

Since the family's conditions will be met in any way, is it reasonable for the donation to go ahead?

This is a difficult issue because to accede to the donor family request for racial donation would be to accept this as a valid reason or condition, which would be contrary to the tenets held by our multicultural and multiracial society. On the other hand, a pragmatic approach would say that since circumstance arising from tissue typing has dictated that the organs will go to the racial group favoured by the donor family, why not go ahead.

Aftermath

The donation goes ahead and the organs are transplanted into Caucasian recipients. Two outcomes are possible:

1. No one hears about the donation and nine patients receive organ transplants, reducing the number of patients awaiting transplant.

Or:

2. The press gets hold of the story. The case is discussed in Parliament. The (fictional) Minister of Health expresses his outrage ('I was appalled when I learnt today the initial details ... I asked my Permanent Secretary to ... ensure it never happens again.' He states that any form of conditional donation of organs from deceased donors is unacceptable. The Department of Health issues a statement that it was totally against any kind of conditions being attached to organ donation; donated organs are a national resource which are available to people regardless of race, religion, age or other circumstance. This is widely supported by his parliamentary colleagues and by the national press.

The second case represents the reality of what would happen as this issue is presented to the public, putting organ donation and transplantation in a bad light. On the positive side issues would be widely discussed both within the media and by the general public, raising the general profile of transplantation and awareness of the ethical dilemmas presented by such rare situations.

Is any form of conditional donation of organs from deceased donors acceptable?

Conditional donation could be considered in more than one form. Firstly, there may be a situation where donation is directed to an individual and is conditional on this being agreed. Secondly, there may be a situation where conditions are imposed which do not direct donation to an individual but which exclude or discriminate against broad sectors of society. It is possible to imagine some situations where directed donation may be acceptable but discriminatory conditional donation is divisive and breaches societal codes and beliefs, and is therefore considered unacceptable.

Example 3

A 19-year-old student with meningitis is declared brainstem dead. She carries an organ donor card which states 'I request that after my death any part of my body be used for the treatment of others.' Her devastated family are seen by the transplant coordinator. They have a friend who has advanced kidney disease. They agree to kidney and liver donation only. They insist that the heart and corneas should not be removed.

Should the donation go ahead in accordance with the statement by the deceased, or in accordance with the wishes of the family to save them further distress?

The kidneys and liver are removed and transplanted into the three patients on the transplant waiting list. The heart and corneas are not removed. This situation is not unusual. Some families stress that certain organs such as the heart or the eyes are sacrosanct and must not be removed.

Is this conditional donation and is it acceptable?

At the time that Example 3 occurred in 1998, the Human Organ Transplant Act[9] required a lack of objection to the use of organs by the family of a deceased donor irrespective of the wishes stated in life by the donor. Given this information, the family were at that time quite entitled under UK law to object to the removal of their daughter's heart and corneas even though she herself had stated a lack of objection in life. In some senses this is a type of conditional donation but importantly, unlike the previously described cases, no conditions are placed on the allocation of organs. With the introduction of the Human Tissue Act 2004[10] and HTA (Scotland) 2006[11] in the UK, greater priority is given to the stated wishes of the deceased donor made during their lifetime. Had this example occurred since this date, the donor medical team would have had the legal right to override objections from the family, giving precedence to the wishes of the deceased daughter stated on her donor card. Organ donation is, however, a sensitive issue and one which relies entirely on public support. To our knowledge there have been no occasions in the UK where this legal principle has been invoked and the family's wishes been overridden. What current UK legislation does do is offer the medical team addressing the family to take on the role of advocate for the deceased donor and this is often sufficient persuasion to adhere to the wishes of the deceased made in life. This whole issue also raises the question of 'ownership' of the human body. If conditional donation is permitted, this would imply that human remains can be considered property and property implies ownership. In legal circles it is generally considered that the body or parts of it should not be considered property nor should any individual or organisation claim ownership of it. Granting property rights and ownership of body parts to the dead has perhaps greater implications for the living. The whole issue of how the body can be considered after death sits uneasily in legal terms.

Example 4

A 3-year-old girl with chronic liver disease is deteriorating on the liver transplant waiting list. Her parents make an emotional appeal on television for an organ donor. Two days later, a 14-year-old boy is hit by a car and suffers a fatal head injury. His parents are seen by the transplant coordinator. They, together with their son, had been touched by the television appeal. They agree to multiorgan donation but ask that his liver be given to the 3-year-old girl.

Is directed donation to a child in need acceptable?

The transplant coordinator states that she cannot guarantee that the liver will be given to this particular child. She explains that there is a national policy that paediatric organs are offered in the first instance to children on the transplant waiting list. The parents agree to donation on the understanding that his organs will only be transplanted into adults if there are no suitable children on the waiting list.

Is it acceptable that children on the national transplant waiting list have priority for paediatric donor organs?

In this case the family have requested that their deceased son's liver be used for a transplant to another child that they have seen on a television appeal. This would be considered to represent directed donation but importantly is not conditional. The parents have requested that organs be used for children if at all possible but, if not, do not object to their use in adults. On a practical level there are good reasons, such as size and quality of function, why it is preferable to allocate childrens' organs to other children. Their opportunities for transplants are often constrained by the size of organs and since their natural premorbid life expectancy is so much greater than adults it seems reasonable to adopt this policy. Regarding the morality of such a policy, there can be few adults who would object to a child receiving a transplant from another child in preference to the organ being used, based on other priorities, in an adult. This is a bias in favour of children which we as a society have endorsed. This bias is considered justifiable because children needing transplants have an inherent bias against them receiving a suitable organ since most of the organ donors are adults.

Example 5

A 63-year-old woman suffers an intracerebral bleed and is declared brainstem dead. She was teetotal. The family are seen by the transplant coordinator. They note the publicity that George Best (a well-known football star who received a liver transplant) received after he returned to drinking after his liver transplant. They agree to multiorgan donation on condition that her organs should not be transplanted into alcoholic patients. The transplant coordinator contacts UK Transplant for advice.

What advice would you give?

- Accept the family's offer?
- Decline the family's offer?
- Accept the donor as a kidney, pancreas and cardiothoracic donor but decline the liver?

UK Transplant advise that this is conditional donation and is not allowed. The organs are declined. The press then gets hold of the story. You are contacted by a radio station. You explain that, following a previous case (organs must go to a white recipient), the Department of Health has outlawed any form of conditional donation. The radio presenter comments that this is ridiculous because:

- the two situations are quite different – racism is abhorrent whereas alcoholic cirrhosis is self-inflicted;
- we should not have denied the organs to desperate patients on the transplant waiting list.

Is it reasonable for donors to refuse donation to patients who have damaged their organs through drug or alcohol abuse, or dietary excess (obesity)?

Following the radio interview there is a listeners' 'phone-in' which overwhelmingly supports the family's conditions in this case.

Should we rely upon public opinion to determine whether one condition is acceptable and another is unacceptable?

This is another example of conditional donation where conditions are placed on the use of organs. In this case there is a condition that none of the organs can be used for alcoholic patients. In terms of principle it is difficult to accept degrees of conditions and so we would view it as an unreasonable request made by the family. The issue of whether individuals could refuse to donate to individuals with self-inflicted disease is again complex. Many individuals with hepatitis C virus-associated cirrhosis or tumours who might be considered for liver transplants contracted their disease through a drug or sexual indiscretion many years before. Should they face similar discrimination? Should cardiologists, vascular surgeons and respiratory medicine specialists refuse to treat patients with smoking-related vascular or lung disease? Should marathon runners be excluded from knee replacements many years later? What is probably needed in this case is a common-sense approach with reassurance to the family that any recipients of any organ other than liver are extremely unlikely to have had their disease caused by alcohol, that only about 20–30% of patients awaiting liver transplantation in the UK have alcohol-related disease and that these patients go through a very rigorous assessment process designed to make the chance of recidivism very remote.

There is little doubt that the public is a powerful lobby, but it would seem unreasonable to make the public entirely responsible for decisions regarding who should receive healthcare. Previous research into different attitudes among the public and doctors about who should receive transplants in various scenarios demonstrated that public decisions are founded on emotive issues, for example favouring children, pregnant women, people of perceived high social worth or productivity, whereas doctors tend to make choices dependent on utilitarian or pragmatic grounds. What is probably required is balance between the two perspectives, with cognisance taken of public opinion guiding utilitarian medical decision-making.

Should anticipated media responses to controversial cases influence clinical decisions?

The media are very powerful in presenting medical innovations and medical controversies to the public. The way in which stories are covered can have a great influence on public perception. The transplantation of the famous retired footballer George Best for alcohol-associated liver disease is a good example of this. Despite satisfying the criteria for suitability for liver transplantation for alcoholic liver disease, the media focus was only on the negative aspects of the case. UK Transplant performed a small study to see whether publicity over the George Best case had an adverse effect on organ donation and concluded that it had not. In spite of this, the view from street level remained that 'most liver transplants were performed for alcoholics ... like George Best'. There have been other celebrity cases in the UK and around the world that have attracted similar media interest and condemnation. This raises the question of whether the interest of the greater good should supersede the interests of an individual if transplantation of that individual may provoke a damaging media-driven response. Doctors have responsibilities to all of their patients and it would be unfair to refuse liver transplantation to an individual (with celebrity) who met acceptable common criteria for listing for fear of adverse publicity. As with most things transparency is probably key and provided that objectivity is maintained, the reasons for proceeding with treatment can be rebutted with honesty and integrity.

Is all publicity good?

There is a philosophy that 'bad news' is good news for the media and that there is little interest or value in a 'good news' story. This slightly cynical perspective is balanced by the philosophy that even bad news can have a positive impact if it raises awareness. For example, a story about the tragedy of a child losing their life for want of a suitable heart is a bad news story with a great potential for positivity in that it is likely to raise awareness of the shortage of organs and the need for public effort toward donation. Debates about paid organ donation, transplant tourism and organ brokering usually have positive aspects in that they highlight the shortage of organs, although not always presenting useful solutions. It is difficult to manipulate the media consistently in a positive way, but by maintaining high standards of integrity and honesty and avoiding clearly dangerous or underhand practices it is possible to portray transplantation and organ donation in a very positive light.

Example 6

A 9-month-old girl with a fatal liver condition is on the liver transplant waiting list. The family have been counselled about liver transplantation in children and the common use of split livers (transplanting a portion of an adult-sized liver into a small child, and transplanting the remainder of the liver into another recipient). Her father is declared brainstem dead at another hospital after a spontaneous intracerebral bleed. The family are seen by the transplant coordinator. They agree to liver donation on the condition that part of it is used to transplant his daughter. The liver can be split and the other half used for another patient on the waiting list. No other organs can be taken.

Is this an exceptional situation in which conditional donation is justified?

The transplant coordinator recognises that this contravenes national policy on conditional organ donation and asks for advice from UK Transplant. The medical director of UK Transplant consults the chair of the Liver Advisory Group to UK Transplant and other experts. They agree that the family's request is reasonable and consult the Department of Health. The Minister of Health's representative takes the view that the public would wish to see this donation go ahead and agrees to the donation.

Who should make decisions in such cases, taking into account that the potential donor is usually medically unstable and there is a short time frame for a decision (less than 2 hours), often late at night?

- A panel of medical experts?
- A panel of lay people?
- A mixed panel of medical experts and lay people?
- The chief executive of the National Transplant Authority?
- A judge in chambers?
- The Minister of Health?

A complex example of directed donation is presented by Example 6. In this case the deceased parent had already been counselled about living liver donation to his child. The likelihood would be that had he lived he would have become a living donor undergoing an ethically and legally acceptable directed donation. The family in this case did not impose conditions on the use of his organs but simply requested that if possible part of his liver should be donated to his son. If the outcome of the case had been that this directed donation had gone ahead and that the remaining liver was successfully transplanted into an adult, then no one would have been disadvantaged by the decision to proceed to directed deceased donation. In this exceptional circumstance this would seem justifiable.

It would seem sensible to establish a policy where such exceptional cases could be reviewed by appropriate individuals who have the ability and experience to make a recommendation about the ethics and legality of a procedure going ahead. The exigencies of transplantation require that such a committee could be convened rapidly and could reach a decision swiftly enough that neither organ donation nor transplantation is prejudiced. A panel might be composed of medical specialists, ethicists, lay people or judges acting in a lay capacity (since they would not have jurisdiction and this is not a legal decision), but importantly would need to be readily accessible at any time of day or night. One practical solution would be to nominate a floating panel of individuals from around the country who have knowledge of transplant legislation and ethics so that an active panel of perhaps three individuals could be selected from regions remote to the donor or potential recipients to avoid a conflict of interest. This panel could then consider the case and give advice to the donor team, UK Transplant and the Department of Health. It is unlikely that such events would occur frequently but a transparent approach to what can be difficult and emotionally charged issues would help to balance the requirements of upholding the principles of justice and equity of access to transplantation with the exceptional societal requests that arise from time to time.

> *"The society which scorns excellence in plumbing as a humble activity and tolerates shoddiness in philosophy because it is an exalted activity will have neither good plumbing nor good philosophy ... neither its pipes nor its theories will hold water."*
>
> John W. Gardiner, US Administrator (1912–2002)

Paid donation and transplant tourism

Example 7

A prominent local businessman originally from Pakistan has renal failure and is on haemodialysis. He presents to the transplant clinic and states that his son has volunteered to donate a kidney to him. The son is resident in Pakistan and will visit for assessment and is asked to bring proof of identity in line with Human Tissue Authority recommendations. The son arrives and his passport confirms the same name as his father. He is unable to provide a birth certificate having never been issued with one.

The potential donor is assessed and found to be fit and blood group compatible; however, HLA typing suggests that there is no genetic relationship whatsoever between the donor and recipient.

The independent assessor notes this apparent discrepancy and delves further. The potential recipient explains that he used the term son in a colloquial rather than literal sense and that the donor is actually his nephew.

Should the transplant go ahead?

Living donors in the UK may be genetically related or unrelated. In the case of genetic relationship there are often few issues related to legitimacy of the reason for donation. In the case of genetically unrelated donation which is not spousal or between partners, a closer enquiry of reasons to proceed with donation is required. In this particular unusual case the donor–recipient pair initially claimed to be genetically related (father/son) but then qualified their relationship when it became clear that they were dissimilar on HLA matching by stating that they were uncle and nephew. In this case this change in apparent relationship created reasonable cause for concern and this was highlighted to the independent medical assessor and also to the Transplants Approval Team of the Human Tissue Authority.

Genetic testing of donor and recipient pairings is no longer performed and HLA mismatch is quite possible between an uncle and his nephew. In view of this and following medical assessment, the donation and transplant go ahead.

Following living kidney donation the donor recovers well and is discharged from hospital. The recipient has initially good graft function but this deteriorates on day 7 postoperatively and he is diagnosed as having severe vascular rejection. He is treated with two pulses of high-dose steroids and graft function improves. A second renal biopsy is taken, following which there is bleeding from the kidney and a decision is taken to perform an angiogram. The angiogram shows that the bleeding has ceased but unfortunately the recipient develops radiological contrast nephropathy and requires a period of renal support.

During his prolonged stay in hospital he is visited by the donor and a conversation is overheard by a nurse in which the donor complains to the recipient that he has not received his money and the recipient is heard to reply 'I am not paying you until I know this kidney is going to work.' The nurse reports the conversation to you.

What should you do?

You challenge the donor and recipient separately and the donor confirms that the recipient agreed to pay him compensation for loss of earnings and that was all that was meant by the overheard conversation. The recipient admits that he was going to pay some money as compensation and states that the donor is permitted to receive compensation for loss of earnings, travel, mortgage, etc. under the Human Tissue Act. You refer the case to the Human Tissue Authority.

In this case there must be a strong suspicion that paid donation has occurred, which is in direct contravention of the Human Tissue Act. While the recipient is correct in his assertion that donors are entitled to reimbursement of legitimate expenses associated with organ donation, the source of reimbursement is the local health board or trust and there is a well-defined process of application for this. Direct payment between donor and recipient remains illegal in the UK and in many other countries around the world.

Example 8

Another patient with renal failure returns from having a kidney transplant abroad and presents to the transplant ward with pain in the right iliac fossa. She has undergone a paid kidney transplant from a living donor but has no details of the donor, the procedure or her perioperative care except the drugs that she has been prescribed. She demands medical care under the auspices of the National Health Service (NHS).

What should you do?

This scenario is increasingly common and there can be few if any transplant units in the UK that do not have similar experiences. The patient is entitled to care under the NHS and the fact that she has a complication of a paid procedure performed in another country is irrelevant to her eligibility for treatment. Data from the USA show that patients having paid transplants overseas have more complications than patients transplanted in the West and are frequently discharged home with little or no information. Common problems are acute rejection, infection related to over-immunosuppression and technical problems. In spite of these problems, what limited data there are suggest that long-term graft function is equivalent to that of patients transplanted in the UK or USA.

What is the legal position of these two cases?

In September 2006, new legislation came into effect in England, Wales and Northern Ireland. The Human Tissue Act 2004 is now the primary legislation regulating transplantation in those countries. Separate legislation applies in Scotland – see the Human Tissue (Scotland) Act 2006.

- Section 32 of the Human Tissue Act 2004 prohibits commercial dealings in human material for transplantation.
- Section 20 of the Human Tissue (Scotland) Act 2006 prohibits commercial dealings in parts of a human body for transplantation.

In the first case, if it could be proved that the patient and donor engaged in a financial relationship which supported the idea of the recipient purchasing the kidney from the donor, then either or both parties could face prosecution under the Human Tissue Act. In the second case, since the transaction occurred in a foreign country the recipient, although an NHS-entitled British citizen, has not broken UK law and is not liable to prosecution in this country. She may be liable to prosecution in Pakistan, which has recently changed its laws to prohibit commercial dealing in organ transplants. Furthermore, the brokers of such transplant contracts may also be liable to prosecution.

"Absurdity, n. A statement or belief manifestly inconsistent with one's own opinion."

Ambrose Bierce (1842–1914),
The Devil's Dictionary

Why does the UK prohibit commercial transplantation?

As a society we are committed to acting within the boundaries of these legislative frameworks. We support the view that organs and tissues should be freely given without exploitative commercial consideration or financial profit. However, we do support the reimbursement of costs incurred and losses attributable to the transplant donation process. We take the view that such reimbursement should be the responsibility of the health services and any system set up to provide such reimbursement should have appropriate safeguards in place that effectively exclude the possibility of exploitation of donors or profit to intermediaries.

What drives the market in organ trading?

1. Demand has outstripped the supply of organs available for clinical transplantation and the number of individuals on the UK Transplant register has increased considerably over recent years.
2. This deficit in the organs available for clinical transplantation in the UK may have prompted many people with end-organ failure to seek a life-saving transplant from sources overseas and this may have involved illegal brokering, exploitation of donors and profit to intermediaries.
3. A position of absolute prohibition on the commercialisation of human material has not to date effectively prevented the trafficking in human material for transplantation.

An illegal market in human body parts for transplantation exists. There is evidence that this illegal market results in the exploitation of (often) vulnerable persons and allows intermediaries to profit financially. This contravenes the principles of equity and justice. That such an illegal market exists is reprehensible.

It is not clear that an 'ethical market' in human body parts is either possible or morally acceptable. In order to give serious consideration to such a market scheme it would need to be established that:

(i) there are circumstances in which human material is property; and
(ii) there are circumstances in which it can be morally legitimate to co-modify this property to be bought and sold.

If both of these conditions were satisfied it would then be necessary to establish that:

(i) it is possible to allow an 'ethical market' to exist in which the risk of contravening the principles of equity and justice can be brought within acceptable limits;
(ii) it is possible to regulate this market in such a way that minimises the risk of coercion of vulnerable persons; and
(iii) implementing this market in combination with other measures of organ donation would save lives and lead to less compromise of accepted moral principles.

Without examining the consequences of a well-regulated and ethically constituted market, there will never be evidence as to whether or not we are able to adequately satisfy these conditions. For this reason it is legitimate to explore how such systems might work cognisant of the fact that they may never be accepted.

Can private organ transplantation coexist within a nationally funded transplant system?

Imagine a situation where private (paying) overseas patients and NHS-entitled patients competed for the same organs. Any individual who received financial gain from the transplant of paying patients would have a huge conflict of interest which would likely be to the detriment of non-paying patients with equal entitlement. The only way of legitimately separating the issue is focused on eligibility to receive an organ. This is the basis of separating entitled or category 1 patients from non-entitled or category 2 patients. Under current legislation category 2 non-entitled patients are only permitted to receive a cadaveric organ transplant in the UK if the organ has been refused for all entitled patients and has been offered to the European Organ Sharing Network and has been similarly declined. This mechanism preserves the better quality organs for entitled citizens and means that private transplantation on the whole utilises organs of poorer quality.

What about the situation where private and NHS patients have equal entitlement to organ transplantation?

Again, where two individuals are equally entitled to receive an organ the fact that one is paying for private healthcare, including fees for surgeon, anaesthetist and hospital costs, creates a huge potential for institutional or individual bias favouring the paying recipient. Without absolute transparency and objectivity this system clearly has the potential to contravene the overarching principle of equity of access to healthcare.

> *"Wisdom is knowing what to do next; virtue is doing it."*
>
> David Starr Jordan (1851–1931), The Philosophy of Despair

References

1. General Medical Council. Tomorrow's doctors. London: GMC, 1993.
2. Gillon R. Philosophical medical ethics. Chichester: Wiley, 1985.
3. Boyd KM, Higgs R, Pinching AJ (eds). The new dictionary of medical ethics. London: BMJ Publishing, 1997.
4. General Medical Council. Good medical practice, 2nd edn. London: GMC, 1998.
5. Davies M. Textbook on medical law, 2nd edn. London: Blackstone Press, 1998.
6. General Medical Council. Seeking patients' consent: the ethical considerations. London: GMC, 1999.
7. Neuberger J, Mayer D. Conditional organ donation: case scenarios and questions. Transplantation 2008; 85(11):1527–9.
8. Neuberger J, Mayer D. Conditional organ donation: the views of the UK general public findings of an Ipsos–Mori Poll. Transplantation 2008; 85(11):1545–7.
9. Human Organ Transplants Act 1989; http://www.opsi.gov.uk/acts/acts1989/ukpga_19890031_en_1.
10. Human Tissue Act 2004; http://www.opsi.gov.uk/ACTS/acts2004/ukpga_20040030_en_1.
11. Human Tissue (Scotland) Act 2006; http://www.opsi.gov.uk/legislation/scotland/ssi2006/20060420.htm.

Further reading

General

Anaise D, Rapaport FT. Use of non heart beating cadavers in clinical organ transplantation – logistics, ethics and legal considerations. Transplant Proc 1993; 25:2153–5.

Andrews LB. My body, my property. Hastings Centre Report 1986; 16:28.

Cahn E. Cited by Calabresi G, Bobbin P. Tragic choices. New York: Norton, 1978.

Dossetor JB. Ethics in transplantation. In: Morris PJ (ed.) Kidney transplantation, 4th edn. Philadelphia: Saunders, 1994; pp. 524–31.

Doyal L. The role of the public in health care rationing. Crit Pub Health 1993; 4:49–53.

Kootsra G. Non-heart beating donor programmes. Transplant Proc 1995; 27:2965.

Mason JK, McCall Smith RA. Law and medical ethics. London: Butterworth, 1994.

Rawls J. A theory of justice. Oxford: Oxford University Press, 1976.

Youngner SJ. Respect for the dead body. Transplant Proc 1990; 22:1014.

Living donation

Department of Health. An investigation into conditional organ donation, 2000; http://www.dh.gov.uk/assetRoot/04/03/54/65/04035465.pdf.

Michielsen P. Medical risk and benefit in renal donors: the use of living donation reconsidered. In: Land W, Dossetor JB (eds) Organ replacement therapy: ethics, justice and commerce. Berlin: Springer-Verlag, 1991.

Najarian JS, Chavers BM, McHugh LE et al. 20 years or more of follow up of living donors. Lancet 1992; 340:807–10.

Russell S, Jacob RG. Living related organ donation: the donor's dilemma. Patient Educ Counselling 1993; 21:89–99.

Paid organ donation and transplant tourism

Bellagio Task Force. Report on transplantation, bodily integrity and the international traffic in organs. Transplant Proc 1997; 29:2739–45.

Bignall J. Kidneys: buy or die. Lancet 1993; 342:45.

Daar AS. Transplantation in developing countries. In: Morris PJ (ed.) Kidney transplantation, 4th edn. Philadelphia: Saunders, 1994; pp. 478–503.

Davies I. Live donation of human body parts: a case for negotiability? Med Legal J 1991; 59:100.

Harris J, Erin C. An ethically defensible market in organs. BMJ 2002; 325:114–15.

Schlitt HJ. Paid non-related living donation: horn of plenty or Pandora's box? Lancet 2002; 359:906–7.

UNOS Ethics Committee, Payment Subcommittee. Financial incentives for organ donation, 1993; http://www.unos.org/resources/bioethics.asp?index=3.

Wigmore SJ, Forsythe JLR. Incentives to promote organ donation. Transplantation 2004; 71:159–61.

Wigmore SJ, Lumsdaine JA, Forsythe JLR. Ethical market in organs – defending the indefensible? BMJ 2002; 325:835.

Wilkinson S, Garrard E. Bodily integrity and the sale of human organs. J Med Ethics 1996; 22:334–9.

2

The acute shortage of deceased donors in the UK – recommendations set in an international context

Christopher J. Rudge

Introduction

Although there has been a shortage of deceased donors for as long as organ transplantation has been part of established clinical practice, that shortage has never been more extreme – or been perceived to be so – than it is now. This is the result of a number of factors, of which perhaps the most important is the very success of transplantation. It would be facile to suggest that the problems of rejection have been overcome – but certainly acute, early rejection has almost disappeared as a cause of graft loss. Chronic graft loss, which may or may not be primarily immunological, remains a major issue but current immunosuppressive regimens are continuing to improve the long-term outcome for transplant recipients (primarily through early outcomes) and transplantation has also become safer. As a result more patients are surviving their first, failed transplant and are in need of a re-transplant. More importantly, the relative success and safety of transplantation means that far more patients are being considered suitable for transplantation – it is not that long ago that, for kidney transplantation, 'older' patients were those over the age of 60, and many centres were reluctant to list such patients. Similarly, patients with diabetes were thought to represent too high a risk in some centres.

Fifteen years ago the deceased organ donor rate in the UK was broadly in line with that of most other western European countries. Sadly that is no longer the case and of those countries with well-developed organ donation and transplant services, we now sit uncomfortably close to the bottom of the league table (**Fig. 2.1**). Of our closer neighbours only Denmark and The Netherlands have a lower donor rate (expressed as donors per million population) and many have rates close to double that of the UK. The rate in Spain is now nearly three times higher.[1]

There have been considerable efforts in recent years to address this problem and in the case of donors after cardiac death (previously known as non-heart-beating donors) these efforts have met with considerable success. The number of such donors is still relatively small (approximately 20% of all deceased donors) but it has risen by 350% in the past 6 years. However, the number of donors after brain death (heart-beating donors) has fallen over the same time period by 14%, and such donors remain the bedrock of liver, heart and lung transplantation. In part to compensate for this lack of deceased donors, the number of living donor kidney transplants has increased markedly (**Fig. 2.2**) and now accounts for 33% of all kidney transplants. Nothing demonstrates the critical shortage of deceased donors more clearly than the willingness of donors, recipients and clinicians to put at risk the life of a fit healthy person in this way.

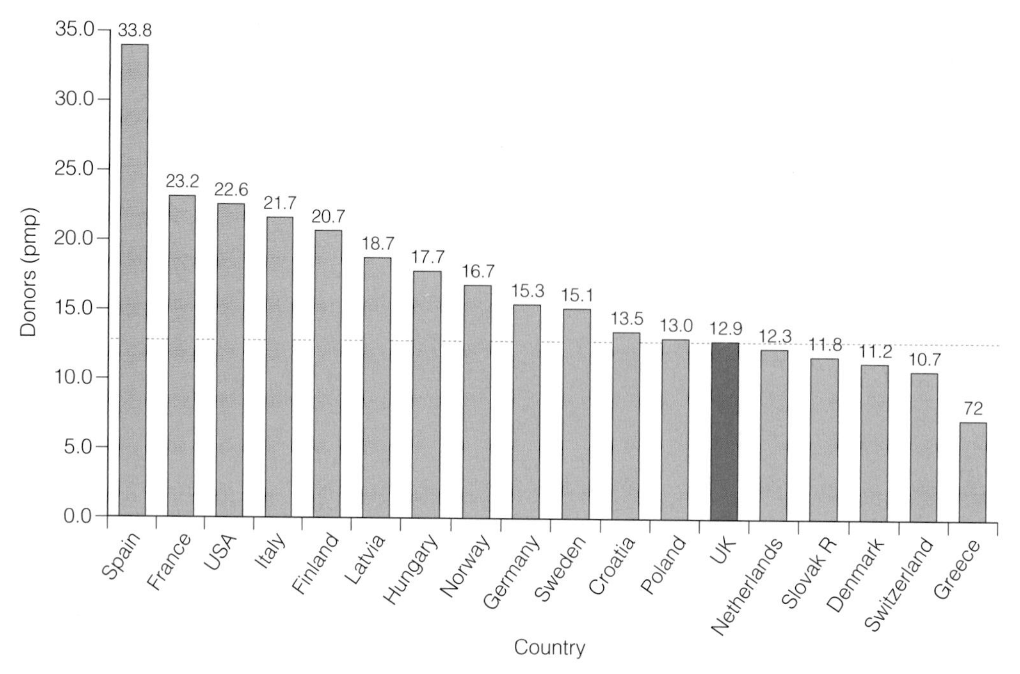

Figure 2.1 • Deceased organ donors rates for Europe and the USA, 2006.

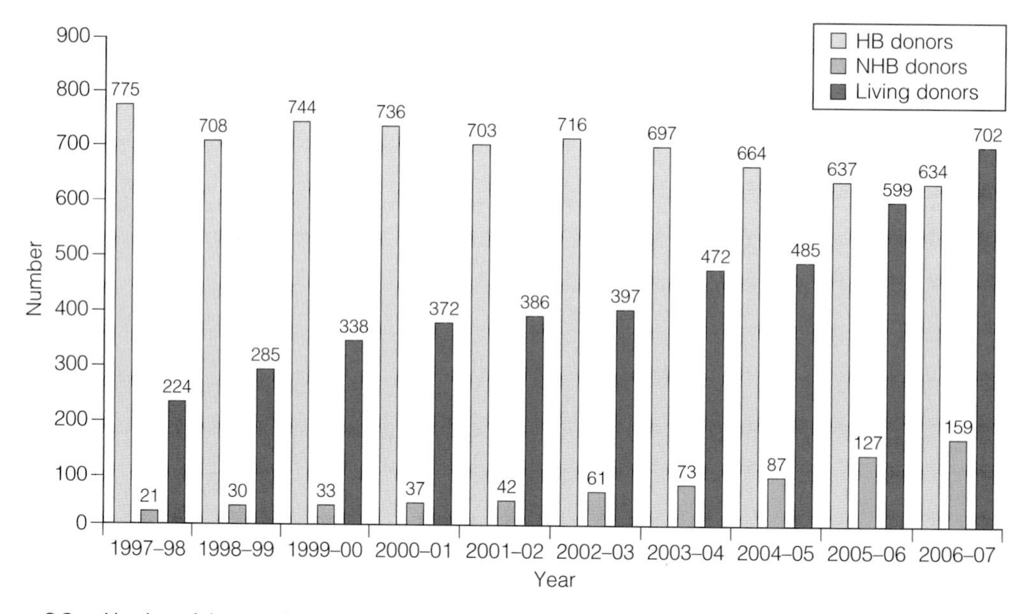

Figure 2.2 • Number of deceased and living donors in the UK, 1 April 1997–31 March 2007. HB, heart-beating; NHB, non-heart-beating.

The need for transplantation

That there is a shortage of suitable organ donors is beyond dispute, but the true extent of the shortage in the UK is far less clearly established. The obvious starting point for any assessment of the need for transplantation would seem to be the number of patients waiting for a transplant – or, more specifically, the number of new patients added to the list each year. However, these figures give only a partial answer to the question because they reflect exactly what they say and – crucially – do *not* reflect the number of patients who may benefit from a transplant that are not added to the waiting list. There is some form of rationing of access to all forms of transplantation but the details differ from organ to organ.

For many years in the UK the number of patients added to the kidney transplant list was relatively static at just over 2000 adults each year (and approximately 100 paediatric patients) but the number increased markedly between 2004 and 2006 and has now reached over 3300 per year. More patients aged over 40 are being registered and the increase is most marked in those over 60 years of age. The number of patients of Asian origin has doubled. As a result, the number of patients actively waiting for a kidney transplant is rising more rapidly than ever before (**Fig. 2.3**).

For the kidney transplant list there are published national assessment criteria modified from the European Best Practice Guidelines.[2] The underlying principle is that patients should have an anticipated survival post-transplant of at least 2 years and this is combined with a series of clinical parameters. Some of these can be assessed accurately but inevitably many are a question of judgement and it is notable that – relatively recently – the proportion of dialysis patients aged under 65 years that were listed for a transplant varied across the UK from 23% to 67%.[3]

Currently in the UK over 23 000 patients are receiving dialysis whilst the active kidney transplant list stands at 6691. There are undoubtedly a number of patients receiving renal replacement therapy for whom a transplant would be inappropriate even if there were a surplus of kidneys available, but it is also highly likely that clinicians would lower their subconscious 'threshold' if more kidneys were available.

When these figures are combined with the increasing number of patients that receive a living-donor kidney before dialysis is needed, and who are never listed for a deceased donor transplant, it is clear that the minimum number of kidney transplants that is needed each year is of the order of 3500–4000, and that if the clinical criteria were to be relaxed the true need could be far greater.

Clinical practice in liver and cardiothoracic transplant centres is more complex. For a patient with renal failure, listing for a transplant offers a possible escape, eventually, from dialysis and some patients are relatively content to dialyse for many years even though that hope is never realised. For patients with end-stage liver, heart or lung disease, however, there is no realistic alternative and many clinicians are reluctant to hold out the hope of a life-saving transplant to more patients than are realistically likely to be able to receive an organ. As a result, the number of patients listed for a liver transplant in the last financial year was 814, for a heart transplant was 194 and for lung transplantation was 204. Despite this practice, between 10% and 15% of these

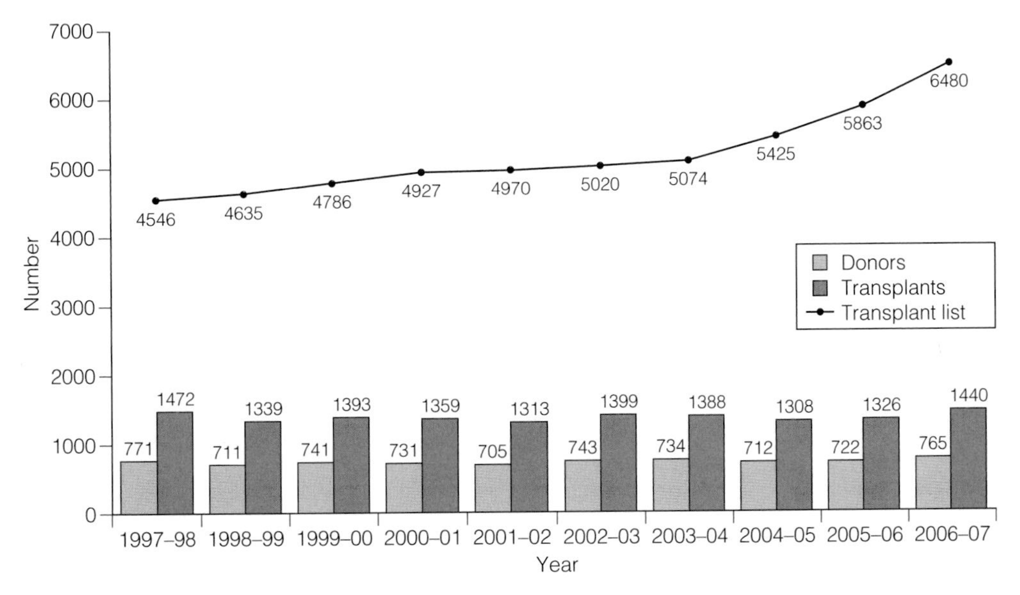

Figure 2.3 • Deceased donor kidney programme in the UK, 1 April 1997–31 March 2007. Number of donors, transplants and patients on the active transplant list at 31 March.

patients die before a transplant becomes available, or are removed from the list because their condition deteriorates beyond the point at which transplantation is appropriate.

Very detailed clinical criteria for listing for liver transplantation were introduced in September 2007, building on earlier guidelines that had been used for a number of years and which were designed to restrict access to the waiting list to patients with an anticipated 5-year survival post-transplant of over 50%. Similar, though less detailed, criteria are also available to assess patients for heart and lung transplantation. However, there are no reliable data to identify accurately the number of patients who could benefit from transplantation and it is very clear that there are many patients whose survival would be greater with a transplant than without that are never listed.

It is apparent from this that the waiting list at any given time is a very poor indication of the need for transplantation, representing as it does not only the balance between new patients listed and those removed from the list following a transplant or their death or deterioration, but also the inevitable rationing of access to a transplant list that is a consequence of the shortage of donors. However, the waiting list continues to rise, and the gap between 'supply' and 'demand' is rising (**Fig. 2.4**). For the record, the active UK transplant lists at 27 December 2007 were:

Kidney	6691
Kidney/pancreas	229
Liver	307
Heart	85
Heart–lung	20
Lung	258
Other	20

It should also be noted that the organ allocation system might have a direct impact on waiting list practice. For example, in the USA for many years a major part of the liver allocation algorithm was the time for which the patient had been listed. As a result clinicians would happily list patients with slowly progressive liver disease who were several years away from needing a liver transplant, in the hope that by the time the patient's liver function had deteriorated the patient would have accumulated enough waiting time to be allocated an organ. At one time the USA (population 300 million) liver waiting list approached 15 000 patients – the highest year-end figure reached in the UK (population 60 million) is 360. The move to a Mayo End-stage Liver Disease (MELD)-based allocation system in the USA (i.e. primarily on the basis of clinical need) has seen the waiting list fall markedly.[4]

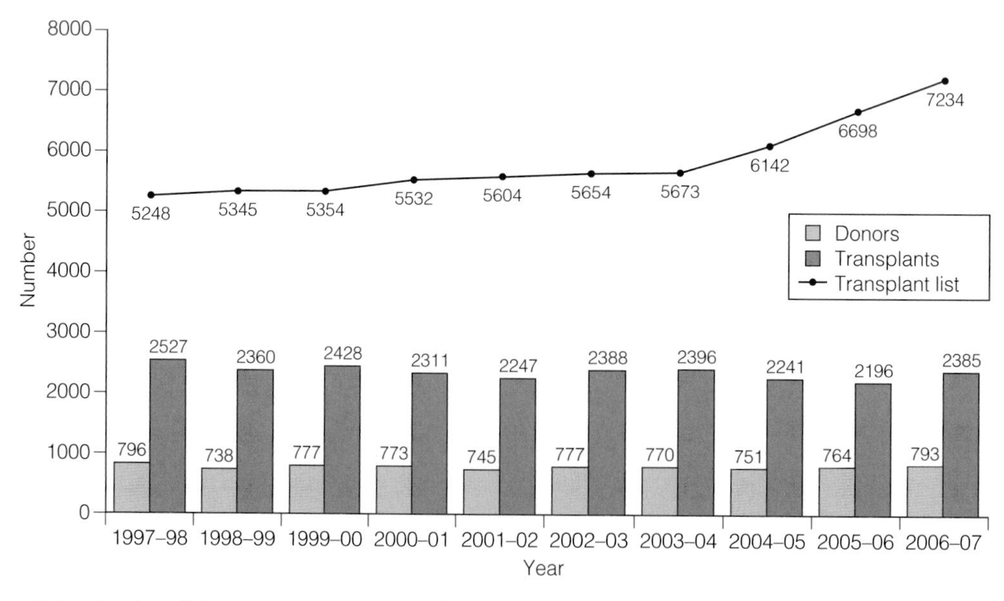

Figure 2.4 • Number of deceased donors and transplants in the UK, 1 April 1997–31 March 2007, and patients on the active transplant lists at 31 March.

There is one further source of information that gives some indication of the need for transplants – the Council of Europe annual Transplant Newsletter.[1] This provides international figures on organ donation and transplantation, although it is the donation figures that are most widely known. The transplant figures are revealing, although international comparisons can be fraught with difficulties. For each organ, Table 2.1 shows the transplant rate per million population (pmp) in 2006 in the UK and the range in five of the western European countries with the highest rates (Spain, France, Belgium, Norway and Sweden).

These figures show not only the poor record of the UK in organ transplant rates, but also give an indication of the number of patients that actually receive a transplant in countries with higher donation rates. It seems inescapable that the need for organ donors is likely to be at least 50% greater – probably double – the rate currently being achieved in the UK and that there is also a need to increase significantly the proportion of donors whose hearts and lungs are suitable for transplantation.

Nor is the need likely to diminish in the foreseeable future. There are several disturbing trends in the UK that will make even greater demands on donation and transplantation in the coming years. There is a general expectation that an 'epidemic' of liver failure associated with hepatitis C is only 5–10 years away, the incidence of diabetes and obesity in the general population are increasing, there are an increasing number of patients with congenital heart disease surviving into adulthood who will require heart transplants, the need for lung transplantation for patients with cystic fibrosis will continue, and (as noted above) the first evidence of the rising need for kidney transplantation amongst the BME (black and minority ethnic) population is now appearing. The acute shortage of donors is very real, and it is almost inevitable that the gap between supply and demand will increase unless radical steps are taken to bring organ donation in the UK very much closer to those in similar western European countries.

Table 2.1 • Transplant rate per million population (pmp) in 2006 in the UK and the range in five western European countries*

	Transplant rate (pmp)	
Organ	**UK**	**Europe**
Kidney	23.2	25.7–46
Liver	10.7	13.2–23.5
Heart	2.6	4.4–7.4
Lung	2.0	3.3–8.4

*Spain, France, Belgium, Norway and Sweden.

Current obstacles to organ donation in the UK

Introduction

There are a number of 'external' factors that will undoubtedly have an impact on organ donation rates – although the extent of that impact is a matter of debate. The overwhelming majority of deceased donors after brain death are patients whose clinical care has been provided in critical care units, although there is a small but increasing number of such patients that are identified in Accident and Emergency departments. It is inescapable that the number of intensive care beds in a local region or country will influence clinical practice in the admission to ICU of patients with catastrophic brain injury and it is notable that Spain, with its donor rate that is three times greater than that of the UK, has many more critical care beds pro rata than the UK. Differences in the number of road traffic accident deaths and the incidence of intracerebral bleeding may also have an impact when making international comparisons, and when appropriate adjustments are made some of the international variation can perhaps be explained.[5,6]

There are also concerns about the definitions used in international comparisons.[7] The donation rate can be expressed most meaningfully in terms of the number of potential donors that become actual donors – i.e. the 'conversion rate'[8] – and the number of potential donors is, at least in part, influenced by some of these 'external' factors.

However, what is essential in any donation system is to maximise the conversion rate, and there is clear evidence that the UK is not succeeding in this regard. The UK Transplant Potential Donor Audit (PDA) was established in April 2003[9] and, whilst it is acknowledged that the format of the PDA can be improved, it provides important information. The data for the past 2 years[10] suggest that over 20% of patients in whom brainstem death was a possible diagnosis were not in fact tested, that only 81%

of patients certified dead after brainstem tests were referred to the donor coordinator network, and that the consent rate when the potential donor's family were approached was only 61%. The overall conversion rate was 48%. Even allowing for uncertainties about the PDA it is clear that at each stage of the process that can lead to organ donation there are patients who could potentially be donors after brain death who do not in fact become donors.

For potential donors after cardiac death the situation is even less satisfactory, although the caveats about the data are perhaps greater. The referral rate for such potential donors was only 32% and the conversion rate was 12%.

The Organ Donation Taskforce

It was in recognition of the poor donation rate in the UK and the increasing gap between the number of patients listed for a transplant and the number of organs available from deceased donors that the ministerial Organ Donation Taskforce was established by the Department of Health in late 2006. The Taskforce was charged with identifying obstacles to donation in the UK and making recommendations to overcome these obstacles. It had a wide-ranging membership including transplant surgeons and coordinators, critical care clinicians, NHS management, and individuals with expertise in ethics, the media and ethnic minority issues. From the outset there was uniform agreement that organ donation and transplantation can only be approached within a UK-wide perspective and the devolved administrations in Scotland, Wales and Northern Ireland were closely involved.

There have been a number of previous reports into donation and transplantation in the UK, none of which has been implemented in full, and in addition to reviewing the available evidence the Taskforce commissioned further work on various aspects of donation. It took evidence from those most involved with the successful Spanish model, which has been introduced in northern Italy and several South American countries, and with the more recent Organ Donation Breakthrough Collaboratives in the USA. The final Taskforce Report was publicly launched in January 2008. As this Report is the most significant structured approach to improving donation in the UK for many years, much of the remainder of this chapter will be based largely on the approach taken by the Taskforce and on its recommendations, although the author takes sole responsibility for this description that in parts goes beyond the Report itself. The report is available at www.dh.gov.uk (http://www.dh.gov.uk/en/Publicationsandstatistics/Publications/PublicationsPolicyAndGuidance/DH_082122).

The key stages in organ donation

There are a number of key stages that may, in appropriate circumstances, lead to organ donation and in identifying the obstacles, and therefore the solutions, it is helpful to consider each stage separately. However, all these stages take place in the wider context of the legal framework and public attitude towards donation, which also need consideration. The key stages are similar for donation after brain death and donation after cardiac death, although some of the issues raised may be specific to one or other type of donor. The common framework is that patients with major cerebral injury from a variety of causes are treated actively in the expectation or hope of survival until such time as either testing for brainstem death is carried out, or further treatment is deemed to be no longer appropriate and a decision to withdraw active treatment is made.

The key stages are:

1. Donor identification.
2. Donor referral.
3. Consent.
4. Donor management.
5. Organ retrieval.

Donor identification

Historically organ donation has been very largely restricted to intensive care units but there is increasing evidence from local audits that as a result of several factors including the wider availability of scanning techniques and – it must be said – the lack of intensive care beds in some areas, there are an increasing number of potential donors after both brain death and cardiac death who are never admitted to intensive care units. Throughout the rest of this chapter the term 'critical care' will be used to describe any part of a hospital where such patients are treated and where decisions are made.

Both steps – brainstem testing or treatment withdrawal – must be made solely in the patient's best interests and should be made regardless of whether organ donation is possible or not. This is a fundamentally important point. There would still appear to be too many situations where the use of brainstem tests to diagnose death are felt to be necessary only when organ donation is being considered and it is the clear view of the Taskforce – and of both the General Medical Council and the British Medical Association – that brainstem death testing should be carried out in all patients where brainstem death is a likely diagnosis even if organ donation is an unlikely outcome.

Although it is not possible from the PDA data to be certain that those patients who were not tested for brainstem death would all have met the criteria, it is likely that a significant proportion would have done so. It is also likely that some patients would have been suitable organ donors. Unless all appropriate patients are tested as part of the routine standard of care there will continue to be potential donors after brain death that are not identified.

For potential donors after cardiac death the issues are somewhat different. It is not uncommon in critical care to reach a stage in a patient's care at which further active treatment is no longer thought to be appropriate even though the patient would not meet the criteria for death by testing the brainstem. When that decision is made, in the patient's best interests, recognition that donation should be a normal part of end-of-life care is the key to the identification of potential donors.

This acceptance that organ donation should be a normal part of end-of-life care for all appropriate patients is the case in some but not all critical care settings at present. To make that shift from donation being unusual to being usual will represent a challenge to some critical care clinicians, despite the support of the relevant Royal Colleges and professional societies.

Donor referral

A number of systems have been introduced in other countries designed to ensure that all possible donors are referred to the donor coordinator network. The use of 'clinical triggers' to achieve this has been the subject of much discussion and in the USA the mandatory notification of all 'imminent deaths' has been introduced in some states. However, discussions on the exact definition of 'imminent death' have not yet reached a national consensus. There is also potential confusion over the difference – if any – between 'notification' of a patient to the donor coordinator (which should mean nothing more than it says) and 'referral', which implies the start of the donation process. In the UK there has been considerable concern that 'premature' notification could result in the notification of patients who subsequently survive. If this were confused with referral it could reinforce the false perception that donation was more important than active treatment of a patient and would be unacceptable to both clinicians and the public. Evidence that would allow the development of appropriate detailed clinical triggers in this context is not available but nevertheless notification criteria can be established. In addition to the potential benefits in terms of increasing donation, the introduction of clear criteria for the notification of patients to the donor coordinator are necessary if the donation process is to be accurately monitored in all hospitals (see below) and this information will provide a robust denominator in measuring the true conversion rate.

Consent

In the UK, the Human Tissue Act 2004 and its Scottish equivalent were both introduced in September 2006 and the Acts, together with the Codes of Practice published by the Human Tissue Authority, set out the requirements for consent for organ donation. The underlying principle is that organ removal cannot take place unless explicit consent (or authorisation) is available. If the individual in life had given consent in the form of, for example, registration on the NHS Organ Donor Register (ODR) or a living will, then that consent is legally valid and is paramount. It cannot be overridden, at least in law. This has changed fundamentally the legal standing of ODR registration from a record of the wishes of the individual to donate organs and/or tissues to legally valid consent. Concerns have been raised that those people who had registered prior to September 2006 had done so without the information that by so doing they were giving consent, and that those who register since that time may still not be given sufficient information to give informed consent. To some extent this is a question of semantics about the term 'informed consent' – normally used in clinical medicine as being required before an operation or other interventional procedure and relatively well defined. However, in practice the process of organ donation requires that detailed discussions about a wide range of medical and behavioural issues takes place with the potential donor's family or friends, and coordinators are very sensitive to the need to ensure that ODR registration truly reflects the wishes of the individual. Currently approximately 15 million people are registered on the ODR and appropriate individuals can consult it at any time to identify whether a potential donor has registered.

If the individual had not given consent but had named a designated individual it is for that individual to give consent. If neither of the above steps had been taken there is a defined hierarchy of individuals in a 'qualifying relationship' with the potential donor who are able to give consent.

The PDA records the outcome of the approach to consent for organ donation and shows that, overall, consent is given by 61% of families whilst 39% do not do so. The consent rate is very much higher when the wishes of the individual to donate have been recorded or made known. When the potential donor comes from any ethnic background other than white (Caucasian) the consent rate drops to approximately 25%.[10] These figures are in stark contrast to the 85% consent rate in Spain, and the UK has one of the lowest consent rates recorded in the Council of Europe Newsletter, although the data are more limited and less comparable than for national donation and transplant rates.

Experience from pilot studies in the UK and from elsewhere in the world suggest that there are two main prerequisites for discussing organ donation with the family or next of kin of the potential donor in a way that ensures that their wishes are respected whilst maximising the consent rate. These are that the individuals concerned should have appropriate and specific training, and should have as much time as is necessary. In the UK consent may be sought by the intensive care clinician who has been involved with the care of the patient, by the donor transplant coordinator, or by the two individuals together. A randomised study is currently under way in the UK to compare these approaches. However, despite the best efforts of all concerned, the two prerequisites set out above are difficult to achieve within the current structure in the UK. Very specific skills are required of those who approach the donor's family and although much detailed training is now available, this is a relatively recent development and has not yet been introduced universally. Even more difficult is the need to devote extensive time to the needs of the donor's family. A donor coordinator has a myriad of tasks to carry out, of which consent is only one, and with the best will in the world this makes the role very difficult indeed.

Donor management

The dramatic physiological consequences of death of the brainstem are well described. Approximately 80% of potential donors become hypotensive, 65% develop diabetes insipidus, 30% have cardiac arrhythmias and 20% develop pulmonary oedema. Although these changes affect all organs to some extent they are most significant in the case of the heart and the lungs. It is, at least in part, as a result of this that cardiothoracic transplant rates are particularly low in the UK, but the early post-transplant function of the liver and the kidneys is also affected adversely. Good donor management can result in more potential donors becoming actual donors, and in more organs being retrieved and transplanted from each donor.

Following the certification of death by brainstem testing, the emphasis of clinical management can change in an attempt to improve organ function (see Chapter 6 for full details). In order to achieve this non-invasive or, more commonly, invasive methods of cardiac monitoring are required and this may include the use of a pulmonary artery catheter and echocardiography – ideally with a transoesophageal scan. The goals of haemodynamic management are to optimise cardiac output, maintaining normal preload and afterload, whilst reducing or eliminating inotropes and vasopressors. In particular, vasopressin may be a much more satisfactory inotrope than adrenaline or noradrenaline. There is some evidence that hormone replacement therapy (pitressin, insulin and tri-iodothyronine) improves organ function and this is now used in most cardiac donors. If the lungs are potentially suitable the ventilator settings should be adjusted to provide a modest level of positive end-expiratory pressure and to minimise the inspired oxygen concentration. Other aspects of donor management include chest physiotherapy and avoidance of infection or hypothermia. On occasions bronchoscopy may be needed to assess the lungs and their suitability for transplantation. It must be recognised that the complex care that is required can be emotionally demanding and stressful to both the nursing staff involved and to the donor's family, all of whom will require support.

Whilst both the principles and practice of donor management are well described and are published,[11] and are all within the skills available in any critical care unit, there are concerns that they are not always implemented.

This results from a lack of clarity as to whether they are the responsibility of the donor hospital clinicians or of the transplant teams, compounded by the inevitable pressures on local clinicians, particularly in smaller units where senior clinicians may well have a number of conflicting responsibilities. The Scottish Organ Retrieval Team was a fully resourced pilot study during 2005–6 and a key part of the success of the team was the availability of an experienced consultant anaesthetist able to provide advice on donor management to local clinicians and also able to travel to the donor hospital some hours in advance of the retrieval team in order to optimise donor management. Other local initiatives to provide support and expertise in donor management have clearly shown the benefits in terms of both the number of organs retrieved per donor and the function of donated organs.

Sufficient evidence is available to support the routine adoption for all donors of a more aggressive approach to donor management than is the norm in the UK, but in order to achieve this it is clear that either more awareness and expertise is needed in many donor hospitals, or must be provided externally through the donor coordinator network or the organ retrieval teams.

Organ retrieval

As with so many aspects of donation, arrangements for organ retrieval have developed over many years in response, largely, to local needs and local solutions. However, a general pattern has emerged in that most abdominal organs – i.e. the liver and the kidneys – are now retrieved by surgical teams based in the seven UK liver transplant units. The recent rapid growth of pancreas transplantation has seen an increasing trend back to the situation where several teams from different centres are needed, as not all liver centres carry out pancreatic transplants and not all pancreas centres are co-located with liver teams. Cardiothoracic organ retrieval remains the responsibility of the six UK transplant units.

Teams vary in their composition and experience, surgeons are frequently involved in elective clinical work whilst on call for retrieval, few teams are truly self-sufficient (i.e. they require a level of anaesthetic and operating theatre support from the donor hospital) and simultaneous retrievals in the same area are often difficult to arrange.

These problems do not all exist in all units – many offer an excellent service – but it is becoming increasingly difficult to sustain the service as training and contractual requirements continue to change.

These difficulties are compounded by changes to the on-call arrangements in many donor hospitals, which are making it increasingly difficult for them to provide support for the organ retrieval process that may take up to 6–8 hours. The donor management requirements outlined above continue during the retrieval operation.

In recognition of these difficulties the Taskforce heard considerable evidence that led it to conclude that the key principles underlying organ retrieval are that teams should be:

1. virtually self-sufficient and not require anaesthetic, theatre or surgical staff from the donor hospital (other than the minimum required for local 'liaison');
2. available 24 hours a day, without elective commitments during their time on-call for retrieval;
3. able to respond appropriately if there is more than one donor on the same day;
4. able to provide opportunities for training.

There will continue for the foreseeable future to be a need for separate cardiothoracic and abdominal retrieval surgeons, but the abdominal team should be fully competent to remove the liver, the kidneys and the pancreas.

There are a number of other key aspects to donation that are clearly essential and have been fundamental to the successes achieved with the Spanish model and in the USA. These are the donor coordinator network, the role of the wider NHS (together with the processes for monitoring, oversight and accountability of donation performance by each hospital), and the legal and ethical framework.

Donor coordination

The donor coordinator system is only part of the overall Spanish model but is widely perceived to be the most important component. Spain has an integrated national, regional and local network of donor coordinators. The 155 hospitals in Spain that are recognised for organ donation have over 300 coordinators, of which approximately two-thirds are doctors and one-third are nurses. All are employed within individual hospitals and are responsible for maximising donation from that hospital. Many of the doctors are relatively senior and they come from either a critical care or a nephrology background. They are typically employed on a part-time basis, whereas more of the nurse coordinators are full-time. The central organisation (ONT) acts to ensure that the whole system works appropriately.

This is in sharp contradistinction to the arrangements currently in place in the UK that have developed and changed over many years with little integrated planning. In the UK donors may come from over 275 hospitals yet there are only approximately 125 coordinators. With the exception of the 20 'in-house coordinators' funded as part of a UK Transplant pilot study over the past 2 years, the donor coordinators are based not in critical care areas of each and every hospital but in 18 teams around the UK, most of which are situated in, or very strongly linked with, transplant units. The coordinators may play a dual role, having some responsibility for recipients as well as for donation. Virtually all are employed by local trusts or transplant units and any national network is therefore subject to the vagaries of local priorities and decision-making.

A careful analysis of the working practice of donor coordinators in the UK reveals further difficulties that demonstrate that the current arrangements are not sustainable. Over and above work to promote donation and teaching/training in general, it is the work involved in the facilitation of each donor that is the raison d'être of the donor transplant coordinator (DTC). The system relies upon the DTC being called to a hospital where there is a potential donor, which may be some hours away from the coordinator's base. There are then three, and possibly four, aspects to the coordinator's role, each of which could be considered to be a job in its own right.

First is consent. This has been described above, but it is easy to underestimate the length and complexity of the process, particularly when the family relationships are themselves not straightforward. Moreover, the needs for detailed and sensitive information about the donor's medical and social history are now extensive and time-consuming.

Second is coordination of donation itself. In addition to arrangements for tissue typing and microbiological screening tests to be carried out, the necessary information about the donor's physiological condition and the varied tests of specific organ function must be obtained and recorded. Individual assessment of the suitability of different organs will involve discussion with local transplant teams and this may be complex if the donor's condition changes over time. Arrangements for organ retrieval must be made with the local hospital's operating theatres and with up to three separate retrieval teams. Throughout this time the family frequently require further support, and if the donor becomes physiologically unstable the DTC may be involved in aspects of donor management or in reconsideration of the suitability of some organs.

Finally the DTC attends and facilitates the retrieval procedure, completes large volumes of necessary paperwork and ensures that necessary samples and documentation accompany the organs. Any last wishes of the family are met and the last offices are administered.

From beginning to end this whole process rarely takes less than 12 hours and may well extend to 18–24 hours from the time of the first notification. For one individual to do all of this is not sustainable.

The role of the NHS

One other general principle that is emphasised in the Taskforce Report is that all parts of the NHS must embrace donation as the norm, and must be held accountable for their performance. Organ donors may come from any one of over 275 hospitals in the UK, and it is important that every hospital does everything possible to maximise the identification and referral of potential donors. Almost every hospital will see, from time to time, patients with severe kidney, liver, heart or lung disease and in due course a number of these patients will be referred to tertiary centres and ultimately listed for a transplant. The limited number of transplant centres cannot make organs available for transplantation as though by magic – the donors come from the same wide range of hospitals as the recipients. Although transplant units and donor coordinators are only too well aware of the shortage of donors, as are many critical care clinicians, work carried out on behalf of the Taskforce showed that some hospital chief executives and other senior managers were less aware of the need, had little information about the donation activity – good or bad – of their own hospital, and felt little or no responsibility for making improvements. However, there was a general readiness to take a more active role if provided with the information and support that would enable them to do so.

A key part of the improvements in the USA in recent years has been the monitoring and oversight of donation performance at individual hospitals that has been introduced with, in some places, the possibility of quite significant financial or regulatory penalties.

In the UK all stages of organ donation are currently measured by the PDA, which is currently being improved in order to provide more detailed and reliable information, and therefore it is possible to provide hospital management with the necessary information and to hold them to account. It must be emphasised that this can never be done in terms of donor numbers alone, but can and should be done in terms of a target that 100% of potential donors are identified and referred to the DTC network, and that the ODR be consulted and the relatives approached. Additionally the conversion rate at individual hospitals can be reported. Comparisons with other local hospitals and with the national picture can also be valuable.

The law

There has been much debate for many years about the possible benefits of moving to a legal system of opting out (presumed consent) rather than the opting in, or explicit consent, which is enshrined in current UK laws. Much of the argument in favour of opting out is based on a simplistic assumption of cause and effect between high donor rates and opting-out legislation in countries such as Spain. A somewhat stronger causal link between legislation and organ donor rates is claimed in the case of Belgium, where one part of the country introduced opting out and saw a rise in donation rates that was not seen in the other part of the country which did not change the law.

However, a critical review of the evidence does not support a clear relationship between the law and donation, particularly when other factors such as the road traffic accident fatality rate and the incidence of fatal intracerebral bleeding are taken into account.[6,12] It is important to note that when the ONT was charged with improving donation in Spain in 1989, the Spanish opting-out legislation had been in place for many years (since 1979). The dramatic rise in the Spanish donor rate has taken place without any change in the law and in his evidence to the Taskforce the architect of the Spanish model, Rafael Matesanz, repeated his view[13] that the improvement in donation has resulted from improvements in the organisation, not changes in legislation.

Moreover, the requirements for detailed donor information that ensure the highest possible quality and safety of donated organs mean that in practice the family of the donor are very closely involved in the discussions and donation is almost impossible without their cooperation, and so regardless of the legal framework the family have a form of de facto veto over donation if they do not wish to cooperate. In Spain, donation does not occur without the agreement of the family.

Public opinion – and that of the profession – is fairly equally divided on the merits of opting out, although there has probably been a modest shift towards a change in recent years. It raises a number of ethical and practical issues, many of which have the potential to harm donation just as much as they may have the potential to increase it. These issues require very serious consideration.

The Taskforce Terms of Reference required its recommendations to be 'within existing operational and legal frameworks' and therefore opting out was not considered. However, the Taskforce published a subsequent report in November 2008 on the issue of presumed consent.[14] A detailed systematic literature review led to the conclusion that "the evidence identified and appraised is not robust enough to provide clear guidance for policy." The Taskforce concluded it was not confident that the introduction of opt-out legislation would increase organ donor numbers, and that there is evidence that donor numbers may go down. Other issues considered included legal, ethical and practical aspects, together with the attitudes of the public and of different faith and belief groups. Whilst acknowledging that this is a finely balanced question the Taskforce reached a clear consensus that an opt-out system should not be introduced in the UK at the present time.

Ethics

The evolving practice of organ donation continues to present ethical and legal problems, and recent legislation has on occasions produced unintended consequences. The diagnosis of death by neurological criteria has been well defined for many years, was re-stated in the Intensive Care Society Guidelines in 2005,[11] and the Academy of Royal Medical Colleges has published an updated guidance document. There are few, if any, major ethical difficulties associated with donation after brain death. However, the increasing use of donors after cardiac death is still associated with concerns in the eyes of some clinicians.

Firstly there is the so-called 'Lazarus' phenomenon – the spontaneous return of cardiac activity some minutes after cardiopulmonary resuscitation has been abandoned. The number of well-described cases is extremely small and they have occurred in a different group of patients than those who are potential donors after cardiac death – i.e. patients who are known to have suffered a catastrophic brain injury and in whom further active treatment is withdrawn. Nevertheless concern has been expressed, and in particular questions have been raised about the period of time for which absence of respiratory or cardiac activity must be documented before death can be certified. The Academy report is, however, very explicit that death can be certified after a period of observation for a minimum of 5 minutes to establish that irreversible cardiorespiratory arrest has occurred.

Secondly there are understandable concerns over the interpretation of the phrase 'in the patient's best interests' as applied to donation after cardiac death. Once the decision has been made that no further treatment options are available or appropriate, it can be argued that any delay in withdrawal of treatment – as may be needed if organ donation after death has been certified is to be arranged – is not in the patient's best interests and is therefore in breach of the Mental Capacity Act 2005. This may be counter-intuitive; most clinicians would accept that such a delay in order to allow other family members to see the patient before death would be understandable, and if it were known that the patient wished to be an organ donor – for example, through registration on the ODR – once again it could easily be argued that the delay is in the patient's best interests. Any such arrangements are only made with the full agreement of the donor's family, but in this litigation-conscious age even clinicians who are fully supportive of donation after cardiac death would welcome clarification that the necessary steps are clearly lawful.

Solutions

In the light of this analysis of the shortcomings of the current organ donation system in the UK the Organ Donation Taskforce Report makes 14 specific recommendations. These are set out below together with a brief commentary where appropriate.

1. **A UK-wide Organ Donation Organisation should be established.** At present no single organisation in the UK has overall responsibility for organ donation. UK Transplant plays a role, as do several other groups, but the Taskforce was clear that a greater degree of responsibility and accountability should be given to a designated authority.
2. **The establishment of the Organ Donation Organisation should be the responsibility of NHSBT.**
3. **Urgent attention is required to resolve outstanding legal, ethical and professional issues in order to ensure that all clinicians are supported and are able to work within a clear and unambiguous framework of good practice. Additionally, an independent UK-wide Donation Ethics Group should be established.** This is crucial to allow the expansion in donation after cardiac death that is so clearly possible. Current results for kidney transplantation suggest that the short- to medium-term results are broadly equivalent to those achieved using kidneys from donors after brain death, although there is more concern about the outcome of livers from such donors. There is considerable experimental evidence to suggest that lung transplantation from donors after cardiac death may even be preferable to donors after brain death and the early results in the UK are extremely encouraging.
4. **All parts of the NHS must embrace donation as a usual, not an unusual event. Local policies, constructed around national guidelines, should be put in place. Discussions about donation should be part of all end-of-life care, when appropriate. Each hospital should have an identified Clinical Donation Champion and a Hospital Donation Committee to help achieve this.** To make that shift from donation being unusual to being usual will represent a challenge to

some critical care clinicians, despite the support of the relevant Royal Colleges and professional societies. In order to help achieve this shift, and to ensure that local policies are in place that build on the extensive national guidance that is now available, the Taskforce recommends that every acute trust (or equivalent) should have an identified Clinical Donation Champion (probably a critical care clinician) supported by a Trust Donation Committee with a non-clinical chairman (perhaps a patient or prominent local figure).

5. **Minimum notification criteria for potential organ donors should be introduced on a UK-wide basis. These criteria should be reviewed after 12 months in the light of evidence of their effect, and the comparative impact of more detailed criteria should also be assessed.** The Taskforce adopted the following approach as a national protocol for the notification of potential organ donors:

   (a) When no further treatment options are available or appropriate, and there is a plan to confirm death by neurological criteria, the donor coordinator should be notified as soon as sedation/analgesia is discontinued or immediately if the patient has never received sedation/analgesia. This notification should take place even if the attending clinical staff believe that donation after death confirmed by neurological criteria might be contraindicated or inappropriate.

   (b) In the context of a catastrophic neurological injury, when no further treatment options are available or appropriate and there is no intention to confirm death by neurological criteria, the donor coordinator should be notified when a decision has been made by a consultant to withdraw active treatment and this has been recorded in a dated, timed and signed entry in the case notes. This notification should take place even if the attending clinical staff believe that death cannot be diagnosed by neurological criteria, or that donation after cardiac death may be contraindicated or inappropriate.

These are felt to be minimum criteria that should be applied in all critical care areas, but there are likely to be clinicians in some trusts who would wish to introduce more specific criteria, and this is to be encouraged. Given the sensitivities that surround this issue it is very important that any such initiatives are introduced in a manner that allows a thorough assessment of their impact on staff and patients and their families in addition to any impact on donation. If more detailed criteria can be shown to be more effective, feasible and acceptable than the minimum criteria outlined above there would be a very strong case for their wider introduction in due course.

For a number of clinicians the introduction of notification criteria that identify possible donors before death has been certified will be a radical change of practice and it is very gratifying that these recommendations have been endorsed by the Intensive Care Society and have also been supported by the Royal College of Anaesthetists.

6. **Donation activity in all trusts should be monitored. Rates of potential donor identification, referral, approach to the family and consent to donation should be reported. The Hospital Donation Committee should report to the hospital board through the clinical governance process and the medical director, and the reports should be part of the assessment of trusts through the relevant healthcare regulator. Benchmark data from other hospitals should be made available for comparison.** The PDA has recently been revised to ensure that this information can be collected in an accurate and timely manner.
7. **Brainstem death testing should be carried out in all patients where brainstem death is a likely diagnosis even if organ donation is an unlikely outcome.**
8. **Financial disincentives to trusts facilitating donation should be removed through the development and introduction of appropriate reimbursement.** The Taskforce was clear that under current legislation financial inducements or rewards for donation are not appropriate. However, the care of an organ donor inevitably incurs costs, in terms of critical care nursing, drugs and equipment and operating theatre time. The increasing move towards a tariff-based funding system in many areas of the NHS offers an opportunity to remove

any financial disincentives that may currently exist and would bring organ donation into line with the funding arrangements for most other aspects of healthcare delivery.

9. **The current network of donor transplant coordinators should be expanded and strengthened through central employment by a UK-wide Organ Donation Organisation. Additional coordinators, embedded within critical care areas, should be employed to ensure a comprehensive highly skilled, specialised and robust service. There should be a close and defined collaboration between donor coordinators, clinical staff and Hospital Donation Champions. Electronic online donor registration and organ offering systems should be developed.** This is perhaps the most fundamental recommendation and is a clear move towards the Spanish model. The Taskforce identified the weaknesses of the current arrangements, as described above, and sees a clear need for a radical redevelopment of the DTC network. Central employment will bring all coordinators together to provide a national service. A significant increase in the number of coordinators will be needed, as the Taskforce believes that the several components of the coordinator role should be more clearly separated. At least two, and possibly three, coordinators may be involved with each donor, with the intention that one person will devote their time exclusively to the donor's family (consent, donor history and family support) whilst a second will be responsible for all the practical aspects involved in coordination of the donation process, organ offering and allocation, liaison with retrieval teams and (possibly) the retrieval procedure itself. Work is currently underway to identify the likely number of coordinators that will be needed, but at a minimum it is likely to double the current number. Most coordinators will spend the majority of their time working within critical care units in individual hospitals or trusts and they will work closely with the identified clinical champions. Through a revised team structure they will in addition provide the on-call service to all hospitals in their area.
10. **A UK-wide network of dedicated Organ Retrieval Teams should be established to ensure timely, high-quality organ removal from all heart-beating and non-heart-beating donors. The Organ Donation Organisation should be responsible for commissioning the retrieval teams and for audit and performance management.** Working closely with the British Transplantation Society, which has recently reviewed current and future possible retrieval arrangements, the existing multiorgan retrieval teams will be strengthened to ensure that they can meet the criteria set out above. Clear standards will be agreed for the retrieval teams themselves, and it is expected that these will include arrangements to optimise donor management and to increase both the viability of organs retrieved and the number of organs retrieved per donor.
11. **All clinical and nursing staff likely to be involved in the treatment of potential organ donors should receive mandatory training in the principles of donation. There should also be regular update training.**
12. **Appropriate ways should be identified of personally and publicly recognising individual organ donors, where desired. These may include national memorials, local initiatives and personal follow-up to donor families.**
13. **There is an urgent requirement to identify and implement the most effective methods through which organ donation and the 'gift of life' can be promoted to the general public, and specifically to the BME population. Research should be commissioned through Department of Health Research and Development funding.** This is of critical importance given the relatively low consent rate that is seen in the UK – despite the general public's wide support in principle for organ donation. The imbalance between the needs for transplantation in the BME population and the low donation rate is of particular concern and whilst much is known about the reasons for non-donation there is a need for more focused research into the active steps that could be taken to resolve this problem.
14. **The Department of Health and the Ministry of Justice should develop formal guidelines for coroners concerning organ donation.** Most coroners are very supportive of organ donation, and it is clearly accepted that in

rare and unusual circumstances the coroner may be unable to agree to donation. However, there are occasions when the coroner's refusal is more difficult to understand and this causes distress and concern – not least to the potential donor's family – out of all proportion to the rarity of its occurrence. The Coroners' Society is extremely supportive of donation.

Conclusion

The recommendations have achieved widespread and senior political support in all parts of the UK and are to be implemented in full.

These recommendations must be taken together in order to ensure that the identification and referral of potential donors and consent to donation are maximised. Improved donor management and organ retrieval will improve the number of organs that can be removed and successfully transplanted. The support of transplant and critical care clinicians, coordinators and NHS management is crucial, as are steps to improve the public's active support for donation. Successful implementation of the recommendations will undoubtedly be a challenge but they offer the most realistic opportunity for many years to make the step-change in UK donation and transplant rates that is so clearly needed.

Key points

Recommendations from the UK Organ Donor Taskforce to bring about change in organ donation

- A UK-wide Organ Donation Organisation should be established.
- The establishment of the Organ Donation Organisation should be the responsibility of NHSBT.
- Urgent attention is required to resolve outstanding legal, ethical and professional issues in order to ensure that all clinicians are supported and are able to work within a clear and unambiguous framework of good practice. Additionally, an independent UK-wide Donation Ethics Group should be established.
- All parts of the NHS must embrace donation as a usual, not an unusual event. Local policies, constructed around national guidelines, should be put in place. Discussions about donation should be part of all end-of-life care, when appropriate. Each hospital should have an identified Clinical Donation Champion and a Hospital Donation Committee to help achieve this.
- Minimum notification criteria for potential organ donors should be introduced on a UK-wide basis. These criteria should be reviewed after 12 months in the light of evidence of their effect, and the comparative impact of more detailed criteria should also be assessed.
- Donation activity in all trusts should be monitored. Rates of potential donor identification, referral, approach to the family and consent to donation should be reported. The Hospital Donation Committee should report to the hospital board through the clinical governance process and the medical director, and the reports should be part of the assessment of trusts through the relevant healthcare regulator. Benchmark data from other hospitals should be made available for comparison.
- Brainstem death testing should be carried out in all patients where brainstem death is a likely diagnosis even if organ donation is an unlikely outcome.
- Financial disincentives to trusts facilitating donation should be removed through the development and introduction of appropriate reimbursement.
- The current network of donor transplant coordinators should be expanded and strengthened through central employment by a UK-wide Organ Donation Organisation. Additional coordinators, embedded within critical care areas, should be employed to ensure a comprehensive highly skilled, specialised and robust service. There should be a close and defined collaboration between donor coordinators, clinical staff and Hospital Donation Champions. Electronic online donor registration and organ offering systems should be developed.

- A UK-wide network of dedicated Organ Retrieval Teams should be established to ensure timely, high-quality organ removal from all heart-beating and non-heart-beating donors. The Organ Donation Organisation should be responsible for commissioning the retrieval teams and for audit and performance management.
- All clinical and nursing staff likely to be involved in the treatment of potential organ donors should receive mandatory training in the principles of donation. There should also be regular update training.
- Appropriate ways should be identified of personally and publicly recognising individual organ donors, where desired. These may include national memorials, local initiatives and personal follow-up to donor families.
- There is an urgent requirement to identify and implement the most effective methods through which organ donation and the 'gift of life' can be promoted to the general public, and specifically to the BME population. Research should be commissioned through Department of Health Research and Development funding.
- The Department of Health and the Ministry of Justice should develop formal guidelines for Coroners concerning organ donation.

References

1. Council of Europe Transplant Newsletter, 12(1). September 2007. http://www.coe.int/t/dg3/health/themes_en.asp#transplantation.
2. Seewww.uktransplant.org.uk http://www.uktransplant.org.uk/ukt/about_transplants/organ_allocation/kidney_(renal)/national_protocols_and_guidelines/protocols_and_guidelines/transplant_list_criteria.jsp.
3. UK Renal Registry Tenth Annual Report, December 2007, Chapter 11; www.renalreg.com
4. United Network for Organ Sharing. See http://www.unos.org/.
5. Cuende N, Cuende JI, Fajardo J et al. Effect of population aging on the international organ donation rates and the effectiveness of the donation process. Am J Transplant 2007; 7(6):1526–35.
6. Coppen R, Friele RD, Marquet RL et al. Opting out systems: no guarantee for higher donation rates. Transplant Int 2005; 18:1275.
7. Roels L. Cohen B, Gachet C. Countries donation performance in perspective: time for more accurate comparative methodologies. Am J Transplant 2007; 7:1439–41.
8. Sheehy E, Conrad SL, Brigham LE et al. Estimating the number of potential organ donors in the United States. N Engl J Med 2003; 349(7):667–74.
9. Barber K, Falvey S, Hamilton C et al. Potential for organ donation in the United Kingdom: audit of intensive care records. BMJ 2006; 332(7550):1124–7.
10. See www.uktransplant.org.uk; http://www.uktransplant.org.uk/ukt/statistics/potential_donor_audit/potential_donor_audit.jsp.
11. Intensive Care Society Guidelines. Donation of Organs for Transplantation, 2005; http://www.ics.ac.uk/icmprof/standards.asp?menuid=7.
12. Rudge C. Organ donation and the law. Transplantation 2006; 82:1140–1.
13. Matesanz R, Rudge CJ. The acute shortage of donors: a UK and European perspective. In Forsythe JLR (ed.) Transplantation (Companion to Specialist Surgical Practice series), 3rd edn, Chap. 2. Elsevier Saunders, 2005.

Immunology of graft rejection

Rajesh Hanvesakul
Simon T. Ball
Margaret J. Dallman

Introduction

The optimal treatment of end-stage renal failure is transplantation, since it enhances both quality of life and survival compared with dialysis. The combination of pancreatic and kidney transplantation has similar benefits whilst transplantation of other organs is truly life saving since no long-term biomechanical support is available.

In clinical practice, organ donor and recipient are genetically distinct except in the case of identical twins (see Box 3.1 for terminology). The genetically encoded immunologically mediated barrier to transplantation was defined over the course of the 20th century. The study of transplantation has played a pivotal role in defining fundamental immunological phenomena and immunology has provided a rationale for the development of clinical transplantation.

Box 3.1 • Transplant terminology

Autograft (autologous transplant)
Transplantation of an individual's own tissue to another site, e.g. the use of a patient's own skin to cover third-degree burns or a saphenous vein femoro-popliteal graft

Isograft (syngeneic or isogeneic transplant)
Transplantation of tissue between genetically identical members of the same species, e.g. kidney transplant between identical twins or grafts between mice of the same inbred strain

Allograft (allogeneic transplant)
Transplantation of tissue between genetically non-identical members of the same species, e.g. cadaveric renal transplant or graft between mice of different inbred strains

Xenograft (xenogeneic transplant)
Transplantation of tissue between members of different species, e.g. baboon kidney into a human

The first successful renal transplant was between identical twins, but the development of transplantation as an important facet of modern medical therapy followed the introduction of immunosuppression to prevent and treat rejection of allogeneic organs. Clinical transplantation developed alongside an appreciation that rejection of foreign tissue belonged 'to the general category of actively acquired immunity' consequent upon differences in genetically encoded histocompatibility antigens. The process of rejection was shown to be caused by infiltrating leucocytes; it exhibits specificity and memory and is prevented by lymphocyte depletion. The major histocompatibility complex (MHC) was identified as encoding the predominant transplantation antigens responsible for acute rejection; these were then shown to be identical to serologically defined leucocyte antigens and subsequently to the elements responsible for self-restriction of immunological responses to conventional antigen.

These fundamental observations form the scientific basis of transplantation, and there is now a highly sophisticated understanding of the molecular

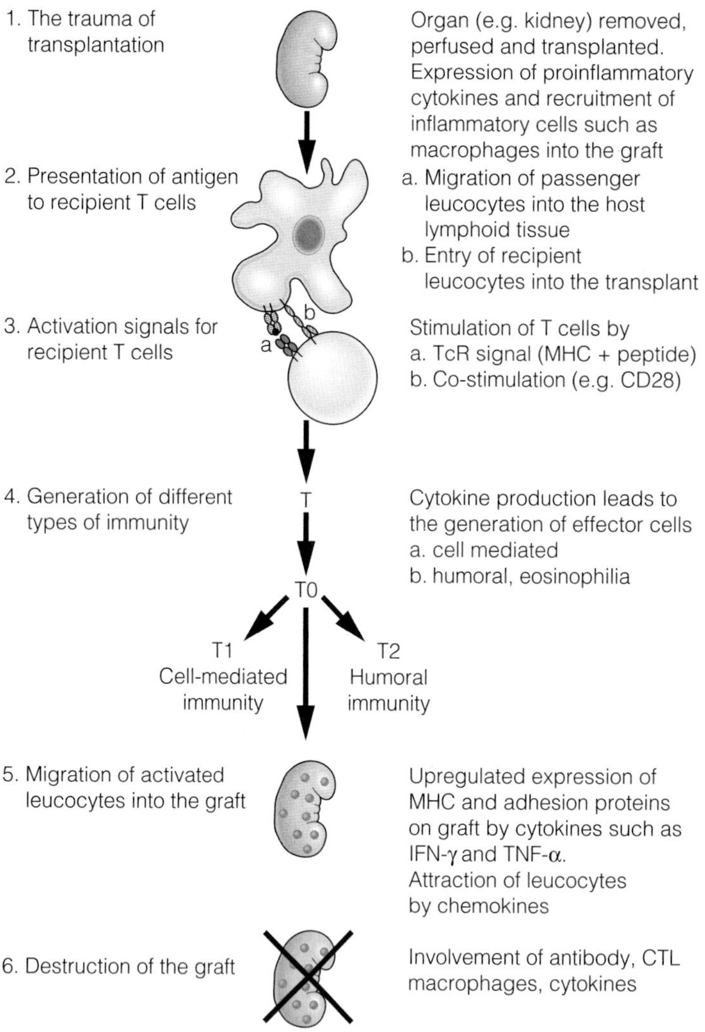

1. The trauma of transplantation — Organ (e.g. kidney) removed, perfused and transplanted. Expression of proinflammatory cytokines and recruitment of inflammatory cells such as macrophages into the graft
2. Presentation of antigen to recipient T cells — a. Migration of passenger leucocytes into the host lymphoid tissue; b. Entry of recipient leucocytes into the transplant
3. Activation signals for recipient T cells — Stimulation of T cells by a. TcR signal (MHC + peptide) b. Co-stimulation (e.g. CD28)
4. Generation of different types of immunity — Cytokine production leads to the generation of effector cells a. cell mediated b. humoral, eosinophilia

T → T0 → T1 Cell-mediated immunity / T2 Humoral immunity

5. Migration of activated leucocytes into the graft — Upregulated expression of MHC and adhesion proteins on graft by cytokines such as IFN-γ and TNF-α. Attraction of leucocytes by chemokines
6. Destruction of the graft — Involvement of antibody, CTL macrophages, cytokines

Figure 3.1 • The evolution of the immune response following kidney transplantation. CTL, cytotoxic T cell; IFN, interferon; MHC, major histocompatibility complex; TcR, T-cell receptor; TNF, tumour necrosis factor.

and cellular events which lead to graft rejection or acceptance, the subjects of this chapter (**Fig. 3.1**).

Before transplantation

Inflammation lies at the centre of the afferent and efferent arms of rejection. This begins prior to transplantation in the haemodynamic and neuroendocrine responses to brainstem death, in the process of multiorgan retrieval and the act of cold preservation. The transplant surgical procedure then adds a period of warm ischaemia and reperfusion. The consequent injury contributes to delayed graft function and generates an inflammatory response, which may itself promote and shape alloantigen-specific immunity.

The importance of these aspects of transplantation is illustrated by:

- the superior outcome of live donor transplants even in the face of significant MHC mismatch;
- the importance of cold ischaemia time in graft outcome;
- reportedly higher rates of rejection observed in individuals with delayed graft function.

Indeed, in experimental syngeneic transplantation, graft histology very similar to that seen in chronic allograft nephropathy may be reproduced by prolonged ischaemia.

Ischaemia–reperfusion injury (IRI)

Early following reperfusion, oxidative stress induced by IRI activates a range of cell types with consequent release of immunologically active soluble proteins such as interleukins 1 and 6 (IL-1 and IL-6) and various chemokines. Vascular endothelium is activated, becoming prothrombotic and expressing adhesion molecules (selectins) that mediate local leucocyte recruitment. The expression of selectins, integrins, cytokines and chemokines promote leucocyte extravasation, via processes that include rolling, formation of tight adhesions and endothelial transmigration. There is an early infiltrate of inflammatory cells even in syngeneic transplants in which alloimmunity is absent. IL-6 released systemically during reperfusion may suppress the generation of regulatory cells (see below), leading to further augmentation of a damaging response.

The severity of this initial injury and the nature of the subsequent inflammatory infiltrate is likely to be important in the stimulation of specific alloimmunity: a maximally damaged organ generates a maximal 'danger signal'[1] which initiates productive immunity manifest as rejection. The relationship between cellular infiltration and outcome is not simple in that this also occurs during the acquisition of tolerance in many animal models of transplantation. The context of such an initial encounter with alloantigen must therefore be important in determining the subsequent nature of any response. The mechanisms by which innate immunity shapes the ensuing responses to alloantigen are becoming increasingly defined. These include Toll-like receptor (TLR) engagement and γ-interferon (γ-IFN) production by natural killer (NK) cells activating dendritic cells (DCs). TLRs expressed by antigen-presenting cells (APCs) are a family of pattern recognition receptors that recognise microbes and endogenous stress proteins such as heat shock proteins whose ligation leads to APC maturation.[2] TLR activation and upregulation has been demonstrated in models of IRI.[3] The relationship between TLR engagement and acute allograft rejection has also been shown in clinical studies of lung and kidney transplantation.[4] IRI also activates other aspects of innate immunity such as complement that may contribute to immediate graft damage and influence adaptive immune responses.

Finally it is also worth noting that non-immunological mechanisms related to IRI directly influence long-term graft outcome. For example, accelerated replicative senescence of parenchymal cells may play a role in the development of chronic allograft nephropathy and this may be promoted by prolonged ischaemia.

Histocompatibility

Histocompatibility antigens were defined on the basis of their preventing transplantation between outbred members of the same species. In vertebrates they can be classified into the major histocompatibility complex (MHC) and numerous minor histocompatibility (miH) systems. Differences between histocompatibility antigens stimulate rejection and by far the most vigorous response arises from discordance at the MHC. This arises in part from the extreme polymorphism of MHC antigens and their physiological role in immune recognition. This role as the self-restricting element in engagement of the T-cell receptor (TCR) by antigenic peptide will be described in the next section.

Although differences in multiple minor histocompatibility antigens alone can result in allograft rejection in rodents, in clinical solid-organ transplantation the MHC is of overwhelming importance. However, this is not the case in bone marrow transplantation in which grafts between human leucocyte antigen (HLA)-identical siblings may be rejected or cause graft-versus-host disease[5] due to differences in only a limited number of minor antigens.

Antigens stimulating rejection by T lymphocytes

MHC class I and II

The MHC is divided into three regions, namely the class I, II and III regions, and these are described in Chapter 4. A large array of genes has been localised to the MHC by sequencing, but two major classes of transplantation antigens were those first mapped by classical genetic techniques. The structure of these molecules is shown in **Fig. 3.2**.

MHC class I proteins are cell-surface glycoproteins comprising a heavy chain (≈45 kDa), which is highly polymorphic, and a less variable light chain, β_2-microglobulin (≈12 kDa), which is encoded outside of the MHC. β_2-Microglobulin is not anchored in the membrane and binds non-covalently to the heavy chain. MHC class I proteins are expressed on most nucleated cells and generally engage T

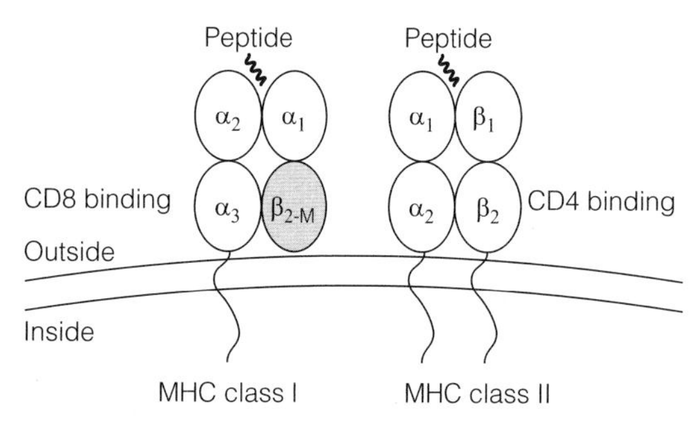

Figure 3.2 • Structure of MHC class I and II molecules. Schematic diagram of MHC class I and II proteins showing their basic structure, membrane association and peptide-binding groove. α_1, α_2, α_3 of MHC class I are membrane integral and form a non-covalent bond with β_2-microglobulin, which is not membrane bound. MHC class II is formed from α and β chains, each with two domains and each membrane integral. $\beta_{2\text{-M}}$, β_2-microglobulin.

lymphocytes bearing the cell-surface protein CD8. CD8 is required for effective TCR engagement of MHC class I, resulting in intracellular signalling.

MHC class II proteins consist of two membrane anchored glycoproteins (α-chain, ≈35 kDa; β-chain, ≈28 kDa), which are both encoded within the MHC. The tissue distribution of MHC class II proteins is more restricted than that of MHC class I. MHC class II is expressed constitutively by B lymphocytes and dendritic cells, which have specialised functions in the generation of the immune response. It is also expressed on some endothelial cells in humans and can be induced by inflammation on many cell types. MHC class II generally engages T lymphocytes bearing the CD4 surface protein. CD4 is required for effective TCR engagement of MHC class II and resultant intracellular signalling.

In humans these transplantation antigens were defined serologically on leucocytes and so are referred to as human leucocyte antigens (HLAs).

Structure of MHC class I and II proteins

MHC class I and II proteins have a similar three-dimensional structure forming a 'peptide binding groove' between two alpha helices (Fig. 3.2). The amino acids that constitute this groove are those which vary most between allotypes. During synthesis and transport of MHC class I and class II proteins to the cell surface the groove binds a wide range of short peptides. The groove in MHC class I has closed ends and the peptides bound are therefore 8–10 amino acids long. The groove in MHC class II is open and can accommodate peptides of considerably greater length. The sequences flanking those bound within the groove whilst not contributing directly to antigen specificity may contribute to the physicochemical properties of the complex. The peptides bound to MHC class I are primarily derived from intracellular proteins (but see 'Indirect/cross-presentation' below) and those from MHC class II primarily from extracellular proteins (**Fig. 3.3**). This difference arises from their distinct intracellular trafficking. In the presence of infection, foreign protein is processed to generate peptides bound to MHC forming a compound antigenic determinant, which engages the TCR and stimulates T-lymphocyte activation. This compound determinant is the structural basis for self-restricted antigen recognition by a T-cell repertoire that is skewed, by thymic selection, toward the engagement of peptide presented in the context of self-MHC. In the absence of foreign protein, all peptides are derived from self-proteins to which the individual is, to varying degrees, tolerant. A significant proportion of self-peptides are derived from MHC proteins themselves and this has consequences for allorecognition.

CD8⁺ T lymphocytes engage antigenic peptide bound to MHC class I and are typically cytotoxic. The normal function of these lymphocytes is to lyse cells infected intracellularly with, for example, virus. $CD8^+$ cells also produce a range of cytokines and some have regulatory properties.

$CD4^+$ T lymphocytes engage antigenic peptide bound to MHC class II. Antigen generally derives from extracellular proteins. These cells have a variety of functions, summarised in their designation as 'helper cells'. These functions include help for cytotoxic T-cell generation, B-cell maturation and promoting delayed-type hypersensitivity (DTH) inflammation. $CD4^+$ T-cell responses are evidently central to many forms of allogeneic rejection and their activation by dendritic cells is an important component of this response. In some rodent models absence of $CD4^+$ T-cell stimulation can abolish rejection despite the presence of $CD8^+$ T cells and fully allogeneic MHC class I mismatches. In such cases tolerance to alloantigen may emerge rather than productive immunity, illustrating the crucial importance of $CD4^+$ T-cell responses in rejection.

The specific structure of the binding groove determines the peptide that can be bound by any given MHC and there are a limited number of peptides from a given protein that form a stable complex.

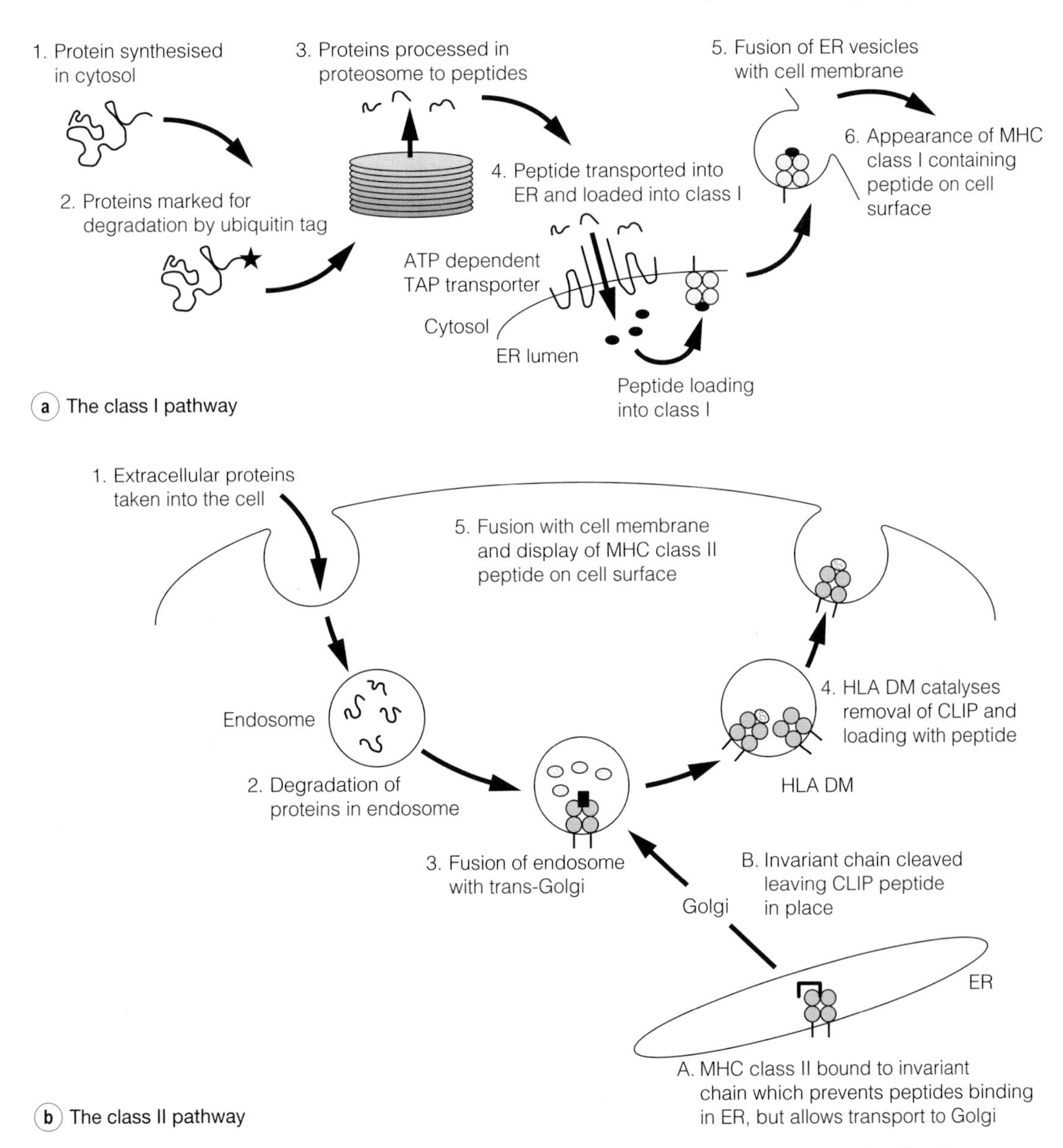

Figure 3.3 • Antigen processing and presentation in the MHC class I and II pathways. **(a)** Processing of endogenous antigens occurs primarily via the class I pathway. Peptides are produced and loaded into MHC class I proteins as shown in steps 1–4. During the synthesis of MHC class I proteins (steps 1–3) the α chain is stabilised by calnexin before β_2-microglobulin binds. Folding of the MHC class I/β_2-microglobulin remains incomplete but the complex is released by calnexin to bind with the chaparone proteins, tapaisin and calreticulin. Only when the TAP transporter delivers peptide to the MHC class I/β_2-microglobulin can folding of this complex be completed and transport to the cell membrane occur (steps 5 and 6). **(b)** Processing of exogenous antigens occurs primarily via the class II pathway. Antigens are taken up into intracellular vesicles where acidification aids their degradation into peptide fragments (steps 1 and 2). Vesicles containing peptides fuse with trans-Golgi, containing CLIP–MHC class II complexes (step 3). HLA-DM aids removal of CLIP and loading of peptide before the class II peptide complex is displayed on the cell surface (steps 4 and 5). MHC class II proteins are synthesised in the endoplasmic reticulum, where peptide binding is prevented by invariant chain. Invariant chain is cleaved leaving the CLIP peptide still in place (steps A and B) before fusing with acidified vesicles containing peptide. In B lymphocytes and epithelial cells of the thymus an atypical class II protein, HLA-DO, is expressed; this is a dimer of HLA-DNα and HLA-DOβ. It, like HLA-DM, is not expressed at the cell surface and inhibits the action of HLA-DM. Its precise role is unknown. ATP, adenosine triphosphate; CLIP, class II-associated invariant chain peptide; ER, endoplasmic reticulum; MHC, major histocompatibility complex; TAP, transporters associated with antigen processing.

The ontogenetic drive for extreme polymorphism in the MHC is therefore likely to reflect the need to counter a wide variety of infectious organisms that rapidly mutate and could evolve protein sequences that do not bind an individual's MHC repertoire. This is unlikely to occur across a whole species if there is a full range of polymorphism as observed in humans. Certain species with limited polymorphism at either MHC class I or II can be devastated by infections that, in closely related species with polymorphic MHC, do not threaten the population.

It is now possible to predict the sequences of peptides likely to bind to a given MHC from its structure and to then confirm this from peptide elution and peptide-binding studies. The peptide can be orientated in the groove and those amino acids in contact with MHC and those in contact with the TCR can be predicted. This is likely to be a powerful tool in future vaccine development.

Assembly of the MHC–peptide complex

The assembly of the MHC–peptide complex (Fig. 3.3) involves sophisticated mechanisms of antigen uptake, processing to peptide and, in the case of MHC class I, transport into the endosomal compartment. The proteosome components (low-molecular-mass polypeptide, LMP) and transporters associated with antigen processing (TAP) responsible for these aspects of processing are encoded within the MHC. In the case of classical MHC class II, assembly requires the initial presence of an invariant chain from which a peptide, class II-associated invariant chain peptide (CLIP), combines with α and β chains to form a nascent complex. CLIP is subsequently exchanged for antigenic peptide.

Non-classical MHC

The MHC locus is large and it encodes a wide range of proteins other than the classical histocompatibility antigens described above. Some of these have a structure similar to classical histocompatibility antigens but are not polymorphic. They may present specialised antigens such as lipids (e.g. mycolic acid and lipoarabinomannan from mycobacterium) or peptides of various sequences but with common characteristics (e.g. with N-formylated amino termini). The relevance of some of these molecules to transplantation is unclear but they do not behave as classical transplantation antigens.

Molecules such as HLA-DM, TAP and LMP that have a role in antigen processing are encoded within the MHC locus as are other molecules more broadly involved in immunity, for example tumour necrosis factors α and β (TNF-α and TNF-β) and complement components C2 and C4. Polymorphisms linked to these genes have been implicated in disease expression and transplantation responses.[6]

HLA-E and HLA-G are expressed in the placenta and are thought to play a role in modulating alloimmune responses to the fetus. There is emerging evidence that the expression of these molecules may also be associated with improved outcome in allogeneic transplantation.[7,8] HLA-E modulates alloimmunity through its interaction with CD94-NKG2A, an inhibitory receptor expressed on NK and cytotoxic T cells. HLA-G interacts with inhibitory receptors ILT-2 and -4 expressed by APCs, NK and T cells. In a large series of renal transplant patients, the formation of soluble HLA-G was strongly correlated with diminished acute rejection rates.[8]

MHC class I-related chain

MHC class I-related chain A (MICA) genes are located within the class I region of the chromosome and are highly polymorphic. Although constitutively expressed at low levels by various cell types, including monocytes, epithelial cells, fibroblasts and endothelial cells, they may be upregulated by stress signals of oxidative stress and sepsis. Expression of MICA in response to stress may activate NK and T cells through its interaction with cytolytic activating receptor NKG2D. MICA is polymorphic and its alloantigenicity has been demonstrated in animal models. There is now good evidence that formation of antibody to MICA is associated with poor graft outcomes in renal transplantation[9] and increased rates of acute rejection in cardiac transplantation.[10]

Minor histocompatibility antigens

The study in rodents of genetic loci determining organ rejection identifies a range of sites encoding transplant antigens that are distinct from the MHC. These minor histocompatibility (miH) antigens are less polymorphic and typically induce a weaker response than disparities in the MHC. It is almost impossible to raise antibodies to these antigens, although they induce a well-defined cellular response.

The crystallographic structure of the MHC with bound peptide provides a structural explanation for these properties of miH antigens. They are peptides derived from proteins of limited polymorphism that, when bound to syngeneic MHC, constitute an antigenic determinant in the same way as any conventional antigen. This explains why it is difficult to raise antibodies to miH antigens, since antibodies typically engage conformational determinants on proteins and the B-lymphocyte receptor is less discriminating of subtle differences in the structure of peptide–MHC composites than is the TCR.

The most easily defined miH antigen is the male-specific H-Y antigen, which is in fact a series of antigenic peptides derived from proteins encoded on the Y chromosome, accounting for its consistent detection across a wide range of background MHC (although in certain strain combinations there is a single dominant antigen[11]). In rodent studies the role of miH proteins as transplantation antigens can be readily demonstrated, particularly when there has been prior sensitisation to the relevant antigen.

In rare cases of rejection in renal transplants between HLA-identical siblings, miH are thought to be the inciting antigens but their importance in other forms of solid-organ transplantation is generally limited. This is not true, however, in bone marrow transplantation, in which rejection is easily stimulated by such differences causing graft vs. host or graft vs. tumour response.[12]

Recognition of alloantigen

The TCR can engage MHC in two distinct ways, either **directly** on APCs derived from the graft, or **indirectly** as processed peptides in the context of self-MHC on self-APCs. A third, **semi-direct** route of antigen presentation has recently been described.

Direct allorecognition

Allogeneic MHC on DCs derived from the graft will be occupied by any number of different endogenous peptides (**Fig. 3.4**). The frequency of alloreactive T lymphocytes stimulated through this pathway is high, of the order of $1/10^2$. Although this appears to contradict the principles of 'self-MHC' restriction resulting from thymic education, it emerges as a consequence of the vast array of endogenous peptides that occupy the MHC generating an equivalent number of compound epitopes. It also arises from the relatively limited differences in the structure of MHC alloantigens outside the peptide-binding groove and the corresponding fact that T lymphocytes can distinguish subtle differences in the kinetics of TCR–peptide–MHC interactions.

These facts apply equally to the response to recall antigens; that is, other than in naive animals any

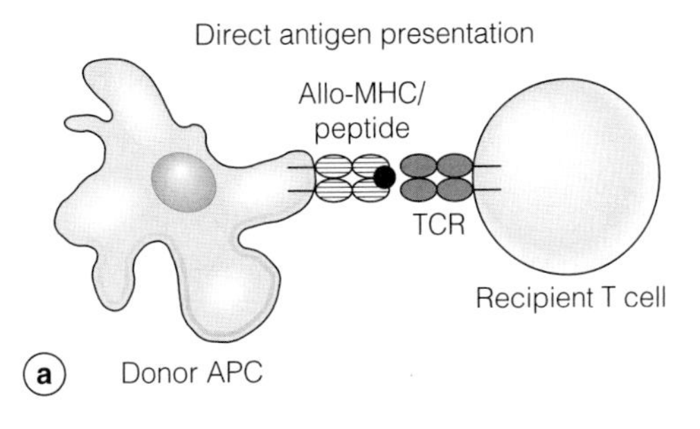

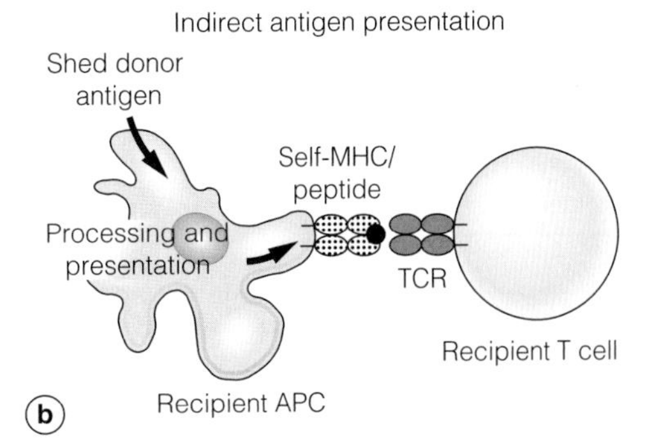

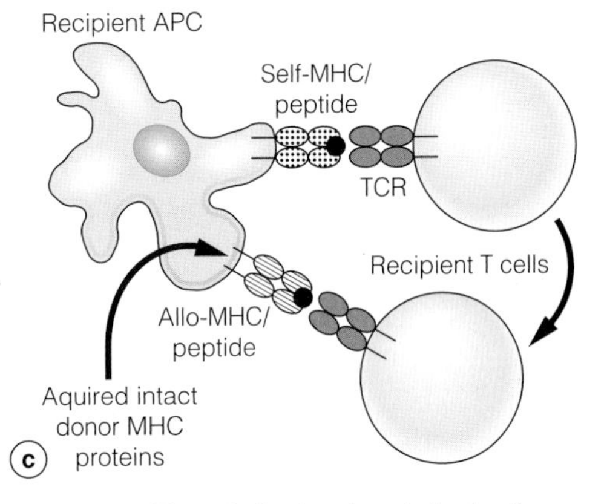

Figure 3.4 • Direct, indirect and semi-direct pathways of antigen presentation. Sensitisation of the recipient can occur by antigen presentation delivered through passenger leucocytes or dendritic cells of donor origin **(a)** or recipient origin **(b,c)**. APC, antigen-presenting cell; MHC, major histocompatibility complex; TCR, T-cell receptor.

direct alloresponse will include cross-reactions with self-restricted secondary responses to conventional antigens.[13–16] As most studies on experimental transplantation, particularly those on the induction of tolerance, use naive rodents these are unlikely to accurately reflect the alloresponse observed clinically. Indeed, strategies to induce transplantation tolerance which succeed in the naive animal are unsuccessful in previously infected animals,[15,16] providing one possible explanation for the difficulties in translating such strategies into preclinical models.

Indirect allorecognition

Proteins from extracellular infectious organisms are generally processed by APCs and presented in the context of MHC class II. As discussed above, only a small proportion of potential peptide sequences from a protein will be processed, bind MHC and generate a response.

In the setting of allogeneic transplantation any donor protein can be processed and presented in the context of MHC by **recipient** APCs. As most proteins have little or no polymorphism within a species, they will not always initiate an alloimmune response, although occasionally transplant rejection is associated with the development of autoimmune responses. However, polymorphic proteins such as those of the MHC can stimulate an immune response via this conventional or 'indirect pathway' of antigen presentation.

The potential importance of this route of allorecognition is demonstrated by experiments in which immunisation with MHC class I peptides prime for subsequent rejection of a skin allograft bearing the intact class I antigens from which the peptides were derived.[17] Studies in other models have demonstrated that priming with MHC class I- and II-derived peptides lead to chronic allograft rejection, with accelerated vasculopathy.[18,19] These peptides are most likely to be presented in the context of recipient class II since they are extracellular for the self-APCs. However, there is now evidence of crossover or 'cross-presentation' between the two pathways, so there may also be presentation of such peptides in the context of MHC class I.[20,21]

The importance of indirect presentation of alloantigen is also illustrated by transplantation of skin allografts from MHC class II deficient onto normal mice. Donor APCs are unable to stimulate recipient $CD4^+$ T cells and yet rejection remains dependent upon $CD4^+$ T cells. It is likely that this reflects recipient $CD4^+$ T-cell stimulation through indirect presentation of donor alloantigens by self-MHC.[20–22] In this latter case, of course, peptides are not artificially introduced into the recipients but produced by normal processing of MHC from apoptosed donor-derived cells.

In long-standing renal transplant recipients there is evidence of donor-specific hyporesponsiveness of peripheral blood lymphocytes to directly presented alloantigen,[27–29] and this may be true even in those who have suffered allograft failure.[27] There is increasing evidence, however, for the association of indirect pathway alloreactivity in peripheral blood lymphocytes with chronic rejection in solid-organ transplants.[28,30–32] The significance[33] and interpretation of peripheral responses as indicative of tissue responses must be guarded, but these observations suggest strategies for monitoring patient responses[32] and assessing the effects of treatment[34] for chronic rejection.

Indirect pathway allorecognition has also been linked to **regulation** of alloimmune responses and this is discussed in a later section.

Semi-direct allorecognition

If host T cells are stimulated by recipient-derived DCs via indirect antigen presentation, the MHC restriction of the effector cell population will be to host rather than donor. A problem could arise if a cytotoxic T cell, once stimulated with self-MHC and allogeneic peptide, comes to lyse its target cell – in the case of graft rejection, the foreign transplanted tissue, which does not express self-MHC molecules. This problem is overcome if the foreign MHC on the target cell cross-reacts with self insofar as the T cell is concerned, if the effector arm of the immune response does not require MHC restriction (e.g. macrophages, DTH) or if the effector population is primed by the donor-derived MHC. How could this latter situation arise if the T cells are primed by recipient-derived DCs? It has been known for several years now that intact proteins can be exchanged between cells in cell culture systems and indeed that MHC proteins transferred in this fashion can stimulate alloreactive responses.[23–25] The importance of this in stimulation of alloreactive responses in the whole animal has been highlighted in recent work,[26] although its importance in inducing graft rejection has yet to be established.

T-cell interactions with antigen-presenting cells: initiation of the immune response

'Passenger leucocytes' in transplantation

In rodent models of transplantation, allogeneic MHC can be more or less immunogenic according to the context in which it is encountered by the recipient immune system. Following transplantation, 'passenger leucocytes' (tissue-resident immature DCs) migrate to secondary lymphoid organs, mature and there deliver a powerful stimulus to the recipient immune system (direct allostimulation). Mature DCs express high levels of MHC class I and II and can therefore stimulate both $CD8^+$ and $CD4^+$ T lymphocytes. They also have other properties including efficient co-stimulatory activity, which render them uniquely powerful stimulators of naive T cells[35–37] and earn them the title 'professional' APCs. A number of experimental systems have been used to dissect the role of donor DCs in stimulating allograft rejection.

Allografts can be depleted of donor-derived passenger leucocytes by irradiation, a period of tissue culture, or by exchange with DCs of recipient type whilst 'parked' in an immunoincompetent animal of recipient origin. These organs when re-transplanted into immunocompetent animals often fail to induce rejection. These experiments suggest that DCs are required for the initiation of the immune response that leads to acute rejection. After such successful transplantation rejection may be induced by the infusion of donor-origin leucocytes.

Yet, clear data implicating the DC as the primary (or only) stimulus of graft rejection in all donor–recipient pairs is lacking. Also, direct application of these findings to clinical transplantation must be cautious, given the fact that in many experimental models of long-term graft survival an active regulatory immune response plays a role in the failure of organ rejection. DCs themselves may be important for the induction of a regulatory response rather than rejection, although this may be most relevant to recipient DCs (indirect pathway) rather than donor DCs (direct pathway). Furthermore, although DCs are very important in the stimulation of the primary T-cell response, they seem less important for secondary responses. In clinical transplantation a significant proportion of the alloresponse is likely to be heterologous[13,15] – that is, primed, self-restricted, pathogen-specific, T-lymphocyte cross-reacting on an allogeneic MHC–peptide complex. These primed cells are significantly less dependent upon DCs for activation. Nevertheless, the stimulation of lymphocytes through the 'direct' pathway of allorecognition by the migration of 'passenger' DCs seems often to provide a powerful early stimulus to acute graft rejection,[38,39] whether this is a primary or secondary response.

Vascular endothelium constitutively expresses MHC class II in humans and it too may play a greater role in stimulating alloimmunity than in rodents, offering a further explanation for the difficulties of translating animal models of transplantation tolerance to the clinic.[40,41]

Activation of dendritic cells

Although the immunostimulatory properties of DCs are most apparent in the experimental data described above, it is now evident that marked phenotypic and functional differences exist between dendritic cells, depending upon the context in which they are studied. The factors that contribute to acquisition of such properties include:

- lineage;
- maturity;
- activation by infection and inflammation;
- interaction with T cells and T-cell-derived cytokines.

The precise mechanisms of dendritic cell differentiation and maturation are currently under intense investigation. For example, DCs are activated by TLR engagement and by other facets of the innate immune system, including complement, NK cells and $\gamma\delta$ T cells.

These set DC phenotype, which may then determine the phenotype of the subsequent adaptive immune response. Generally 'immature' DCs play a role in maintaining peripheral tolerance by both deletional and regulatory mechanisms, and these properties may themselves be regulated by T lymphocytes. DCs therefore bidirectionally connect innate and acquired immunity, and by doing so link the nature of antigen-specific memory to the context in which antigen is first encountered.[42,43]

Co-stimulation

Engagement of the TCR can result in a range of outcomes depending upon context, as already alluded to in the discussion of DCs. This context is established by the engagement of co-stimulatory molecules on APCs and soluble mediators including cytokines. The outcome for the T lymphocyte includes proliferation, apoptosis or anergy, acquisition of different effector

Table 3.1 • Members of the CD28 and CD40–ligand (CD154) families of molecules expressed on T lymphocytes and their corresponding ligands

	CD28–B7 family					
	CD28	**CTLA4**	**ICOS**	**PD-1**	**TLT-2**	**BTLA**
Expression	Constitutive on T cells	Induced on T cells	Induced on T cells	Induced on T, B and dendritic cells Constitutive on NK cells	Induced on T cells	Induced on DCs, monocytes, T cells and B cells
Ligand	B7-2 > B7-1	B7-1 > B7-2	B7RP-1	PD-L1 and PD-L2	B7-H3	HVEM (TNFR Family)
Activity	Blockade promotes allograft survival and tolerance in some models	Blockade inhibits tolerance induction and ligation inhibits activation	Complex effects, improved graft survival after blockade in some models	Activation promotes and blockade inhibits allograft survival	Various effects on T-cell activation reported	Negatively regulates T-cell responses

	CD40 ligand–CD40 family (TNF receptor–TNF superfamily)				
	CD154	**CD137 (4-1 BB)**	**CD134 (OX-40)**	**CD27**	**LIGHT**
Expression	Induced on T cells, NK cells and eosinophils	Induced on T cells	Induced on T cells	Constitutive on T cells and on APCs	
Ligand	CD40	CD137 ligand	CD134 ligand	CD70	HVEM
Activity	Blockade promotes graft survival/tolerance	Blockade inhibits $CD8^+$ T-cell-mediated rejection	Blockade inhibits CD28/CD154-independent rejection, inhibition of primed responses	Effector and memory cell generation (B cells and T cells)	Positively regulates T-cell responses

phenotypes, acquisition of different regulatory phenotypes or differentiation into a memory cell.

The importance of co-stimulation was first evident from experimental models in which its inhibition was achieved by various means, including the fixation of APCs, presentation by purified MHC in lipid bilayers and by 'non-professional' APCs transfected with MHC class II. In these studies $CD4^+$ T cells that engaged with cognate antigen in the absence of co-stimulation did not proliferate. Furthermore, they often acquired a stable change in phenotype: they became 'anergic', failing to proliferate on re-stimulation with competent APCs. This generated a model in which TCR was said to provide antigen-specific 'signal 1' and co-stimulatory molecule engagement 'signal 2'. Together, these signals resulted in proliferation but signal 1 in the absence of signal 2 resulted in anergy. In many in vitro systems signal 2 was found to be the engagement of CD28 by B7-1 or B7-2 (CD80 or CD86).

The field of co-stimulation has since become considerably more complex, in that large numbers of molecules with signal 2 properties have been identified in different experimental systems, and some of these act to provide 'negative co-stimulation' – that is, they have a dominant negative effect on T-cell activation and promote the acquisition of anergic/regulatory phenotypes or cell death. The most studied examples of such negative co-stimulation include CTLA4 and PD1.

Co-stimulatory interactions can be broadly separated into those of the CD28–B7 family and those of the tumour necrosis factor receptor (TNFR)/tumour necrosis factor (TNF) family, as summarised in Table 3.1. The relative importance of different co-stimulatory pathways and the way in which they interact remains to be fully elucidated. In alloimmunity it is not yet clear whether different pathways are functionally redundant or, more likely, that they play specialised roles relevant to different aspects of successful transplantation.

CD28–B7

The role of CD28 has been the most intensively investigated in the field of co-stimulation.[44–46] CD28 is a homodimeric glycoprotein present on the surface of T cells, which interacts with two counter-receptors, CD80 and CD86, expressed on the surface of APCs. CD86 is expressed constitutively at low level by APCs and is upregulated rapidly following interaction with the T lymphocyte. It has a rather low affinity for CD28 whereas CD80, which is not constitutively expressed, is upregulated with slower kinetics but has an approximately 10-fold greater binding affinity for CD28 than does CD80. The result of ligation of CD28 by either CD86 or CD80 appears to be increased cytokine synthesis and proliferation, and they do not appear to be qualitatively distinct. The different kinetics of expression and affinity for other ligands such as CTLA4 (CD152) may result in different effects on cellular phenotype.

CD28 ligation by CD80/86 promotes initial T-cell activation but also plays a role in CTLA4 (CD152) regulatory T-cell homeostasis. Blocking the CD28 pathway in rodents can have dramatic effects on the generation of primary immune responses and may result in prolonged graft survival or even tolerance of grafts in some experimental models,[47,48] but mice with a disrupted CD28 gene can make productive immune responses, albeit with sometimes altered kinetics. This may be due to redundancy of co-stimulatory pathways, including other members of the B7 and CD28 families.

CTLA4 ligation by CD80/86 inhibits T-cell activation. The severe phenotype of CTLA4 –/– mice, which die from uncontrolled lymphoid proliferation shortly after birth,[49] suggests that there is less redundancy in this aspect of immune regulation.

Other CD28 family members, e.g. ICOS, PD-1 and BTLA (B- and T-lymphocyte attenuator), are inducible on T cells and seem to have roles regulating previously activated T cells.

There are **five other B7 family members**, ICOS ligand, PD-L1 (B7-H1), PD-L2 (B7-DC), B7-H3 and B7-H4 (B7x/B7-S1), that are expressed on various cell types and are likely to provide further control of T-cell activation and tolerance in the periphery. B7 molecules seem to be able to bidirectionally deliver signals into B7-expressing cells as well as those expressing their ligands.

The most widely used method to block CD28–B7 interactions has been through the use of CTLA4-Ig but this will also block CTLA4–B7 interaction, and inhibition of negative CTLA4 signalling may account for diverse observations on immune responses in different experimental systems. A high-affinity analogue of CTLA4-Ig has been developed as an immunosuppressive for use in clinical practice,[50] and the development of reagents that preferentially inhibit the CD28–B7 interaction is also ongoing.

CD154–CD40

Upon activation T cells express CD154 (CD40 ligand, gp39). Interaction of this protein with its counter-receptor, CD40, appears to be critical for the activation of B cells, dendritic cells and monocytes. This activation is itself crucial to subsequent T-cell stimulation. In dendritic cells CD40 ligation upregulates IL-12 production; in macrophages this ligation produces a range of proinflammatory cytokines including TNF-α and IL-8 whilst in B cells CD40–CD154 provides signals for proliferation, maturation, isotype switching and generation of memory.

CD40–CD154 blockade prolongs graft survival in a murine cardiac transplant model,[51] and combined CD28 and CD40 blockade induces indefinite survival of an allogeneic skin graft in mice with no long-term deterioration of graft integrity.[52] The latter is a rigorous marker of non-responsiveness but interestingly these animals did not exhibit tolerance to a second allograft. In short this combination seems to be highly effective immunosuppression in the setting of experimental allogeneic transplantation but without evidence of tolerance induction.

CD154–CD40 promotes DC maturation in the presence of $CD4^+CD25^+$ regulatory lymphocytes which otherwise maintain DC immaturity. This illustrates the importance of DC maturation not only by innate immune mechanisms but also by activated T lymphocytes.

These observations underlie significant beneficial effects for anti-CD154 seen in preclinical models of transplantation;[53] unfortunately, its clinical application has as yet been prevented by prothrombotic effects – a probable consequence of the expression of CD154 on platelets.

The T-cell synapse

On TCR engagement, multiple cell-surface proteins (**Fig. 3.5**) become highly organised within the two dimensions of the cell membrane, forming a structure referred to as the immunological synapse. This depends upon cytoskeletal–cell membrane interactions and is in part orchestrated by the coordinated movement of proteins congregated within lipid rafts. The orderly congregation of molecules has been referred to as a supramolecular activation complex (SMAC).[54] The nature of the synapse in vivo requires further investigation, as there is some evidence to suggest considerably greater motility of T cells in lymph nodes than would be suggested from the kinetics of T-cell–DC interaction observed in vitro.

Although the signal 1/signal 2 model of co-stimulation proved useful it is increasingly apparent that the distinction between adhesion molecules and co-stimulatory molecules is not strict. Thus the kinetics of TCR-binding, non-antigen-specific interactions between T cells and APCs, the presence of cytokines, and the state of T-cell maturation all contribute to determining cell fate on TCR engagement.

TCR signalling

The majority of T cells bear a TCR of two chains – the α and β chains – complexed with various other proteins, in particular those of the CD3 complex (γ, δ, ε) and ζ chains. The phosphorylation of ζ chains results in zap-70 recruitment and subsequent congregation of other signalling elements around a membrane scaffold protein, linker for activation of T cells (LAT), which is located in lipid rafts.

Diverse signalling pathways are activated which ultimately lead to transcriptional activation and the de novo expression of a range of genes, including those encoding cytokines and new cell-surface proteins. These pathways are highly complex but, in broad terms, those relevant to T-cell activation include:

- phospholipase C-γ1 (PLC-γ1), which through Ca^{2+} influx activates nuclear factor of activated T cells (NFAT);
- protein kinase C-θ (PKC-θ), which acts through nuclear factor-κB (NF-κB) and activator protein 1 (AP1);

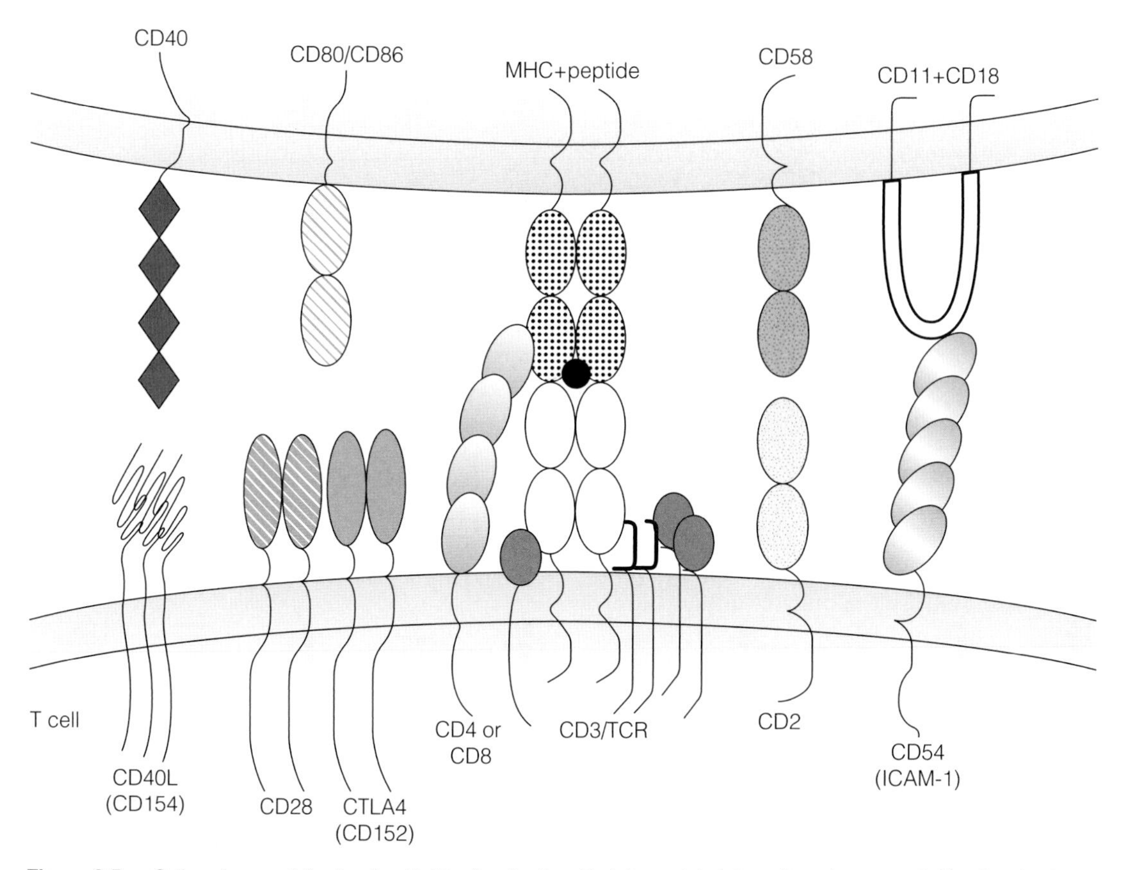

Figure 3.5 • Cell-surface proteins involved in T-cell activation. Protein–protein interactions important in T-cell activation. Receptors on the T cell are shown with their ligands on the APC. Ig domains are common amongst cell-surface proteins of leucocytes and are denoted in this diagram by ovals. Several of these interactions are important in delivering signals to both T cells and APCs.

- RAS-guanosine-releasing protein (GRP) and the growth factor receptor-bound protein 2 (GRB2)–SOS complex, which act via RAS.
- VAV1–SLP76 complex activation through RHO family GTPases.

These signalling pathways coordinate to drive proliferation and differentiation of the T lymphocyte and are therefore targets or potential targets of pharmacological intervention.[55] The activation of all these intracellular signalling pathways can be influenced by various co-stimulatory molecules.

T-cell differentiation, cytokine production and regulation

As described above, the interaction of T cells and APCs is central to initiation of the alloimmune response. The default for such a response following transplantation is the generation of productive immunity manifest as rejection mediated through various effector pathways. A productive immune response generally results in proliferation and differentiation of 'helper' T cells, which drive and direct antigen-specific immune responses through various effector pathways. They are mostly $CD4^+$ T lymphocytes but in certain situations $CD8^+$ T lymphocytes perform similar functions. The type of effector response is determined in part by the elaboration of particular patterns of cytokines determined by the signals delivered to cells of the innate immune system. These effector mechanisms are thereby orchestrated to deliver effective host defence but at the same time maintain self-tolerance in the face of an adaptive immune system of immense potential diversity (**Fig. 3.6**).

In alloimmune responses, naive helper T cells are induced to commit to various cell lineages including Th1, Th2, Th17 and T-regulatory cells (Treg) by different cytokines; the state of maturation of APC and antigen load T-cell differentiation, cytokine production and subsequent immune deviation to particular cell lineages ultimately determine the type of effector immune response, whether it be malign or benign for the transplanted organ. Th1 differentiation is promoted by IL-12, Th2 by IL-4, and Th17 by IL-6/TGF-β; regulatory T cells of various types

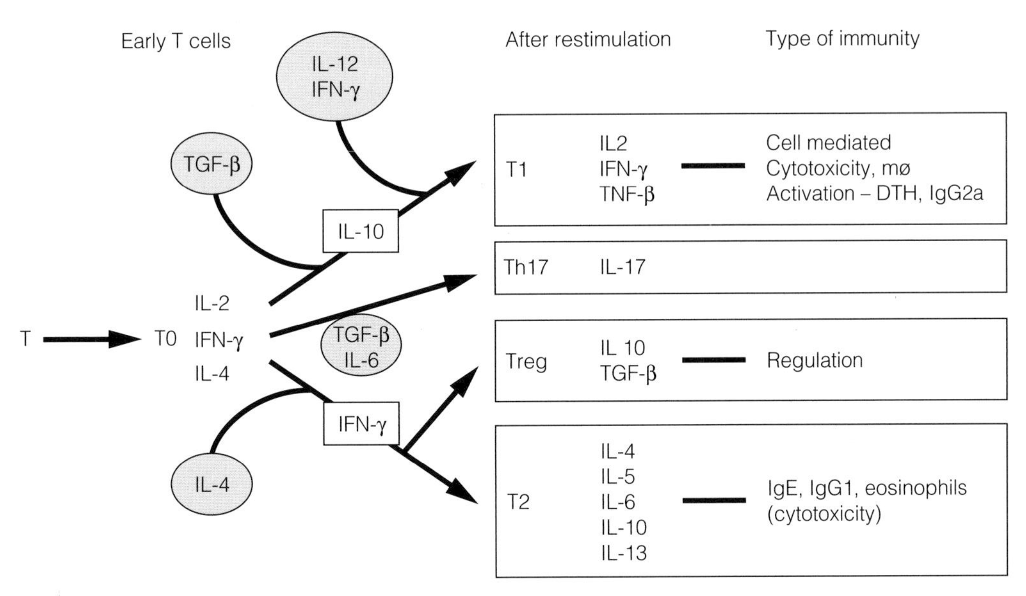

Figure 3.6 • Peripheral T-cell differentiation and cytokine production. Cytokines produced at various stages during the evolution of immune responses are shown. Cytokines that positively or negatively influence the divergence of Th1 and Th2 cells are shown in circles or squares respectively. Treg cells are those that negatively regulate the immune response and are important in controlling self-reactivity as well as limiting immune responses to pathogens. Their divergence from the Th2 pathway is based on data from many groups indicating that T cells can co-express IL-10 and IL-4 under many conditions. The role of Th17 in mediating allograft damage remains uncertain but its contribution to destructive immunity is well established in autoimmune diseases. DTH, delayed-type hypersensitivity.

are promoted by cytokines such as TGF-β and IL-10. Cell lineages are characterised by the synthesis of specific transcription factors (Th1: T-bet; Th2: GATA-3; Th17: RORgammat; CD4⁺CD25^hi^Treg: Foxp3).

Animal models of transplantation which generally rely on early, acute rejection as a 'readout' of allogeneic responsiveness have generally associated this with a Th1 response[56–61] and it is likely that Th17 responses will also be implicated in autoimmunity.[62] Th2 and Treg induction are, by contrast, associated with protection from acute rejection; however, the relevance of such a simple model to clinical transplantation is uncertain.

There is increasing recognition that regulatory cells are important determinants of self-tolerance, that they may play an important role in models of transplantation tolerance and that an ability to induce or expand such cells in the clinical setting may be of therapeutic value.[63] The presence of regulatory or suppressor T cells has been suggested by various observations.

The absence of specific T-cell subpopulations can result in autoimmune disease and their repletion prevents the development of autoimmunity, suggesting that tolerance is maintained by a '**dominant negative mechanism**'.

There is good evidence that, following their induction, regulatory cells not only suppress activation but can confer a regulatory phenotype on naive cells, a process termed '**infectious tolerance**'.[64]

Also, tolerance to alloantigen of one type can spread to alloantigen of another type when the two are expressed on the same cell, in a process termed '**linked suppression**'.

The presence of regulatory cells can be inferred from the aforementioned depletion experiments and has been more directly demonstrated in cell transfer experiments; however, interpretation of these experiments must be cautioned by the fact that lymphopenic or otherwise manipulated recipients are generally used and this has significant effects on the behaviour of lymphocytes through homeostatic proliferation.

The presence of regulatory cells in allogeneic transplant models, not involving transfer into lymphopenic hosts, has in most cases involved only minor histocompatibility antigens[65] or inhibition of direct pathway stimulation by prior depletion of DCs.[66] Some groups have expanded allospecific regulatory cells ex vivo with a view to overcoming some of the limitations encountered in models of major histoincompatibility.[63] These regulatory cells could have either direct[67] or indirect allospecificity.[68]

The most highly investigated group of regulatory cells are CD4⁺CD25⁺ regulatory cells. Neonatal rodents undergoing early thymectomy develop autoimmune disease which can be prevented by the infusion of a CD4⁺CD25⁺ subpopulation of T lymphocytes from non-thymectomised syngeneic donors. A role for such CD4⁺CD25⁺ regulatory cells has been demonstrated in various model systems, including thyroiditis,[69] diabetes mellitus,[70] encephalitis[71,72] and colitis.[73] The role of such cells in various models of transplantation tolerance is now extensively described,[74] although many models involve lymphopaenic hosts,[75] with consequent caveats with regard to their relevance to the clinical setting.

The antigen specificity of CD4⁺CD25⁺ regulatory cells is also variously reported.[76,77] There are at least two relevant specificities: that restricting the regulatory cell and that of the regulated cell. This relates to definition of the precise mechanism of action of these cells. In vivo imaging studies have demonstrated that regulatory cells can interact specifically with antigen-bearing DCs and prevent stable contacts between effector T cells and DCs, thereby inhibiting effective priming.[78] This may account for activity against T cells specific for antigens not related to the specificity of the regulatory cell.[79]

There is also evidence in vivo for regulation by CD4⁺CD25⁻,[80] CD8⁺ T-lymphocyte subpopulations and NK T cells.[81,82] The mechanisms identified are incomplete, but they point to relevant pathways. For example, the induction of non-responsiveness in naive cells has been inhibited by neutralising specific cytokines such as IL-10[83–85] and TGF-β, and cells whose regulatory activity are so inhibited are referred to as Tr1 and Th3 cells respectively. The relationship between CD4⁺CD25⁺ regulatory cells and these cells defined in other systems has yet to be clearly defined.[86]

Interaction between innate immunity and Treg function has also been established. TLR engagement on DCs has been shown to prevent suppression by Treg in an IL-6-dependent fashion. Furthermore, IL-6 released during IRI has been shown to inhibit the recruitment of Treg into the graft and enhance the differentiation of Th17 cells.

In summary, these observations on T regulatory cells offer hope for the induction of dominant tolerance which may benefit transplantation but the limitations of the experimental models involved need careful consideration before extrapolating to the clinical scenario.

The effector arm of the immune response

The immune system generates many different effector mechanisms depending on the challenge it meets. In certain infections a single mechanism is essential for the clearance of the organism and the absence of that mechanism renders the host susceptible to disease. For example, in the clearance of lymphocyte choriomeningitis virus infections in mice, cytotoxic cells are absolutely required and disabling this arm of immunity by disrupting the perforin gene leads to death of infected animals. In general, factors such as activation of innate immunity, antigen dose and prior immune stimuli determine which immune effector mechanisms predominate, but in clinical transplantation a range of different responses are activated. Unfortunately most of these are capable of damaging a graft such that the obliteration of any single mechanism often has little benefit on graft survival, which is why the prevention of rejection is difficult. It also accounts for the importance of the $CD4^+$ T-cell response since this is involved in the orchestration of many of these mechanisms and for interest in the role of regulatory cells, because if it were possible to commandeer a mechanism evolved to control autoimmunity then abrogation of alloimmunity is likely to be significantly easier than inhibiting multiple different effector pathways.

Migration of activated leucocytes into the graft

In order to enter a site of inflammation, leucocytes migrate across vascular endothelium. This is mediated through a variety of chemokines and by cell–cell interactions between leucocyte and the endothelium. Activated cells and memory cells express a range of chemokine receptors and adhesion molecules that promote migration into peripheral tissues.

Cell–cell interactions

The adhesion of leucocytes to the endothelium is a multistep process, involving a series of interactions between the leucocyte and endothelial cell.[87] The proteins involved fall into three groups: selectins, integrins and immunoglobulin (Ig) superfamily members along with mucins on a range of glycoproteins. The initial step is attachment and rolling; the cell may then detach or be activated, adhere and transmigrate. Rolling of leucocytes along the endothelium allows the leucocyte to sample the endothelial environment whilst maintaining its ability to detach and travel elsewhere. Attachment is largely mediated by selectin binding. Under appropriate conditions leucocytes are activated and adhesion will be followed by passage through the endothelium in steps largely mediated by integrins and Ig superfamily members – intercellular adhesion molecules (ICAMs) and vascular cell adhesion molecules (VCAMs).

The expression of adhesion molecules involved in these interactions is upregulated by proinflammatory cytokines. The retrieval of a donor organ upregulates expression of various cytokines with subsequent upregulation of selectins, ICAM-1 and VCAM-1.[88,89] Before any alloimmune response has been elicited the transplanted organ is consequently attractive to circulating leucocytes. Although naive cells tend not to enter non-lymphoid sites this is not true of activated and memory cells;[90] there may therefore be early ingress of specific T lymphocytes in the clinical scenario with implications for therapeutic strategies aimed at reducing trafficking into the graft.[91–93]

Chemokines

Chemokines are small soluble proteins (8–11 kDa), which are responsible for leucocyte recruitment, homeostatic trafficking and play an important role in the activation and effector functions of leucocytes. Approximately 50 chemokines have been described and are classified into four groups based on specific sequence motifs: **CC chemokines**, which are important in mononuclear cell recruitment (e.g. macrophage inflammatory protein-1α/β (MIP-1α/β), RANTES and macrophage chemoattractant protein-1 (MCP-1); **CXC chemokines**, which are important in neutrophil recruitment (e.g. CXCL-9, CXCL-10); **CX3C chemokines**; and **fractalkine**.

In transplantation a variety of chemokines are involved in orchestrating graft infiltration during an alloimmune response.[94] Although modest reduction in rejection rates following antagonism of specific chemokines (e.g. CXCR3 and CCR5 receptor antagonist) have been achieved in mouse models of transplantation,[95] this has not been so in larger animal models. This is not surprising, as in human, multiple chemokines are likely to interact, compensating for loss of individual pathways. Chemokines also play an important role in memory cell, regulatory cell (CCR4) and dendritic cell (CCR7) trafficking to and from lymphoid tissue. Currently clinical trials on chemokine therapy have focused on diseases such as HIV and autoimmunity, and there is as yet no clear evidence for its successful use in human transplantation.

Specificity of rejection

The nature of tissue destruction during rejection tells us about the processes involved – graft destruction can show exquisitely fine specificity for cells carrying donor alloantigens. Mintz and Silvers demonstrated the specificity of donor cell lysis in experiments using allophenic mice as tissue donors. Allophenic or tetraparental mice are bred by fusing the embryos from mice of two different genetic origins. The tissues of the resulting mosaic offspring are made up of patches of cells from each parental type. If skin from an allophenic donor is grafted to mice of either parental origin only the cells of non-identical type are rejected, leaving cells of recipient type intact. A similar level of specificity has been observed in other experimental situations.

Nevertheless, 'bystander' destruction of tissue may be observed following specific immune responses to foreign antigens. In the aforementioned experiments of donation from tetraparental animals, if the majority of cells in the graft were allogeneic to the recipient the overwhelming inflammatory response actually led to destruction of the entire tissue. Thus, although the immune system can exhibit fine specificity at the cellular level, once initiated, ongoing inflammation may result in non-directed mechanisms of cell death and this is likely to be relevant in clinical, acute rejection.

The effector systems that can damage tissue and their roles in hyperacute, acute and chronic rejection are described below.

Humoral mechanisms

Antibody causes tissue damage through the fixation of complement, but also through antibody-dependent cellular cytotoxicity, although the role of this pathway in allograft rejection has not been fully elucidated.[96]

Patients exposed to MHC antigens through transplantation, blood transfusion or pregnancies often develop antibodies against MHC antigens. These preformed antibodies can cause **hyperacute** rejection, a process in which the organ fails immediately following revascularisation.[97,98] There is deposition of antibody and complement and the accumulation of polymorphonuclear leucocytes within the graft.[97] This is no longer a significant cause of graft failure because of the success of screening for MHC-directed cytotoxic antibodies and the use of the lymphocyte crossmatch with increasing levels of sensitivity. However, in the long term the formation of antibody in waiting-list patients may be the greatest barrier to successful transplantation by creating a group of patients who become virtually 'untransplantable' or in whom transplantation is so delayed as to materially affect outcome,[40] or who require novel and potentially hazardous immunosuppressive protocols.[41,99–101]

Anti-HLA antibody may develop in the course of acute rejection and this is associated with vascular involvement and a relatively poor response to conventional therapy.[102–106]

The appearance of donor-specific anti-HLA antibodies following the acute phase of transplantation is not necessarily associated with acute rejection and this state has been termed accommodation. The mechanisms of accommodation are poorly understood but are of considerable interest, particularly in the fields of ABO incompatibility and xenotransplantation.[107–109]

The presence of donor-specific antibodies is usually associated with poor long-term graft outcome[106] and their deposition in the kidney has been inferred from peritubular capillary staining for C4d.[110] The causal significance of such antibodies is also supported by observations in mice with severe combined immunodeficiency in which the infusion of donor-specific antibody causes lesions similar to chronic allograft vasculopathy.[111] On an individual basis these findings may therefore be considered as good evidence for an immunological component to chronic allograft nephropathy.[110] Even in the absence of donor-specific antibody the presence of HLA antibodies is a risk factor for allograft loss,[112] perhaps because of low-level B-cell priming not reflected by detectable donor-specific antibody or because the presence of HLA antibody reflects the priming of T lymphocytes specific for indirectly presented HLAs.[113]

Interestingly, liver transplantation appears to be an exception to the rule not to transplant across a positive crossmatch. In fact, liver transplants are carried out with little regard at all to alloimmunity and are performed successfully not only across positive crossmatches but also without MHC matching of donor and recipient (see below).

Cellular mechanisms

Natural killer (NK) cells

NK cells, although derived from the same lineage as T cells (lymphoid), are important components of the innate immune system, with a major role in the first line of defence against virus and tumour cells.[114] More recent experiments have recognised mechanisms that may be involved in alloreactivity.[115] The ability to mediate allogenicity was first described following the demonstration of the 'hybrid resistance' phenomenon whereby F1 mice would reject parent haematopoetic cells. Parent to F1 progeny transplantation provides a reliable model for NK-mediated rejection as it excludes T- and B-cell-mediated responses. NK cells recognise and react to the absence of autologous MHC class I products on the surface of target cells; this is the 'missing self' hypothesis.[116] Effector functions are further modulated by a complex array of inhibiting and activating receptors that determine the threshold for activation. The ligands for these receptors are poorly defined but include HLA class I molecules and stress-induced molecules such as MICA and ULBP (UL-16 binding protein), unusual members of the extended MHC class I superfamily.

There is now good evidence that NK cells play an important role in stem cell transplantation. Alloreactivity induced by donor NK cells when the recipient is lacking cognate HLA ligands is associated with enhanced engraftment, a reduction in acute graft-versus-host disease, eradication of malignant cells and successful reconstitution of the immune system.[117] Human trials are ongoing to address the clinical applicability of these mechanisms in bone marrow transplantation.

The contribution of NK cells to rejection in solid-organ transplantation remains uncertain.[118] SCID and RAG knockout mice that are T- and B-cell deficient accept organ allografts indefinitely, suggesting that NK cells do not play a direct role in rejection in this setting. In a mouse model of cardiac allograft vasculopathy the role of NK cells is partly CD4 T-lymphocyte dependent, offering one explanation for these observations.[119]

The human NK-cell repertoire is though more complex. Human genetic correlation studies have demonstrated that certain NK-cell receptor HLA-ligand polymorphisms modulate the incidence of acute rejection in liver transplantation,[120] but do not impact on outcomes in renal transplantation.[121]

As previously discussed, NK cells not only act as effector cells but interact with components of the innate and adaptive immune systems, promoting DC maturation and T-cell differentiation.[42,115,122]

Antigen-specific cytotoxic T cells

In cell culture systems, MHC-mismatched lymphocytes proliferate and produce cytokines in response to one another in the mixed lymphocyte reaction (MLR). The resulting cytokine production allows the differentiation of precursor cytotoxic T lymphocytes into effector cells (cytotoxic T cells, CTLs) that lyse target cells bearing mismatched MHC antigens. The identification of a powerful antigen-specific response, which could be quantified in the MLR, made the CTL a prime suspect as an important effector mechanism of acute graft rejection, and evidence for this is considerable:

- The majority of class I MHC-specific CTLs express CD8 and graft rejection may be delayed in the absence of $CD8^+$ cells in some animal models.[43,123–126]
- CTLs can be recovered from allografts that are undergoing rejection; however, they may be present only at low levels in grafts of animals that have been treated with ciclosporin to prevent rejection.[127,128]
- Cloned populations of CTLs can cause the type of tissue damage associated with rejection.[129,130]

Conversely, graft destruction may occur in the absence of demonstrable CTL activity and the presence of such cells within a graft may not always lead to graft destruction. Alloantigen-specific CTL activity may be recovered from a non-rejected graft and high alloantigen-specific CTL activity has been demonstrated in the splenocytes of rats in which prolonged allograft survival has been induced by donor-specific preoperative blood transfusion.[131,132]

This suggests that in vitro cytotoxic activity does not necessarily correlate with the destruction of allogeneic cells.

CTLs kill their targets in a variety of ways, including through the action of perforins and granzymes that attack membrane integrity, through Fas-mediated apoptosis and the secretion of TNF-α. Although direct cytotoxicity may not be necessary to mediate rejection it is likely to play a role in most cases, and CD8+ T cells are likely to contribute through the elaboration of cytokines that recruit and activate cells involved in more generalised inflammatory responses during the acute phases of rejection.

Macrophages and DTH reactions

T lymphocytes can generate responses that are both exquisitely specific at a cellular level and that non-specifically generate widespread local inflammation (e.g. the delayed-type hypersensitivity reaction (DTH), first described by Koch in the cutaneous reaction to tuberculin). The afferent phase of DTH is antigen specific but the efferent phase is not.

DTH involves the production of a wide range of cytokines by CD4+ T lymphocytes, which activate effector cells, the most important of which are macrophages. These cells themselves elaborate a variety of substances, such as oxygen-derived intermediates and TNF-α, that mediate tissue damage. The importance of DTH is illustrated by acute rejection of H-Y disparate grafts in which cell-mediated cytotoxicity does not correlate with graft rejection but the DTH response to H-Y antigens does.[133] Furthermore, the reconstitution of irradiated rats with CD4+ T lymphocytes can cause allograft rejection in the absence of detectable cytotoxic T-cell activity.[43]

Chronic allograft nephropathy is likely, at least in part, to be due to an alloimmune response in which cytokines, CD4+ T lymphocytes and macrophages play a part.[134,135] Cytokines such as IL-1, TNF-α, TGF-β and platelet-derived growth factor (PDGF) lead to smooth muscle proliferation and increased synthesis of extracellular matrix protein. They are therefore candidate mediators of the vasculopathy and interstitial fibrosis observed in chronic allograft nephropathy.

Eosinophils

Acute and chronic kidney allograft rejection are associated with a varying level of eosiniphilia,[136,137] but their importance in rejection is uncertain. In an experimental model of acute mouse cardiac allograft rejection in which the depletion of CD8+ T lymphocytes results in a dominant T2 response, rejection appears to be mediated by eosinophils. In another model, in which acute rejection of MHC class II disparate mouse skin grafts was studied, IL-5-dependent infiltration with eosinophils was observed. In this model, when Fas/FasL interactions were absent, neutralising antibodies to IL-5 blocked eosinophilia and rejection, implicating the eosinophil as an effector cell in this system.[138] In another experimental model of skin allograft rejection, the same group has shown a role for IL-5 and eosinophilia in chronic rejection, but in this system not all of the pathology could be attributed to eosinophils.[139] In situations where classical pathways of graft rejection are absent or which are dominated by a T2-type response, the eosinophil appears therefore to be critical in graft destruction.[140]

Mast cells

There is new evidence linking mast cells to the development of Treg-dependent transplant tolerance. In a mouse model of allogeneic skin transplantation, mast cell deficiency was associated with the inability to induce tolerance associated with a possible role of IL-9 in Treg generation.[141]

Cytokines

The primary role of cytokines is in the initiation and regulation of immunity as described above, but cytokines may also play an important role in the effector phase of the alloimmune response. This may be the case in both acute and chronic rejection and in non-immune mechanisms of allograft damage such as calcineurin inhibitor toxicity. Thus, TGF-β may play an important intermediate role in alloimmune regulation, chronic rejection and in the nephrotoxicity of calcineurin inhibitors.[142–144]

Privileged sites

Privileged sites are those in which tissue allografts appear to elicit a weak immune response and there may consequently be prolonged allograft survival. These sites include the anterior chamber of the eye, the cornea, the brain and the testis.[145,146] They are typified by absent or limited lymphatic drainage. The degree of 'privilege' seems to vary depending on the nature of the transplanted tissue as well as the site of transplantation.

The liver is an unusual vascularised graft in that, despite its extensive blood supply and high immune cell content, it often fails to elicit rejection and may protect co-transplanted organs from rejection despite their usual immunogenicity. Outbred pigs often fail to reject orthotopic liver allografts[147] and simultaneous renal allografts from the same donor show prolonged survival despite the fact that they would otherwise have been rejected. In rat strain combinations in which an orthotopic liver allograft is not rejected the graft may even abrogate an existing state of sensitisation of the host against donor histocompatibility antigen.[148,149] Similarly, although crossmatching has been demonstrated to be of some relevance in liver transplantation outcome,[150] urgently implanted grafts survive well despite a positive crossmatch.

The mechanisms underlying these remarkable properties of the liver are still to be fully explained. They may depend on the antigenic load, given the liver's size, and its Promethean properties of regeneration; however, if specific mechanisms could be elucidated then they could be applied therapeutically in other solid organs.[151–153] Those so far identified include the induction of regulatory CD4+ T cells and partial activation followed by apoptosis of CD8+ T cells as they encounter antigen on liver sinusoidal epithelial cells.

Chronic rejection

The clinicopathological entity of chronic rejection is perhaps most easily considered from the perspective of renal transplantation. It is apparent that some of the changes associated with deteriorating graft function in established transplants are multifactorial and may relate to donor and recipient factors such as age,[154] peritransplant variables such as cold ischaemia time,[155] nephrotoxic treatment such as calcineurin inhibitors,[156] and hypertension. The relevance of these factors is acknowledged in the frequent use of the designation chronic allograft nephropathy or chronic allograft injury (CAI) rather than chronic rejection. However, it is apparent that immunological factors do impinge on the long-term vascular and tubulointerstitial pathology described as CAI. A history of acute rejection is associated with CAI and whilst this could be a consequence solely of damage established at the time and subsequent changes in calcineurin inhibitor dosing, it is now apparent that CAI is associated with enhanced responses of T lymphocytes through the indirect pathway.[28,30,157] This association has been demonstrated with antigen from disrupted cells and with specific donor-derived HLA-DR peptides.[32] It is apparent that modern immunosuppressive therapy that has markedly improved early allograft survival and freedom from acute rejection has only had a limited impact on long-term graft attrition. This may be because conventional immunosuppression has relatively little impact on indirect pathway activation[32] and on B-cell activation, which may both be important.[106,110,158]

A recent report of lung transplant recipients with evidence of humoral chronic rejection failed to detect anti-HLA antibodies, leading to speculation that tissue-specific antibodies may develop in this setting,[159] but these data are limited.

In animal models of chronic rejection there is evidence of upregulation of a wide variety of cytokines, growth factors and lipid mediators of inflammation, but their relative importance has yet to be established. There is therefore not a single candidate pathway which might be targeted for therapeutic purposes.

In summary, chronic rejection is likely to largely involve the indirect pathway of allorecognition, resulting in the elaboration of cytokines that interact with non-immunological factors to produce the vasculopathy and tubulointerstitial atrophy and fibrosis typical of chronic allograft nephropathy. This immune response is relatively resistant to conventional immunosuppressive regimes and is therefore a potential target for future prophylactic therapy. However, this must be safe given the relatively good results of renal transplantation.

Key points

- The immune response to a transplanted allogeneic organ is multifaceted – and the associated literature is becoming increasingly complex. There is increasing evidence for the importance of indirect allorecognition in chronic rejection and appreciation of the importance of heterologous memory T-cell responses in clinical allorecognition.
- A range of recently characterised molecules, including co-stimulators and chemokines and their receptors, offers potential new targets for therapy.
- The actions of immunoregulatory cells are now the subject of intense study, and commandeering such mechanisms perhaps offers the most attractive prospect of improving graft outcome.
- Finally, it is notable that considerable effort has been expended in attempts to improve graft outcome but perhaps the most important group of patients are those who are never or rarely transplanted because they have formed alloreactive antibodies. If mechanisms could be identified to prevent or reverse the formation of such antibodies safely and easily, this would benefit a subgroup of patients with end-stage organ failure who suffer markedly reduced survival because of this failure to be transplanted.

References

1. Matzinger P. Tolerance, danger, and the extended family. Annu Rev Immunol 1994; 12:991–1045.
2. Iwasaki A, Medzhitov R. Toll-like receptor control of the adaptive immune responses. Nat Immunol 2004; 5(10):987–95.
3. Peng Y, Gong JP, Liu CA et al. Expression of toll-like receptor 4 and MD-2 gene and protein in Kupffer cells after ischemia–reperfusion in rat liver graft. World J Gastroenterol 2004; 10(19):2890–3.
4. Goldstein DR, Tesar BM, Akira S et al. Critical role of the Toll-like receptor signal adaptor protein MyD88 in acute allograft rejection. J Clin Invest 2003; 111(10):1571–8.
5. den Haan JM, Meadows LM, Wang W et al. The minor histocompatibility antigen HA-1: a diallelic gene with a single amino acid polymorphism. Science 1998; 279(5353):1054–7.
6. Turner D, Grant SC, Yonan N et al. Cytokine gene polymorphism and heart transplant rejection. Transplantation 1997; 64(5):776–9.
7. Lila N, Amrein C, Guillemain R et al. Human leukocyte antigen-G expression after heart transplantation is associated with a reduced incidence of rejection. Circulation 2002; 105(16):1949–54.
8. Qiu J, Terasaki PI, Miller J et al. Soluble HLA-G expression and renal graft acceptance. Am J Transplant 2006; 6(9):2152–6.
9. Zou Y, Stastny P, Susal C et al. Antibodies against MICA antigens and kidney-transplant rejection. N Engl J Med 2007; 357(13):1293–300.
10. Suarez-Alvarez B, Lopez-Vazquez A, Gonzalez MZ et al. The relationship of anti-MICA antibodies and MICA expression with heart allograft rejection. Am J Transplant 2007; 7(7):1842–8.
11. Scott DM, Ehrmann IE, Ellis PS et al. Identification of a mouse male-specific transplantation antigen, H-Y. Nature 1995; 376(6542):695–8.
12. Marijt WA, Heemskerk MH, Kloosterboer FM et al. Hematopoiesis-restricted minor histocompatibility antigens HA-1- or HA-2-specific T cells can induce complete remissions of relapsed leukemia. Proc Natl Acad Sci USA 2003; 100(5):2742–7.
13. Lombardi G, Sidhu S, Daly M. Are primary alloresponses truly primary? Int Immunol 1990; 2(1):9–13.
14. Brehm MA, Markees TG, Daniels KA et al. Direct visualization of cross-reactive effector and memory allo-specific CD8 T cells generated in response to viral infections. J Immunol 2003; 170(8):4077–86.
15. Adams AB, Williams MA, Jones TR et al. Heterologous immunity provides a potent barrier to transplantation tolerance. J Clin Invest 2003; 111(12):1887–95.

   This paper demonstrates that heterologous immunity, which is virally induced, is a potent barrier to conventional models of tolerance developed in pathogen-free rodents. Although not a new concept this demonstration is both elegant and important, coming from a laboratory that has been at the forefront of investigations in transplantation tolerance.
16. Williams MA, Onami TM, Adams AB et al. Cutting edge: persistent viral infection prevents tolerance induction and escapes immune control following CD28/CD40 blockade-based regimen. J Immunol 2002; 169(10):5387–91.

17. Fangmann J, Dalchau R, Fabre JW. Rejection of skin allografts by indirect allorecognition of donor class I major histocompatibility complex peptides. J Exp Med 1992; 175(6):1521–9.

18. Lee RS, Yamada K, Houser SL et al. Indirect recognition of allopeptides promotes the development of cardiac allograft vasculopathy. Proc Natl Acad Sci USA 2001; 98(6):3276–81.

19. Vella JP, Magee C, Vos L et al. Cellular and humoral mechanisms of vascularized allograft rejection induced by indirect recognition of donor MHC allopeptides. Transplantation 1999; 67(12):1523–32.

20. Lee RS, Grusby MJ, Glimcher LH et al. Indirect recognition by helper cells can induce donor-specific cytotoxic T lymphocytes in vivo. J Exp Med 1994; 179(3):865–72.

21. Lee RS, Grusby MJ, Laufer TM et al. CD8+ effector cells responding to residual class I antigens, with help from CD4+ cells stimulated indirectly, cause rejection of "major histocompatibility complex-deficient" skin grafts. Transplantation 1997; 63(8):1123–33.

22. Auchincloss H Jr, Lee R, Shea S et al. The role of "indirect" recognition in initiating rejection of skin grafts from major histocompatibility complex class II-deficient mice. Proc Natl Acad Sci USA 1993; 90(8):3373–7.

23. Bedford P, Garner K, Knight SC. MHC class II molecules transferred between allogeneic dendritic cells stimulate primary mixed leukocyte reactions. Int Immunol 1999; 11(11):1739–44.

24. Russo V, Zhou D, Sartirana C et al. Acquisition of intact allogeneic human leukocyte antigen molecules by human dendritic cells. Blood 2000; 95(11):3473–7.

25. Harshyne LA, Watkins SC, Gambotto A et al. Dendritic cells acquire antigens from live cells for cross-presentation to CTL. J Immunol 2001; 166(6):3717–23.

26. Herrera OB, Golshayan D, Tibbott R et al. A novel pathway of alloantigen presentation by dendritic cells. J Immunol 2004; 173(8):4828–37.

27. Mason PD, Robinson CM, Lechler RI. Detection of donor-specific hyporesponsiveness following late failure of human renal allografts. Kidney Int 1996; 50(3):1019–25.

28. Baker RJ, Hernandez-Fuentes MP, Brookes PA et al. Loss of direct and maintenance of indirect alloresponses in renal allograft recipients: implications for the pathogenesis of chronic allograft nephropathy. J Immunol 2001; 167(12):7199–206.

29. Ng WF, Hernandez-Fuentes M, Baker R et al. Reversibility with interleukin-2 suggests that T cell anergy contributes to donor-specific hyporesponsiveness in renal transplant patients. J Am Soc Nephrol 2002; 13(12):2983–9.

30. Ciubotariu R, Liu Z, Colovai AI et al. Persistent allopeptide reactivity and epitope spreading in chronic rejection of organ allografts. J Clin Invest 1998; 101(2):398–405.

31. SivaSai KS, Smith MA, Poindexter NJ et al. Indirect recognition of donor HLA class I peptides in lung transplant recipients with bronchiolitis obliterans syndrome. Transplantation 1999; 67(8):1094–8.

32. Najafian N, Salama AD, Fedoseyeva EV et al. Enzyme-linked immunosorbent spot assay analysis of peripheral blood lymphocyte reactivity to donor HLA-DR peptides: potential novel assay for prediction of outcomes for renal transplant recipients. J Am Soc Nephrol 2002; 13(1):252–9.

This paper demonstrates indirect alloreactivity of peripheral blood lymphocytes from renal transplant recipients to HLA-DR-derived peptides corresponding to the donor mismatch. The technique used to demonstrate responsiveness was an elispot to γ-interferon. The frequency of responsive cells was greater in those that had undergone an episode of acute rejection (and worse renal function: 'high-risk' patients), than those who had not. It is proposed that this form of 'immunological monitoring' could be used to tailor immunosuppression to the individual but it also suggests that HLA-derived peptides could be used as therapeutic agents.

33. Waanders MM, Heidt S, Koekkoek KM et al. Monitoring of indirect allorecognition: wishful thinking or solid data? Tissue Antigens 2008; 71(1):1–15.

34. Benichou G, Tam RC, Soares LR et al. Indirect T-cell allorecognition: perspectives for peptide-based therapy in transplantation. Immunol Today 1997; 18(2):67–71.

35. Larsen CP, Morris PJ, Austyn JM. Migration of dendritic leukocytes from cardiac allografts into host spleens. A novel pathway for initiation of rejection. J Exp Med 1990; 171(1):307–14.

36. Steinman RM, Gutchinov B, Witmer MD et al. Dendritic cells are the principal stimulators of the primary mixed leukocyte reaction in mice. J Exp Med 1983; 157(2):613–27.

37. Austyn JM, Weinstein DE, Steinman RM. Clustering with dendritic cells precedes and is essential for T-cell proliferation in a mitogenesis model. Immunology 1988; 63(4):691–6.

38. Benichou G, Valujskikh A, Heeger PS. Contributions of direct and indirect T cell alloreactivity during allograft rejection in mice. J Immunol 1999; 162(1):352–8.

39. Illigens BM, Yamada A, Fedoseyeva EV et al. The relative contribution of direct and indirect antigen recognition pathways to the alloresponse and graft rejection depends upon the nature of the transplant. Hum Immunol 2002; 63(10):912–25.

40. Meier-Kriesche HU, Kaplan B. Waiting time on dialysis as the strongest modifiable risk factor for renal transplant outcomes: a paired donor kidney analysis. Transplantation 2002; 74(10):1377–81.

41. Zachary AA, Montgomery RA, Ratner LE et al. Specific and durable elimination of antibody to donor HLA antigens in renal-transplant patients. Transplantation 2003; 76(10):1519–25.

42. Yu G, Xu X, Vu MD et al. NK cells promote transplant tolerance by killing donor antigen-presenting cells. J Exp Med 2006; 203(8):1851–8.

43. Lowry RP, Gurley KE, Forbes RD. Immune mechanisms in organ allograft rejection. I. Delayed-type hypersensitivity and lymphocytotoxicity in heart graft rejection. Transplantation 1983; 36(4):391–401.

44. Harding FA, McArthur JG, Gross JA et al. CD-28 mediated signalling co-stimulates murine T cells and prevents the induction of anergy in T cell clones. Nature 1992; 356:607–9.

45. Lenschow DJ, Walunas TL, Bluestone JA. CD28/B7 system of T cell costimulation. Annu Rev Immunol 1996; 14:233–58.

46. Acuto O, Michel F. CD28-mediated co-stimulation: a quantitative support for TCR signalling. Nat Rev Immunol 2003; 3(12):939–51.

This is an excellent review of the role of CD28, which discusses the quantitative and qualitative effects of co-stimulation on the consequences of TcR engagement.

47. Pearson TC, Alexander DZ, Winn KJ et al. Transplantation tolerance induced by CTLA4-Ig. Transplantation 1994; 57(12):1701–6.

48. Turka LA, Linsley PS, Lin H et al. T-cell activation by the CD28 ligand B7 is required for cardiac allograft rejection in vivo. Proc Natl Acad Sci USA 1992; 89(22):11102–5.

49. Tivol EA, Borriello F, Schweizer AN et al. Loss of CTLA-4 leads to massive lymphoproliferation and fatal multiorgan destruction, revealing a critical negative regulatory role of CTLA-4. Immunity 1995; 3:541–7.

50. Dharnidharka VR. Costimulation blockade with belatacept in renal transplantation. N Engl J Med 2005; 353(19):2085–6; author reply 2085–6.

51. Larsen CP, Alexander DZ, Hollenbaugh D et al. CD40–gp39 interactions play a critical role during allograft rejection: Suppression of allograft rejection by blockade of the CD40–gp39 pathway. Transplantation 1996; 61(1):4–9.

52. Larsen CP, Elwood ET, Alexander DZ et al. Long-term acceptance of skin and cardiac allografts after blocking CD40 and CD28 pathways. Nature 1996; 381:434–8.

53. Kirk AD, Burkly LC, Batty DS et al. Treatment with humanized monoclonal antibody against CD154 prevents acute renal allograft rejection in nonhuman primates. Nat Med 1999; 5(6):686–93.

54. Monks CR, Freiberg BA, Kupfer H et al. Three-dimensional segregation of supramolecular activation clusters in T cells. Nature 1998; 395(6697):82–6.

55. Gummert JF, Ikonen T, Morris RE. Newer immunosuppressive drugs: a review. J Am Soc Nephrol 1999; 10(6):1366–80.

56. Sayegh MH, Akalin E, Hancock WW et al. CD28–B7 blockade after alloantigenic challenge in vivo inhibits Th1 cytokines but spares Th2. J Exp Med 1995; 181(5):1869–74.

57. Strom TB, Roy-Chaudhury P, Manfro R et al. The Th1/Th2 paradigm and the allograft response. Curr Opin Immunol 1996; 8(5):688–93.

58. VanBuskirk AM, Wakely ME, Orosz CG. Transfusion of polarized TH2-like cell populations into SCID mouse cardiac allograft recipients results in acute allograft rejection. Transplantation 1996; 62(2):229–38.

59. Orosz CG, Wakely E, Sedmak DD et al. Prolonged murine cardiac allograft acceptance: characteristics of persistent active alloimmunity after treatment with gallium nitrate versus anti-CD4 monoclonal antibody. Transplantation 1997; 63(8):1109–17.

60. Sirak JH, Orosz CG, Roopenian DC et al. Cardiac allograft tolerance: failure to develop in interleukin-4-deficient mice correlates with unusual allosensitization patterns. Transplantation 1998; 65(10):1352–6.

61. Bickerstaff AA, VanBuskirk AM, Wakely E et al. Transforming growth factor-beta and interleukin-10 subvert alloreactive delayed type hypersensitivity in cardiac allograft acceptor mice. Transplantation 2000; 69(7):1517–20.

62. Afzali B, Lombardi G, Lechler RI et al. The role of T helper 17 (Th17) and regulatory T cells (Treg) in human organ transplantation and autoimmune disease. Clin Exp Immunol 2007; 148(1):32–46.

63. Jiang S, Camara N, Lombardi G et al. Induction of allopeptide-specific human CD4+CD25+ regulatory T cells ex vivo. Blood 2003; 102(6):2180–6.

64. Qin S, Cobbold S, Pope H et al. 'Infectious' transplantation tolerance. Science 1993; 259:974.

65. Wise MP, Bemelman F, Cobbold SP et al. Linked suppression of skin graft rejection can operate through indirect recognition. J Immunol 1998; 161(11):5813–16.

66. Yin D, Fathman CG. CD4-positive suppressor cells block allotransplant rejection. J Immunol 1995; 154(12):6339–45.

67. Sanchez-Fueyo A, Domenig CM, Mariat C et al. Influence of direct and indirect allorecognition pathways on CD4+CD25+ regulatory T-cell function in transplantation. Transplant Int 2007; 20(6):534–41.

68. Golshayan D, Buhler L, Lechler RI et al. From current immunosuppressive strategies to clinical tolerance of allografts. Transplant Int 2007; 20(1):12–24.

69. Seddon B, Mason D. Peripheral autoantigen induces regulatory T cells that prevent autoimmunity. J Exp Med 1999; 189(5):877–82.

70. Stephens LA, Mason D. CD25 is a marker for CD4+ thymocytes that prevent autoimmune diabetes in rats, but peripheral T cells with this function are found in both CD25+ and CD25– subpopulations. J Immunol 2000; 165(6):3105–10.

71. Kohm AP, Carpentier PA, Anger HA et al. Cutting edge: CD4+CD25+ regulatory T cells suppress antigen-specific autoreactive immune responses and central nervous system inflammation during active experimental autoimmune encephalomyelitis. J Immunol 2002; 169(9):4712–6.

72. Kohm AP, Carpentier PA, Miller SD. Regulation of experimental autoimmune encephalomyelitis (EAE) by CD4+CD25+ regulatory T cells. Novartis Found Symp 2003; 252:45–52; discussion 252–4, 106–14.

73. Mottet C, Uhlig HH, Powrie F. Cutting edge: cure of colitis by CD4+CD25+ regulatory T cells. J Immunol 2003; 170(8):3939–43.

74. Karim M, Kingsley CI, Bushell AR et al. Alloantigen-induced CD25+CD4+ regulatory T cells can develop in vivo from CD25-CD4+ precursors in a thymus-independent process. J Immunol 2004; 172(2):923–8.

75. Kingsley CI, Karim M, Bushell AR et al. CD25+CD4+ regulatory T cells prevent graft rejection: CTLA-4- and IL-10-dependent immunoregulation of alloresponses. J Immunol 2002; 168(3):1080–6.

76. Cobbold SP, Nolan KF, Graca L et al. Regulatory T cells and dendritic cells in transplantation tolerance: molecular markers and mechanisms. Immunol Rev 2003; 196:109–24.

77. Bushell A, Karim M, Kingsley CI et al. Pretransplant blood transfusion without additional immunotherapy generates CD25+CD4+ regulatory T cells: a potential explanation for the blood-transfusion effect. Transplantation 2003; 76(3):449–55.

78. Tadokoro CE, Shakhar G, Shen S et al. Regulatory T cells inhibit stable contacts between CD4+ T cells and dendritic cells in vivo. J Exp Med 2006; 203(3):505–11.

79. Kang SM, Tang Q, Bluestone JA. CD4+CD25+ regulatory T cells in transplantation: progress, challenges and prospects. Am J Transplant 2007; 7(6):1457–63.

80. Lin CY, Graca L, Cobbold SP et al. Dominant transplantation tolerance impairs CD8+ T cell function but not expansion. Nat Immunol 2002; 3(12):1208–13.

81. Chang CC, Ciubotariu R, Manavalan JS et al. Tolerization of dendritic cells by T(S) cells: the crucial role of inhibitory receptors ILT3 and ILT4. Nat Immunol 2002; 3(3):237–43.

82. Wong KK, Carpenter MJ, Young LL et al. Notch ligation by Delta1 inhibits peripheral immune responses to transplantation antigens by a CD8+ cell-dependent mechanism. J Clin Invest 2003; 112(11):1741–50.

This Notch signalling pathway is highly conserved from *Drosophila* to humans and controls cell fate decisions during development. Notch receptors and ligands are widely distributed throughout the haematopoietic system, including on mature leucocytes of most if not all lineages. In this paper, overexpressing of one of the Notch ligands, Delta-like1, on alloantigen-bearing cells could, when introduced into animals, inhibit the immune response to the same antigen delivered subsequently in the absence of this ligand – this even allowed prolongation of heart allograft survival in an antigen-specific fashion. Interestingly, the immune response was skewed away from the normal T1 type of immunity seen following transplantation towards an IL-10-dominated response, suggesting the activation of Treg in this system, thus providing a possible new opportunity for manipulation of the immune system.

83. Asseman C, Mauze S, Leach MW et al. An essential role for interleukin 10 in the function of regulatory T cells that inhibit intestinal inflammation. J Exp Med 1999; 190(7):995–1004.

84. Sundstedt A, O'Neill EJ, Nicolson KS et al. Role for IL-10 in suppression mediated by peptide-induced regulatory T cells in vivo. J Immunol 2003; 170(3):1240–8.

85. Dieckmann D, Bruett CH, Ploettner H et al. Human CD4(+)CD25(+) regulatory, contact-dependent T cells induce interleukin 10-producing, contact-independent type 1-like regulatory T cells [corrected]. J Exp Med 2002; 196(2):247–53.

86. Roncarolo MG, Gregori S, Levings M. Type 1 T regulatory cells and their relationship with CD4+CD25+ T regulatory cells. Novartis Found Symp 2003; 252:115–27; discussion 127–31, 203–10.

87. Butcher EC, Picker LJ. Lymphocyte homing and homeostasis. Science 1996; 272(5258):60–6.

88. Koo DD, Welsh KI, McLaren AJ et al. Cadaver versus living donor kidneys: impact of donor factors on antigen induction before transplantation. Kidney Int 1999; 56(4):1551–9.

89. Schwarz C, Regele H, Steininger R et al. The contribution of adhesion molecule expression in donor kidney biopsies to early allograft dysfunction. Transplantation 2001; 71(11):1666–70.

90. Valujskikh A, Lakkis FG. In remembrance of things past: memory T cells and transplant rejection. Immunol Rev 2003; 196:65–74.

91. Pinschewer DD, Ochsenbein AF, Odermatt B et al. FTY720 immunosuppression impairs effector T cell peripheral homing without affecting induction, expansion, and memory. J Immunol 2000; 164(11):5761–70.

92. Troncoso P, Ortiz M, Martinez L et al. FTY 720 prevents ischemic reperfusion damage in rat kidneys. Transplant Proc 2001; 33(1–2):857–9.

93. Kobayashi H, Koga S, Novick AC et al. T-cell mediated induction of allogeneic endothelial cell chemokine expression. Transplantation 2003; 75(4):529–36.

94. Inston NG, Cockwell P. The evolving role of chemokines and their receptors in acute allograft

rejection. Nephrol Dial Transplant 2002; 17(8):1374–9.

95. Akashi S, Sho M, Kashizuka H et al. A novel small-molecule compound targeting CCR5 and CXCR3 prevents acute and chronic allograft rejection. Transplantation 2005; 80(3):378–84.
96. Tilney NL, Strom TB, Macpherson SG et al. Surface properties and functional characteristics of infiltrating cells harvested from acutely rejecting cardiac allografts in inbred rats. Transplantation 1975; 20(4):323–30.
97. Williams GM, Hume DM, Hudson RP Jr et al. "Hyperacute" renal-homograft rejection in man. N Engl J Med 1968; 279(12):611–8.
98. Patel R, Terasaki PI. Significance of the positive crossmatch test in kidney transplantation. N Engl J Med 1969; 280(14):735–9.
99. Glotz D, Antoine C, Julia P et al. Intravenous immunoglobulins and transplantation for patients with anti-HLA antibodies. Transplant Int 2004; 17(1):1–8.
100. Glotz D, Antoine C, Julia P et al. Desensitization and subsequent kidney transplantation of patients using intravenous immunoglobulins (IVIg). Am J Transplant 2002; 2(8):758–60.
101. Warren DS, Zachary AA, Sonnenday CJ et al. Successful renal transplantation across simultaneous ABO incompatible and positive crossmatch barriers. Am J Transplant 2004; 4(4):561–8.
102. Crespo M, Pascual M, Tolkoff-Rubin N et al. Acute humoral rejection in renal allograft recipients: I. Incidence, serology and clinical characteristics. Transplantation 2001; 71(5):652–8.
103. Mauiyyedi S, Crespo M, Collins AB et al. Acute humoral rejection in kidney transplantation: II. Morphology, immunopathology, and pathologic classification. J Am Soc Nephrol 2002; 13(3):779–87.
104. Bohmig GA, Regele H, Exner M et al. C4d-positive acute humoral renal allograft rejection: effective treatment by immunoadsorption. J Am Soc Nephrol 2001; 12(11):2482–9.
105. Halloran PF, Wadgymar A, Ritchie S et al. The significance of the anti-class I antibody response. I. Clinical and pathologic features of anti-class I-mediated rejection. Transplantation 1990; 49(1):85–91.
106. Worthington JE, Martin S, Al-Husseini DM et al. Posttransplantation production of donor HLA-specific antibodies as a predictor of renal transplant outcome. Transplantation 2003; 75(7):1034–40.
107. Salama AD, Delikouras A, Pusey CD et al. Transplant accommodation in highly sensitized patients: a potential role for Bcl-xL and alloantibody. Am J Transplant 2001; 1(3):260–9.
108. Delikouras A, Fairbanks LD, Simmonds AH et al. Endothelial cell cytoprotection induced in vitro by allo- or xenoreactive antibodies is mediated by signaling through adenosine A2 receptors. Eur J Immunol 2003; 33(11):3127–35.
109. Delikouras A, Dorling A. Transplant accommodation. Am J Transplant 2003; 3(8):917–8.
110. Mauiyyedi S, Pelle PD, Saidman S et al. Chronic humoral rejection: identification of antibody-mediated chronic renal allograft rejection by C4d deposits in peritubular capillaries. J Am Soc Nephrol 2001; 12(3):574–82.
111. Russell PS, Chase CM, Winn HJ et al. Coronary atherosclerosis in transplanted mouse hearts. II. Importance of humoral immunity. J Immunol 1994; 152(10):5135–41.
112. Hourmant M, Cesbron-Gautier A, Terasaki PI et al. Frequency and clinical implications of development of donor-specific and non-donor-specific HLA antibodies after kidney transplantation. J Am Soc Nephrol 2005; 16(9):2804–12.
113. Hanvesakul R, Maillere B, Briggs D et al. Indirect recognition of T-cell epitopes derived from the alpha3 and transmembrane domain of HLA-A2. Am J Transplant 2007; 7(5):1148–57.
114. Hamerman JA, Ogasawara K, Lanier LL. NK cells in innate immunity. Curr Opin Immunol 2005; 17(1):29–35.
115. Raulet DH. Interplay of natural killer cells and their receptors with the adaptive immune response. Nat Immunol 2004; 5(10):996–1002.
116. Ljunggren HG, Karre K. In search of the 'missing self': MHC molecules and NK cell recognition. Immunol Today 1990; 11(7):237–44.
117. Ruggeri L, Aversa F, Martelli MF et al. Allogeneic hematopoietic transplantation and natural killer cell recognition of missing self. Immunol Rev 2006; 214:202–18.
118. Vilches C, Parham P. Do NK-cell receptors and alloreactivity affect solid organ transplantation? Transplant Immunol 2006; 17(1):27–30.
119. Uehara S, Chase CM, Kitchens WH et al. NK cells can trigger allograft vasculopathy: the role of hybrid resistance in solid organ allografts. J Immunol 2005; 175(5):3424–30.
120. Moya-Quiles MR, Alvarez R, Miras M et al. Impact of recipient HLA-C in liver transplant: a protective effect of HLA-Cw*07 on acute rejection. Hum Immunol 2007; 68(1):51–8.
121. Tran TH, Mytilineos J, Scherer S et al. Analysis of KIR ligand incompatibility in human renal transplantation. Transplantation 2005; 80(8):1121–3.
122. Obara H, Nagasaki K, Hsieh CL et al. IFN-gamma, produced by NK cells that infiltrate liver allografts early after transplantation, links the innate and adaptive immune responses. Am J Transplant 2005; 5(9):2094–103.
123. Tilney NL, Kupiec-Weglinski JW, Heidecke CD et al. Mechanisms of rejection and prolongation of vascularized organ allografts. Immunol Rev 1984; 77:185–216.

124. Cobbold SP, Jayasuriya A, Nash A et al. Therapy with monoclonal antibodies by elimination of T-cell subsets in vivo. Nature 1984; 312(5994):548–51.

125. Madsen JC, Peugh WN, Wood KJ et al. The effect of anti-L3T4 monoclonal antibody treatment on first-set rejection of murine cardiac allografts. Transplantation 1987; 44(6):849–52.

126. Madsen JC, Superina RA, Wood KJ et al. Immunological unresponsiveness induced by recipient cells transfected with donor MHC genes. Nature 1988; 332:161.

127. Mason DW, Morris PJ. Inhibition of the accumulation, in rat kidney allografts, of specific – but not nonspecific – cytotoxic cells by cyclosporine. Transplantation 1984; 37(1):46–51.

128. Bradley JA, Mason DW, Morris PJ. Evidence that rat renal allografts are rejected by cytotoxic T cells and not by nonspecific effectors. Transplantation 1985; 39(2):169–75.

129. Engers HD, Sorenson GD, Terres G et al. Functional activity in vivo of effector T cell populations. I. Antitumor activity exhibited by allogeneic mixed leukocyte culture cells. J Immunol 1982; 129(3):1292–8.

130. Tyler JD, Galli SJ, Snider ME et al. Cloned LYT-2+ cytolytic T lymphocytes destroy allogeneic tissue in vivo. J Exp Med 1984; 159(1):234–43.

131. Armstrong HE, Bolton EM, McMillan I et al. Prolonged survival of actively enhanced rat renal allografts despite accelerated cellular infiltration and rapid induction of both class I and class II MHC antigens. J Exp Med 1987; 165(3):891–907.

132. Dallman MJ, Wood KJ, Morris PJ. Specific cytotoxic T cells are found in the nonrejected kidneys of blood-transfused rats. J Exp Med 1987; 165(2):566–71.

133. Liew FY, Simpson E. Delayed-type hypersensitivity responses to H-Y: characterization and mapping of Ir genes. Immunogenetics 1980; 11(3):255–66.

134. Chen J, Myllarniemi M, Akyurek LM et al. Identification of differentially expressed genes in rat aortic allograft vasculopathy. Am J Pathol 1996; 149(2):597–611.

135. Paul LC, Saito K, Davidoff A et al. Growth factor transcripts in rat renal transplants. Am J Kidney Dis 1996; 28(3):441–50.

136. Kormendi F, Amend WJ Jr. The importance of eosinophil cells in kidney allograft rejection. Transplantation 1988; 45(3):537–9.

137. Nolan CR, Saenz KP, Thomas CA 3rd et al. Role of the eosinophil in chronic vascular rejection of renal allografts. Am J Kidney Dis 1995; 26(4):634–42.

138. Le Moine A, Surquin M, Demoor FX et al. IL-5 mediates eosinophilic rejection of MHC class II-disparate skin allografts in mice. J Immunol 1999; 163(7):3778–84.

139. Le Moine A, Flamand V, Demoor FX et al. Critical roles for IL-4, IL-5, and eosinophils in chronic skin allograft rejection. J Clin Invest 1999; 103(12):1659–67.

140. Goldman M, Le Moine A, Braun M et al. A role for eosinophils in transplant rejection. Trends Immunol 2001; 22(5):247–51.

141. Lu LF, Lind EF, Gondek DC et al. Mast cells are essential intermediaries in regulatory T-cell tolerance. Nature 2006; 442(7106):997–1002.

142. Khanna AK, Cairns VR, Becker CG et al. TGF-beta: a link between immunosuppression, nephrotoxicity, and CsA. Transplant Proc 1998; 30(4):944–5.

143. Khanna A, Cairns V, Hosenpud JD. Tacrolimus induces increased expression of transforming growth factor-beta1 in mammalian lymphoid as well as nonlymphoid cells. Transplantation 1999; 67(4):614–19.

144. Khanna AK, Hosenpud JS, Plummer MS et al. Analysis of transforming growth factor-beta and profibrogenic molecules in a rat cardiac allograft model treated with cyclosporine. Transplantation 2002; 73(10):1543–9.

145. Streilein JW, Takeuchi M, Taylor AW. Immune privilege, T-cell tolerance, and tissue-restricted autoimmunity. Hum Immunol 1997; 52(2):138–43.

146. Streilein JW. Immune privilege as the result of local tissue barriers and immunosuppressive microenvironments. Curr Opin Immunol 1993; 5(3):428–32.

147. Calne RY, Sells RA, Pena JR et al. Induction of immunological tolerance by porcine liver allografts. Nature 1969; 223(205):472–6.

148. Kamada N, Brons G, Davies HS. Fully allogeneic liver grafting in rats induces a state of systemic nonreactivity to donor transplantation antigens. Transplantation 1980; 29(5):429–31.

149. Kamada N, Davies HS, Roser B. Reversal of transplantation immunity by liver grafting. Nature 1981; 292(5826):840–2.

150. Doyle HR, Marino IR, Morelli F et al. Assessing risk in liver transplantation. Special reference to the significance of a positive cytotoxic crossmatch. Ann Surg 1996; 224(2):168–77.

151. Kamada N, Wight DG. Antigen-specific immunosuppression induced by liver transplantation in the rat. Transplantation 1984; 38(3):217–21.

152. Farges O, Morris PJ, Dallman MJ. Spontaneous acceptance of liver allografts in the rat. Analysis of the immune response. Transplantation 1994; 57(2):171–7.

153. Farges O, Morris PJ, Dallman MJ. Spontaneous acceptance of rat liver allografts is associated with an early downregulation of intragraft interleukin-4 messenger RNA expression. Hepatology 1995; 21(3):767–75.

154. Meier-Kriesche HU, Cibrik DM, Ojo AO et al. Interaction between donor and recipient age in determining the risk of chronic renal allograft failure. J Am Geriatr Soc 2002; 50(1):14–7.

155. McLaren AJ, Jassem W, Gray DW et al. Delayed graft function: risk factors and the relative effects of early function and acute rejection on long-term survival in cadaveric renal transplantation. Clin Transplant 1999; 13(3):266–72.

156. Meier-Kriesche HU, Kaplan B. Cyclosporine microemulsion and tacrolimus are associated with decreased chronic allograft failure and improved long-term graft survival as compared with sandimmune. Am J Transplant 2002; 2(1):100–4.

157. Hornick PI, Mason PD, Baker RJ et al. Significant frequencies of T cells with indirect anti-donor specificity in heart graft recipients with chronic rejection. Circulation 2000; 101(20):2405–10.

158. Martin L, Guignier F, Mousson C et al. Detection of donor-specific anti-HLA antibodies with flow cytometry in eluates and sera from renal transplant recipients with chronic allograft nephropathy. Transplantation 2003; 76(2):395–400.

159. Magro CM, Klinger DM, Adams PW et al. Evidence that humoral allograft rejection in lung transplant patients is not histocompatibility antigen-related. Am J Transplant 2003; 3(10):1264–72.

4

Testing for histocompatibility

Philip A. Dyer, Alison Logan, Kay Poulton, Karen Wood, Judith Worthington

Introduction

The immune system has evolved specifically to recognise and destroy hazardous infective agents such as bacteria and viruses. These same mechanisms function to reject non-self-allogenic tissues, which are an irritant to surgical transplantation. Successful organ transplants occur between monozygotic twins (syngeneic transplants) or between genetically related or unrelated individuals (allogeneic transplants) – but only when the immune system is hindered by effective immunosuppression. The degree of immunosuppression needed, which reflects the frequency and strength of the alloimmune response, will be determined by the immunogenetic disparity between the donor and the recipient. The ability of a recipient to respond to allogeneic tissue will reflect their own immunogenetic constitution – their immune responsiveness. There are two major genetic systems determining human allogenicity, each of which can convey a biological veto or a biological impediment to effective clinical transplantation.

Veto

- antibody to incompatible ABO blood group; or
- mismatched donor human leucocyte antigen (HLA) evidenced by IgG antibody present at the time of transplantation.

Impediment

- low-titre antibody specific for ABO blood group; or
- recipient previously exposed to HLA antigens present in the donor.

If transplantation proceeds, in the biological veto situation then hyperacute rejection of the transplant is highly probable.

In the situation of biological impediment, effective clinical transplantation can only be achieved through use of special protocols to remove circulating antibody (see Chapter 7).

In this chapter we explain the immunology and genetics of the HLA system in the context of clinical organ transplantation, current techniques used to identify the extensive HLA polymorphism at the gene (allele) and protein (specificity) levels, and will detail techniques used to establish recipient allosensitisation. The application of these techniques to attain effective clinical transplantation will be highlighted.

HLA genetics

The human leucocyte antigen (HLA) complex is located on the short arm of chromosome 6 at 6p21.3. This region is also known as the major histocompatibility complex (MHC). This collection of highly polymorphic genes codes for HLA molecules, which are cell-surface proteins that play a pivotal role in antigen presentation and recognition, and hence the survival of transplanted organs and tissues. For the purposes of this chapter the emphasis will be placed on elements of the HLA system that are characterised on a routine basis in the histocompatibility and immunogenetics laboratory.

There are two classes of HLA genes that are involved in the immune response to transplanted organs, class I and class II. These classes are structurally and functionally distinct. Class I molecules are involved in the processing and presentation of intracellular peptide to $CD8^+$ T cells. Class II molecules process and present extracellular peptide to $CD4^+$ T cells. This process is described in more detail in Chapter 3 on transplantation immunology.

Potential recipients and donors are tested for HLA class I (HLA-A, HLA-B and HLA-Cw) and HLA class II (HLA-DR, HLA-DQ and in some instances HLA-DP) to facilitate organ allocation (HLA matching), crossmatching and antibody screening. There are two means of determining the HLA type of an individual: by serological methods or DNA polymerase chain reaction (PCR)-based techniques.

HLA nomenclature

Nomenclature for both accepted and novel HLA alleles is regulated by the World Health Organisation (WHO) Nomenclature Committee for factors of the HLA system.[1] HLA sequences are officially recorded on the IMGT/HLA Sequence Database (www.ebi.ac.uk/imgt), which is updated quarterly and is part of the International Immunogenetics Project (IMGT). HLA genes are highly polymorphic. In January 2008, this database contained sequences for 2047 class I alleles and 944 class II alleles.

Over time, major revisions of HLA nomenclature are necessary due to the ever-increasing numbers of alleles identified. A guide to the most recent nomenclature for HLA antigens and alleles is summarised in Table 4.1, where resolution of HLA alleles to the four-digit level is shown. In practice matching beyond this level, even in stem cell transplantation, is impractical. Resolution to eight digits can be carried out and in some instances an alphabetical suffix is used to describe the biological expression of an encoded molecule.

HLA typing

Serologically based techniques: complement-dependent cytotoxicity (CDC)

In a CDC test, antisera with specificity for HLA antigens are incubated with peripheral blood lymphocytes in the presence of complement. The antiserum used is usually monoclonal. If the target lymphocytes carry HLA antigen(s) to which the antiserum has specificity the test leads to killing of the target cells. In order to visualise the cell death a cocktail of fluorescent dyes is added to distinguish between live cells (negative test) and dead cells (positive test). CDC serological HLA typing identifies HLA protein polymorphisms as expressed at the cell surface.

Molecular techniques

Molecular techniques for HLA typing of DNA sequence polymorphisms have largely replaced serology since they offer flexibility of resolution, much improved reproducibility and greater accuracy. The invention of the polymerase chain reaction (PCR)[2] has revolutionised HLA typing techniques by facilitating identification of HLA polymorphisms at the single nucleotide polymorphism (SNP) level.

Principle of the polymerase chain reaction

Use of the PCR allows amplification of selected regions of interest within a length of target DNA. The technique involves heat separation of double-stranded DNA, primer annealing and extension, resulting in exponential amplification of the template DNA. The essential agent is Taq polymerase, which is a thermostable enzyme that facilitates nucleotide extension from primer pairs, constructing a DNA copy of the template DNA strand. Primers are chemically synthesised oligonucleotides, usually 17–30 nucleotides in length. The primers are designed to flank the region of interest by binding to complementary sequences on the target DNA. Adenine (A) binds to thymine (T) and guanine (G) to cytosine (C) via hydrogen bonds. The PCR mixture contains the target DNA, the primers, the four deoxyribonucleotide triphosphate building blocks (A, T, G and C), Taq polymerase enzyme and reaction buffer.

Table 4.1 • HLA nomenclature: basic overview of the levels of HLA typing performed in the histocompatibility laboratory

WHO nomenclature	Interpretation
HLA-B	Identification of HLA locus
HLA-B44	HLA defined by serology-based technique
HLA-B*44	Asterisk denotes HLA alleles defined by analysis of DNA
HLA-B*44 two-digit resolution	Denotes the allele family Corresponds where possible to the serological group Often termed 'low resolution' Level used for matching in solid organ transplants
HLA-B*4402 four-digit resolution	Allele sequence variation results in amino acid substitutions, coding variation or non-synonymous changes Level of matching used in haemopoietic stem cell transplantation

This process is performed in a thermocycler, which creates rapid (millisecond) changes in temperature in a controlled environment.

- Double-stranded DNA is denatured by heating to 90–95 °C.
- Cooling to 40–60 °C allows the primers to anneal to their complementary sequences on the target DNA.
- Taq polymerase synthesises new strands from the primed sequences using target DNA as template.
- Denaturation, annealing and extension steps are repeated through many cycles and at each cycle new strands can act as templates for the next round of synthesis. This allows exponential increases in product.

PCR sequence-specific primers (PCR-SSP)

PCR-SSP is currently the HLA typing system of choice in most histocompatibility and immunogenetics laboratories for typing deceased organ donors. A result can be generated in 3 hours. There are various commercially available PCR-SSP kits in use, such as those manufactured by Invitrogen (www.invitrogen.com), Olerup (www.alphahelix.co.uk) and BAG Health Care GmbH (www.bag-germany.com). These kits are constantly updated by the manufacturer to incorporate new WHO-recognised alleles. Some laboratories construct in-house SSP trays using their own design primers; however, as more alleles are defined it has become almost impossible for individual laboratories to keep their own primer design up to date.

The underlying concept of PCR-SSP typing of HLA alleles is based on the fact that Taq polymerase lacks 3′ to 5′ exonuclease proofreading activity. Therefore only primers that are matched to the 3′ end of the template will facilitate DNA extension.

To perform HLA typing on one individual, multiple PCRs are performed simultaneously using different combinations of sequence-specific primers. The combinations utilised should allow amplification of all known HLA alleles.

The assignment of alleles is based on the presence or absence of amplified product. To ensure that the absence of a specific product is due to the individual lacking the corresponding sequence and is not simply due to a technical error, a control is included in each of the SSP reactions. The control is an invariant region of a gene that is constant between all individuals, such as human growth hormone. The PCR-SSP product is visualised by size differences using agarose gel electrophoresis, as shown in **Fig. 4.1.** Electrophoresis through agarose relies on the movement of negatively charged DNA (due to the phosphate backbone) towards the anode. Fragments of DNA differentially migrate and thus can be identified according to their size. DNA is visualised on the gel by staining with ethidium bromide, which intercollates between the strands of DNA and fluoresces under ultraviolet light. In Fig. 4.1, control bands are visible in each of the SSP reactions and the specific product is visible as a second band in the well. By knowing which wells contain which SSP, the HLA type of an individual can be allocated.

PCR sequence-specific oligonucleotide probes (PCR-SSOP)

PCR-SSOP was the first PCR-based technique used for detecting HLA polymorphisms.[3,4] The technique has advantages over PCR-SSP, in particular a large sample throughput can be achieved.

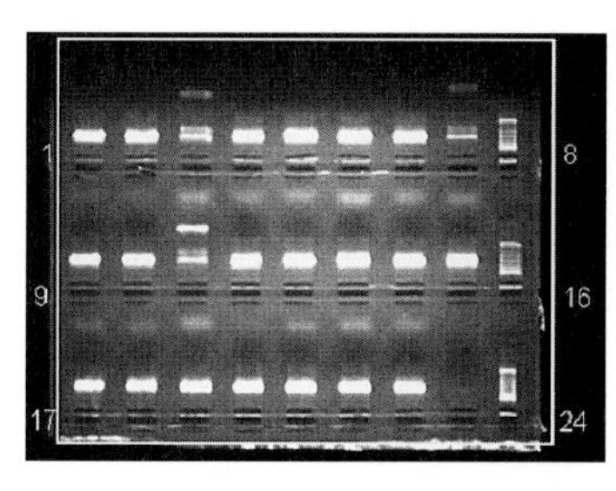

Figure 4.1 • A PCR sequence-specific primer (PCR-SSP) reaction. A control band (DNA) can be seen in each well and specific product bands are visible in wells 3, 8 and 11. By knowing the specific primers in each well, the HLA type of an individual can be determined.

Until recently the methodology and interpretation of results was complex and time-consuming, which meant that it was not suitable for deceased donor HLA typing.

The development of luminex technology which incorporates a reverse sequence-specific oligonucleotide (RSSO) HLA typing system in conjunction with an analyser has meant that RSSO is a viable alternative to PCR-SSP.

Microbead array technology

Polystyrene beads (5.6 µm diameter) are colour coded internally using red and infrared fluorophores. Different amounts of the two dyes are added to each bead set. In this way 100 different bead sets can be constructed. Each bead set will have a unique identifiable spectral signature.

Each set of beads is then coated with either antibodies, peptides, oligonucleotides or receptors depending on the target to be detected. For histocompatibility testing, coated beads are used that allow the detection of HLA-specific antibodies present in patients' serum, and which can HLA type patients and donors (detailed below).

Microbead array technology is based on the principle of flow cytometry where coated beads are passed in a fluid phase through the path of a laser beam. The laser excites the internal dyes of each bead and also any reporter dye that is bound to a target molecule on the bead surface. The resulting fluorescent signal is detected within the dedicated cytometer analyser. Computer software is then used to interpret and convert the signal information into the corresponding HLA type.

Further details, along with a step-by-step animation of the principles of this technology, can be viewed at www.luminex.corp.

RSSO in conjunction with microbead array technology

Target DNA is amplified in the PCR using primers that are labelled with biotin. The double-stranded DNA is then denatured to leave single-stranded DNA. At this point SSO probes bound to fluorescent microspheres are added. The probes determine the specificity of the reaction and different numbers of probes are used depending on the locus to be tested. The SSO probes bind to allele-specific regions of DNA. When the beads are washed any unbound probes will be removed. The addition of PE-conjugated streptavidin then causes a chemical reaction with biotin to release fluorescence, which is detected by a dedicated analyser (e.g. LABscan™ 100). The analyser in conjunction with an XY platform and sheath delivery system means that 96-well trays can be read automatically. Software is then used to analyse the fluorescence pattern (probes bound) and matches them to the patterns of known alleles, hence defining the HLA type. Several manufacturers provide RSSO kits and technology, including LABType™ produced by OneLambda (www.onelambda.com) and Lifematch™ produced by Tepnel Life Sciences.

The throughput of RSSO coupled with microbead array technology is greater than that obtained by conventional PCR-SSO and PCR-SSP, and for this reason the typing system is commonly used in histocompatibility laboratories. In most instances the resolution is intermediate; however, high-resolution HLA typing kits are being trialled. Currently PCR-SSP is still routinely used for HLA typing deceased donors due to time constraints, but it is envisaged that RSSO will supersede this technique.

Sequence-based typing (SBT)

SBT allows a greater resolution of HLA typing than both PCR-SSP and most PCR-SSOP techniques, but it is rarely employed for solid organ transplantation as the criteria used for HLA matching do not require resolution to the HLA allele level. SBT is mainly used for haemopoietic stem cell transplantation (HSCT), where HLA typing to allele level is necessary due to the risk of graft vs. host disease.

SBT involves typing of HLA genes at the nucleotide or allele level. The technique utilises the elegant way in which DNA is copied by a DNA polymerase enzyme. Bases are incorporated into an extending DNA strand in a PCR through the formation of a phosphodiester bridge between the 3′ hydroxyl group at the growing end of a primer and the 5′ phosphate group of the nucleotide that is being incorporated.[5] In SBT, DNA polymerases incorporate analogues of A, G, T and C that do not have a 3′ hydroxyl group. This results in termination of the strand.[6] The analogues of A, G, T and C are labelled with a fluorescent dye. Detection of the dye then reveals which base is incorporated at each position. In this way the entire sequence can be determined. There are various SBT techniques and software options available from manufacturers such as Abbott Diagnostics (www.abbottmolecular.com) and Protrans (www.protrans.com) for HLA typing.

Clinical application of HLA typing

HLA typing and evaluation of HLA mismatching are essential prerequisites to kidney and stem cell transplantation.

The level of HLA typing employed for solid organ transplantation is generally low to medium resolution (two digits). HLA matching is not applied in pancreas, heart, liver or lung transplantation unless the recipient is sensitised to donor HLA antigens. In the case of haemopoietic stem cell transplants (HSCTs), medium- to high-resolution (four digits) HLA typing is performed. The level of matching is increased for HSCTs due to the risk of graft vs. host disease, when immunocompetent T cells transferred with the graft attack the recipient's own cells and tissues.

Causes of HLA-specific alloimmunisation

HLA-specific antibodies can be produced in response to blood transfusions, pregnancy or transplantation.[7–9]

The method of antigen presentation and the dose of the antigen influence the nature of the immune response, which will differ for each of these stimuli.

Blood transfusion

The antigen source in a blood transfusion is the lymphocyte content of the transfused blood.[10] A number of studies have indicated that a low number of transfusions are a minimal risk for sensitisation whilst multiple transfusions are required to cause alloimmunisation.[11,12] These data suggest that one or a few – usually less than 10 – blood transfusions induce minimal clonal expansion with limited proliferation of T and B cells. This is usually a transient IgM antibody, or IgM antibody followed by IgG antibody production, which is promptly downregulated. Multiple transfusions induce significant clonal expansion where antibody production is more likely and may result in a sustained level of IgG antibody production.[13]

Pregnancy

HLA-specific antibodies can be produced during pregnancy as a result of stimulation by paternal HLA antigens expressed on the fetus. Stimulation occurs when high levels of antigen enter maternal circulation during the birth of the first child. Subsequent pregnancies may induce a secondary response. Multiparous female patients are at greater risk of sensitisation following blood transfusions, suggesting that a degree of clonal expansion has occurred during pregnancy so that antibodies are more easily induced by subsequent blood transfusions.[14]

Previous transplantation

An allograft recipient is exposed to those HLA antigens of the donor that were mismatched and consequently may develop antibodies to those antigens.[9] However, transplant recipients are immunosuppressed and the immune response can be immature.

Available evidence suggests that the immune system of all allograft recipients undergoes clonal expansion.

This process develops further if the graft is lost or immunosuppressive therapy is discontinued, resulting in detectable donor-specific HLA antibodies in the serum.

Detection of alloantibodies

Complement-dependent cytotoxicity

The long-established method used to screen patient sera for HLA-specific antibodies is the CDC assay outlined above. A patient's serum sample is incubated with a panel of lymphocytes of known HLA phenotype. If the patient's serum contains antibodies specific for HLA molecules on the target, cell death occurs. By using a selected panel of target cells of known HLA phenotype, it is possible to assign specificity to the antibody detected.[15] The CDC assay is also used for recipient/donor crossmatching, which was introduced as an essential pretransplant test in the mid-1960s.[16]

CDC enables the detection and definition of complement-fixing IgG and also IgM antibodies, which may be directed against HLA or non-HLA targets.

Despite its widespread use in antibody screening the CDC assay has a number of inherent problems, not least that non-complement-fixing antibodies, which have been shown to be relevant to graft outcome, cannot be detected.[17] A number of solid-phase laboratory techniques using flow cytometry, luminex and enzyme-linked immunosorbent assay (ELISA) are now available for the definition of recipient sensitisation. The advantages of these alternate technologies is that they detect only HLA-specific antibodies and non-complement-fixing as well as complement-fixing antibodies; they can be semiautomated and they remove the need for viable target cell panels.

Enzyme-linked immunosorbent assay (ELISA)

One of the first solid-phase assays to be introduced was ELISA-based testing. Whilst these have largely been superseded by Luminex technology they remain a useful component of laboratory antibody detection and definition repertoire. Commercially available kits can be divided into antibody detection kits which act as pre-screen assays, simply determining the presence or absence of HLA-specific antibody, and antibody definition kits, which enable definition of HLA specificity. Antibody detection kits include Quikscreen (GTI Inc., www.gtidiagnostics.com) for the detection of IgG HLA class I-specific antibodies, B-Screen (GTI Inc., www.gtidiagnostics.com) for the detection of IgG HLA class II-specific antibodies and LATM (One Lambda Inc., www.onelambda.com) used to detect the presence or absence of HLA class I- and class II-specific antibodies in a single assay.[18–21]

ELISA kits are also commercially available for the definition of HLA-specific antibodies. Quik-ID (GTI Inc.) enables the definition of IgG class I HLA-specific antibodies. It uses microtitre trays precoated with a panel of HLA antigens derived from platelets. In the same manner as CDC the pattern of reactivity can be examined to determine antibody specificity. Class II ID (GTI Inc.) is also designed to define specificity and the antigen source is EBV-transformed cell lines; it defines class II HLA-specific antibodies. The Lambda Antigen Trays (One Lambda Inc.) also feature purified HLA class I and class II antigens attached to a Terasaki format tray, and are designed for the definition of HLA-specific IgG antibodies.

Flow cytometry assays

Techniques that use a flow cytometer to detect antibody binding have been developed for the identification of IgG antibody binding to lymphocytes.[22] A flow cytometer is a machine in which particles, such as target cells, pass in a fluid phase through a narrow channel at which a laser beam is focused. The particle size and fluorescence are detected by photomultipliers feeding data to a computer. Software is used to interpret the captured data, which are displayed for interpretation.

FlowPRA beads ('Flow Panel Reactive Antigen'; One Lambda Inc.) are microparticles coated with purified HLA class I or class II antigen.[23] After incubation of the beads with serum and staining with fluorescence-labelled antihuman IgG antibody, positive sera are detected in the flow cytometer.

A schematic representation of a single FlowPRA bead is illustrated in **Fig. 4.2**. The HLA class I beads are non-fluorescent particles whereas the class II coated beads are internally fluorescent, allowing the

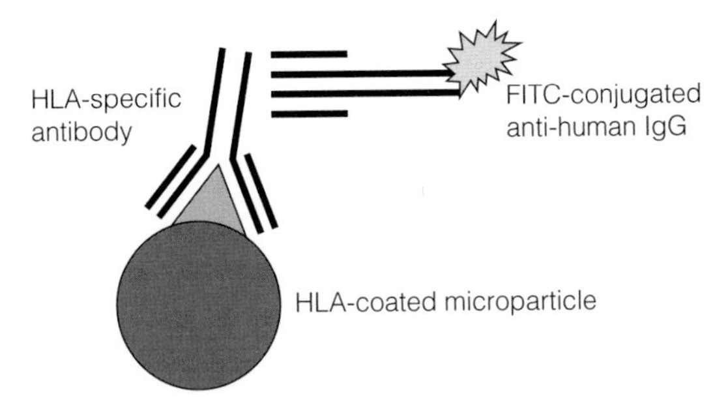

Figure 4.2 • Flow Panel Reactive Antigen (PRA) beads. The bead or microparticle is coated with HLA antigen and is incubated with patient serum. HLA-specific antibody present in the patient serum binds to the corresponding target antigen on the bead. A fluorescently labelled second antibody then binds to the antigen–antibody complex. The fluorescently labelled bead is then detected using a flow cytometer.

FlowPRA class I beads to be separated from the FlowPRA class II beads when tested in the same assay. Consequently it is possible to distinguish HLA class I-specific antibodies from HLA class II-specific antibodies in a single test. In addition to beads used for detecting the presence of antibodies, FlowSpecific (One Lambda Inc.) beads are available for the definition of antibody specificity. There are separate tests for the determination of HLA class I and class II specificities. The class I assay involves testing the serum with four pools, each containing eight beads with six antigens per bead. The class II specificity test similarly involves four pools, each of eight beads, but with four antigens per bead. The eight individual bead populations can be distinguished on the basis of their internal fluorescence, and antibody binding is detected by using a fluorescein isothiocyanate (FITC)-conjugated antihuman IgG. A flow cytometer with several photomultipliers fitted can detect multiple fluorescence spectra. A further development was the introduction of FlowPRA® Class I or Class II Single Antigen kits; these kits consist of 32 single antigen beads in four groups. Each bead is coated with single HLA antigens produced by recombinant technology. The use of single antigen particles enables the definition of 'hidden' specificities in multispecific sera in which specificities can be masked in highly sensitised patients.

Luminex technology

Microbead array technology was described previously (see 'HLA typing' section) and the same principles are applied to antibody detection and definition. Luminex technology allows for simultaneous measurement of multiple analytes in a single reaction. After incubation of the test serum with microspheres any bound antibodies are labelled with a fluorescent antihuman immunoglobulin. The luminex analyser uses a flow cytometer to detect the fluorescence intensity of the antihuman immunoglobulin on each bead. Analysis software determines whether HLA-specific antibody has bound to a specific bead population, thus assigning both positivity and HLA specificity.[24] Commercially available assays include LABScreen® products (One Lambda Inc.www.onelambda.com) which utilise the Lambda Array Beads Multi-Analyte System® (LABMAS). The range of products available includes the Mixed pre-screen assay, the PRA tests which enable specificity definition and the Single Antigen assay. The Lifematch HLA Antibody Detection System (Tepnel Life Sciences PLC, www.tepnel.com) includes pre-screen, specificity definition and single antigen formats. Additionally the Lifecodes range includes a donor-specific antibody monitoring assay which involves the purification and capture of donor HLA antigens onto beads. These beads can then be incubated with patient serum and antibody binding detected by the luminex analyser after the addition of antihuman IgG conjugated with PE.

Clinical significance of HLA-specific antibodies

The humoral response to an allograft is predominantly an initial IgM response, with low affinity for antigen, followed by a secondary response that is IgG and is of much greater affinity. There have also been reports that IgA may be involved.[25]

The majority of alloantibodies are directed against cell-surface antigens encoded by genes within the MHC at the class I (HLA-A, -B and -Cw) and class II (HLA-DR and -DQ) loci. Such antibodies, if donor specific, cause rapid rejection of vascularised grafts.

Hyperacute rejection (HAR) is apparent within minutes of vascular anastomoses being complete and always within the first 24 hours. Preformed antibodies react with antigens on the vascular endothelial cells of the graft and initiate the complement cascade causing inflammation, blocking the vessels of the graft and causing its death.[26] The introduction of the pretransplant crossmatch has virtually eliminated hyperacute rejection.

Acute rejection occurs within a few weeks or months following transplantation. It is caused by the primary response of the recipient to HLA molecules expressed by the transplanted organ. Helper $CD4^+$ T cells stimulate alloreactive B cells to make alloantibody. These antibodies bind to the graft endothelium and initiate rejection in the same way as in hyperacute rejection. Chronic rejection occurs after months or years of good function. There are well-defined pathological changes in the blood vessels, glomeruli and interstitium but the pathogenesis of chronic rejection is less well understood than that of acute rejection (see more detail in Chapter 3).

A review by McKenna et al. cited more than 23 studies showing that de novo production of donor-specific HLA antibodies is associated with increased acute and chronic rejection and decreased graft survival in kidney, heart, lung, liver and corneal transplants.[27] This was reiterated in a more recent review by Terasaki.[28]

Avoidance of positive crossmatches

The screening of potential transplant recipients for HLA-specific antibodies is an important role of the histocompatibility laboratory. Definition of the HLA specificity of antibodies in patients awaiting kidney transplantation avoids not only graft failure but also unnecessary crossmatching as the HLA antigens that must be avoided in a donor can be identified in advance. Furthermore, a comprehensive screening programme will reduce the positive crossmatch rate and subsequent unnecessary shipping of organs and prolonged ischaemia times.

Screening strategies

The definition of HLA-specific antibodies in patients awaiting transplantation is time-consuming and labour intensive. With the introduction of flow cytometry and ELISA-based techniques the range of technical options available for the laboratory has increased. Flow cytometry and ELISA have generally been considered as alternatives but they have limitations. Consequently a number of laboratories have devised screening strategies that employ each method in a stepwise manner to maximise the information obtained whilst minimising effort. The main aim of these strategies is to screen out antibody-negative sera, which are the majority, and focus specificity definition for only those sera known to contain HLA-specific antibodies. By implementing these strategies the workload of the laboratory is reduced whilst improving the level of specificity definition.[29]

Other antigenic targets

HLA-DP is expressed on the kidney and the advances in antibody screening technology have made it possible to detect and define DP-specific antibodies. The incidence of DP-specific antibodies in patients waiting for a kidney transplant is 7%.[30] Whilst the influence of DP mismatching on graft survival remains to be clarified, it has been suggested that DP-specific antibodies may play a role in rejection, particularly in re-transplants.[31] Testing for DP-specific antibodies in sensitised patients and the selection of DP-compatible donors may prevent positive crossmatches and/or transplant rejection and aid interpretation of crossmatches.

MHC class I-related chain (MIC) genes are located in close proximity to the HLA-B locus on chromosome 6 and encode a 62-kDA cell-surface glycoprotein, which shares limited homologies with HLA class I. There are five MIC genes and two are expressed: MICA and MICB. These antigens are detected on endothelial cells but not on lymphocytes.[32] The presence of MICA-specific antibodies has been correlated with rejection episodes and graft loss.[33] Recently an MIC pool was added to the LabScreen mixed assay and single antigen beads are available for the definition of MIC specificities.

Post-transplant monitoring and diagnosis of antibody-mediated rejection

A number of studies have demonstrated that the development of donor HLA-specific antibodies following transplantation is associated with poor graft outcome and antibody-mediated rejection (AMR) is now recognised as a histopathological entity.[34] AMR most commonly occurs in patients who have received a high-risk transplant such as one that required the use of an antibody-reducing protocol or had a historical positive crossmatch. However, de novo AMR can also occur in previously unsensitised patients. Antibody-mediated activation of the classical complement pathway results in the generation of the complement fragment C4d that covalently binds to the surface of endothelial cells.[35] Studies have demonstrated a correlation between deposition of C4d in the peritubular capillaries of kidney transplant biopsies and the development of donor-specific antibodies.[36,37] It is now recognised that AMR should be diagnosed on the basis of allograft dysfunction, characteristic features of histology and C4d immunohistology, and the presence of donor-specific antibodies.[38] Post-transplant monitoring of HLA-specific antibodies enables the identification of those patients at risk of AMR and their

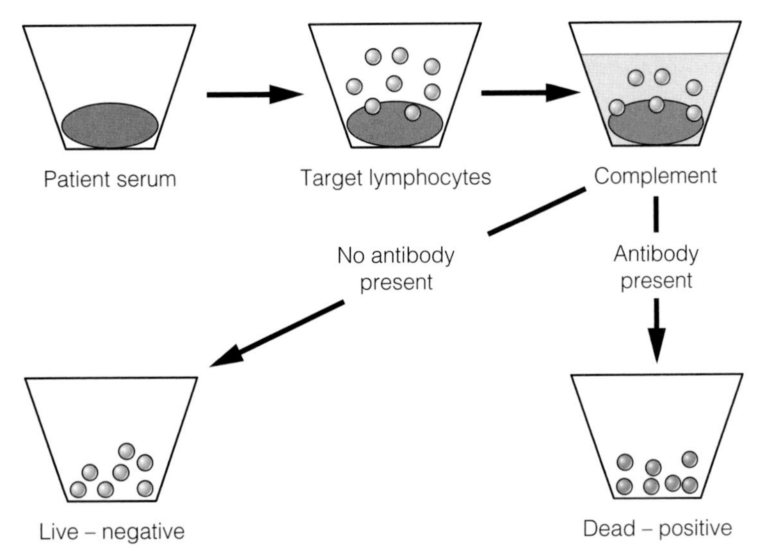

Figure 4.3 • Schematic representation of the complement-dependent cytotoxicity (CDC) assay. Patient serum and target lymphocytes are incubated and any specific antibodies present in the patient serum will bind to target antigen. Upon addition of complement, target cells with bound antibody are lysed. Cell death is visualised and indicates that the patient serum contains antibody. In the case of a crossmatch, cell death is referred to as a 'positive crossmatch' and therefore a contraindication to kidney transplantation.

immunosuppressive therapy can be tailored accordingly.[39–41] Importantly, by collecting patient samples post-transplant, a comprehensive antibody history is available should the graft fail and further transplantation become necessary.

Crossmatching

The complement-dependent cytotoxicity crossmatch (CDC-XM)

This assay follows a protocol similar to that used for cytotoxic antibody screening and is outlined in **Fig. 4.3**. In the CDC-XM, recipient serum is incubated with donor lymphocytes and the subsequent addition of complement results in lysis of lymphocytes that have been bound by antibody. The mechanism of complement-dependent cytotoxicity is detailed in Chapter 3. Cell viability is visualised by staining the cells and viewing them with a microscope. Cell death indicates a positive crossmatch and is a contraindication to successful kidney transplantation. A negative crossmatch indicates that the kidney transplant can proceed. The introduction of the CDC-XM eliminated antibody-mediated hyperacute rejection.

Although the CDC-XM is an essential prerequisite before transplantation there is still debate about the clinical relevance of a positive crossmatch report. The crucial factors in determining clinical relevance are the specificity, immunoglobulin class and the time of occurrence of the antibody causing the positive results.

It is accepted that IgG antibodies specific for donor HLA specificities will result in hyperacute rejection. The role of IgM alloantibodies in graft outcome is less clear.

IgM antibodies may be directed to HLA or non-HLA targets. It is known that IgM antibodies without specificity that react with a patient's own cells in an auto CDC-XM are not clinically relevant and may be an in vitro artefact. The establishment of the presence of such reactivity is important to allow apparently sensitised patients access to transplantation. Other technical problems associated with the CDC-XM include the need for viable target cells, and that only complement-fixing antibodies are detected. Despite its age and simplicity, the CDC-XM is highly effective in the prevention of hyperacute rejection.

The flow cytometry crossmatch (FC-XM)

There is now increasing evidence that more sensitive crossmatch techniques have a role to play in predicting transplant failure. Sensitivity of the crossmatch should equal sensitivity of the screening test used. The first study using a flow cytometer to detect antibody binding in a crossmatch test for kidney transplantation showed the technique to be more sensitive than CDC-XM for the detection of donor-specific antibody.[42]

Selected patient sera are incubated with donor lymphocytes at 21 °C and afterwards the target cells are then washed to remove any unbound antigen. Fluorochrome-tagged antibodies are added in a two-step incubation. Antihuman IgG–FITC in the first step binds with any IgG antibody from the patient sera that may be bound to the donor cells, then CD3-RPE and CD19-RPE segregate lymphocyte population into T and B cells respectively. This three-colour FC-XM is then visualised in a flow cytometer; a positive crossmatch results from an increase in fluorescence of the target cells contrasted with a negative control.[43]

The association of a positive FC-XM with reduced transplant outcome and the increased sensitivity of this test have now been demonstrated in a number of studies.[44,45]

Most centres have adopted the use of FC-XM for selected cohorts of patients considered to be at high risk, such as repeat transplants, recipients with multiple antibody sensitisation and in other situations where time permits, such as transplantation of organs from living donors.

There are few reports on the clinical significance of the FC-XM for other than kidney transplants, although one study of liver and heart transplants showed that recipients with donor-specific IgG antibodies detected by FC-XM had a poorer prognosis compared with crossmatch-negative recipients.[46]

Patient pathway

There now follows a description of a typical patient pathway from the view of the histocompatibility and immunogenetics laboratory.

The laboratory performs HLA typing for each patient referred with a view to entering the transplant list. In addition, HLA-specific antibody screening is performed to identify unacceptable HLA antigens. **Figure 4.4** illustrates the way in which kidney failure patients are investigated in our centre as an example. Inter-laboratory differences exist.

A frequently asked question concerns the time which each laboratory procedure will take. An approximate timescale is given in Table 4.2 to indicate the length of time needed to HLA type a potential organ donor and then to crossmatch the donor with potential kidney recipients.

Allocation

The degree of HLA matching used in successful solid organ transplantation is dependent on the type of organ and the policy of the transplanting centre.

Evidence-based reviews have demonstrated the effectiveness of HLA matching in improving outcomes, but waiting times may increase as a result.[47]

Kidney transplantation

Kidneys donated after brain death (DBD)

Following evidence-based review and wide discussion with users, the UK national scheme for the allocation of kidneys donated after brain death was revised in April 2006 with the aim of maximising outcomes and access to donated kidneys.[47] Patients suitable for kidney transplantation must be identical or, in specified circumstances, compatible with the ABO blood group of the donor. Patients are HLA typed and screened for the presence of HLA-specific antibody sensitisation. These details are entered into local and national 'waiting' lists to which potential donors will be matched. ABO blood group-identical recipients can be selected and ranked according to agreed allocation rules. From April 2006 in the UK the rules shown in Table 4.3 apply. Discriminatory points are assigned to benefit:

- patients who have waited longest;
- younger adults with least HLA mismatch with the donor;
- closer age match.
- minimisation of transport and hence cold storage time;
- patients with low-frequency HLA types or who are ABO-B, which are harder to match.

Kidneys donated after cardiac death (DCD)

UK practice currently mainly supports 'controlled' asystolic donation, which has now reached about 20% of all kidney transplants from deceased donors. Because cold storage time is perceived to be a major influence on outcome after kidney transplantation from DCD donors, most are allocated to recipients within the organ retrieval

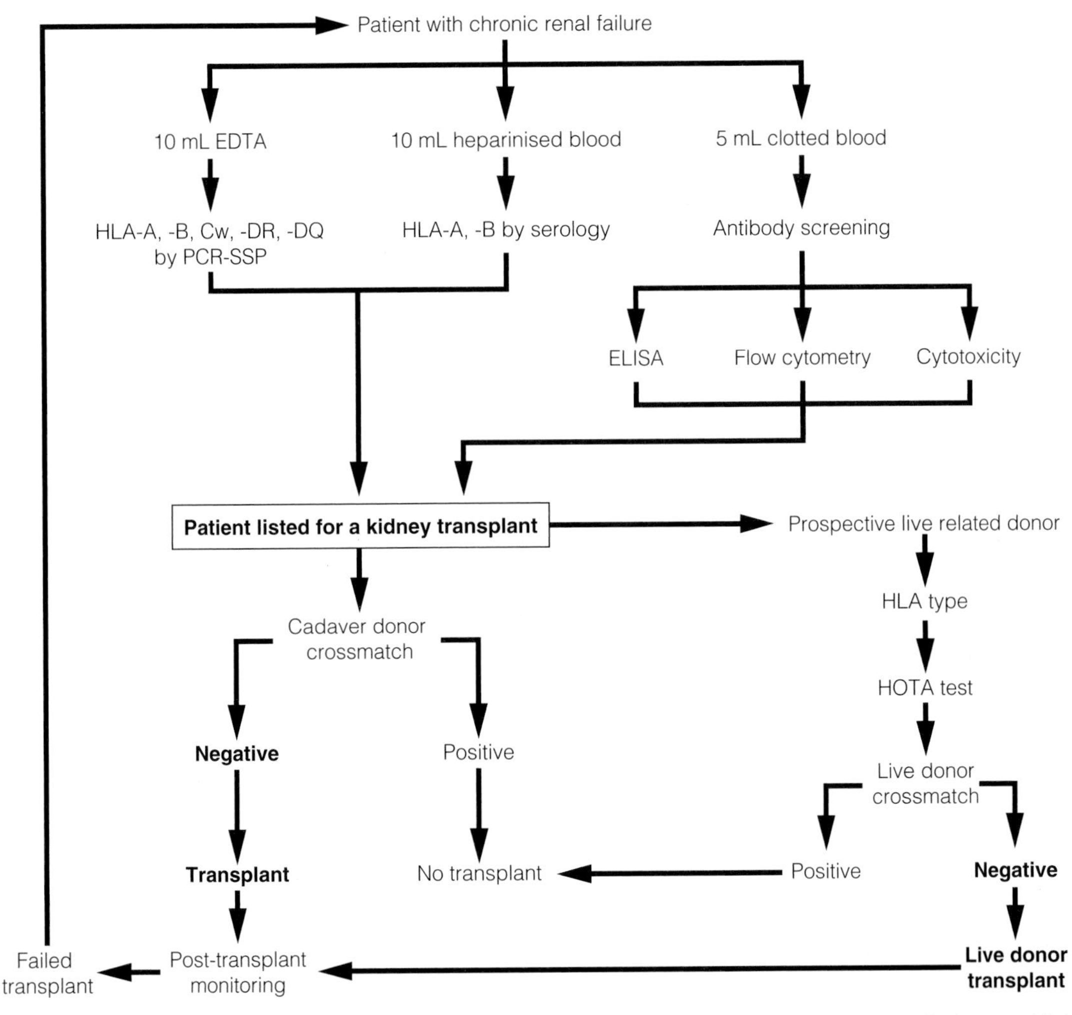

Figure 4.4 • Overview of the kidney patient process within the Manchester Transplantation Laboratory. Patients are HLA typed and unacceptable antigens are determined prior to entry onto the kidney transplant list. Other histocompatibility laboratories generally follow a similar process.

team area. It is common practice to allocate one kidney from a DCD donor to a recipient who is ABO identical, crossmatch negative and has waited for a prolonged time for a kidney transplant. The other donated kidney is allocated to an ABO-identical, crossmatch-negative and minimally HLA-mismatched recipient. In this way access to DCD kidneys is maintained for all listed patients and outcome can be maximised.

Pancreas transplantation

Since 2004 transplantation of a whole pancreas, simultaneously (SPK) or after (PAK) kidney transplantation for diabetic patients with kidney failure, has become increasingly common in the UK. Not all transplant centres perform pancreas transplants but those that do allocate the organs according to a local policy. At the start of 2008, the UK introduced an exchange scheme to benefit sensitised patients. Again, all recipients must be ABO identical (or compatible) with the donor and a crossmatch is always done to assess the risk of any donor HLA-specific sensitisation. So far too few pancreas transplants have been done in the UK to assess any influence of HLA mismatching, but reports from the USA have encouraged allocation ignoring HLA mismatch, so waiting time has become the main discriminator.

Heart and lung transplantation

Although minimising HLA mismatches has been shown to significantly improve survivals, the

Table 4.2 • Approximate timescale for HLA typing of a potential organ donor followed by completion of a crossmatch of donor and potential kidney recipient(s)

0 hours	Lab notified of local donor by transplant coordinator Blood sample sent to the histocompatibility laboratory
After 5 hours	HLA type of donor interpreted UKT run matching programme
After 6 hours	UKT provides shortlist of patients for crossmatch
After organ retrieval	Crossmatch performed
5 hours later	Crossmatch results reported to clinicians

UKT = UK Transplant.

irreversible damage caused by prolonged cold storage time discourages exchange of hearts and lungs over large distances simply to minimise HLA mismatches. Recipients in need of a thoracic organ transplant who are sensitised to HLA antigens fare particularly poorly since there is often a prolonged wait for an HLA-compatible, crossmatch-negative donor organ. During 2007 there have been discussions in the UK to establish an exchange scheme to benefit sensitised patients.

Liver transplantation

Retrospective reviews of HLA mismatching in liver transplantation have shown little impact on graft survival.

The reason for this remains unclear; however, occurrence of chimerism detected after successful transplantation suggests that tolerance mechanisms may occur in liver transplantation.

Corneal transplantation

In corneal transplantation, the degree of HLA mismatch is not a consideration prior to grafting. The cornea is usually avascular and is therefore considered an 'immunologically privileged' site. If the graft bed becomes vascularised as a consequence of primary graft failure, HLA matching has been shown to be beneficial in reducing the risk of rejection in a re-graft.

Conclusion

The seminal development of the CDC-XM to prevent hyperacute rejection caused by preformed donor HLA-specific sensitisation by IgG antibodies has saved many thousands of kidneys from immediate failure.

In deceased donor kidney transplantation, avoiding HLA antigen mismatches between the donor and recipient has shown improved survivals at 1 year after transplantation of between 5% and 10%.

The clinical value of HLA matching in allocation of kidneys from deceased donors varies throughout the world. In countries where there is no exchange system or where there are only a few patients on a waiting list, then avoiding HLA mismatches might not be practicable. The clinical value to be gained by allocating kidneys to minimise HLA mismatching can be contrasted with the experience of immunosuppressive drugs trials that have taken place since the introduction of ciclosporin in the mid-1980s. Nearly all of these trials, using new drugs or drug combinations, have failed to achieve any improvement in graft survivals at 1 year. For selection of living kidney donors most centres aim to minimise HLA mismatching when selecting a donor from several potential donors. When only one donor is available, HLA mismatching is not

Table 4.3 • Kidney donation rules of allocation

Tier (priority)	Recipient	HLA-A, -B, -DR mismatch	HLA-specific sensitisation, reaction frequency	Discriminator within tier
A	Paediatric	Zero	≥85%	Time waited
B	Paediatric	Zero	<85%	Time waited
C	Adult	Zero	≥85%	Points score
D	Adult	Zero	<85%	Points score
E	Any eligible	Minimised	<85%	Points score

a veto to effective transplantation unless there is HLA-specific sensitisation to donor HLA antigens.

Recent advances in effective identification of recipient sensitisation to HLA antigens and definition of the specificity of sensitisation has opened important new approaches to effective allocation of organs for transplantation to avoid antibody-mediated rejection. A further advantage of the effective definition of recipient sensitisation is transplantation without the need to perform a crossmatch when sensitisation records will influence organ allocation and unacceptable HLA mismatches can be ruled out. It will remain essential that HLA typing of both donor and recipient is performed to facilitate definition of sensitisation.

Studies identifying improved survivals when HLA mismatches are minimised (by chance) in heart and lung transplantation have been published.

In thoracic organ transplantation, HLA mismatching is not considered as a major factor in allocation except when the recipient is sensitised to donor-mismatched HLA antigens. Such patients may be passed over for a non-sensitised recipient and so will wait much longer, with the real possibility of never receiving a transplant.

The true value of histocompatibility testing resides in supporting clinically effective donor organ allocation systems because of ongoing shortages in donor organ supply and the important need to ensure optimal survival of this rare resource.

Key points

- Mismatching of ABO blood groups is a veto to effective organ transplantation, unless special procedures are used.
- Mismatching of HLA antigens to which a patient is sensitised is a veto to effective organ transplantation, unless special procedures are used.
- The genetics and nomenclature of the HLA system are complex and are defined by a WHO committee.
- Techniques to define HLA types and sensitisation to HLA antigens are precise. Their expert performance and interpretation are vital.
- Application of these techniques to effective transplantation of individual patients requires expert advice and interaction with clinical staff.
- HLA matching, sensitisation and crossmatching are important components of transplant organ allocation.

References

1. WHO HLA nomenclature report; www.ebi.ac.uk/imgt/hla.
2. Mullis K, Faloona F. Specific synthesis of DNA in vitro via a polymerase catalysed chain reaction. Methods Enzymol 1987; 155:335.
3. Saika RK, Bugawan TL, Horn GT et al. Analysis of enzymatically amplified β-globin and HLA-DQA DNA with allele specific oligonucleotide probes. Nature 1986; 324:163.
4. Bugawan TL, Horn GT, Long CM. Analysis of HLA-DP allelic sequence polymorphism using the in vitro enzymatic DNA amplification of DP-A and DPB loci. J Immunol 1988; 141:4024.
5. Watson JD, Hopkins NH, Roberts JW et al. Molecular biology of the gene, 4th edn. Menlo Park, CA: Benjamin-Cummings.
6. Sanger F, Nicklen S, Coulson AR. DNA sequencing with chain-terminating inhibitors. Proc Natl Acad Sci USA 1977; 74:5463–7.
7. Opelz G, Graver B, Mickey MR et al. Lymphocytotoxic antibody responses to transfusions in potential kidney transplant recipient. Transplantation 1981; 32:177–83.
8. Nymand G, Heron I, Jensen KG et al. Occurrence of cytotoxic antibodies during pregnancy. Vox Sang 1971; 21:21.

9. Scornick JC, Salomon DR, Lim PB et al. Post-transplant antidonor antibodies and graft rejection. Transplantation 1989; 47:287–90.
10. Martin S, Dyer PA, Harris R. Successful renal transplantation of patients sensitised following deliberate unrelated blood transfusions. Transplantation 1985; 39:256–8.
11. Pfaff WW, Fennell RS, Howard RJ et al. Planned random blood transfusions in preparation for transplantation. Transplantation 1984; 38:701.
12. Reisner EG, Kostyu DD, Phillips G et al. Allo-antibody responses in multiply transfused sickle cell patients. Tissue Antigens 1972; 2:415.
13. Scornick J, Brunson M, Howard R et al. Allo-immunization, memory and the interpretation of crossmatch results for renal transplantation. Transplantation 1992; 54:389–94.
14. Sanfilippo F, Vaughan WK, Bollinger RR et al. Comparative effects of pregnancy, transfusion and prior graft rejection in sensitisation and renal transplant results. Transplantation 1982; 34:360–6.
15. Martin S, Class F. Antibodies and crossmatching for transplantation. In: Dyer PA, Middleton D (eds) Histocompatibility testing – a practical approach. Oxford: IRL Press, 1993; pp. 81–104.
16. Patel R, Terasaki PI. Significance of the positive crossmatch test in kidney transplantation. N Engl J Med 1969; 280:735–9.
17. Nanni-Costa A, Scolari MP, Lannelli S et al. The presence of post transplant HLA specific IgG antibodies detected by enzyme-linked immunoabsorbent assay correlates with specific rejection pathologies. Transplantation 1997; 63:167–9.
18. Kao KJ, Scornick JC, Small SJ. Enzyme-linked immunoassay for anti-HLA antibodies – an alternative to panel studies by lymphocytotoxicity. Transplantation 1993; 55:192.
19. Worthington JE, Thomas AA, Dyer PA et al. GTI Quikscreen for the detection of HLA class I specific antibodies. Eur J Immunogenet 1995; 22:110.
20. Wang G, Tarsitani C, Takemura S et al. ELISA assay for the detection of specific antibodies to class I HLA antigens in transplant patients with panel reactive antibodies. Hum Immunol 1996; 49:109.
21. Worthington JE, Sheldon S, Langton A et al. Evaluation of LATM assay for the detection of HLA class I and II specific antibodies. Eur J Immunogenet 1999; 26:69.
22. Harmer AW, Sutton M, Bayne A et al. A highly sensitive rapid screening method for the detection of antibodies directed against HLA Class I and Class II antigens. Transplant Int 1993; 6:277.
23. Pei R, Wang C, Tarsitani C et al. Simultaneous HLA class I and class II antibodies screening with flow cytometry. Hum Immunol 1998; 59:313.
24. Susskind B. Methods for histocompatibility testing in the early 21st century. Curr Opin Organ Transplant 2007; 12:393–401.
25. Karuppan SS, Ohlam S, Moller E. The occurrence of cytotoxic and non-complement fixing antibodies in the crossmatch serum of patients with early acute rejection episodes. Transplantation 1992; 54:839–43.
26. Feucht HE, Opelz G. The humoral immune response towards HLA class II determinants in renal transplantation. Kidney Int 1996; 50:1464–75.
27. McKenna RM, Takemoto S, Terasaki PI. Anti HLA antibodies after solid organ transplantation. Transplantation 2000; 69:319–26.

A seminal report and review.

28. Terasaki PI. Humoral theory of transplantation. Am J Transplant 2003; 3:665–73.

A synthesis of data supporting an ongoing role for antibody-mediated loss after transplantation.

29. Worthington JE, Langton A, Liggett H et al. A novel strategy for the detection and definition of HLA specific antibodies in patients awaiting renal transplantation. Transplant Int 1998; 11:372–6.
30. Pfeiffer K, Vogeler U, Albrecht KH et al. HLA-DP antibodies in patients awaiting renal transplantation. Transplant Int 1995; 8:180–4.
31. Buckingham M, McDonald L, Barabanova Y et al. The relevance of HLA-DP antibodies in kidney allograft outcome. Hum Immunol 2005; 66:5.
32. Zwirner NW, Marcos CY, Mirbaha F et al. Identification of MICA as a new polymorphic allo-antigen recognized by antibodies in sera of organ transplant recipients. Hum Immunol 2000; 61:917–24.
33. Mizutabi K, Terasaki PI, Rosen A et al. Serial ten year follow up of HLA and MICA antibody production prior to kidney failure. Am J Transplant 2005; 5:2265–72.
34. Racusen LC, Colvin RB, Solez K et al. Antibody mediated rejection criteria – an addition to the Banff 97 classification of renal allograft rejection. Am J Transplant 2003; 3:708.
35. Feucht HE, Schneeberger EE, Hillebrand G et al. Capillary deposition of C4d complement fragment and early graft loss. Kidney Int 1993; 43:1333.
36. Collins AB, Schneeberger EE, Pascual MA et al. Complement activation in acute humoral renal allograft rejection: diagnostic significance of C4d depositis in peritubular capillaries. J Am Soc Nephrol 1999; 10:2208.
37. Worthington JE, McEwen A, McWilliam LJ et al. Association between C4d staining in renal transplant biopsies, production of donor specific HLA antibodies and graft outcome. Transplantation 2007; 83:398–403.
38. Montgomery RA, Hardy MA, Jordan SC et al. Consensus opinion from the antibody working group

on the diagnosis, reporting and risk assessment for antibody mediated rejection and desensitisation protocols. Transplantation 2004; 78:181.

39. Bohmig GA, Regele H. Diagnosis and treatment of antibody-mediated kidney allograft rejection. Transplant Int 2003; 16:773–87.

40. Scornik JC, Guerra G, Schold JD et al. Value of post transplant antibody tests in the evaluation of patients with renal graft dysfunction. Am J Transplant 2007; 7:1808–14.

41. Lefaucher C, Nochy D, Hill GS et al. Determinants of poor graft outcome in patients with antibody mediated acute rejection. Am J Transplant 2007; 7:832–41.

42. Garovoy MR, Rheinschmidt MA, Bigos M. Flow cytometry analysis: a high technology cross-match technique facilitating transplantation Transplant Proc 1985; 15:1939–44.

43. Robson A, Martin S. T and B cell crossmatching using three-colour flow cytometry. Transplant Immunol 1996; 4:203.

44. Cook DJ, Terasaki PI, Iwaki Y et al. An approach to reducing early kidney transplant failure by flow cytometry crossmatching. Clin Transplant 1987; 1:253–6.

45. Ogura K, Terasaki PI, Johnson C et al. The significance of a positive flow cytometric crossmatch test in primary kidney transplantation. Transplantation 1993; 56:294–8.

46. Talbot D. Flow cytometric crossmatching in human organ transplantation. Transplant Immunol 1994; 2:138–9.

47. http://www.uktransplant.org.uk/ukt/about_transplants/organ_allocation/kidney_(renal)/renal_organ_sharing_principles/kidney_organ_allocation_scheme_2006.jsp.

A report detailing the current allocation system for DBD donor kidneys in the UK.

5

Immunosuppression with the kidney as paradigm

Amy F. Rosenberg
Herwig-Ulf Meier-Kriesche

Introduction

While the successful first kidney transplant was performed without immunosuppression, long-term graft survival after kidney transplantation requires maintenance immunosuppression except for those few fortunate situations of transplants between identical twins. Over the decades immunosuppressive regimens have become more sophisticated, allowing for longer-term graft survival and establishing kidney transplantation as the treatment modality of choice for patients with end-stage renal disease. Nevertheless immunosuppressive medications have significant toxicities and side-effects, and it has been the transplant community's mission to find efficacious yet tolerable immunosuppressive combination regimens for transplant recipients. Undoubtedly the long-term goal would be to provide long-term graft acceptance without immunosuppression, but in clinical practice some degree of long-term immunosuppression is needed in most renal transplant recipients. In this chapter we will describe the commonly used immunosuppressive agents and how their combination has evolved in kidney transplantation. **Figure 5.1** illustrates the mode of action of immunosuppressive drugs used in clinical practice.

Azathioprine

Azathioprine (AZA), along with corticosteroids, was the earliest immunosuppressive medication used for the prevention of acute cellular rejection in kidney transplant recipients. Until the published reports by Murray et al. in 1963, total body or selective tissue irradiation was the primary method used to prevent rejection immediately after implantation of the transplanted kidney.[1] This treatment resulted in unacceptable toxicity and was not effective long term. Transplant physicians began to employ pharmacological agents (AZA and corticosteroids) solely for the prevention of acute rejection.[2] This resulted in acceptable graft survival rates and, after determining the appropriate doses of these medications, less toxicity.

AZA is a prodrug of 6-mercaptopurine (6-MP). Once absorbed into the bloodstream, AZA is completely converted to 6-MP. 6-MP exerts its immunosuppressive activity through purine synthesis inhibition. Purines (adenine, guanine and hypoxanthine) are necessary for DNA formation in new lymphocytes. 6-MP is transformed via hypoxanthine phosphoribosyltransferase (HGPRT) and a series of kinases into 6-thioguanine (6-TGN) nucleotides. It is then incorporated into elongating DNA chains, thus halting DNA chain elongation.[3] By competing for HGPRT, 6-MP also renders this enzyme less available for true purines to be incorporated into DNA.

AZA is usually dosed at 1–3 mg/kg/day in a single daily dose, and undergoes multiple metabolic pathways. Once AZA is converted into 6-MP in the bloodstream, it may then be metabolised by three separate pathways: the thiopurine methyltransferase

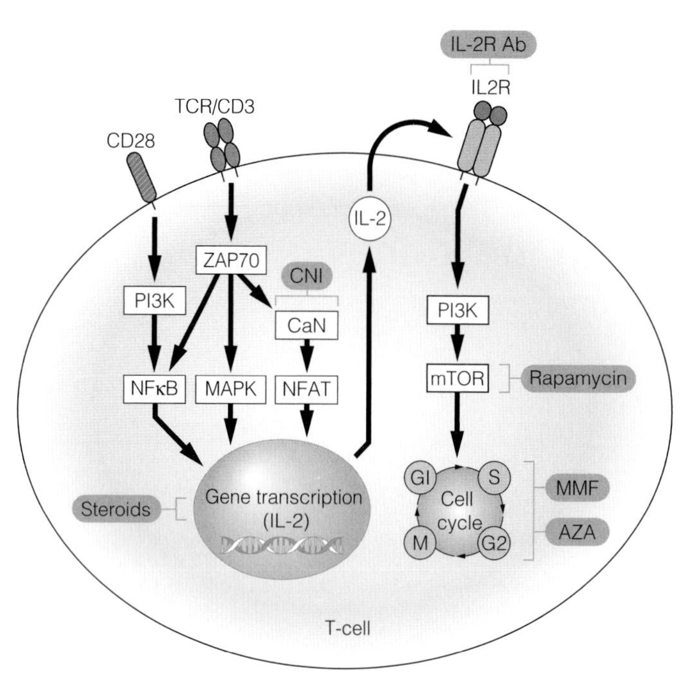

Figure 5.1 • Schematic representation of mode of action of commonly used immunosuppressive drugs. Reproduced from Demirkiran A, Hendrikx TK, Baan CC et al. Impact of immunosuppressive drugs on CD4+CD25+FOXP3+ regulatory T cells: does in vitro evidence translate to the clinical setting? Transplantation 2008; 85:783–9. With permission from Lippincott, Williams & Wilkins.

(TPMT) pathway, the HGPRT pathway or the xanthine oxidase (XO) pathway. Both the TPMT and XO pathways result in inactive metabolites. The HGPRT pathway results ultimately in the production of 6-TGN, which is incorporated into the elongating DNA chain or methylmercaptopurine nucleotide, which also inhibits de novo purine synthesis. The TPMT pathway is one of the best-described medication metabolic pathways subject to pharmacogenomic alterations. Approximately 10% of the population exhibits lower than normal TPMT activity, thereby forming fewer inactive metabolites and more active metabolites. These individuals tolerate only low doses of AZA. An important drug–drug interaction with AZA involves the coadministration of allopurinol, an XO inhibitor, usually used to prevent gouty arthritis attacks. By inhibiting XO, allopurinol eliminates or decreases one of 6-MP's two metabolic pathways leading to inactive metabolites. This results in the formation of fewer inactive metabolites and more active metabolites. For patients who require treatment with allopurinol, substitution of AZA with an alternative immunosuppressive medication such as mycophenolate mofetil (MMF) is a possibility. The primary side-effect noted with AZA is bone marrow suppression.

In 1963, Murray et al. published the first case series using only medications for immunosuppression after kidney transplant.[1] In 13 cadaveric and live-donor kidney transplant cases, they used AZA or 6-MP as their primary immunosuppressive agent. The first five patient cases had allograft function for less than 30 days, much of this due to AZA or 6-MP drug toxicity, which necessitated discontinuation of these medications. Discontinuation of these medications then resulted in rejection. Several patients also died due to complications from profound bone marrow suppression. In these early cases, AZA doses ranged from 6 to 12 mg/kg. The researchers quickly recognised that these doses resulted in profound toxicity. For the eight subsequent patients, doses were reduced to 2–4 mg/kg daily. Of these eight patients, at the time of publication, five had functioning grafts. The duration of these functioning grafts ranged from 40 to 365 days.

Shortly after Murray et al. published their results, Starzl and colleagues from the University of Colorado reported their experiences with an AZA-based immunosuppressive regimen.[2] They published the results of 45 live-donor kidney transplant recipients, 32 of which received kidneys from related donors. These patients were treated with 2–4 mg/kg/day of AZA for 7–10 days prior to their kidney transplant, and daily thereafter. Doses were adjusted based on white blood cell indices. At the time of publication, 27 of the 45 transplanted patients were alive with functioning grafts. In these early days of transplantation, the balance between prevention and treatment of rejection and infectious complications was not well established. This was the cause of most post-transplant mortality. Despite the high rejection and mortality rates seen in these early case series, they represented a major milestone in the possibility of successful long-term kidney transplantation. These published series lead to the routine use of AZA as a primary maintenance immunosuppressive agent in kidney transplantation between the mid-1960s and the 1980s, when ciclosporin was studied.

The introduction of ciclosporin

The introduction of ciclosporin (CYA) resulted in dramatically improved short- and long-term outcomes in kidney transplantation. CYA was originally approved for the prevention of rejection in kidney allografts in 1983. It represented a new class of medications, the calcineurin inhibitors (CNIs), which would become the backbone of solid-organ transplant immunosuppression. CYA inhibits the production of key cytokines for early T-lymphocyte activation, such as interleukin-2 (IL-2) and interferon-γ. Ciclosporin does this by binding to the intracellular protein cyclophilin. The CYA–cyclophilin complex then inhibits the nuclear factor of activated T cells (NFAT). NFAT is a key enzyme involved in the signal transduction of important lymphokines for early T-cell activation such as IL-2 and interferon-γ. These actions ultimately result in the inability of T lymphocytes to progress from the G0 to the G1 phase of activation.

CYA was originally approved in the 1980s as the non-modified version. As a large cyclic polypeptide, CYA is insoluble in water and is only absorbed in the upper part of the small intestine. In addition, its absorption in this non-modified form is highly influenced by food, the rate of bile flow and gastrointestinal motility. These factors contributed to the high inter- and intrapatient variability in bioavailability. For CYA, a critical dose medication, this unpredictable bioavailability proved clinically problematical. In 1995, a microemulsion (also called modified CYA) of CYA was approved. This microemulsion resulted in better solubility in the aqueous fluids of the gastrointestinal tract, and ultimately better and more consistent oral bioavailability.

After initial studies in healthy volunteers,[4,5] followed by a pilot study in 18 renal transplant recipients,[6] the new modified CYA was studied in 55 stable renal transplant recipients.[7] The purpose of this study was to determine if trough CYA concentrations better predicted systemic exposure (area under the curve, AUC) with CYA microemulsion compared with non-modified CYA. The other purpose of the study was to determine if intraindividual variability was lower with the CYA microemulsion formulation compared with the non-modified formulation when studied in a larger stable kidney transplant population. This pharmacokinetic study included four study periods with two crossover time points. This analysis showed an increase of 59% and 30% in maximum concentration (C_{max}) and AUC respectively with CYA microemulsion, when compared with non-modified CYA. There was also stronger correlation between CYA trough concentrations and AUC with the microemulsion (r^2 = 0.806) compared with non-modified product (r^2 = 0.492). In addition, there was less intrapatient variability for every measured pharmacokinetic parameter (t_{max}, C_{max}, AUC and peak-to-trough fluctuation percentage, PTF%), except for trough concentrations. Although the intrapatient coefficient of variation (CV) between CYA microemulsion and non-modified CYA for trough concentrations was not statistically different, there was a strong trend towards decreased variability in this parameter as well (CV CYA microemulsion 14% vs. CV non-modified CYA 20%, $P = 0.069$).

Following this initial crossover study in renal transplant patients, a longer-term double-blind study was performed to compare pharmacokinetics between patients on non-modified and microemulsion CYA.[8] This study included short-term (8 and 12 weeks) and longer-term (1 year) analyses. Fifty-nine stable renal transplant patients were randomised in a 1:4 fashion to either remain on their twice daily non-modified CYA regimen ($n = 12$) or switch (1 mg:1 mg) to CYA microemulsion ($n = 47$). A baseline pharmacokinetic evaluation was performed 1 week prior to randomisation, then again at weeks 8 and 12. For 46 patients who remained in the study for long-term analysis, a pharmacokinetic evaluation was performed again at 1 year. Similarly to the previous study, this study showed better correlation between trough concentrations and AUC with the microemulsion formulation (baseline/non-modified CYA: r^2 = 0.507; CYA microemulsion week 8: $r^2 = 0.748$; CYA microemulsion week 12: $r^2 = 0.822$; CYA microemulsion 1 year: $r^2 = 0.576$). Also, less intraindividual variability was shown with the microemulsion product in all pharmacokinetic parameters (t_{max}, C_{max}, AUC and PTF%, $P < 0.001$), except trough concentrations.

Due to these more predictable pharmacokinetic parameters, the modified product replaced the non-modified as the preferred product. Both modified and non-modified CYA dosing is most routinely monitored through whole-blood trough concentrations. Target trough concentrations depend upon patient-specific factors, other immunosuppressive medications used in combination and timing post-transplant. Although most centres monitor CYA trough concentrations, some monitor concentrations 2 hours after the dose is administered (C2 monitoring). CYA concentrations 2 hours after dosing have been shown to correlate better with AUC than do trough concentrations. However, this strategy may be less practical for many patients and transplant centres, and since most clinical trials have utilised trough concentration monitoring, this strategy is still the standard for most transplant centres.

CYA is primarily metabolised by the cytochrome P450 3A4 isoenzyme found both in the liver and intestinal brush border. It is also a substrate of P glycoprotein and subject to drug interactions via this enzyme as well. Addition or discontinuation of any medication that inhibits or induces this enzyme system should result in a CYA dosage change and temporarily increased blood concentration monitoring.

Non-modified CYA was originally studied in two large randomised controlled trials published in the early 1980s. One study was conducted in Europe and the other in Canada.[9,10] Both studies enrolled between 200 and 300 patients receiving cadaveric kidney transplants. The European study compared single-agent CYA treatment to AZA and corticosteroids. The Canadian study compared CYA and corticosteroids to best local therapy, which primarily consisted of AZA and prednisone. Some centres, however, did use antilymphocyte antibodies as part of their induction immunosuppression regimen in the control group. The European study at 1 and 5 years showed improved graft survival in the CYA-treated patients.[11,12] At 1 year, graft survival was 72% in the CYA group and 52% in the AZA group ($P = 0.001$). This benefit was maintained at 5 years with graft survival of 55% in the CYA group and 40% in the AZA/prednisone group ($P < 0.01$). Patient survival was no different between the groups at either time point in the European study. The Canadian study showed both improved patient and graft survival in the CYA group compared to the AZA group at 3 years post-kidney transplant.[13] Graft survival was 69% in the CYA group and 58% in the AZA group ($P = 0.05$), and patient survival was 90% in the CYA group and 82% in the AZA group ($P = 0.04$). This improvement in short- and intermediate-term graft survival with a new immunosuppressive medication would later prove to be one of the most significant contributions to kidney transplantation to date.

While the improvement in graft survival with CYA was a major advance for kidney transplantation, this was tempered by CYA's primary side-effect of nephrotoxicity. In both the European and the Canadian multicentre studies, for the patients with surviving grafts, the CYA treatment groups had worse renal function, defined by serum creatinine (SCr) concentrations or calculated creatinine clearances (CrCl), than did the control groups. At 5 years, CYA-treated patients in the European trial had SCr concentrations of 2.3 mg/dL and AZA/prednisone patients had SCr concentrations of 1.5 mg/dL ($P < 0.001$). Similarly, a 3-year analysis of the Canadian trial showed a mean calculated CrCl of 48 mL/min in the CYA group compared with 63 mL/min in the AZA group ($P = 0.001$). Both of these trials used high doses of CYA compared with more recent regimens, and only the Canadian study adjusted doses based on serum concentrations. Although nephrotoxicity has become a well-recognised CYA side-effect, subsequent studies using lower doses of CYA have not shown renal function to be as poor at these time points. In addition to nephrotoxicity, other important side-effects noted with CYA use were hirsutism, gingival hyperplasia, tremor and hypertension. Despite CYA's nephrotoxicity and numerous other side-effects, the improvement in graft survival for all forms of transplant was a major contribution to the success of modern-day transplantation.

The introduction of tacrolimus and mycophenolate

After the introduction of CYA in the early 1980s, there were no new maintenance immunosuppressive medications approved for a little over 10 years. The mid-1990s, however, saw two new maintenance immunosuppressive medications that brought about substantial improvements in acute rejection rates in renal transplantation. Tacrolimus (TAC) and mycophenolate mofetil (MMF) would both prove to decrease acute rejection rates further when compared with traditional CYA-based immunosuppression. This added two new medications for transplant clinicians to use in maintenance immunosuppressive regimens.

TAC, the second CNI approved for clinical use, was approved in 1994 for the prevention of acute allograft rejection in liver transplant patients. TAC inhibits calcineurin by binding to FK binding protein 12 (FKBP12). This TAC–FKBP12 complex then binds with calcineurin, inhibiting the phosphatase activity of calcineurin. The inhibition of calcineurin further inhibits the dephosphorylation and translocation of NFAT. This ultimately inhibits the production of key cytokines involved in cell-mediated immune function, such as IL-2 and interferon-γ.[14] TAC is usually administered orally every 12 hours. Its oral bioavailability ranges from 17% to 22%.[14] Therapeutic drug monitoring for TAC is usually accomplished through 12-hour serum trough concentration monitoring, similar to CYA. Stated therapeutic doses for TAC range from 5 to 15 ng/mL, but depend largely upon patient-specific factors and concomitant immunosuppressive therapy. TAC is primarily metabolised by hepatic and intestinal cytochrome P450 3A4, and is subject to many drug interactions with other medications that either inhibit or induce this microenzyme.[14] It is also a substrate of P glycoprotein and subject to drug interactions via this enzyme as well. Any addition or discontinuation of other medications that inhibit or induce these enzyme systems must be followed by TAC dosage adjustments and more intensive serum concentration monitoring.

TAC was first studied in the liver transplant population as part of a dual-drug regimen along with corticosteroids, and shown to be superior to CYA at preventing episodes of acute rejection. Two randomised open-label multicentre trials conducted in the early 1990s compared TAC/corticosteroid immunosuppression with CYA/AZA/corticosteroid immunosuppression in liver transplant recipients. Both the US and the European multicentre randomised trials concluded that at 1 year following a first-time liver transplant, TAC resulted in fewer episodes of acute rejection and refractory rejection.[15,16]

Shortly after being studied in liver transplant patients, TAC was studied in large numbers of kidney transplant patients, with similar results. Two similarly designed multicentre randomised, open-label, clinical trials were conducted in the USA and Europe.[17,18] Each were large studies, with the US study randomising 412 patients and the European study 448 patients. Each of these trials compared regimens of TAC/AZA/corticosteroids to CYA (non-modified)/AZA/corticosteroids. The US study investigators used antibody induction therapy in all patients in addition to the above-described maintenance regimens.[17] The primary outcome in each study was incidence of biopsy-proven acute rejection (BPAR) 1 year after cadaveric renal transplantation. Target CYA and TAC concentrations were similar in each study, with the US study targeting slightly higher CYA concentrations than the European study. Both studies targeted TAC troughs between 10 and 25 ng/mL for the first 3 months, then 5–15 ng/mL thereafter. CYA target concentrations in the two studies ranged between 100 and 400 for the first three months and 100–300 thereafter. Although open label in design, each study's pathologists were blinded to treatment assignment to minimise bias in determination of rejection presence.

These two multicentre randomised trials in kidney transplantation showed similar positive results as had the earlier liver transplant studies. The US investigators reported BPAR rates of 30.7% in the TAC group and 46.4% in the CYA group ($P = 0.001$).[17] The European investigators reported this same outcome in 25.9% of patients in the TAC arm and 45.7% of patients in the CYA arm ($P < 0.001$).[18] At 1 year, there was no difference in either trial between groups for either patient survival or graft survival.

A 5-year follow-up of the US multicentre trial showed that improved outcomes with TAC proved durable.[19] Although the intent-to-treat analysis failed to show any difference in patient or graft survival at 5 years, crossover analyses revealed improved outcomes in patients who were receiving TAC at year 5. Throughout the 5-year follow-up, 27.5% of patients crossed over from CYA to TAC, while only 9.3% crossed over from TAC to CYA ($P < 0.001$). The majority of these crossovers occurred in the first year of treatment; therefore the crossover analyses should represent long-term treatment on each medication. The majority (68.4%) of CYA randomised patients crossed over to TAC due to refractory rejection. Treatment failure, defined as graft loss or study drug discontinuation, was higher in the CYA group (56.3%) compared with the TAC group (43.8%, $P = 0.008$). In addition, graft survival was higher in African-Americans receiving TAC (65.4%) compared with CYA (42.6%, $P = 0.013$). Primary side-effects noted in most of the clinical trials are tremor, elevated blood glucose, nephrotoxicity and hypertension.

One possible weakness of the above studies was that the non-modified formulation of CYA was used. As previously discussed, CYA microemulsion demonstrates more reliable pharmacokinetics, which likely translates to better post-transplant outcomes. Since publication of these original comparisons between CYA (non-modified) and TAC in kidney transplants, comparison between TAC and CYA microemulsion has also been conducted. Johnson et al. studied three different immunosuppressive regimens in patients receiving a first cadaveric kidney transplant.[20] They randomised 223 patients to open-label treatment with either CYA/MMF ($n = 75$), TAC/MMF ($n = 72$) or TAC/AZA ($n = 76$). Target trough concentrations for CYA and TAC were similar to the previously discussed studies. Primary end-points were again patient and graft survival, BPAR, development of insulin dependence, and renal function.

The 1-year analysis of this study showed no difference in the primary outcome measures, but did reveal some noteworthy findings. As expected, at 1 year, there was no difference in patient or graft survival amongst the three groups. Although BPAR was higher in the CYA/MMF group (20%) than the TAC/MMF group (15.3%), this difference was not statistically significant. However, of the patients who experienced rejection, significantly more patients in the CYA/MMF group (69%) required antilymphocyte antibody therapy to treat refractory rejection than in the TAC/MMF group (27%, $P = 0.05$). In addition, a higher proportion of patients in the CYA group had serum creatinine >1.5 compared with patients treated with TAC ($P = 0.03$).

The primary outcomes of this 1-year analysis remained the same in a 3-year follow-up analysis.[21] The authors additionally noted that, after 3 years, patients who initially experienced delayed graft function and received TAC/MMF survived at a higher rate (84.1%) than those who received CYA/MMF (49.9%, $P = 0.02$). This finding suggests that higher risk patients may derive benefit from receiving TAC over CYA. In the present study, for TAC efficacy analysis, the only appropriate groups to compare are the CYA/MMF and the TAC/MMF groups, since MMF has been shown to be more effective than AZA at preventing rejection. This study, which used modified CYA, did not find a difference in acute rejection rates as had the previously discussed European and US studies. The numbers of patients compared in the TAC/MMF and CYA/MMF groups were much smaller than those studied in the US and European randomised controlled trials. This may have resulted in a lack of power to detect a difference in acute rejection rates between modified CYA- and TAC-treated groups.

Taking all of these findings together, it appears that TAC is at least as effective as modified CYA and may be more effective at preventing acute rejection.

TAC, even though possibly less nephrotoxic than CYA at the target concentrations commonly used, also has important side-effects. In addition to its intrinsic nephrotoxicity, neurotoxicity, diabetes and alopecia are common side-effects.

Mycophenolate mofetil (MMF)

MMF was approved in 1995 for the prevention of acute rejection in kidney transplant patients. MMF is a prodrug which is hydrolysed immediately in the bloodstream to its active form of mycophenolic acid (MPA). In 2004, an enteric-coated dosage form of mycophenolate sodium was also approved. MPA is the active moiety of both preparations. MPA prevents lymphocyte activation through its inhibition of inosine monophosphate dehydrogenase, an important enzyme in the production of guanosine nucleotides for lymphocyte DNA formation.[22] Since B and T lymphocytes require de novo synthesis of purines, such as guanosine, MPA has potent cytostatic effects on these cell lines. When administered orally, MMF exhibits approximately 94% bioavailability in healthy volunteers. In kidney transplant recipients, bioavailability improves as patients progress further from the time of transplant. By month 3 following kidney transplant, bioavailability is similar to that of a healthy volunteer.

MMF was originally approved based on several randomised double-blind trials, all of which showed decreased treatment failure (biopsy-confirmed acute rejection, graft loss, patient death or discontinuation of study drug) at 6 months post-kidney transplantation with MMF when compared with either placebo or AZA.[23–25] The US MMF Study Group and the Tricontinental MMF Study Group both enrolled approximately 500 cadaveric renal transplant recipients and randomised them to receive either MMF 2 g/day, MMF 3 g/day or AZA 1–2 mg/kg/day. All patients also received CYA and corticosteroids as their maintenance immunosuppression. The primary end-point of each study was BPAR or treatment failure. The initial reports from these studies analysed outcomes at 6 months.

In each case, both MMF groups were shown to be superior to AZA, with substantially fewer patients in the MMF groups reaching the primary end-point when compared to AZA.

The Tricontinental study showed 34.8% of patients in the MMF 3 g group experienced treatment failure compared with 38.2% in the MMF 2 g group and 50% in the AZA group (MMF 3 g vs. AZA: $P = 0.0045$; MMF 2 g vs. AZA: $P = 0.0287$). The US study showed that 31% of patients in either MMF group reached the primary end-point as compared with 47.6% of AZA-treated patients ($P = 0.0015$ in the MMF 2 g group and $P = 0.0021$ in the MMF 3 g group).

Similar in design to the Tricontinental and US MMF studies, the European MMF Cooperative Study Group showed positive results for MMF-treated patients, but compared them to placebo rather than AZA active control. A total of 491 patients was enrolled. Like in the previous two trials, MMF-treated patients were randomised to either 2 g/day or 3 g/day, and a third group was randomised to placebo. As expected, this trial too showed superior results, with both MMF-treated groups having significantly fewer patients reaching the primary end-point of treatment failure. Compared with placebo, the MMF 2 g/day group demonstrated a relative risk (RR) of 0.535 (97.5% CI 0.399–0.718) of reaching the primary end-point and the MMF 3 g/day group an RR of 0.658 (97.5% CI 0.494–0.875).

The European MMF Cooperative Group, US MMF Study Group and Tricontinental Study Group all reported 3-year follow-up results with particular interest in patient and graft survival.[26–28] Neither of the active control trials (US trial or the Tricontinental trial) reported differences in patient or graft survival at 3 years. However, the European trial did report improved 3-year graft survival (excluding death with a functioning graft) in the MMF 2 g group compared with the placebo group. The graft loss rate in the placebo group was 16% compared with 8.6% in the MMF 2 g group ($P = 0.03$).

With the exception of the Tricontinental placebo-controlled trial MMF 2 g group, in each of the above randomised trials, although MMF was shown to produce superior outcomes with fewer patients experiencing treatment failure when compared with either AZA or placebo, these studies showed no improvement in graft survival. This is likely due to lack of power in the above studies to detect such a difference. In an effort to determine whether or not MMF use was related to improved graft survival, Ojo et al. examined the United States Scientific Transplant Registry and compared patients treated with MMF with those treated with AZA between 1992 and 1997.[29] The primary end-point is this study was late (more than 6 months post-transplantation) renal allograft failure. A multivariate analysis including potential risk factors for chronic allograft failure was performed on 66 774 patients who received a CNI, corticosteroids and either AZA or MMF; 48 436 patients received azathioprine, 8435 received MMF and 9903 received a mixture of different maintenance regimens. Patients were analysed based on their medication assignment at the time of discharge from transplant surgery. Although not a primary aim of the study, the analysis detected a higher rate of acute rejection in AZA-treated patients (24.7%) compared with MMF-treated patients (15.5%, $P < 0.001$). Graft survival (death censored) at 4 years was higher among the MMF group (85.6%) compared with the AZA group 81.9% ($P < 0.0001$) (**Fig. 5.2**). Patient survival at 4 years was also higher in the MMF group (91.4%) compared with 89.9% in the AZA group ($P = 0.002$). The risk factor most strongly associated with late graft loss was the presence of acute rejection, with a risk ratio of 2.41 ($P < 0.001$).

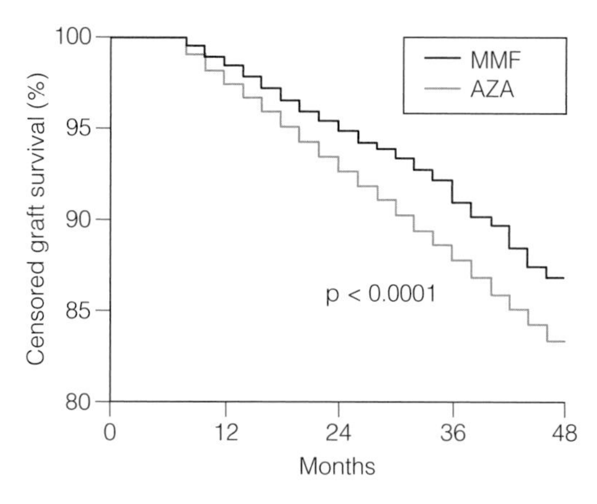

Figure 5.2 • Comparison of Kaplan-Meier 4-year censored graft survival beyond 6 months after transplantation. Reproduced from Ojo AO, Meier-Kriesche HU, Hanson JA et al. Mycophenolate mofetil reduces late renal allograft loss independent of acute rejection. Transplantation 2000; 69:2405–9. With permission from Lippincott, Williams & Wilkins.

After controlling for acute rejection, treatment with MMF was found to protect from late graft loss with a risk ratio of 0.73 compared with AZA treatment ($P < 0.001$). In addition to controlling for acute rejection in the overall analysis, the authors analysed the risk of developing chronic allograft failure between patients treated with MMF and AZA in patients who had never experienced an acute rejection episode.

This analysis revealed that the risk of chronic allograft failure was 20% lower in patients treated with MMF compared with AZA (risk ratio 0.8, $P < 0.001$) even if they had never experienced an acute rejection episode. While other randomised controlled trials lacked the power to detect graft survival differences over short periods of time, this analysis of a large database confirmed that rejection rates in MMF-treated patients were lower than in AZA-treated patients, but more importantly that MMF-treated patients had better graft survival even when analysed independently of rejection episodes. Therefore MMF possibly exerts a beneficial effect beyond its ability to prevent acute rejection episodes.

Sirolimus

Sirolimus, the first of a new class of immunosuppressive medications, was approved in 1999. It belongs to the therapeutic class, mammalian target of rapamycin (mTOR) inhibitors. Like other immunosuppressive agents, sirolimus inhibits T-lymphocyte activation and proliferation, but does so through a unique mechanism. Sirolimus, like TAC, binds to the intracellular FKBP12, but the sirolimus–FKBP12 complex results in different downstream effects than the TAC–FKBP12 complex. The sirolimus–FKBP12 complex binds and inhibits mTOR. mTOR is a key enzyme, allowing the T lymphocyte to respond to cytokine-initiated cell proliferation and move from the G_1 to the S phase of the cell cycle.[30]

Sirolimus exhibits important pharmacokinetic differences from other commonly used immunosuppressive medications, and also undergoes clinically important drug–drug interactions. Sirolimus is administered once daily and exhibits bioavailability of approximately 14% as the oral liquid formulation and approximately 40% as oral tablets. Despite this difference in bioavailability it is recommended that 2 mg of sirolimus oral solution and 2 mg sirolimus administered as tablets are clinically equivalent and may be interchanged 1:1 when formulation change is necessary.[30] This has not been confirmed for higher doses of sirolimus. When administered with CYA, it is recommended that sirolimus be separated from CYA administration by 4 hours to avoid an approximately 45% increase in sirolimus exposure.[31] This interaction has not been shown when used in combination with TAC. Although serum concentration monitoring was not initially thought to be important with sirolimus, this practice is now routine. This therapeutic drug monitoring is accomplished through 24-hour trough concentration monitoring. Like CYA and TAC, the therapeutic range of sirolimus trough concentrations depends largely on other immunosuppressive medications and patient-specific factors, but in general the therapeutic range is 5–15 ng/mL. Sirolimus is primarily metabolised by cytochrome P450 3A4. Concomitant administration with medications that inhibit or induce this microenzyme should result in dosage adjustments and increased trough concentration monitoring of sirolimus.

Sirolimus was approved for the prevention of acute rejection in kidney transplant recipients based on the results of two randomised, double-blind trials.[32,33] Each trial randomised patients to one of three groups: a 2 mg/day sirolimus group, a 5 mg/day sirolimus group or an AZA group (US trial) or a placebo group (worldwide trial). Each trial also used CYA and prednisone as additional maintenance immunosuppression. Neither study allowed antibody induction treatment at the time of transplantation. The primary composite end-point of each study was the rate of efficacy failure (BPAR, graft loss or death) at 6 months post-transplantation.

The US study showed decreased efficacy failure rates for both sirolimus groups as compared to the AZA-treated group; 719 patients were randomised to receive sirolimus 2 mg/day (n = 284), 5 mg/day (n = 274) or AZA 2–3 mg/kg/day (n = 161). The primary end-point analysis showed RR values of efficacy failure of 0.61 (95% CI 0.44–0.84) and 0.58 (95% CI 0.41–0.81) for the sirolimus 2 mg and 5 mg groups respectively when compared to the AZA-treated group at 6 months. BPAR alone was also less likely in both of the sirolimus groups than the AZA group at 6 months; RR was 0.58 (95% CI 0.41–0.81) in the 2 mg group and 0.49 (95% CI 0.34–0.71) in the 5 mg group. There were no differences in death or graft loss between the groups at 1 year.

The worldwide study, which compared sirolimus to placebo rather than AZA, also showed decreased efficacy failure in the sirolimus-treated groups. In this trial 576 patients were randomised to receive either sirolimus 2 mg (n = 227), 5 mg (n = 219) or placebo (n = 130). Efficacy failure occurred in 30% of the patients receiving sirolimus 2 mg and 25.6% of the patients receiving sirolimus 5 mg, compared with 47.7% of the patients receiving placebo. When examined alone, BPAR also occurred less often in the sirolimus 2 mg/day group (24.7%) and 5 mg/day group (19.2%) compared with placebo-treated patients (41.5%) (P = 0.003 and P < 0.001 respectively). There was also no difference in graft or patient survival at 1 year.

Despite better efficacy at preventing acute rejection episodes with sirolimus, both the US and the Worldwide Rapamune study groups showed that at 6 months and 1 year sirolimus-treated patients had worse renal function compared to either the AZA or placebo groups. At 6 months in the US study, the 2 mg sirolimus group had an SCr of 1.7, the 5 mg group had an SCr of 1.8 and the AZA group an SCr of 1.4 (P < 0.001 for both sirolimus groups compared with AZA). These SCr values remained worse in the sirolimus groups at 1 year: 1.8 mg/dL for the 2 mg sirolimus group, 1.9 mg/dL for the 5 mg sirolimus group and 1.5 mg/dL in the AZA group (P < 0.01 for both sirolimus groups compared with AZA). This finding was also replicated in the worldwide trial comparing sirolimus to placebo. While differences in SCr were not found to be statistically different between the groups in the worldwide trial, CrCl was worse at 6 months in the sirolimus 5 mg group when compared to placebo-treated patients (56.42 mL/min vs. 62.58 mL/min, P = 0.05).

This finding of apparent synergistic nephrotoxicity with full-dose CNI and sirolimus has been replicated in another randomised controlled trial comparing TAC/sirolimus vs. TAC/MMF.[34]

In addition, analyses of the United States Scientific Renal Transplant Registry showed lower 4-year death-censored graft survival with CYA/sirolimus immunosuppression as compared with CYA/MMF immunosuppression, and 3-year death-censored graft survival with a TAC/sirolimus immunosuppression as compared with TAC/MMF in deceased donor transplants[35,36] (Table 5.1). These outcomes lead to the practice of no longer using sirolimus with full-dose CNI, but rather exploring the use of sirolimus in various CNI-sparing regimens. These regimens will be discussed later in this chapter.

In addition to the decreased renal function seen when sirolimus is used with full-dose CNIs, several other common side-effects have been noted in clinical trials with sirolimus use. The most common side-effects seen are elevated lipid parameters, especially triglycerides, leucopenia and thrombocytopenia.

Also, since the initial pivotal trial publication, problems with impaired wound healing and increased incidence of lymphoceles have been noted with sirolimus.

Contribution of antibody products for induction immunosuppression and for treatment of acute rejection episodes

Antibody therapies are generally used for two indications following kidney transplantation: induction treatment and treatment of acute rejection episodes. Induction immunosuppressive treatment occurs at the time of transplantation and refers to the use of

Table 5.1 • Overall and death-censored graft survival in tacrolimus/mycophenolate mofetil and tacrolimus/sirolimus regimens

Population*	Regimen†	Overall graft survival		Death-censored graft survival	
		3 years (%)	*P* value‡	3 years (%)	*P* value‡
All	TAC/SRL	80.3		87.4	
	TAC/MMF	85.9	<0.001	92.0	<0.001
DDTx	TAC/SRL	74.5		83.6	
	TAC/MMF	82.8	<0.001	90.4	<0.001
Living transplants	TAC/SRL	87.5		92.1	
	TAC/MMF	90.4	0.015	94.4	0.044
Caucasian recipients	TAC/SRL	83.6		90.4	
	TAC/MMF	87.4	<0.001	93.6	<0.001
African-American recipients	TAC/SRL	72.5		80.4	
	TAC/MMF	81.4	<0.001	87.7	<0.001
ECD Tx	TAC/SRL	57.5		64.8	
	TAC/MMF	74.5	<0.001	84.6	<0.001

*DDTx, deceased-donor transplants; ECD Tx, expanded criteria donor transplants.
†MMF, mycophenolate mofetil; SRL, sirolimus; TAC, tacrolimus.
‡*P* value for log-rank test comparing TAC/sirolimus and TAC/MMF survival.
Reproduced from Meier-Kriesche H, Schold J, Srinivas T et al. Sirolimus in combination with tacrolimus is associated with worse renal allograft survival compared to mycophenolate mofetil combined with tacrolimus. Am J Transplant 2005; 5:2273–80. With permission from Blackwell Publishing.

an antibody preparation to provide increased immunosuppression during and immediately after transplantation. Antibody therapy for the treatment of acute rejection is generally selected after high-dose corticosteroid failure.

Antibody preparations can be divided into depleting antibodies and non-depleting antibodies. Depleting antibody administration results in the removal of targeted cells from the peripheral blood. Non-depleting antibodies alter the function of targeted cells while maintaining their numbers in peripheral blood. Depleting antibody preparations were in use earlier, beginning in the early 1980s with equine antithymocyte globulin and the late 1980s with muromonab CD3. These agents have been approved for the treatment of acute rejection, but are also used off-label for induction therapy. The first non-depleting antibody, daclizumab, was introduced in 1997 followed by basiliximab, introduced in 1998. These two non-depleting antibodies are generally used for induction therapy only. As antibody induction agents, all of these products have been associated with decreased early acute rejection rates; however, their effects on long-term outcomes is less clear. In addition to being used as induction therapy along with immediate triple-drug maintenance immunosuppression, these agents have also been used to allow for delayed initiation of CNI treatment, especially in patients with delayed graft function. This section will highlight the approval and contributions of each antibody agent to the treatment and/or prevention of acute rejection, beginning with depleting antibody agents.

Depleting antibody preparations

Depleting antibody preparations may be used both for the treatment and the prevention (as induction immunosuppression given at the time of transplant) of acute rejection in kidney transplantation. There are four primary depleting antibody preparations in use to date. These agents can be further subdivided into monoclonal and polyclonal agents. The two available polyclonal products are equine antithymocyte globulin (E-ATG) and rabbit antithymocyte globulin (R-ATG). The two monoclonal products are muromonab CD3 and alemtuzumab. Muromonab CD3 is a chimeric monoclonal antibody, whereas alemtuzumab is a humanised monoclonal antibody. These four depleting antibody agents will be discussed in the order in which they became available for use, with E-ATG approved in 1981, muromonab CD3 in 1986, R-ATG in 1999 and alemtuzumab (used off-label in organ transplantation) approved for treatment of chronic lymphocytic leukaemia in 2001.

Equine antithymocyte globulin

E-ATG was the first depleting or non-depleting antibody to receive approval in kidney transplantation. It is currently approved for the management of allograft rejection in renal transplant patients.[37] E-ATG is a polyclonal IgG product produced by immunising horses with human T lymphocytes. E-ATG is thought to exert its antirejection effects by clearing T lymphocytes from peripheral blood, and modulation of T-lymphocyte activity. When used for treatment of acute rejection, E-ATG is usually administered in doses of 15 mg/kg intravenously daily for 14 days. E-ATG should be administered into a high-flow central vein through a 0.2–1 μm in-line filter. The manufacturer of E-ATG recommends administration of an intradermal test dose of 0.1 mL of a 1:1000 dilution (5 μg horse IgG). Although still recommended by the manufacturer, this practice has not been shown to accurately predict which patients may experience allergic reactions to E-ATG.

E-ATG was originally studied in 358 renal transplant recipients in an open-label study design as induction immunosuppression.[38] Patients were randomly assigned to receive either E-ATG (n = 183) for 14 days at the time of transplant or no E-ATG (n = 175) along with their maintenance immunosuppression consisting of AZA and prednisone. E-ATG doses were between 10 and 30 mg/kg/day. The outcomes of interest were: time to first rejection episode, steroid dosage requirements, functional graft survival and patient survival. When all four study lots of E-ATG were combined for data analysis, there was no difference between patients who received E-ATG and those who did not in any parameter except time to first rejection episode. For approximately the first 28 days after transplant, the E-ATG group experienced fewer rejection episodes than did the control group. After this period of time the occurrence of rejection became similar between the groups. Thus, active dosing of E-ATG conferred protection from acute rejection, but not long after dosing ended, rejection rates in the treatment group resembled those of the control group.

Several months later, Shield et al. published their experience using E-ATG for treatment rather than prevention of rejection.[39] Twenty patients with established rejection were randomised to receive E-ATG 15 mg/kg (doses were then adjusted based on presence of T-cell function measured by sheep red blood cell rosetting levels) for 14 days or methylprednisolone 1000 mg i.v. daily for up to 5 days. As in the previous study, maintenance immunosuppression consisted of AZA and prednisone in both groups. Both treatments successfully reversed rejection episodes in these 20 patients, with E-ATG reversing rejection on average in 3 days and methylprednisolone in 6 days. This early experience, although underpowered to detect any difference in efficacy between the two agents, seemed to prove that E-ATG was at least as effective as high-dose methylprednisolone and might reverse rejection episodes more quickly.

Given that E-ATG appeared at least as effective at treating rejection as high-dose corticosteroids, Hardy et al. investigated in a small group of patients the effect of E-ATG on allograft rejection that was resistant to standard high-dose corticosteroids.[40] The authors' typical treatment for acute rejection was methylprednisolone 15 mg/kg/day (not to exceed 750 mg) plus graft irradiation on days 1, 3 and 5 after rejection diagnosis. Ten patients who experienced acute rejection but did not respond to this typical treatment were treated with E-ATG 15 mg/kg/day for 21 days. Their response to this treatment was then described and compared to that of 10 other patients who were treated with and responded to the typical rejection protocol. The authors were particularly interested in whether or not the experimental group showed a return of SCr to baseline, the frequency of other rejection episodes following E-ATG treatment, 1-year graft survival and patient mortality. Although the patients in the experimental group did not respond initially to therapy as did the control group, seven of these patients' SCr returned to pre-rejection concentrations compared to only three patients in the control group. Only two patients in the experimental group experienced a second rejection episode within 1 year and no patient in the experimental group experienced a third rejection episode. In contrast, four patients in the control group experienced a second rejection episode and two of these patients experienced rejection a third time within 1 year. SCr concentrations also appeared to be lower in the experimental group compared with the control group 1 year later. Five patients in the experimental group had SCr less than or equal to 1.5 mg/dL compared with two patients in the control group. Graft failure appeared to occur equally among the two groups, with five graft failures in the control group and four in the experimental group over the 1-year follow-up. No patient in either group died during the year after rejection diagnosis. Later, this same group of investigators published their experience with 12 additional patients receiving E-ATG for their first acute rejection and 20 control patients receiving the standard therapy for their first acute rejection.[41] In this series graft survival at 1 year was 73% in patients who received E-ATG for rejection treatment compared with 46.6% in patients who received standard therapy. Mortality was 3% in the E-ATG group compared with 20% in the standard therapy group. This follow-up study appeared to suggest that treatment of acute rejection episodes with E-ATG rather than high-dose corticosteroids may confer better graft survival when used on the background of AZA/corticosteroid maintenance immunosuppression. Not long after E-ATG was approved for use, however, CYA was also approved as maintenance immunosuppressive therapy. As such, these same outcomes with E-ATG may not exist when stronger maintenance immunosuppressive therapy, including a CNI, is used.

Although small in patient numbers, and used with less than optimal maintenance immunosuppression, these early published experiences with E-ATG established antilymphocyte antibody therapy as effective for the prevention and treatment acute rejection. Although large randomised controlled trials were lacking, these experiences also established antilymphocyte antibodies as being effective against rejection that did not respond to the standard of high-dose corticosteroid treatment.

Muromonab CD3

Muromonab CD3 was the first monoclonal antibody approved for use in kidney transplantation. It was approved in 1986 for the treatment of acute rejection in kidney transplant recipients. It is typically reserved for rejection that is refractory to standard high-dose corticosteroid treatment or severe acute rejection. Muromonab CD3 is an entirely murine monoclonal antibody directed against the CD3 surface antigen present on mature T lymphocytes. When given for the treatment of acute rejection, muromonab CD3 is typically administered as a 5-mg dose daily for 10–14 days. After one to two doses, peripheral blood T-lymphocyte numbers are dramatically reduced. In many cases, no T lymphocytes are measurable in peripheral blood. Since muromonab CD3 is monoclonal, it is specific for T lymphocytes and does not result in depletion of other cellular elements in the blood. Development of human antimurine neutralising antibodies has also been observed, and is thought to be responsible for the occasional observation of rising CD3-positive T-cell concentrations towards the end of treatment. The most notable side-effect of muromonab CD3 treatment is the cytokine release syndrome. Initial binding of muromonab CD3 to the CD3 receptor on T lymphocytes results in transient activation of these cells. This results in a transient release of cytokines, such as tumour necrosis factor-α, IL-2, IL-6 and interferon-γ. Clinically, this manifests as symptoms ranging from mild and flu-like to severe shortness of breath from life-threatening pulmonary oedema. Risk of the more severe manifestations of the cytokine release syndrome can be minimised by avoiding muromonab CD3 administration in patients who are fluid overloaded.

Muromonab CD3 was initially studied in an open-label randomised trial comparing high-dose corticosteroids with muromonab CD3 for the treatment of acute rejection in 122 renal transplant patients.[42] Patients were equally randomised to receive either 500 mg methylprednisolone for three doses or 5 mg muromonab CD3 for 10–14 doses. Those randomised to muromonab CD3 received pretreatment with methylprednisolone 1 mg/kg, acetaminophen and an antihistamine. Ninety-four percent of patients treated with muromonab CD3 experienced successful reversal of rejection compared with 75% of steroid-treated patients ($P = 0.004$). When these patients were followed for 1 year after rejection treatment, there was no difference in patient survival, but a significantly higher graft survival rate was present in the muromonab CD3 group (62%) compared with the steroid treatment group (45%, $P = 0.029$). The cytokine release syndrome, including the mildest of symptoms, was noted in a high percentage of patients. Seventy-three percent of patients experienced fever, 57% chills, 21% dyspnea, 14% chest pain and tightness, while 10–11% experienced wheezing, nausea and tremors. Infectious complications were similar between the two groups during the first 45 days after treatment and during the first year after treatment.

The previous study documented the effectiveness of muromonab CD3 as treatment for acute rejection, but as mentioned earlier, since approval in 1986 it has also been used as induction treatment at the time of transplantation. Norman et al. published a multicentre randomised trial evaluating muromonab CD3 as an induction agent when given along with AZA, and corticosteroids at the time of transplant, followed by CYA beginning on day 11 after transplantation.[43] Two hundred and seven renal transplant recipients were randomised to receive either CYA, AZA and corticosteroids at the time of transplantation or AZA, corticosteroids and muromonab CD3 (5 mg i.v. daily) for 14 days, then CYA beginning on day 11 after transplantation. The primary outcomes of the study were to compare the safety and efficacy of the muromonab CD3 induction regimen with that of conventional therapy. Analysis at 1 year revealed a lower acute rejection rate in the experimental group (51%) compared with 66% in the conventional therapy group ($P = 0.032$). Fewer patients in the experimental group had more than one episode of acute rejection (17%), compared with 32% in the conventional group ($P = 0.025$). In addition, time to rejection was longer (45.5 days) in the experimental group compared with the conventional treatment group (8 days, $P = 0.001$). Despite the decrease in acute rejection rates, there was no difference in graft survival at 6 months, 1, 2 or 5 years. Although delaying initiation of CNI dosing is sometimes undertaken to decrease the risk of delayed graft function, interestingly in this study, the rate of delayed graft function was identical between the two groups ($n = 35$ experimental group, $n = 36$ conventional group). Patients in the experimental group did report more frequent side-effects. Most commonly reported were fever, tachycardia, nausea, chills, hypertension and hypotension. Although overall rates of infection did not differ between the groups, the experimental group reported more cytomegalovirus (CMV) infections and more herpes simplex infections compared with the conventional therapy group (CMV: experimental 13% vs. conventional 5%, $P = 0.055$; herpes simplex: experimental 31% vs. conventional 19%, $P = 0.06$).

These two studies demonstrated muromonab CD3's arguably better efficacy in the treatment and prevention of acute rejection when compared with standard therapy. However, side-effects, most notably the cytokine release syndrome and increased viral infections when used for induction treatment, likely limited its use to high-risk patients or patients who did not respond to conventional treatment of acute rejection episodes.

Rabbit antithymocyte globulin

Although used in European countries extensively prior to approval in the USA, R-ATG was approved in the USA in 1999 for the treatment of acute rejection in renal transplant recipients.[44] R-ATG is also used off-label as induction immunosuppression at the time of transplantation. It is a polyclonal IgG antibody product, produced through immunisation of rabbits with human T lymphocytes. R-ATG is comprised of multiple different antibodies directed at various T- and some B-lymphocyte surface antigens, such as CD2, CD3, CD4, CD8, CD11a, CD18, CD25, human leucocyte antigen (HLA)-DR and HLA class I.[44,45] R-ATG, like E-ATG, is thought to exert its antirejection effects by clearing T lymphocytes from peripheral blood, and modulation of T-lymphocyte activity. When used for treatment of acute rejection, R-ATG is usually administered in doses of 1.5 mg/kg intravenously daily for 7–14 days. It is recommended to be infused through a high-flow vein and through an in-line 0.22 μm filter. Primary side-effects of R-ATG are divided into side-effects noted during the infusion and primarily haematological side-effects that are noted days to weeks later. During infusion, patients may experience fever, chills, rigors, headache, dyspnoea, hypertension or hypotension. Although these side-effects may still occur with corticosteroid, acetaminophen and antihistamine premedication, this practice reduces their severity and is recommended. R-ATG also results in haematological side-effects such as leucopenia and thrombocytopenia. These may persist for several weeks after the last dose of R-ATG. Complete blood counts should be monitored during therapy to ensure that blood cell indices do not fall too low. The manufacturer of R-ATG recommends that the dose be halved if the white blood count (WBC) falls to between 2000 and 3000 cells/mm^3 or if the platelet count falls to between 50 000 and 75 000 cells/mm^3.[44] The manufacturer also recommends to consider stopping R-ATG treatment if the WBC falls below 2000 cells/mm^3 or the platelet count falls below 50 000 cells/mm^3.[44] Many groups administer this agent according to the absolute T-cell count with daily monitoring of this parameter.[46]

R-ATG was studied in a randomised, double-blind, multicentre trial in the USA to evaluate its efficacy at reversing acute rejection in renal transplant recipients compared to E-ATG.[47] Patients (n = 163) with mild to severe acute rejection were randomised to either receive R-ATG (1.5 mg/kg i.v. daily for 7–14 days) or E-ATG (15 mg/kg i.v. daily for 7–14 days). In the study, both groups received a mean of 10 days of treatment. The primary outcome was successful treatment of acute rejection, defined as the return of SCr to baseline or below on two consecutive measurements taken at least 2 days apart at the end of treatment. This outcome was reached by 88% of patients treated with R-ATG compared with 76% of patients treated with E-ATG (P = 0.027) (**Fig. 5.3**). The difference in treatment effect was primarily seen in patients with moderate and severe rejection. There was no difference between the groups in graft survival at 1 year, SCr (as % of baseline) 30 days after treatment or improvement in biopsy results after treatment. However, repeated episodes of rejection were less common in the R-ATG group than the E-ATG group (P < 0.05). Five study centres (26 patients) performed T-cell subset analysis during therapy and found that T-cell depletion was more profound and more prolonged with R-ATG compared to E-ATG (median CD3 values during treatment: 5 cells/mm^3 for R-ATG and 147 cells/mm^3 for E-ATG, P = 0.001; median CD3 values 30 days after treatment: 180 cells/mm^3 for R-ATG and 722 cells/mm^3 for E-ATG, P = 0.016). Related to this more profound lowering of T cells, more patients in the R-ATG group required

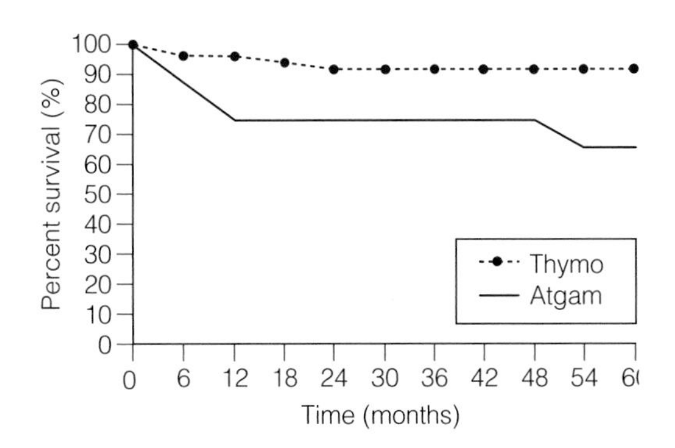

Figure 5.3 • Freedom from acute rejection. At 5 years after transplantation, freedom from acute rejection occurred in 92% of thymoglobulin-treated patients and 66% of antithymocyte globulin (Atgam)-treated patients (P = 0.0073). Reproduced from Hardinger K, Scnitzler M, Miller B et al. Five-year follow up of thymoglobulin versus ATGAM induction in adult renal transplantation. Transplantation 2004; 78:136–41. With permission from Lippincott, Williams & Wilkins.

dosage reduction due to leucopenia (34% vs. 10%, P < 0.001).

This study showed that R-ATG appears to eliminate rejection at least more quickly than E-ATG, and also resulted in fewer patients experiencing repeated rejection episodes. R-ATG also resulted in more leucopenia than E-ATG, but in this study an increase in infectious complications as a consequence of more leucopenia was not noted.

As mentioned earlier, R-ATG is also used for induction immunosuppression at the time of kidney transplantation. This practice was studied in a 1999 single-centre randomised, double-blind, prospective trial performed to compare the efficacy of R-ATG and E-ATG when used for induction therapy.[48] Seventy-two kidney transplant recipients were randomised in a 2:1 fashion (2 R-ATG to 1 E-ATG) to receive either R-ATG (1.5 mg/kg/day i.v.) or E-ATG (15 mg/kg/day i.v.) for at least 7 days. The primary end-points were rejection at 6 and 12 months, and patient and graft survival at 12 months. Maintenance immunosuppression for all patients was similar and consisted of CYA, AZA and prednisone. At 6 months post-transplantation, there were fewer acute rejection episodes in the R-ATG group (4%) compared with the E-ATG group (17%, P = 0.038). This difference in treatment groups remained at 1 year, with no more rejection episodes in the R-ATG group between months 6 and 12, and two more rejection episodes in the E-ATG group (overall 1-year rejection rates: R-ATG 4% vs. E-ATG 25%, P = 0.014). Although graft survival was reported as better in the R-ATG group, only one graft loss in the E-ATG group was thought to be due to immunological factors. The other graft losses were attributed to surgical complications. Consistent with previous observations, this study also noted a higher rate of leucopenia in the R-ATG group (56%) compared with the E-ATG group (4%, P < 0.001). Although the overall rate of infections did not differ between the groups, interestingly the R-ATG group experienced fewer episodes of CMV disease (12.5%) compared with the E-ATG group (33.3%, P = 0.056). This may be related to fewer rejection episodes in the R-ATG group, thus less of a need for additional immunosuppression to treat rejection. These positive benefits of R-ATG induction treatment as compared with E-ATG induction were also maintained in a 5-year follow-up analysis.[49]

Both E-ATG and R-ATG have been used for the treatment or prevention (as induction therapy) of acute rejection in kidney transplantation. Improved long-term graft survival has not yet been demonstrated for either agent, although it appears that R-ATG more rapidly and effectively treats and prevents rejection episodes when compared with E-ATG. For this reason R-ATG is used more frequently than E-ATG for these indications. When compared with muromonab CD3, which has also been used for these same indications, the two polyclonal preparations (R-ATG and E-ATG) have not been shown to result in the production of neutralising antibodies as has muromonab CD3. In addition, while infusion-related side-effects do occur with the two polyclonal products, severe cytokine release syndrome occurs more commonly with muromonab CD3.

Alemtuzumab

Alemtuzumab was approved in 2001 for the treatment of chronic lymphocytic leukaemia, but has since been used off-label for prevention and treatment of rejection in organ transplantation. Alemtuzumab is a humanised monoclonal antibody directed against the CD52 antigen present on malignant and normal B lymphocytes, T lymphocytes, monocytes, macrophages and natural killer cells.

When used as induction treatment, alemtuzumab has most commonly been part of a regimen using lower doses of other traditional immunosuppressive agents in an effort to decrease toxicities of these agents. It has also been used early after transplantation in the hopes of inducing tolerance. Although small case series and numerous pilot studies in kidney transplantation exist, there is an absence at this time of large randomised controlled trials comparing alemtuzumab with other agents for induction immunosuppression.

Ciancio et al. randomised 90 first-time renal transplant recipients to receive either induction treatment with daclizumab (group A), alemtuzumab (group B) or R-ATG (group C). Maintenance immunosuppression in groups A and C consisted of TAC with target trough concentrations of 8–10 ng/mL, MMF 1g twice daily and prednisone at 0.3 mg/kg/day until month 3, when the prednisone dose was halved. Group B received less intense maintenance immunosuppression, with TAC target trough concentrations between 4 and 7 ng/mL for 1 month, then between 4 and 6 ng/mL thereafter. This group also received a lower MMF dose of 500 mg twice daily and no planned steroids after 1 week post-transplantation. Patients were followed for a median of 15 months (range 4–25 months). At 12 months there was no difference in BPAR noted between the groups, with 18% in group A, 18% in group B and 19% in group C ($P = 0.99$). There was also no difference in graft failure between the groups at 1 year (group A 3%, group B none and group C none, $P = 0.38$). There was also no difference in renal function as defined by calculated CrCl between the groups at 1 year (group A 80 mL/min, group B 72.8 mL/min and group C 81 mL/min, $P = 0.63$). The presence of new-onset diabetes after transplant was also similar amongst the groups, with 23% of patients in group A, 11% in group B and 14% in group C developing post-transplant diabetes ($P = 0.51$). Although there did not appear to be any differences in outcomes between these three groups, the authors note that 80% of patients assigned to group B remained free of prednisone treatment at 1 year despite similar rejection rates and graft survival. While these outcomes with alemtuzumab induction treatment appear promising, larger trials with longer-term outcome measures are needed before alemtuzumab's true place in therapy is known.

Non-depleting antibody preparations

Daclizumab

In 1997, daclizumab was the first of the two non-depleting antibody preparations to be approved for the prevention of acute rejection in renal transplantation when used along with conventional maintenance immunosuppressive regimens. Daclizumab is a monoclonal humanised IgG immunoglobulin directed against the Tac subunit of the IL-2 receptor present on the surface of activated T lymphocytes.[50] Daclizumab is approved for administration intravenously at 1 mg/kg on the day of transplantation, then every 2 weeks thereafter for a total of five doses. Although daclizumab was originally studied to be administered in a five-dose regimen, it has since been studied in regimens as short as single-dose regimens. Daclizumab's terminal elimination half-life is approximately 20 days, similar to naturally occurring human immunoglobulins. Daclizumab does not appear to result in any significant drug–drug interactions.

Daclizumab is well tolerated, with no significant side-effects reported in renal transplants. A transient elevation in blood glucose was noted in patients who received daclizumab on the day of administration. There was no difference seen in infectious complications in renal transplant patients who received daclizumab; however, in heart transplant patients who received daclizumab in combination with CYA, MMF and corticosteroids, mortality was increased when compared with those patients who did not receive daclizumab.[51] This mortality was primarily attributed to infection-related deaths in patients who also required treatment with depleting antibodies for refractory rejection.

Daclizumab was approved for use based on the outcome of two randomised, double-blind, placebo-controlled trials. One trial took place in the USA, Canada and Sweden, and utilised triple-drug therapy (CYA, AZA, corticosteroids) as maintenance immunosuppressive therapy.[52] The other trial, which enrolled patients in Europe and Australia, utilised double-drug therapy (CYA and corticosteroids) as maintenance immunosuppressive therapy.[53] Both trials examined BPAR at 6 months as the primary outcome. Secondary outcomes were also identical in these two trials: number of acute rejection episodes in the first 6 months, time to acute rejection, graft function at 6 months, the need for antilymphocyte antibody treatment due to refractory rejection, and patient and graft survival.

Both studies showed decreased acute rejection episodes at 6 months in the daclizumab-treated patients. The triple-therapy (TT) study showed a 22% acute rejection rate in the daclizumab arm versus a 35% acute rejection rate in the placebo arm ($P = 0.03$). The double-therapy (DT) study showed 28% acute rejection in the daclizumab arm versus 48% in the placebo arm ($P = 0.001$). In the secondary end-point analysis, both studies also showed a decreased number of acute rejection episodes per patient and a longer time to first acute rejection episode in the daclizumab groups. The DT trial also showed better graft function at 6 months in the daclizumab group, SCr = 1.7 mg/dL and glomerular filtration rate (GFR) 58 mL/min, compared with placebo, SCr = 1.9 mg/dL and GFR 51 mL/min ($P = 0.04$ and 0.02 for SCr and GFR respectively). In addition, fewer patients in the DT trial who received daclizumab required antilymphocyte antibody treatment for refractory rejection (8% in the daclizumab group and 16% in the placebo group, $P = 0.02$). Surprisingly, the DT trial showed better survival at 1 year in the daclizumab-treated patients (99%) when compared with placebo-treated patients (94%, $P = 0.01$). The reason for this increased survival was unclear, but did not appear to relate to any of the study's outcome measures. Despite the apparent improved immunosuppressive efficacy of daclizumab when compared to placebo, there was no difference in infectious complications between the two groups. This was particularly encouraging, as a possible consequence of improved immunosuppression may be increased infectious complications.

A combined 3-year analysis of these two studies reported the durability of daclizumab's positive effects on acute rejection rates.[54] However, in this longer-term analysis there was still no difference in graft survival or renal function between groups in either study.

Development of malignancy was also no different between treatment groups in either study. The increased survival seen in the initial report of the DT trial was maintained at 3 years, with 88% surviving in the placebo group compared with 96% in the daclizumab group ($P = 0.017$). This increased survival still did not appear to be related to any of the study end-points. The authors commented that this may represent differences in cardiovascular risk between the two study groups, as more patients in the placebo group died of cardiovascular events.

Basiliximab

Basiliximab is a chimeric monoclonal antibody, targeting the α subunit (CD25) of the IL-2 receptor.[55] It was approved for the prevention of renal allograft rejection in 1998. Basiliximab is similar to daclizumab in its mechanism of action; while daclizumab is considered a humanised monoclonal antibody, basiliximab is chimeric. Basiliximab's entire variable region is mouse derived, whereas only daclizumab's hypervariable region is mouse derived. While this difference may account for a slightly longer terminal half-life with daclizumab, none of these differences have resulted in clinically important differentiation between the two agents.

Basiliximab is administered in a two-dose regimen, with 20 mg given intravenously preoperatively followed by another 20 mg on day 4 after kidney transplant surgery. Basiliximab has a terminal half life of 7.2 ± 3.2 days.[55] When administered as the two-dose regimen it provides a 30- to 45-day blockade of the IL-2 receptor in adults.[56,57]

Basiliximab was originally studied in two identical randomised, double-blind, placebo-controlled trials, one conducted in the USA and the other in Europe and Canada.[58,59] The US trial enrolled 348 patients and the European/Canadian trial enrolled 380 patients. Each trial evenly randomised patients to receive either basiliximab or placebo as induction treatment on the day of transplant and 4 days later. Maintenance immunosuppression in each trial was CYA and corticosteroids. Each trial used BPAR at 6 months as the primary study end-point. Secondary end-points studied were incidence of steroid-resistant acute rejection, patient and graft survival, and graft function.

Both studies showed that fewer patients in the basiliximab group experienced BPAR at both 6 months and 1 year, while differences in most secondary outcomes were not demonstrated. At 6 months the US study showed BPAR of 32.9% in the basiliximab group compared with 49.1% in the placebo group (P = 0.017), and the European/Canadian study showed 29.8% (basiliximab group) compared with 44% (placebo group, P = 0.012). This benefit was also maintained at 1 year in both studies: the US study showed BPAR of 35.3% in the basiliximab group and 49.1% in the placebo group (P = 0.009) and the European/Canadian study showed 37.9% in the basiliximab group and 54.8% in the placebo group (P = 0.002). Neither study demonstrated a difference in graft function 1 year after transplant. CrCl in both studies at 1 year ranged from 52 to 58 mL/min. Patient and graft survival were no different between groups in either study. Both studies showed that fewer patients in the basiliximab treatment groups experienced steroid-resistant acute rejection, requiring treatment with a depleting antibody agent. In the US study, 41.6% of placebo-treated patients experienced steroid-resistant acute rejection compared with 25.4% of basiliximab-treated patients (P = 0.001). In the European/Canadian trial much lower numbers of patients experienced steroid-resistant acute rejection, with 20.8% in the placebo group versus 5.4% in the basiliximab group (P < 0.001). As with the daclizumab multicentre studies, despite the improvement in acute rejection rates and decreased severity of rejection in the basiliximab group, this did not result in higher rates of infectious complications or malignancy in either study.

Both of the original trials with basilixmab were performed with maintenance immunosuppression consisting only of CYA and corticosteroids. By this point in time, triple immunosuppression was the standard at many centres; therefore, an additional trial was performed to determine whether or not the benefits of basiliximab remained when used with triple-drug maintenance immunosuppression, and whether or not the same safety profile existed with stronger maintenance immunosuppression. Three hundred and forty patients from Europe, Israel, Mexico and South Africa were randomised to receive basiliximab or placebo.[60] Their maintenance immunosuppression consisted of CYA (modified), AZA and corticosteroids. Acute rejection episodes at 6 months were lower in the basiliximab group (20.8%) compared with the placebo group (34.9%, P = 0.005). Also, fewer patients in the basiliximab group (4.2%) experienced two or more rejection episodes when compared with the placebo group (7%, P = 0.026). There were no differences at 1 year in patient or graft survival or in graft function.

These results were similar to those seen in the daclizumab studies, both agents appearing to reduce acute rejection rates by approximately 13–16% within the first year when used with a double- or triple-medication maintenance regimen of CYA and corticosteroids with or without AZA.

While all of the antibody induction agents showed decreased acute rejection rates, they were all initially studied in patients receiving maintenance immunosuppression with CYA-based regimens with or without AZA. However, during the time that each of these agents were being studied as induction therapy, two new maintenance immunosuppressive agents, MMF and TAC, were approved. Each of these agents also produced decreases in acute rejection episodes when compared with the standard CYA-based therapy with or without AZA. Clinical trials would therefore need to be performed with newly available agents in combination in order to determine the most efficacious regimen for different patient groups.

As mentioned earlier, polyclonal antibody therapy such as R-ATG is also used as induction treatment. The question of whether or not this therapy is more efficacious was studied by Brennan et al. in a comparison of basiliximab induction to R-ATG induction treatment in patients at high risk for delayed graft function and acute rejection.[61] High-risk patients (n = 278) receiving a cadaveric renal transplant were randomised 1:1 in an open-label fashion to receive either R-ATG (1.5 mg/kg, first dose intraoperatively, then once daily for three more doses) or basiliximab (20 mg intraoperatively, then once again 4 days later). Both groups received CYA (modified), MMF and steroids as maintenance immunosuppression. The primary end-point was a composite of first BPAR, delayed graft function, graft loss, or death at 1 year. There was no difference in the primary end-point between the groups, with 50.4% of the R-ATG and 56.2% of the basiliximab group experiencing the composite end-point ($P = 0.34$). However, BPAR was lower in the R-ATG group (15.6%) compared with the basiliximab group (25.5%), as was steroid-resistant rejection requiring antibody treatment (1.4% R-ATG group vs. 8% basiliximab, $P = 0.005$) (Table 5.2). There was no difference between the groups in the rates of delayed graft function (40.4% R-ATG vs. 44.5% basiliximab, $P = 0.54$), graft loss (9.2% R-ATG vs. 10.2% basiliximab, $P = 0.68$) or patient death (4.3% R-ATG vs. 4.4% basiliximab, $P = 0.09$). The R-ATG group experienced an overall higher rate of infectious complications (85.8%) compared with the basiliximab group (75.2%, $P = 0.03$) and a higher rate of leucopenia (33.3% R-ATG vs. 14.6% basiliximab, $P < 0.001$).

In this case it appears that while R-ATG treatment did not result in lower numbers of patients reaching the composite outcome, it did result in lower acute rejection rates. These lower rejection rates, however, came unsurprisingly at the price of higher rates of infectious complications and haematological side-effects.

Induction therapy with either daclizumab or basiliximab has not been shown in clinical trials to improve allograft survival. Webster et al. undertook a meta-analysis to determine whether or not a study involving larger numbers of patients may uncover graft survival benefit with these induction agents.[62]

After rigorous inclusion criteria the group analysed 2786 renal transplant patients in clinical trials of either basiliximab or daclizumab induction therapy compared with placebo. This analysis did not detect any difference in overall graft survival in patients receiving one of the IL-2 receptor antagonists compared with those receiving placebo at 1 (RR 0.84, CI 0.64–1.1) or 3 years (RR 1.08, CI 0.71–1.64).

The hope of lower toxicity immunosuppressive regimens

While the advances in immunosuppressive medications in the 1980s and 1990s greatly decreased rejection rates and improved graft survival (compared with the pre-CYA era), long-term immunosuppressive toxicity remains an issue today. Corticosteroids are associated with myriad side-effects such as weight gain, decreased bone mineral density, hyperglycaemia, hyperlipidaemia, hypertension and cosmetic side-effects (**Fig. 5.4**). CNIs, the backbone of the modern immunosuppressive regimen, unfortunately cause some of these same side-effects, along with chronic nephrotoxicity. Since the introduction of newer immunosuppressive agents such as MMF, sirolimus and newer antibody induction agents, the possibility of lower toxicity regimens has been and remains currently under study. These lower toxicity regimens are generally aimed at reducing or eliminating corticosteroids or CNIs from the immunosuppressive regimen in hopes of decreasing these side-effects.

Corticosteroid-sparing regimens

Corticosteroid minimisation regimens have taken three forms: steroid-withdrawal, steroid-avoidance and steroid-free regimens. Steroid withdrawal is accomplished by maintaining patients on steroids for several months to a year, then discontinuing steroids. Steroid avoidance is accomplished by removing steroids from the immunosuppressive regimen within days after transplantation. Complete steroid-free regimens use no steroids at all.

Early attempts at corticosteroid withdrawal occurred in the 1980s and early 1990s with CYA-based immunosuppression. Of these experiences, the largest randomised trial was published by Sinclair et al.[63] They randomised 523 renal transplant patients with well-functioning grafts at 90 days to either stay on low-dose prednisone (0.3 mg/kg every other day) or to have prednisone replaced by placebo in a blinded fashion. Patients' other immunosuppression consisted only of CYA with target levels between 75 and 200 ng/mL. Patient and graft actuarial survival were analysed at 5 years. Graft survival was 73% in the placebo group compared with 85% in the prednisone group ($P = 0.03$). There was no difference between the groups

Table 5.2 • Efficacy end-points 12 months after transplantation

End-point	No. of patients (%)		*P* value
	Antithymocyte globulin (n = 141)	Basiliximab (n =137)	
Composite of acute rejection, delayed graft function, graft loss and death	71 (50.4)	77 (56.2)	0.34
Biopsy-proven acute rejection	22 (15.6)	35 (25.5)	0.02
No. of black recipients/total no. of such recipients	8/41 (19.5)	13/39 (33.3)	0.14
No. of non-black recipients/total no. of such recipients	14/100 (14.0)	22/98 (22.4)	0.07
No. of recipients in USA/total no. of such recipients	13/91 (14.3)	21/92 (22.8)	0.07
No. of recipients in Europe/total no. of such recipients	9/50 (18.0)	14/45 (31.1)	0.12
Antibody-treated acute rejection	2 (1.4)	11 (8.0)	0.005
Delayed graft function	57 (40.4)	61 (44.5)	0.54
Graft loss	13 (9.2)	14 (10.2)	0.68
From death	4 (2.8)	3 (2.2)	
From acute rejection	1 (0.7)	1 (0.7)	
From primary non-function	1 (0.7)	4 (2.9)	
From graft thrombosis	0	4 (2.9)	
From chronic rejection	2 (1.4)	1 (0.7)	
From infection	1 (0.7)	1 (0.7)	
From toxic effects of ciclosporin	1 (0.7)	0	
From recurrent disease	1 (0.7)	0	
From hypertension	1 (0.7)	0	
From urinary fistula	1 (0.7)	0	
Death	6 (4.3)	6 (4.4)	0.90
From cardiovascular disease	2 (1.4)	5 (3.6)	
From pulmonary disease	1 (0.7)	1 (0.7)	
From gastrointestinal disease	2 (1.4)	0	
From unknown cause	1 (0.7)	0	

Reproduced from Brennan D, Daller J, Lake K et al. Rabbit antithymocyte globulin versus basiliximab in renal transplantation. N Engl J Med 2006; 355(19):1967–77. With permission from the Massachusetts Medical Society.

in patient survival. Importantly, this study did not show any difference in graft survival until approximately 2 years after transplant. It was at this time point that graft survival curves began to diverge. This demonstrated that longer follow-up may be necessary to detect differences in steroid-withdrawal regimens. Renal function remained similar between the groups in patients who maintained functioning grafts.

Unfortunately, some of the expected benefits of steroid withdrawal were not recognised in this study. There was no difference in mean blood glucose, serum cholesterol, or triglycerides. While there was also no difference in blood pressure, the authors noted that patients in the prednisone group received more antihypertensive medications. The absence of a difference between the two groups in these parameters may be partially explained because patients were allowed to resume steroid treatment if rejection episodes occurred or if the physician deemed it necessary to restart steroids. This occurred in approximately 50% of patients randomised to the placebo group. These results were not encouraging as not only did patients not gain

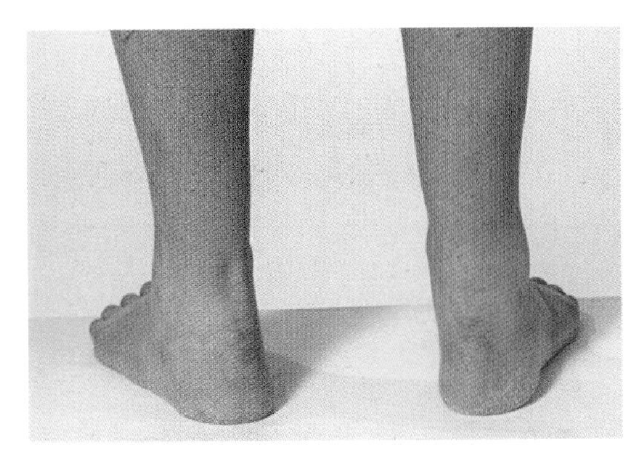

Figure 5.4 • Achilles tendon rupture in a renal transplant patient given long-term steroids.

substantial benefits from steroid discontinuation, but graft survival suffered.

While the previous study used CYA-only maintenance immunosuppression, the question remained: would results of steroid withdrawal be different if a more efficacious maintenance regimen were used? Ahsan et al. conducted a multicentre, double-blind, placebo-controlled trial of steroid withdrawal at 90 days post-transplant, in patients receiving CYA and MMF as their other maintenance immunosuppression.[64] Similar to the low-risk population enrolled in the previous trial, patients were only eligible if they had no acute rejection episodes within the first 3 months after transplant. After 266 patients were randomised, the study was halted early by the data safety monitoring board due to a statistically significant difference in acute rejection rates between the steroid-withdrawal group and the control group (withdrawal 15% vs. control 5%, $P = 0.004$). The majority of rejection episodes in the withdrawal group occurred after steroids had been withdrawn. In addition to this increased risk of rejection, improvements in known steroid side-effects were modest at best in the withdrawal group. Total cholesterol levels were 13.7 mg/dL lower in the withdrawal group and fewer patients in the withdrawal group required antihypertensive medications (83% withdrawal group vs. 93% maintenance group, $P = 0.01$).

While there have been multiple other uncontrolled, retrospective studies conducted, based on the results of these two large randomised controlled trials, it appears that while steroid withdrawal may be possible in select patients, overall, rejection rates and graft survival are likely to be poorer when this strategy is applied to the entire population of low-risk kidney transplant recipients. This was seen even when more efficacious agents such as MMF were used in the remaining immunosuppressive regimen.

Steroid avoidance differs from steroid withdrawal in that steroid agents are only used for several days following transplantation. Many single-centre, short-term studies have been conducted and showed short-term efficacy with this rapid steroid-withdrawal approach. However, long-term results of this strategy in randomised controlled trials have yet to be published. Matas et al. reported their single-centre results after following patients for 4 years;[65] 477 live-donor and cadaveric-donor kidney transplant recipients had steroids withdrawn on postoperative day 6. Patients received either CYA/MMF or TAC/sirolimus as maintenance immunosuppression. At the time of this publication, the authors noted no difference in rejection rates or graft survival between patients receiving CYA/MMF and those receiving TAC/sirolimus, so these patient groups were reported together. All patients received 5 days of R-ATG induction therapy. At 4 years, actuarial graft survival was 90%, acute rejection-free graft survival was 86%, chronic rejection-free survival was 95%, and SCr concentrations were a mean of 1.6 mg/dL. When compared with historical control patients, all of these outcomes were found to be more favourable in the steroid-withdrawal group. An additional 5-year follow-up was also published and showed graft survival of 84%, acute rejection-free graft survival of 84%, chronic rejection-free graft survival of 87%, and a mean SCr of 1.7 mg/dL.[66] While this study published the largest number of patients undergoing rapid steroid withdrawal while receiving more modern immunosuppressive regimens, its use of historical controls is one of its key limitations.

Woodle et al. have reported 4-year outcomes in their randomised, double-blind, placebo-controlled trial comparing 191 early steroid-withdrawal patients (day 7 post-transplant) to patients who received chronic steroid maintenance therapy (n = 195).[67] All patients received TAC and MMF as maintenance immunosuppression, and received induction therapy with either R-ATG or an IL-2 receptor antibody. The study remains blinded to date, but 4-year outcomes have been reported in abstract form. The primary end-point of this study was treatment failure defined as death, graft loss, or acute rejection. To date, no difference has been shown in the primary outcome between groups (steroid-withdrawal group 16.8% and continuation group 12.3%). However, a trend was noted for increased BPAR in the steroid-withdrawal group (17.3%) compared with the steroid-continuation group (10.8%, *P* = 0.08). In addition, chronic allograft nephropathy at 4 years in patients who had for-cause biopsies was higher in the steroid-withdrawal group (9.9%) compared with 4.1% in the steroid-continuation group (*P* = 0.03). Patients in the steroid-withdrawal group experienced fewer bone disorders such as avascular necrosis and fractures (3.7%) compared with steroid-maintenance patients (9.7%, *P* = 0.02). Other potential benefits of steroid withdrawal, such as improved lipid parameters, less weight gain and less hypertension, were not found. The findings of increased chronic allograft nephropathy and potentially higher acute rejection rates are concerning, and it again appears that early steroid withdrawal may suffer from the same outcomes seen previously with later steroid withdrawal; however, the final outcomes of this trial have yet to be reported.

After the introduction of newer immunosuppressive agents such as the IL-2 receptor antagonists and MMF transplant, clinicians became interested in the possibility of completely steroid-free immunosuppressive regimens. Steroid-free immunosuppression, unlike the previous two strategies discussed, uses no steroids at all. Rostaing et al. studied this in a randomised, open-label, multicentre study in Europe by comparing patients who received two-dose daclizumab induction treatment, TAC and MMF with those who received TAC, MMF and corticosteroids;[68] 551 patients were randomised 1:1 to the two groups, and the primary end-point of this study was BPAR at 6 months after kidney transplantation. At 6 months BPAR rates were virtually identical between the two groups: 51% of the patients in the daclizumab/TAC/MMF experienced BPAR and 50% in the TAC/MMF/corticosteroid group. In addition, graft survival and renal function were no different between groups at 6 months. New onset of diabetes post-transplant was defined as the new need for insulin for more than 30 days after transplant; 0.4% of patients in the daclizumab/TAC/MMF group experienced new-onset diabetes, while this was present in 5.4% of patients in the TAC/MMF/corticosteroid group (*P* = 0.001). There was also a difference in bone mineral density decline between the two groups; 93 patients had bone mineral densities performed before transplant and 6 months later. The daclizumab/TAC/MMF group experienced no changes in bone density scores in either the femoral or lumbar regions, whereas the TAC/MMF/corticosteroid group showed a 0.15 decrease in *t* score and 0.13 decrease in *z* score in the femoral region (*P* = 0.03). No change in the lumbar region was noted in the TAC/MMF/corticosteroid group. There was no difference noted between the two treatment groups with regard to low-density lipoprotein concentrations at 6 months. While this study certainly presents optimistic results for a steroid-free regimen, the obvious weakness of these results is the relatively short follow-up period. As mentioned previously, 6 months is not likely to be adequate to fully evaluate efficacy differences between these two regimens.

The FREEDOM trial seeks to compare all three steroid-minimisation strategies in a single prospective trial. One-year results of the FREEDOM trial have been reported.[69] It is an open-label, multicentre trial which randomised 335 patients to receive no steroids, short-term steroids (up to day 7) or standard steroids. The remainder of the immunosuppressive regimen consists of CYA (modified), enteric-coated mycophenolic acid and basiliximab induction.

While the 12-month analysis showed no difference in renal function between groups as measured by GFR, there was a significantly higher number of patients in the steroid-free group experiencing BPAR (31.5%) and in the steroid-withdrawal group (26.1%), compared with the standard steroid group (14.7%) (*P* = 0.004 steroid free vs. standard; *P* = 0.046 steroid withdrawal vs. standard) (**Fig. 5.5**). Of the expected benefits from less steroid exposure there were several modest benefits noted in this 1-year analysis. Fewer patients in the steroid-free group (4.5%) required new use of antihypertensive medications compared with patients in the standard steroids group (14.7%, *P* = 0.01). The steroid-withdrawal group at 1 year had a decrease in body mass index of 0.88 kg/m^2 whereas the standard steroid group showed an increase of 1.88 kg/m^2. The mean triglyceride level in the steroid withdrawal group was 141 mg/dL compared with 168 mg/dL in the standard steroid group (*P* = 0.03). There was also a decreased need for lipid-lowering medications in the steroid-withdrawal group (36.5%) compared with the standard steroid group (52.3%, *P* = 0.018).

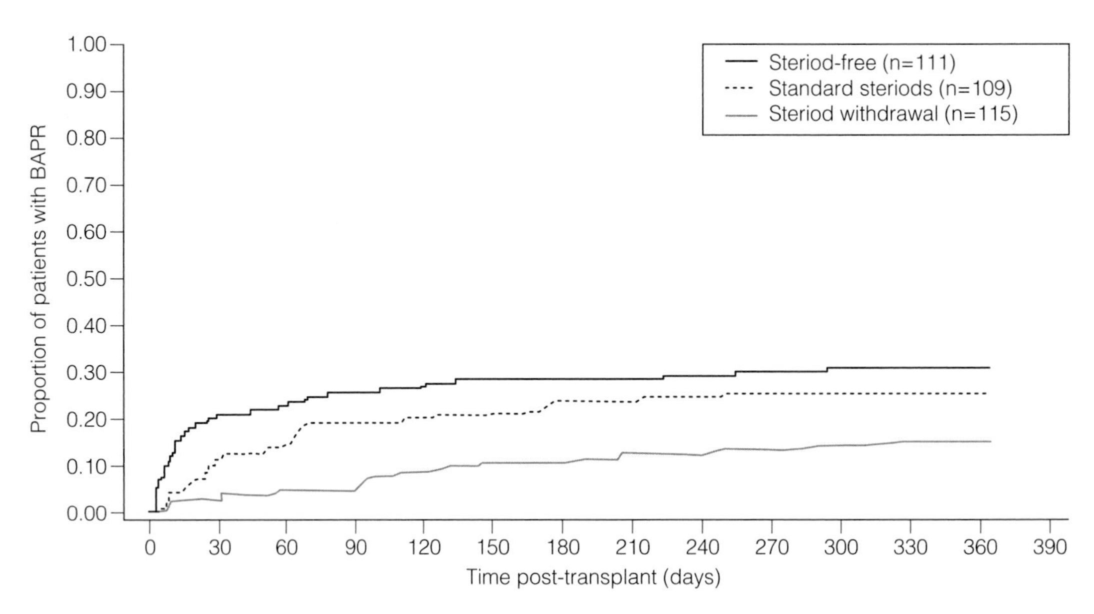

Figure 5.5 • Cumulative incidence of biopsy-proven acute rejection (BPAR) in patients randomised to steroid-free therapy, steroid withdrawal at day 7, or standard steroid (Kaplan–Meier, intention-to-treat population). Reproduced from Vincenti F, Schena F, Paraskevas S et al. A randomized multicenter study of steroid avoidance, early steroid withdrawal or standard steroid therapy in kidney transplant recipients. Am J Transplant 2008; 8:307–16. With permission from Blackwell Publishing.

While this is only a 1-year analysis of the FREEDOM trial, these results are particularly concerning for steroid-free regimens.

The two largest randomised controlled trials of steroid avoidance have shown so far significantly higher acute rejection rates with somewhat concerning secondary findings of possibly worse renal function in the CYA-based and more chronic allograft nephropathy in the TAC-based steroid-avoidance arms. While some patients might be maintained safely on steroid-free regimens, the risks seem to outweigh the benefits when looking at large randomised study outcomes.

Calcineurin inhibitor-sparing regimens

CNIs, like corticosteroids, are associated with many side-effects, such as hypertension, diabetes and tremor. However, probably the most concerning side-effect of CNIs is long-term nephrotoxicity. Recently, with the introduction of MMF and sirolimus, regimens that either use low-dose CNIs or eliminate CNIs totally from the immunosuppressive regimen have been studied with the hopes of eliminating nephrotoxicity and other side-effects.

When sirolimus was originally studied in combination with CYA and steroids, this combination produced lower rejection rates than CYA and steroids alone. However, it also resulted in worse renal function. Not known to be nephrotoxic by itself, sirolimus appeared to potentiate the nephrotoxicity of CYA. Following the outcomes of these initial studies of the CYA/sirolimus combination, the Rapamune Maintenance (RMR) study was designed to determine if the use of concentration-controlled sirolimus maintenance therapy would allow for the early discontinuation of CYA, thereby improving renal function while maintaining acceptably low acute rejection rates. Four-year outcomes of the RMR study have been published;[70] 430 patients received CYA, sirolimus and steroids for the first 3 months after transplant were then randomised 1:1 to either remain on this triple-drug regimen (CYA goal trough concentrations 75–200 ng/mL and sirolimus goal trough concentrations >5 ng/mL), or to have CYA removed from their regimen and remain on just sirolimus (goal trough concentration 20–30 ng/mL) and steroids. The primary end-point was graft survival, with other end-points such as patient survival, incidence and severity of acute rejection, allograft function (determined by calculated glomerular filtration rate and SCr concentrations), and rates of study drug discontinuation. After 4 years of follow-up, patients in the sirolimus/steroids group had better graft survival (96.1%) as compared with the CYA/sirolimus/steroids group (90.6%, $P = 0.026$), and better renal function (SCr 1.38 vs. 1.87 mg/dL, $P < 0.001$). While BPAR rates overall did appear higher in the sirolimus/steroids group (6.5% vs. 10.2%), this difference was not statistically different ($P = 0.223$). There was also a higher treatment assignment discontinuation rate (60.9%) in the CYA/sirolimus/steroids group compared with the sirolimus/steroids group (44%, $P < 0.001$). This was primarily due to side-effects, the most frequent of which was worsening renal function. This 4-year analysis of the RMR study reinforced the nephrotoxicity of the CYA/sirolimus combination, and showed that when CYA was removed from the regimen renal function and graft survival were better. However, the control group in this study (CYA/sirolimus/prednisone) received a regimen that was known to be nephrotoxic from previous studies, therefore the improved outcomes in this study may simply reflect the comparison of sirolimus/steroids to a regimen known to produce worse renal function. Therefore comparison of sirolimus-based regimens to other regimens, such as MMF-based regimens or low-dose CNI regimens, would need to be performed before determining which regimen is superior.

Although several smaller trials have studied the practice of low-dose CNIs, or CNI elimination, the Ciclosporin Avoidance Eliminates Serious Adverse Renal Toxicity (CAESAR) trial most recently reported 1-year outcomes of these approaches.[71] The CAESAR trial randomised 535 first-time renal transplant patients to receive either standard-dose CYA (trough concentrations 150–300 ng/mL through month 4, then 100–200 ng/mL thereafter), low-dose CYA (trough concentrations 50–100 ng/mL) or low-dose CYA (trough concentrations 50–100 ng/mL) with CYA withdrawal at month 4. MMF and corticosteroids were given to patients in all groups, and patients in the low-dose CYA and CYA-withdrawal groups received daclizumab induction treatment. Although no difference between the groups was noted in GFR at 1 year (the primary end-point), significantly more patients experienced at least one episode of BPAR in the CYA-withdrawal group. BPAR was experienced by 38% of patients in the CYA-withdrawal group, 25.4% in the low-dose CYA group and 27.5% in the standard-dose CYA group. This difference was statistically different between both the CYA-withdrawal and low-dose CYA groups ($P = 0.027$) and the CYA-withdrawal and standard-dose CYA groups ($P = 0.04$). There were no differences between the groups in adverse events overall, and in particular no difference in known CYA metabolic side-effects such as hypertension and dyslipidaemia.

These results showed that removal of the CNI from the regimen entirely, even when induction with daclizumab and MMF are used, results in higher rejection rates, without improvement in renal function or metabolic side-effects, and therefore may forecast long-term decline in renal function. However, the CAESAR study also showed that targeting lower CYA trough concentrations did not result in higher acute rejection rates when used with daclizumab induction therapy and MMF.

Given the increase in acute rejection rates previously seen with CNI elimination, but similar rejection rates with low-dose CYA, the Efficacy Limiting Toxicity Elimination (ELITE) Symphony study sought to extend the findings of the CAESAR study by also examining not only low-dose CYA, but low-dose TAC and sirolimus as well. The ELITE Symphony study compared low-dose CYA (trough concentrations 50–100 ng/mL), low-dose TAC (trough concentrations 3–7 ng/mL), low-dose sirolimus (trough concentrations 4–8 ng/mL) and standard-dose CYA (trough concentrations 150–300 ng/mL for 3 months, then 100–200 ng/mL thereafter).[72] All groups received MMF and all groups except for the standard CYA group received daclizumab induction treatment; 1645 patients were randomised in a 1:1:1:1 fashion to one of the four treatment groups and followed for 1 year. The primary end-point was renal function as measured by calculated GFR. Secondary end-points included BPAR, allograft survival, patient survival and treatment failure. Treatment failure was defined as any of the following: additional immunosuppressive agents needed, discontinuation of study medication for more than 14 consecutive days or 30 total days, allograft loss, or death. GFR at 1 year was better in the low-dose TAC group (65.4 mL/min) compared with low-dose CYA (59.4 mL/min), low-dose sirolimus (56.7 mL/min) and standard-dose CYA (57.1 mL/min) ($P < 0.001$ for all comparisons). Significantly less BPAR was experienced in the low-dose TAC group (15.4%) compared with low-dose

CYA (27.2%), low-dose sirolimus (40.2%) and standard-dose CYA (30.1%) groups ($P < 0.001$ for all comparisons). In addition, allograft survival was higher in the low-dose TAC group (94.2%) compared with low-dose sirolimus (89.3%) and standard-dose CYA (89.3%) ($P = 0.007$ for both comparisons). However, there was no difference in allograft survival between the low-dose TAC group and the low-dose CYA group. The low-dose sirolimus group experienced the highest overall rate of side-effects (53%, compared with 43.4–44.3% in other groups, $P < 0.05$ for all comparisons). This group also had the most treatment withdrawals due to side-effects (7.8%, compared with 1.8–3.1% in other groups). The low-dose TAC group experienced more new-onset diabetes (10.6%, compared with 4.7–7.8% in the other groups, $P = 0.02$). This group also experienced more diarrhoea (27.4%, compared with 14.4–24% in the other groups, $P < 0.001$). Overall, opportunistic infections were highest in the standard-dose CYA group (33%, compared with 26.3–28.1% in the other groups, $P = 0.03$). Based on these findings, it appears that when low-dose CNI regimens are used, TAC along with MMF and daclizumab induction therapy results in better renal function and fewer rejection episodes than low-dose CYA along with MMF and daclizumab induction therapy at 1 year. In fact, the low-dose TAC regimen was even superior to the standard-dose CYA regimen at preventing BPAR, preserving renal function, and allograft survival at 1 year. Low-dose TAC was also superior to low-dose sirolimus in preventing rejection episodes, preserving renal function and in tolerability. Longer-term follow-up is necessary to determine if these positive results remain after years of treatment.

The future

The immediate future in renal transplantation will see an ongoing quest for lower toxicity immunosuppressive regimens. Studies will continue to explore whether it is possible to maintain efficacy while reducing short- and long-term toxicities. Given the excellent results in renal transplantation, it is become increasingly difficult to show safety in studies. Hard end-points such as patient and graft loss and specific side-effects are too rare to be detected reliably even in larger studies. In order to safely develop new immunosuppressive strategies, reliable surrogate end-points for graft and patient survival will have to be developed. These surrogate end-points will have to undergo the same scrutiny in randomised prospective trials as new medications.

There are several new immunosuppressive medications in the pipeline. Many of these new agents are being developed within steroid or CNI avoidance strategies in order to further the quest for adequate postapproval market share. As these scenarios have proven rather difficult with current immunosuppressive agents, vigilant scrutiny must be the advice for new immunosuppressive agents developed in this context. The ultimate goal is graft acceptance without long-term immunosuppression; whether the long road of development towards this goal is through refinement of pharmacological approaches or other immunological strategies remains to be seen.

Key points

- Acute rejection has been largely overcome as a significant cause of allograft loss.
- Multiple drugs are now used to maximise immunosuppression balanced with minimisation of side effects.
- The "old" end-points of patient and graft loss are too rare to be detected even in larger studies. Surrogate end-points will have to be developed.
- After a number of years where there are few new choices, there are a number of new immuno suppression medications in the pipeline.

References

1. Murray J, Merrill J, Harrison H et al. Prolonged survival of human-kidney homografts by immunosuppressive drug therapy. N Engl J Med 1963; 268(24):1315–23.
2. Starzl T, Marchioro T, Rifkind D et al. Factors in successful renal transplantation. Surgery 1964; 58(2): 296–318.
3. Imuran® prescribing information. Prometheus Corporation.
4. Mueller E, Kovarik J, van Bree J et al. Improved dose linearity of cyclosporine pharmacokinetics from a microemulsion formulation. Pharm Res 1994; 11:301.
5. Kovarik J, Mueller E, van Bree J et al. Reduced inter- and intraindividual variability in cyclosporine pharmacokinetics from a microemulsion formulation. J Pharm Sci 1994; 83:444.
6. Mueller E, Kovarik J, van Bree J et al. Pharmacokinetics and tolerability of a microemulsion formulation of cyclosporine in renal allograft recipients: a concentration-controlled comparison with the commercial formulation. Transplantation 1994; 57:178.
7. Kovarik J, Mueller E, van Bree J et al. Cyclosporine pharmacokinetics and variability from a microemulsion formulation – A multicenter investigation in kidney transplant patients. Transplantation 1994; 58(6):658–63.
8. Wahlberg J, Wilczek H, Fauchald P et al. Consistent absorption of cyclosporine from a microemulsion formulation assessed in stable renal transplant recipients over a one-year study period. Transplantation 1995; 60(7):648–52.
9. European Multicentre Trial Group. Cyclosporine A as sole immunosuppressive agent in recipients of kidney allografts from cadaver donors. Lancet 1982; 00:57–60.
10. The Canadian Multicentre Transplant Study Group. A randomized clinical trial of cyclosporine in cadaveric renal transplantation. N Engl J Med 1983; 309:809–15.
11. European Multicentre Trial Group. Cyclosporin in cadaveric renal transplantation: one-year follow up of a multicentre trial. Lancet 1983; 2(8357):986–9.
12. European Multicentre Trial Group. Cyclosporin in cadaveric renal transplantation: 5-year follow-up of a multicentre trial. Lancet 1987; 2(8557):506–7.
13. The Canadian Multicentre Transplant Study Group. A randomized clinical trial of cyclosporine in cadaveric renal transplantation: analysis at three years. N Engl J Med 1986; 314:1219–25.
14. Prograf® prescribing information. Astellas, Inc., Nutley, NJ.
15. European FK506 Multicentre Liver Study Group. Randomized trial comparing tacrolimus (FK506) and cyclosporin in prevention of liver allograft rejection. Lancet 1994; 344:423–8.
16. US Multicenter FK506 Liver Study Group. A comparison of tacrolimus (FK506) and cyclosporine for immunosuppression in liver transplantation. N Engl J Med 1994; 331: 1110–15.
17. Pirsch JD, Miller J, Deierhoi MH et al. A comparison of tacrolimus (FK506) and cyclosporine for immunosuppression after cadaveric renal transplantation. Transplantation 1997; 63(7): 977–83.
18. Mayer DA, Dmitrewski J, Squifflet JP et al. Multicenter randomized trial comparing tacrolimus (FK506) and cyclosporine in the prevention of renal allograft rejection: a report of the European Tacrolimus Multicenter Renal Study Group. Transplantation 1997; 64(3):436–43.
19. Vincenti F, Jensik SC, Filo RS et al. A long-term comparison of tacrolimus (FK506) and cyclosporine in kidney transplantation: evidence for improved allograft survival at five years. Transplantation 2002; 73:775–82.
20. Johnson C, Ahsan N, Gonwa T et al. Randomized trial of tacrolimus (Prograf) in combination with azathioprine or mychophenolate mofetil versus cyclosporine (Neoral) with mycophenolate mofetil after cadaveric kidney transplantation. Transplantation 2000; 69(5):834–41.
21. Gonwa T, Johnson C, Ahsan N et al. Randomized trial of tacrolimus + mycophenolate mofetil or azathioprine versus cyclosporient + mycophenolate mofetil after cadaveric kidney transplantation: results at three years. Transplantation 2003; 75(12):2048–53.
22. Cellcept® prescribing information. Nutley, NJ.
23. European Mycophenolate Cooperative Study Group. Placebo-controlled study of mycophenolate mofetil combined with cyclosporine and corticosteroids for prevention of acute rejection. Lancet 1995; 345:1321–5.
24. Sollinger HW et al. Mycophenolate mofetil for the prevention of acute rejection in primary cadaveric renal allograft recipients. Transplantation 1995; 60(3):225–32.
25. Tricontinental Mycophenolate Mofetil Renal Transplantation Study Group. A blinded, randomized clinical trial of mycophenolate mofetil for the prevention of actue rejection in cadaveric renal transplantation. Transplantation 1996; 61(7): 1029–37.

   MMF was seen to be superior to AZA.
26. Behrend M, Grinyo J, Vanrenterghem Y et al. (European Mychophenolate Mofetil Cooperative Study Group). Mycophenolate mofetil in renal transplantation: 3-year results from the

placebo-controlled trial. Transplantation 1999; 68(3):391–6.

27. US Renal Transplant Mycophenolate Mofetil Study Group. Mychophenolate mofetil in cadaveric renal transplantation. Am J Kidney Dis 1999; 34(2):296–303.

28. Mathew T, for the Tricontinental Mycophenolate Mofetil Renal Transplantation Study Group. A blinded, long-term randomized multicenter study of mycophenolate mofetil in cadaveric renal transplantation: results at three years. Transplantation 1998; 65(11):1450–4.

29. Ojo A, Meier-Kriesche H, Hanson J et al. Mycophenolate mofetil reduces late renal allograft loss independent of acute rejection. Transplantation 2000; 69(11):2405–9.

A large registry study indicating the benefit of MMF.

30. Rapamune® prescribing information. Wyeth Pharmaceuticals, Philadelphia, PA.

31. Kaplan B, Meier-Kriesche H, Napoli K et al. The effects of relative timing of sirolimus and cyclosporine microemulsion coadministration on the pharmacokinetics of each agent. Clin Pharmacol Ther 1998; 63:48–53.

32. Kahan BD. Efficacy of sirolimus compared with azathioprine for reduction of acute renal allograft rejection: a randomized multicentre study. Lancet 2000; 356(9225):194–202.

33. MacDonald A. A worldwide phase III, randomized controlled, safety and efficacy study of a sirolimus/cyclosporine regimen for prevention of acute rejection in recipients of primary mismatched renal allografts. Transplantation 2001; 71(2):271–80.

34. Mendez R, Gonwa T, Yang H et al. A prospective randomized trial of tacrolimus in combination with sirolimus or mycophenolate mofetil in kidney transplantation: results at one year. Transplantation 2005; 80(3):303–9.

This study shows the synergistic nephrotoxicity with CNI drugs.

35. Meier-Kriesche HU, Steffen BJ, Chu AH et al. Sirolimus with Neoral versus mycophenolate mofetil with Neoral is associated with decreased renal allograft survival. Am J Transplant 2004; 4:2058–66.

36. Meier-Kriesche H, Schold J, Srinivas T et al. Sirolimus in combination with tacrolimus is associated with worse renal allograft survival compared to mycophenolate mofetil combined with tacrolimus. Am J Transplant 2005; 5:2273–80.

37. ATGAM prescribing information. Pharmacia & Upjohn, New York, NY.

38. Wechter W, Brodie J, Morrell R et al. Antithymocyte globulin (ATGAM) in renal allograft recipients. Transplantation 1979; 28(4):294–302.

39. Shield C, Cosimi A, Tolkoff-Rubin N et al. Use of antithymocyte globulin for reversal of acute allograft rejection. Transplantation 1979; 28(6):461–4.

40. Hardy M, Nowygrod R, Elberg A et al. Use of ATG treatment of steroid-resistant rejection. Transplantation 1980; 29:162–4.

41. Nowygrod R, Appel G, Hardy M. Use of ATG for reversal of acute allograft rejection. Transplant Proc 1981; 13(1):469–72.

42. Goldstein G, Schindler J, Tsai H et al. A randomized clinical trial of OKT3 monoclonal antibody for acute rejection of cadaveric renal transplants. N Engl J Med 1985; 313(6):337–42.

43. Norman D, Kahana L, Stuart F et al. A randomized clinical trial of induction therapy with OKT3 in kidney transplantation. Transplantation 1993; 55(1):44–50.

44. Thymoglobulin prescribing information.

45. Bonnefoy-Bernard N, Flacher M, Revillard JP. Antiproliferative effect of antilymphocyte globulins on B-cells and sB-cell lines. Blood 1992; 79:2164–70.

46. Clark KR, Forsythe JL, Shenton BK et al. Administration of ATG according to the absolute T lymphocyte count during therapy for steroid-resistant rejection. Transplant Int 1993; 6:18.

47. Gaber O, First M, Tesi R et al. Results of the double-blind randomized multicenter, phase III clinical trial of thymoglobulin versus ATGAM in the treatment of acute graft rejection episodes after renal transplantion. Transplantation 1998; 66(1):29–37.

R-ATG was more effective than antithymocyte globulin in treating acute rejection.

48. Brennan D, Flavin K, Lowell J et al. A randomized, double-blinded comparison of thymoglobulin versus atgam for induction immunosuppressive therapy in adult renal transplant recipients. Transplantation 1999; 67(7):1011–18.

49. Hardinger K, Scnitzler M, Miller B et al. Five-year follow up of thymoglobulin versus ATGAM induction in adult renal transplantation. Transplantation 2004; 78:136–41.

50. Daclizumab prescribing information. Roche Pharmaceuticals, Nutley, NJ.

51. Hershberger R, Starling R, Eisen H et al. Daclizumab to prevent rejection after cardiac transplant. N Engl J Med 2005; 352(26):2705–13.

52. Vincenti F, Kirkman R, Light S et al. Interleukin-2 receptor blockade with daclizumab to prevent acute rejection in renal transplantation. N Engl J Med 1998; 338(3):161–5.

53. Nashan B, Light S, Hardie I et al. Reduction of acute renal allograft rejection by daclizumab. Transplantation 1999; 67(1):110–15.

54. Bumgardner G, Hardie I, Johnson R et al. Results of 3-year phase III clinical trials with daclizumab prophylaxis for prevention of acute rejection after renal transplantation. Transplantation 2001; 72(5):839–45.

55. Simulect® prescribing information. Novartis Pharmaceuticals, East Hanover, NJ.

56. Amlot P, Rawlings E, Fernando O et al. Prolonged action of a chimeric interleukin-2 receptor (CD25) monoclonal antibody used in cadaveric renal transplantation. Transplantation 1995; 60:748.
57. Kovarik J, Rawlings E, Sweny P et al. Prolonged immunosuppressive effect and minimal immunogenicity from chimeric (CD25) monoclonal antibody SDZ CHI 621 in renal transplantation. Transplant Proc 1996; 28:913.
58. Kahan B, Rajagopalan P, Hall M et al. Reduction of the occurrence of acute cellular rejection among renal allograft recipients treated with basiliximab, a chimeric anti-interleukin-2 receptor monoclonal antibody. Transplantation 1999; 67(2):276–84.
59. Nashan B, Moore R, Amiot P et al. Randomised trial of basiliximab versus placebo for control of acute cellular rejection in renal allograft recipients. Lancet 1997; 350:1193–8.
60. Ponticelli C, Yussim A, Cambi V et al. A randomized, double-blind trial of basiliximab immunoprophylaxis plus triple therapy in kidney transplant recipients. Transplantation 2001; 72(7):1261–7.
61. Brennan D, Daller J, Lake K et al. Rabbit antithymocyte globulin versus basiliximab in renal transplantation. N Engl J Med 2006; 355(19): 1967–77.

    R-ATG gives lower rejection rates at the cost of higher rates of infection.
62. Webster A, Playford G, Higgins G et al. Interleukin 2 receptor antagonists for renal transplant recipients: a meta-analysis of randomized trials. Transplantation 2004; 77(2):166–76.

    The effect of IL-2 receptor blockers on long-term results in kidney patients.
63. Sinclair N et al. Low-dose steroid therapy in cyclosporine-treated renal transplant recipients with well-functioning grafts. Can Med Assoc J 1992; 147(5):645–57.
64. Ahsan N, Hricik D, Matas A et al. Prednisone withdrawal in kidney transplant recipients on cyclosporine and mychophenolate mofetil, a prospective randomized study. Transplantation 1999; 68(12):1865–74.
65. Matas A, Kandaswamy R, Humar A et al. Long-term immunosuppression, without maintenance prednisone, after kidney transplantation. Ann Surg 2004; 240(3):510–17.
66. Matas A, Kandaswamy R, Gillingham K et al. Prednisone-free maintenance immunosuppression – A 5-year experience. Am J Transplant 2005; 5:2473–8.
67. Woodle ES et al. A randomized double blind placebo-controlled trial of early corticosteroid cessation versus chronic corticosteroids: four year results. Am Transplant Congr 2007; abstract 1704.
68. Rostaing L, Cantarovich D, Mourad G et al. Corticosteroid-free immunosuppression with tacrolimus, mycophenolate mofetil, and daclizumab induction in renal transplantation. Transplantation 2005; 79(7):807–14.
69. Vincenti F, Schena F, Paraskevas S et al. A randomized multicenter study of steroid avoidance, early steroid withdrawal or standard steroid therapy in kidney transplant recipients. Am J Transplant 2008; 8:307–16.

    A randomised, multicentre trial of 335 patients to no steroids, reduced steroids or standard steroids.
70. Oberbauer R, Segoloni G, Campistol J et al. Early cyclosporine withdrawal from a sirolimus-based regimen results in better renal allograft survival and renal function at 48 months after transplantation. Transplant Int 2005; 18:22–8.
71. Ekberg H, Grinyo J, Nashan B et al. Cyclosporine sparing with mycophenolate mofetil, daclizumab, and corticosteroids in renal allograft recipients: the CAESAR study. Am J Transplant 2007; 7: 560–570.
72. Ekberg H, Tedesco-Silva H, Demirbas A et al. Reduced exposure to calcineurin inhibitors in renal transplantation. N Engl J Med 2007; 357:2562–75.

    Two large studies aiming to examine the effect of reducing or eliminating CNI drugs from the immunosuppressive regimen.

6

The donor procedure

Keith P. Graetz
Nick Inston
Keith M. Rigg

Introduction

The majority of organs transplanted over the last 40 years have come from donation after brain death (DBD) donors, but over the last decade the number of suitable donors has declined. As a result the number of living donors, particularly kidney, has significantly increased and now accounts for 35% in the UK and over 50% in the USA. In addition there are now an increasing number of expanded criteria donors (ECDs) and donation after cardiac death (DCD) donors. The terminology used for the different types of deceased donors is summarised in Table 6.1.

The success of transplantation is in part determined by the quality of the donor procedure. Each donor type brings with it its own challenges and the donor procedure has the potential to result in an irretrievably damaged organ or one with a poorer outcome. Equal importance should be given to both the donor and the recipient operation, and although this has certainly been the case for living donation it has not been as true for deceased donation.

This chapter will first consider the general issues that underpin the donor procedure – legislation, organ retrieval, organ preservation and organ damage. It will then consider in more detail the practical aspects of all types of deceased and living donation.

General issues

Before the donor procedure can take place it is important that there is appropriate selection of the donor and that legal requirements have been met. Donor selection will be considered in more detail later in this chapter. The UK legal framework will be highlighted, but mention will be made of legislation elsewhere in the world.

Legislation

The Human Tissue Act 2004 (HT Act) gives a framework for the removal, storage and use of tissues and organs from the deceased, and the storage and use of tissues and organs from the living, for specified health-related purposes, which includes donation for transplantation. The Human Tissue Act 2004[1] covers England, Wales and Northern Ireland, whilst the Human Tissue (Scotland) Act 2006[2] covers Scotland. Consent (or authorisation in Scotland) is the fundamental principle that underpins the Act. In deceased donation the wishes of the deceased in life should take precedence after death. Their wishes may have been expressed in life by being registered on the Organ Donor Register or by letting a family member know. If the wishes of the deceased are not known then consent should be obtained from a nominated person or a person in a qualifying relationship. It is important that trained individuals, be

Table 6.1 • Terminology used in deceased donation

Preferred term	Other terms
Deceased donor	Cadaveric donor
Donation after brain death (DBD)	Heart-beating donor Brainstem-dead donor
Donation after cardiac death (DCD)	Non-heart-beating donor Asystolic donor
Extended criteria donor (ECD)	Marginal donor

they intensive care staff or transplant coordinators, make the approach to the family. Further details can be found in the Codes of Practice on Consent[3] and Donation of organs, tissues and cells for transplantation[4] produced by the Human Tissue Authority (HTA). Internationally there are different legal approaches. In North America, as in the UK, it is the consent of the individual that is key, although there have been a number of modifications to the legal framework to allow an increase in organ donors. In Europe many countries have a legal framework based around presumed consent or opting out, and this approach is currently being debated in the UK.[5]

Consent for the removal of organs from living donors in the UK is covered by the common consent law, although consent for the storage and use is covered by the HT Act. Under the HT Act, the HTA regulates all living donor transplantation. Independent Assessors (IAs) are senior professionals trained and accredited by the HTA who act as representatives of both the donor and the HTA. They assess donor–recipient pairs to ensure the requirements of the HT Act are met and then recommend whether or not HTA approval should be given. Based on the online report submitted by the IA, the HTA executive is able to give approval for all straightforward directed genetically or emotionally related living-donor transplants, whilst a panel of HTA members need to approve all complex cases.

Organ retrieval

There are significant differences in the way in which deceased and living organ retrieval are performed and perceived. In the UK deceased-donor organ retrieval has generally been an out-of-hours activity performed by trainee surgeons that is given little priority on an emergency operating list; whilst living-donor organ removal is a daytime elective activity performed by senior surgeons.

There are a number of logistical problems in providing the existing deceased-donor organ retrieval service in the UK which include:

- shortage of transplant surgeons (consultant and trainees) compounded by European Working Time Directive requirements;
- availability of surgeons because of other conflicting elective and emergency commitments;
- availability of theatres, theatre staff and anaesthetists in donor hospitals;
- ability to provide a retrieval team during daylight hours;
- ability to provide retrieval teams for simultaneous donors;
- other identified delays in the system.

Organ retrieval is a stressful procedure for teams involved and specific stressors identified include emotional, technical, environmental/organisational, communication and long hours (Lorna Marson, Edinburgh, personal communication). In general, organ retrieval is not popular amongst consultant surgeons and consequently training and supervision is usually delegated to other trainees. This trend needs to be reversed in order to improve the quality of organs retrieved and training initiatives include an annual UK Transplant Organ Procurement Workshop and a separate module in the new surgical curriculum.[5]

Working groups from UK Transplant and the British Transplantation Society have produced recommendations for a national organ retrieval service and these are included within the report from the Organ Donation Taskforce.[6] The chief principles of such a service are summarised in Box 6.1. These changes will come at a price, but if implemented will lead to a properly resourced and funded service that will benefit all involved (see Chapter 2). A pilot scheme has already been run in Scotland with good effect.[7]

Organ preservation

Preservation of organs is an important factor in ensuring viability of the organ and in optimising outcomes. Once an organ no longer has a supply of oxygenated blood, cell damage will occur due to a depletion of adenosine triphosphate (ATP) and a failure of the sodium–potassium pump. This leads to swelling of the cell and anaerobic metabolism with the accumulation of lactic acid. This cellular acidosis leads to lysosomal damage and subsequent cell

Box 6.1 • Principles of a national organ retrieval service

1. The team should be virtually self-sufficient and not require theatre, anaesthetic or surgical staff from the donor hospital
2. A team should be available 24 hours a day, not have elective commitments during their time on call and be able to attend the donor hospital within 3 hours where possible
3. The team should be able to respond appropriately if there is more than one donor on the same day
4. The ability to support retrieval of all adult and paediatric organs
5. The provision of opportunities for training
6. The retrieval and recipient services should be separated contractually and financially

death. More detailed descriptions of this process of cell damage can be found in previous editions of this volume[8,9] and elsewhere.[10] Historically static cold storage has been the method of choice, but this may be suboptimal for the increasing number of ECDs and DCD donors. Alternative techniques such as hypothermic machine perfusion and normothermic perfusion have been investigated as means of improving preservation and consequently outcomes.[10]

Static cold storage

Cold storage is the preservation technique of choice in the majority of abdominal organ transplant units. This requires rapid intravascular flushing with cold preservation fluid to promote cooling of the organ, washout of blood components and rapid equilibration of the fluid with the tissues. Cooling results in a 50% reduction of the metabolic rate for every 10 °C fall in temperature, so that at a storage temperature of between 0 and 4 °C the metabolic rate is less than 10%. However, hypothermia alone will not completely prevent cell swelling and damage and the subsequent reperfusion injury, hence the use of a preservation fluid. There are a wide range of preservation fluids whose key components are an osmotic agent or impermeant to prevent cell swelling, a buffer to counteract the effect of intracellular acidosis, and an antioxidant to prevent the formation of reactive oxygen species. Preservation fluids currently in clinical use include Marshall's or Hyperosmolar Citrate (Soltran®, Baxter Healthcare), University of Wisconsin (UW) solution (ViaSpan™, Bristol Myers Squibb), Histidine Tryptophan Ketoglutarate (HTK) solution (Custodial®, Kohler Medical Limited), EuroCollins (Fresenius) and Celsior® (SangStat Medical). A summary of the properties of each is detailed elsewhere.[10] Further solutions are in development, which may help to reduce the ischaemia–reperfusion injury, but also benefit the marginal or extended criteria donor. These include Polysol, IGL-1 and AQIX® RS-1.[10–12]

For deceased donors and kidney donation Marshall's solution is still the preferred preservation solution for DBD donors in the UK, although for DCD donors UW solution is increasingly being used. Elsewhere in the world UW has become the preferred solution.

A large international multicentre analysis of kidney preservation in over 90 000 deceased donors showed that when the cold ischaemic time was less than 18 hours, there was no significant difference in risk of failure for the type of preservation fluid (UW, Marshall's, HTK or EuroCollins) used. However, UW solution showed superior results when the cold ischaemic time was greater than 24 hours.[13]

UW solution is the preferred choice of preservation solution in liver, pancreas and intestinal donors throughout the world. Preservation solutions used in heart and lung donation are discussed in more detail in Chapter 11.

Hypothermic machine perfusion

Machine perfusion was developed and used widely in the 1970s as a means of storing and transporting kidneys. More recently it has come back into favour with the increasing use of ECDs and DCD donors. There are two machines currently in use, the LifePort (Organ Recovery Systems) and the RM3 Renal Preservation System (Waters Medical Systems). The LifePort is most commonly used in the UK and requires little supervision after set-up, but can only be used for one kidney at a time. The RM3 machine does require ongoing supervision but can perfuse two kidneys at once. The machines have become more compact and the potential advantages are that the incidence of delayed graft function (DGF) is reduced, there may be potential for real-time viability testing of the organ and there is the opportunity to add therapeutic agents to the perfusate. Many studies comparing machine perfusion to cold storage have been retrospective in nature.

A systematic and meta-analysis of the effectiveness of machine perfusion to cold storage demonstrated a 20% reduction in risk for DGF, but with no difference in 1-year graft survival. The need for studies of high methodological quality and sufficient size were recommended.[14]

In the multicentre Collaborative Transplant Study (CTS) analysis of kidney preservation, pulsatile machine perfusion was not superior to cold storage, although the numbers in the former group were small (n = 2200, or 2.6%).[13] There are currently two multicentre randomised controlled studies taking place in Europe comparing cold storage with the LifePort machine. One study based in Belgium, The Netherlands and Germany is looking at DBD donor kidneys, and the other based in the UK (the PPART study) is looking at DCD donor kidneys. Recruitment is complete and the findings are due to be reported soon.

Normothermic perfusion

The normothermic perfusion of abdominal organs using a cardiopulmonary bypass system should allow the successful preservation of ECDs as well as permitting viability testing prior to transplantation. This has already been well proven in animal models, although there are logistical issues of using this in humans. A number of centres in the USA and Spain have used veno-arterial extracorporeal membrane oxygenation (ECMO) after cardiac arrest in the DCD group of patients to increase the numbers of kidneys and livers for transplantation.[15,16] This technique restores the flow of warm oxygenated blood during the interval between death and organ procurement. This allows viability assessment and improves early function.

Preconditioning

One further evolving strategy for using ECDs and reducing the ischaemia–reperfusion injury is that of preconditioning, which can be physical or pharmacological. Physical techniques include ischaemic and thermal preconditioning, whilst pharmacological techniques include drugs, cytokines and gene transfer.

A randomised controlled trial of ischaemic preconditioning in liver transplantation improved post-transplant liver function, demonstrating that the technique protects against ischaemia–reperfusion injury.[17]

However, the use of pharmacological techniques raises ethical considerations as treatments are required in the donor that are not of direct benefit to the donor, but rather to the recipient.[18]

Organ damage

Organ damage can be multifactorial and includes pre-existing donor organ disease, surgical damage during retrieval or bench preparation, and warm or cold ischaemia.

Donor organ disease increases with age and in the kidney includes changes due to hypertension and diabetes, whilst in the liver changes due to fatty infiltration are common. With a shortage of deceased-donor organs marginal organs are increasingly being considered, whereas not so long ago they would not have been. In addition infection and malignancy can be transmitted in the donor organ, making quality and safety one of the three key challenges facing transplantation as described in a recent European Union Policy document.[19]

Surgical damage is a potentially avoidable event, which can often be salvaged, but adds to the complexity of the recipient operation and can lead to complications. The largest published study looked at over 9000 kidneys retrieved in the UK.[20] Repairable damage was recorded in 19% of cases and unsalvageable damage in 1%. One- and 3-year graft survival rates were not significantly different between repaired and undamaged organs. The lowest rates of damage were found in those retrieved by multiorgan retrieval teams performing more than 50 retrievals per year. This suggests that surgical training and experience are important and the implementation of a properly resourced national retrieval service should result in fewer surgically damaged organs. Further damage can also occur during benchwork of the organ and this has not been quantified.

The third cause of organ damage is ischaemia, which is classified as warm or cold ischaemia. There are different nomenclatures in use, but the following is suggested for universal use:[21]

- Asystolic warm period (or first warm ischaemic time), which is the time from cessation of circulation of oxygenated blood to the perfusion and cooling of the organ.
- Cold storage period (or cold ischaemic time), which is the time the organ is kept cold by either static cold storage or by hypothermic machine preservation.
- Anastomosis period (or second warm ischaemic time), which is the time taken from removal of the organ from cold storage to reperfusion.

During this time of vascular anastomosis the organ will slowly rewarm despite various possible manoeuvres to keep it cold.

A large international multinational analysis of kidney preservation showed that cold ischaemia times up to 18 hours were not detrimental to graft outcome, but that the risk of graft failure increased beyond that with increasing ischaemic times. Within the first 18-hour period the time of ischaemia had no significant effect upon graft survival.[13]

Similar data from the UK show that the cut-off time is 20 hours, after which there is an increased risk of kidney transplant failure (UK Transplant, unpublished data). In liver transplantation longer cold ischaemic times lead to poorer graft outcomes, although an analysis of the cut-off time has not yet been determined (**Fig. 6.1**). In DCD donors cold ischaemia times should be minimised and implantation performed as soon as possible.

Deceased donation

This section will consider the broad principles of donor identification and assessment, the diagnosis of death and donor management, before considering in more detail the surgical procedure. In addition particular reference will be made to the paediatric donor, the ECD and the DCD donor. Although organ retrieval from the deceased donor is a well-established procedure with an extensive literature, the level of evidence is generally low. A comprehensive review of publications on multiorgan donor/donation, retrieval technique and procurement was performed to look at levels of evidence and grades of recommendation.

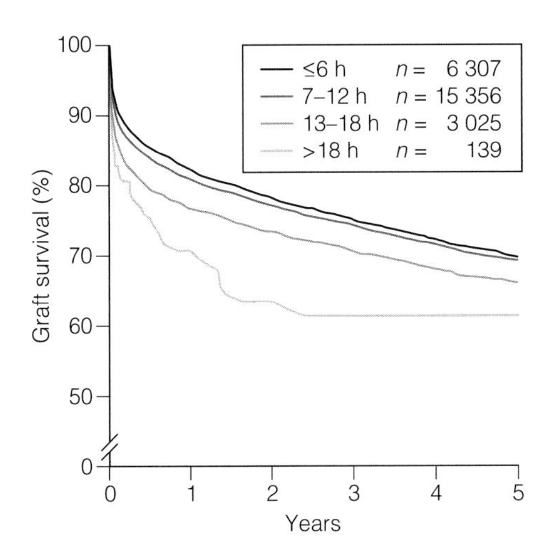

Figure 6.1 • Cold ischaemic times for first deceased-donor liver transplants, 1996–2005. Reproduced with permission of Prof. G. Opelz, Collaborative Transplant Study.

The conclusions of the review were that multi-organ donation need not compromise the outcome of individual organ transplants, dissection of abdominal organs is better performed following cold perfusion, abdominal organs should be rapidly removed en bloc and separation performed on the back table in the cold; single aortic cannulation is superior to combined aortic and portal cannulation and increased aortic perfusion improves outcome.[22]

Donor identification and assessment

The identification and selection of potential donors and the subsequent approach to their family for consent varies between regions and countries. In the UK senior critical care staff have traditionally approached relatives prior to involving the transplant coordinator. Recently, in an attempt to decrease the refusal rate from donor families the concept of collaborative requesting has been introduced. A randomised controlled trial of Assessment of Collaborative Requesting (ACRE) is currently under way and is due to be completed in 2009.

The donor transplant coordinators fulfil a vital role in the early assessment of the donor (Table 6.2) and in liaison with the individual specialists assess the donor with regards to organ suitability for transplantation.

All potential donors should be considered and discussed with the coordinator.[6] There are general contraindications for donation as well as organ-specific criteria (Table 6.3). With the shortage of organs and successful use of organs from ECDs, some of these contraindications may be relative and have been challenged.

Diagnosis of death

There is no legal definition of death in the UK, although a generally accepted definition is the irreversible loss

Table 6.2 • Role of the donor transplant coordinator

Standard		Aim
1	Education	Develop and deliver educational strategies and clinical initiatives in order to raise and maintain the profile and awareness of organ/tissue donation to relevant healthcare professionals
2	Donor identification and referral	To ensure identification and referral of all potential organ/tissue donors to the DTC at the earliest opportunity in order to maximise donor management and outcome
3	Donor management/assessment process	To maintain respect and care of the donor ensuring that optimal donor management is implemented To minimise the risk of transmission of infections/diseases from the donor ensure that all relevant donor information is obtained to enable recipient centres to make an informed decision regarding organ suitability
4	Care of the donor family	To ensure the donor family is given clear and accurate information about organ/tissue donation enabling them to make a fully informed decision
5	The organ/tissue referral process	To ensure that organs/tissues are allocated in line with national policies
6	Care of the donor during the theatre process	To maintain the dignity and respect of the donor whilst continuing the coordination process including donor management to optimise the suitability of the donor organs and tissues being removed
7	Healthcare professional follow-up	To ensure that all healthcare professionals are given adequate and appropriate support

Table 6.3 • Contraindications to organ donation

General absolute contraindications to donation	
Infective	HIV/AIDS or HIV high-risk category (even if HIV antibody negative) Severe sepsis Viral encephalitis Recipients of human growth hormone or risk of nvCreutzfeldt–Jacob disease
Malignant	Concurrent or recent malignancy (excluding CNS malignancy except glioblastoma multiforme) (Recent generally refers to solid-organ malignancy within the last 5 years)
Organ-specific contraindications	
Heart	Established cardiac disease Previous cardiac surgery Prolonged cardiac arrest Myocardial infarction Irreversible poor cardiac output Age >60 Abnormal chest X-ray (minor changes acceptable and occur in 27% of donors) Abnormal 12-lead ECG High inotrope requirement (above <10 μg/kg/min dopamine to maintain systolic blood pressure >90 mmHg if euvolaemic)
Lungs	Age >60 Established lung disease Bronchopneumonia Heavy smoker Pulmonary aspiration and tracheal colonisation with bacteria and fungi Parenchymal trauma Previous lung surgery Irreversible poor gas exchange Abnormal chest X-ray Inadequate gas exchange (P_aO_2 <50 kPa on 100% O_2 with 5 mmHg peak end-expiratory pressure)

Table 6.3 • (*cont.*) Contraindications to organ donation

Liver	Established liver disease Inborn error of metabolism (liver based) History of alcohol abuse (note: liver function tests can be deranged secondary to hypotension/asystole)
Kidneys	Age not strictly limited (although <75 years recommended) Established chronic kidney disease
Pancreas	Age <50 years No history significant cardiovascular disease No history alcohol abuse No major obesity No type II diabetes

of the capacity for consciousness, combined with irreversible loss of the capacity to breathe.[23] Details on the diagnosis of brainstem death and cardiac death can be found in the Code of Practice for the Diagnosis and confirmation of Death.[23]

Donor management

Physiology of brain death

The events preceding brain death, for example trauma or intracranial surgery, may impact on the quality of the retrieved organs. There are also significant physiological changes associated with brain death that affect each of the organs. This has been investigated both in controlled animal models and in human donors. Initially at the point of brain death the brainstem compression associated with coning results in hypertension and bradycardia, the Cushing reflex, which lasts about 15 minutes. This is followed by the autonomic storm in which massive catecholamine drive occurs leading to a transient hyperdynamic state,[24] although end-organ hypoperfusion can arise due to blood shunting. The pituitary hormones vasopressin and adrenocorticotropic hormone (ACTH) decline from 15 to 45 minutes, as do tri-iodothyronine, thyroxine and glucagon. This results in transient hypertension and also in hypoperfusion of organs. Insulin and lactate dehydrogenase show a moderate increase and diabetes insipidus is common.[25]

The effect on the cardiovascular physiology is to increase left and right ventricular end-diastolic pressure, increase cardiac output, and decrease systemic and pulmonary vascular resistance. This is transient and after 2–4 hours right and left ventricular systolic function deteriorates significantly, with the right ventricle being affected more than the left.[26]

The liver is also affected by hypotension and hypoperfusion; and human studies support the principle that brainstem death causes activation of inflammatory mediators within the liver and subsequently results in increased injury upon reperfusion.[27]

In the kidney cardiovascular events may result in tubular injury alone, although studies in human donor organs have implied that brain death produces both non-specific endothelial damage and also increases organ immunogenicity.[28] These effects may be reduced with proper donor management.[29]

Clinical management

The principles of donor management are to provide cardiovascular stability and maintain organ function. The first principle is adequate fluid management and although inotropes are often required their use during donor maintenance should be minimised.[30] The goal of management is to maintain blood pressure at a mean arterial pressure of >60–70 mmHg, with the inotrope of choice being vasopressin.[31] It acts as a vasopressor with no β-adrenergic response and also has an antidiuretic effect. This is also of benefit as diabetes insipidus (DI) occurs in two-thirds of brainstem-dead donors and further contributes to fluid imbalances, although 1-D-amino-8-D-arginine vasopressin (DDAVP) may be required if profound DI occurs. Other inotropes are limited in their use. High-dose adrenaline can be detrimental to donor organs as local vasoconstriction occurs, and the vasodilator effects of dobutamine may lead to undesirable hypotension and tachycardia.[32] The use of thyroxine replacement is controversial, although most guidelines support its use as it increases myocardial contractility due to β-adrenergic receptor-independent processes.[33] Glycaemic control is often deranged in the brain-dead donor and insulin infusion may be required to maintain the blood glucose within a normal range.

A summary of the medical management of the organ donor based on the Intensive Care Society guidelines for Adult Organ and Tissue Donation[34] is shown in Table 6.4.

Surgical management and operative procedure

In most circumstances the DBD donor is indistinguishable from other critical care patients and it is essential that the lead surgeon identify the patient prior to surgery. The preoperative check must also ensure that documentation of brainstem death tests, consent and donor blood group is complete and accurate. It is also important to thoroughly review the patient's notes and charts to confirm the history and review any events that may impact on the procedure and the subsequent outcome of retrieved organs, in particular assessment of the history, the laboratory investigations and radiological studies.

Table 6.4 • Summary of donor management (based on Intensive Care Society guidelines[34])

Clinical problem	Management
Endocrine	
Diabetes insipidus	Maintain sodium at 155 mmol/L with 5% dextrose Maintain urine output at 1–2 mL/kg/h with vasopressin If vasopressin fails to control diuresis, intermittent desmopressin (DDAVP) may occasionally be required
Hyperglycaemia	Insulin infusion to maintain plasma glucose 4–9 mmol/L
Hypothyroidism	Tri-iodothyronine (T3)
General stability and reduction of inflammatory response	Methylprednisolone
Cardiovascular support	
Decreased mean arterial pressure with increased cardiac output	Preload optimisation and vasopressor to increase afterload
Decreased mean arterial pressure with decreased cardiac output	Preload optimisation, vasopressor to increase afterload and inotrope to increase contractility
Increased mean arterial pressure with decreased cardiac output	Preload optimisation and vasodilator to increase afterload ± inotrope to increase contractility
Respiratory	
Maintenance of normocapnia ($P_aCO_2 \approx 5.0–5.5$ kPa)	Consider pressure control ventilation Modes that allow patient triggered ventilation are not appropriate
Ventilation with the lowest F_iO_2 to maintain $P_aO_2 > 10.0$ kPa Peak end-expiratory pressure >5 cmH_2O may reduce cardiac output and is rarely required High inspiratory pressures should be avoided	Very sensitive ventilatory triggers may allow cardiac cycle-induced pressure changes to trigger the ventilator. This may cause diagnostic confusion by giving the appearance of a spontaneous breath
Haematological support	
Haemoglobin concentration should be maintained over 9 g/dL	Blood transfusion
Deranged coagulation	Fresh frozen plasma and platelets

The procedure must be exercised in a professional manner. Currently the donor procedure is often performed out of hours with teams that may not work regularly together, in unfamiliar hospital environments. Separate teams for the thoracic and abdominal organ retrieval are normal and multiple teams including pancreas and renal teams may also be involved.

It is imperative that members of the team communicate effectively with each other and also with the host hospital staff including the anaesthetist, who may already have been involved with the patient prior to brain death. The procedure should be discussed and the sequence of events agreed prior to starting.

The operating surgeons should be experienced in retrieval surgery and in dealing with the challenges of difficult retrievals. Damage to retrieved organs is not uncommon, with studies estimating injuries to be up to 19%.[20] Donors who are obese, paediatric and with intra-abdominal trauma or previous surgery provide significant surgical challenges, calling on a high level of surgical expertise if injury is to be avoided.

Ideally the surgical equipment for the retrieval is packed at the transplant centre and the team should be self-sufficient with respect to instruments. A large insulated icebox, sterile ice and fluids are packed and travel with the team.

The patient is transferred to theatre in the same manner as a normal critical care transfer. The patient is placed supine on the operating table, prepared using aseptic technique and draped leaving the midline exposed from the suprasternal notch to the symphysis pubis. The drapes need go no wider than the midclavicular line bilaterally.

Techniques vary between surgeons and regions but often the procedure will begin with a full midline laparotomy with a thorough inspection of the intra-abdominal viscera to exclude pathology, particularly occult neoplasms (**Fig. 6.2**). Following this the intra-abdominal team proceed to mobilise the right colon and small bowel to expose the retroperitoneum. The dissection extends to the level of the superior mesenteric artery. This exposes the vena cava and renal veins and allows careful assessment of accessory renal vessels. The iliac arteries or lower aorta are then carefully encircled and controlled with heavy ties. These are used to secure the cannulas prior to cold perfusion. The ureters may be identified at this stage but care must be taken to ensure they are not devascularised if dissected. The portal vein may also be cannulated indirectly via the superior or inferior mesenteric veins, although this is not practised universally and may be unnecessary or detrimental if the pancreas is also retrieved.[35,36]

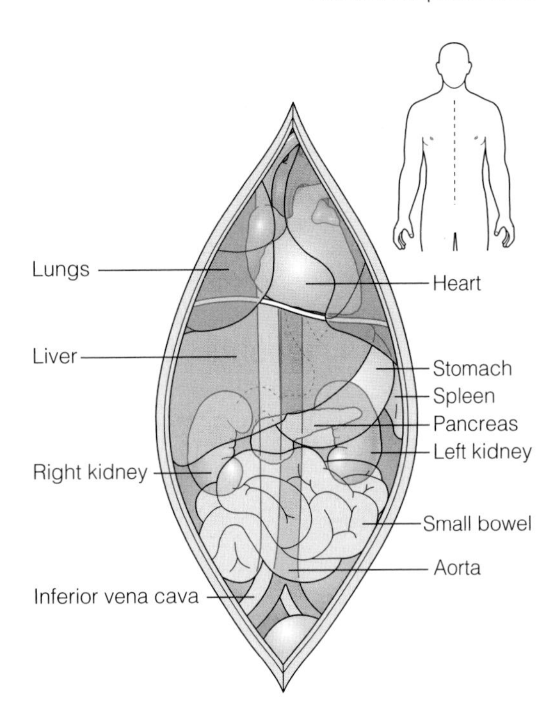

Figure 6.2 • Incision to expose abdominal and thoracic organs.

The duodenum is then mobilised medially in a plane anterior to the kidney to gain exposure to the portal structures. Palpation in the foramen of Winslow for an accessory right hepatic artery is initially performed. The common bile duct is encircled low down, the distal duct ligated and the proximal duct divided with sharp dissection. Care must be exercised to avoid devascularising the duct by dissecting its medial aspect. The portal vein also lies immediately posterior and injury can be avoided by remaining in a plane close to the bile duct. Most centres advocate washing out the gallbladder and duct.

Following division of the bile duct the arterial anatomy of the liver is determined. This is often easier following a midline thoracotomy. The gastroduodenal artery should be dissected and the common hepatic artery followed back to the coeliac trunk. Accessory left hepatic arteries arise from the left gastric artery and may be seen or palpated passing from the lesser curve of the stomach to the liver. The arterial anatomy of the liver and coeliac axis is notoriously variable and up to 40% of donors will have aberrant anatomy. Hence great care should be taken to identify anatomical variation.[37]

Preferences exist as to the timing of the dissection, with some surgeons preferring a long warm dissection to minimise the time taken postperfusion, whereas others favour dissection during the cold phase. Both kidneys should be exposed both on their anterior and posterior surfaces. If pancreas retrieval is to be performed, duodenal sterilisation may be requested via a nasogastric tube (see Chapter 9) and the pancreas is exposed via the lesser sac. Further arterial dissection might be necessary particularly if aberrant anatomy such as an accessory right hepatic artery arising from the superior mesenteric artery occurs. If a cardiothoracic team is present mobilisation of the great vessels in preparation for heart and lung removal is performed. The thoracic organs are retrieved individually or as a heart–lung bloc. Following thoracotomy the pericardium is opened and the pleura opened. The organs are inspected to determine suitability. The superior and inferior venae cavae (SVC and IVC) are dissected and the ascending aorta prepared for cross-clamp.

Upon agreement between teams, heparin is administered and primed perfusion catheters inserted and tied in place, thus isolating the intra-abdominal segment of the aorta (**Fig. 6.3**). On discussion between the teams and anaesthetist, cardioplegia is commenced through the aortic root and the aorta cross-clamped. If the lungs are being removed a perfusion catheter is placed in the pulmonary artery. The time is noted and the patient is exsanguinated, usually via the right atrium and/or the IVC. The left heart is decompressed by incising the right superior pulmonary vein. Simultaneously the infradiaphragmatic aorta is clamped or tied and cold perfusion fluid instilled rapidly via the intra-arterial and/or portal venous catheters. Topical cooling is performed with intra-abdominal sterile ice, ensuring ice is packed around the pancreas and kidneys.

If lung retrieval is performed anaesthetic input is required to maintain ventilation and to keep the lungs fully expanded, otherwise the anaesthetist can end ventilation. The heart is retrieved by division of the aorta, pulmonary artery, SVC and IVC, and pulmonary veins. If a bloc is taken the great vessels are divided and just before excision the endotracheal tube is pulled back and the trachea stapled with the lungs kept inflated. If the organs are to be used separately backbench preparation is performed, prior to packing in sterile bags and placing in a transport cooler packed with ice.

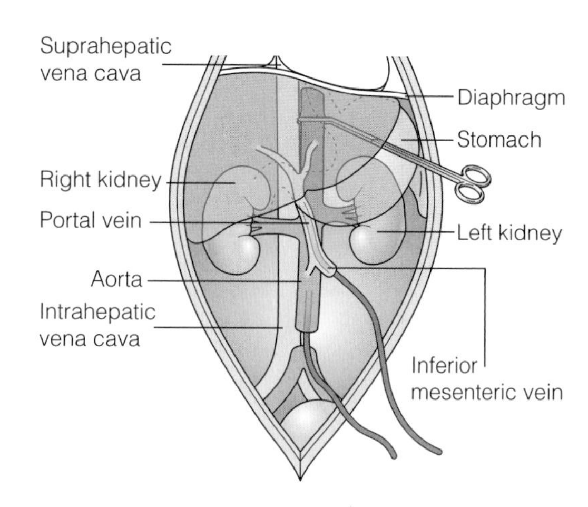

Figure 6.3 • Cannulation of abdominal aorta and inferior mesenteric vein in preparation for perfusion.

Following removal of the intrathoracic organs the coeliac trunk is removed as a patch from the anterior aorta and the splenic artery, left gastric and gastroduodenal arteries divided. This should be done in cooperation if separate liver and pancreas teams are present as conflict over the lengths of vessels must be avoided. The liver is then mobilised with a cuff of diaphragm around the suprahepatic IVC. The infrahepatic IVC is divided above the renal veins and the portal vein divided, usually at the level of the splenic vein, although concomitant pancreas retrieval may necessitate a slightly higher division.

Following hepatectomy the liver is flushed on the backbench prior to being packed in a bowl of fluid in sterile double bags. This is then transported in an insulated cool-box full of ice.

The procedure continues with the pancreatectomy, which requires division of the superior mesenteric artery. For further details and variation of pancreatic retrieval, see Chapter 9. As the renal artery origins may lie in close proximity to the superior mesenteric artery (SMA) the patch should not extend laterally into the aorta. Dissection of the pancreas from the posterior abdominal wall is best performed using a minimal touch technique using the spleen as a retractor to avoid pancreatic trauma. Division of the proximal and distal duodenum using a linear stapler completes the procedure (**Fig. 6.4**).

The kidneys are removed by dividing the IVC in the midline or more usefully by dividing the left renal vein laterally from the IVC, thus providing a caval conduit to lengthen the shorter right renal vein. The

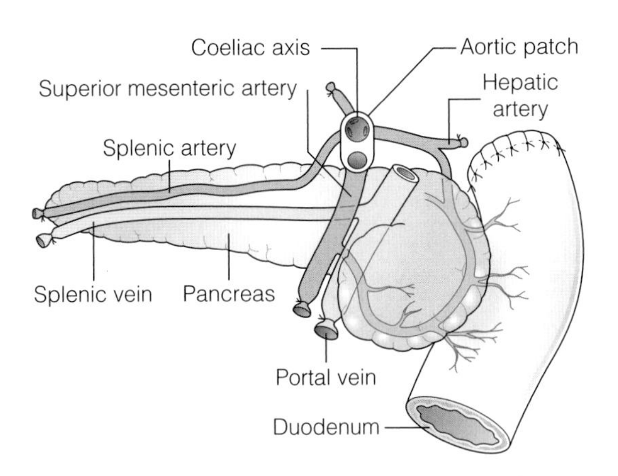

Figure 6.4 • Excision of pancreas with aortic patch.

left renal artery lies immediately posterior to the vein and is at risk of injury at this site. The veins are mobilised laterally and accessory vessels can be identified within the lumen of the IVC. The aorta is divided in the midline and accessory vessels identified, recognising that lower polar renal arteries can sometimes arise from the distal aorta or even common iliac artery. The ostia of all arteries should be taken on a patch of aorta that can be fashioned into appropriate Carrel patches by the recipient surgeon. Some centres advocate en bloc removal of the kidneys and ex situ division of the aorta and IVC (**Fig. 6.5**).

The organs are all flushed and packaged, and lymph nodes and spleen for tissue typing and full-length iliac vessels to be used as conduits if required are removed.

The operation is completed by careful suturing of the wounds, application of dressings and patient cleaning. It is not uncommon for relatives to want to see the donor following the procedure, and throughout the procedure dignity and respect are of paramount importance.

It is the responsibility of the lead surgeon to complete all documentation with careful description of anatomy and any injuries. An operation note should be completed and in some cases a summary for the coroner may be required.

The extended criteria donor

The definition of the ECD is wide ranging and varies according to the organ to be retrieved. Criteria may include age, organ-specific disease such as excess alcohol, generalised disease such as diabetes, cause of death such as a cerebrovascular accident, and donor cardiovascular instability.

With the limited number of organs available many units will consider donors that would previously have been declined, although it has been shown that ECD kidneys have poorer long-term outcome than both standard DBD and DCD kidneys.[38] This can be improved by careful selection and matching,[39] although many clinicians would not consider using organs that represent significant risk to the recipient if other options of management exist. This is particularly true with regard to kidneys as renal replacement therapy in the form of dialysis is a suitable alternative.[40] In patients with end-stage liver, heart or lung disease the risk/benefit balance shifts due to the lack of options remaining to enable survival.

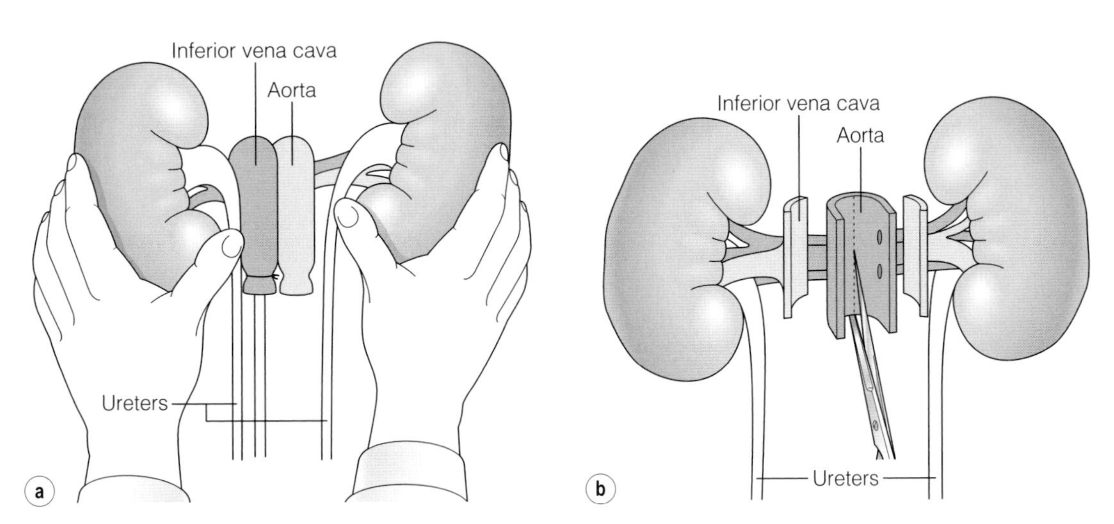

Figure 6.5 • **(a)** En bloc dissection of kidneys. **(b)** Separation of kidneys prior to excision.

The risks of using marginal donors are both organ specific and general. In considering liver transplants the risk of primary non-function is increased in patients with severe steatosis, whilst other aetiological factors are ill defined.[41,42]

Delayed graft function may occur in renal transplant recipients with the risk increased in marginal donors, although this is usually recoverable.[43]

Systemic risks from donors exist primarily due to the transmission of infection or malignancy. The risk of blood-borne viral diseases, particularly hepatitis B, C and human immunodeficiency virus (HIV), are always present when allogeneic tissue transfer is undertaken. The pre-donation screening and microbiological testing of donor blood attempts to identify at-risk donors.[44] The risks of transmitting malignancy are very low. Donors with a history of a non-central nervous system malignancy or a high-grade glioblastoma multiforme, coupled with a thorough laparotomy, will help to minimise this risk.

In the UK the ultimate responsibility is with the surgeon performing the transplant and the risk of disease transmission requires careful balancing with the recipient's overall clinical risk. As part of the recipient consent process it is always important to mention the risk of disease transmission, particularly if this risk is above the baseline.

The paediatric donor

The donor operation in larger paediatric donors is usually identical to that performed in adults. The small donor size provides surgical challenges and although the procedure follows the main principles as outlined above it should be performed by an experienced retrieval surgeon. Often multiple teams will be involved as most paediatric donors have few contraindications to donation and on size matching the organs represent a precious resource.

Surgical dissection progresses as described above; however, in smaller donors the kidneys should be retrieved en bloc as they may be used as a dual transplant.[45,46]

Paediatric donors are often emotionally testing for the teams involved, particularly the staff of the host hospital, and a high level of professionalism and sensitivity is required.

The DCD donor

Over the last 15 years, in an attempt to expand the number of available organs, DCD programmes have developed and have increased organ supply by over 20% in some units.[47] The accepted classification for DCD donors is the Maastricht criteria[48] (Box 6.2). The majority of DCD donors in the UK are category 3, although category 1 and 2 donors have been successfully used. Transplantation has been successful with kidneys, liver, lungs and pancreas from DCD donors.[49–51]

'Guidelines relating to solid-organ transplants from non-heart-beating donors', produced by the British Transplantation Society, cover all aspects of DCD donors, but of particular relevance to this chapter are the sections dealing with donor selection, immediate post-mortem management, the process of organ retrieval and preservation of organs.[52]

The techniques of procurement from DCD donors vary, although two main techniques exist. The first of these is similar to the open technique as described for the DBD donor. A rapid-access laparotomy is performed and the aortic bifurcation cannulated. The IVC is vented and the aorta is clamped either below the diaphragm or in the chest via a rapid thoracotomy. The alternative technique relies on the use of an intra-aortic double-balloon catheter which is placed via the femoral artery through a groin incision. Venous exsanguination is performed via the femoral vein.

Both techniques have their exponents and the principle in both techniques is to flush the organs in as short a time as possible and provide rapid cooling. The open technique has the advantages of allowing topical cooling to be applied and visual assessment of the quality of perfusion, although it does require an experienced surgeon to achieve timely organ perfusion without injury. The double-balloon catheter has the advantages of a minimal approach which can be performed in the emergency department in a short time and can be easily taught. Balloon malposition can be a cause for poor perfusion and it is recommended that they are checked using X-ray.[52] Intraperitoneal cooling can be performed by inserting a cannula into the peritoneal cavity at the same time.

Both techniques require significantly more fluid to fully flush the donor circulation and studies of DCD

Box 6.2 • The modified Maastricht criteria for DCD donors

These donors can be divided into categories based principally on work from the Maastricht group. This is important both for the logistics of retrieval and outcome following transplantation.

- **Category 1**: Dead on arrival at hospital
  For these individuals to be potential donors the moment of sudden death needs to have been witnessed and the time that it occurred documented, as well as pre-admission resuscitation.
- **Category 2**: Unsuccessful resuscitation
  These are individuals in whom cardiopulmonary resuscitation is commenced following collapse. These patients are usually in an Accident and Emergency Department, in which the interval of resuscitation and the efficiency of resuscitation has been well documented.
- **Category 3**: Awaiting cardiac arrest
  These are a group of patients for whom death is inevitable but they do not fulfil brainstem-dead criteria. These patients are cared for in many areas within hospitals but most commonly are identified in Neurosurgical Intensive Care Units, General Intensive Care Units, Coronary Care Units, Accident and Emergency Departments and Medical Wards.
- **Category 4**: Cardiac arrest in a brainstem-dead individual
  This is an individual in whom death has been diagnosed by brainstem criteria who then suffers an unexpected cardiac arrest. On some occasions these cases will be awaiting the arrival of an organ retrieval team.
- **Category 5**: Unexpected cardiac arrest in a patient in an ITU or critical care unit
  This has recently been suggested as an addition to the other four categories.

From: Koostra G, Daemoin J, Oomen A. Categories of non-heart beating donors. Transplant Proc 1995; 27:2893. With permission from The Transplantation Society/Elsevier. Also available at http://www.bts.org.uk/Forms/Guidelines_Nov_2004.pdf. British Transplantation Society. Guidelines relating to solid organ transplants from non-heart beating donors. London, 2004.

kidneys have also shown some benefit from the addition of fibrinolytics to the perfusion fluid.[53]

Living donation

This section will concentrate on the donor part of the live donor kidney transplant. Donor work-up, the operative approaches and associated morbidity and mortality will be covered. Live donation in other solid-organ systems will also be discussed briefly.

Live donor kidney transplantation

Donor selection

Assessment of patients with chronic kidney disease for dialysis and potential transplantation should start early and ideally be in place within 6 months of the patient requiring dialysis. Discussion of the option of living-donor transplantation should be raised with the patient and their family at this stage. Patients need to be aware of its benefits, which include the ability to plan the time of surgery for the recipient, reduced waiting times, and improved graft and patient survival. The living donor coordinator has a key role to play in these early discussions.

In the UK, evidence-based and best-practice donor work-up is driven by the 'United Kingdom guidelines for living donor kidney transplantation' produced by the British Transplantation Society and Renal Association (BTS/RA).[54]

In the USA, the Organ Donor Procurement and Transplantation Network has also produced some less detailed guidelines.[55] The details of donor work-up are dealt with in more detail in the guidelines above, but should include the following aspects.

Assessment of donor premorbid health

All efforts should be made to minimise the risk to the donor and this starts with an assessment of their comorbidity and current health status. Significant comorbidity should halt the work-up process at this point.

Uncontrolled or multiagent controlled hypertension and diabetes mellitus should be considered absolute contraindications because of the risk of deterioration of donor kidney function following nephrectomy. Up to 25% of type II diabetic patients will develop overt diabetic nephropathy within 5 years of diagnosis.[56] The significance of uncontrolled hypertension lies with the increased perioperative anaesthetic risk, whereas hypertension as a risk factor for future kidney function is less clear. The argument follows the theory that uninephrectomy results in hyperfiltration which will progress to glomerulopathy and albuminuria. This is aside from the increased risk from donor vascular disease. Interestingly, one study followed 18 hypertensive donors and eight became normotensive following uninephrectomy.[57]

A history of malignancy and infection does not completely preclude donation but every effort should be made to prevent transmission of malignant cells or microbiological agents to the recipient. Screening should be fully completed before the donor procedure goes ahead.

Obesity is not a contraindication to donation but can lead to increases in the technical difficulty of surgery. There is also a greater incidence of postoperative complications, particularly atelectasis and pneumonia, venous thromboembolic disease and wound infections. This leads to longer length of stay and increased recovery times. Most units in the UK set an upper body mass index (BMI) limit around 30 kg/m^2.

All donors should have routine haematology and biochemistry tests and an electrocardiogram performed. Urine samples should be tested for blood, with any evidence of albuminuria or haematuria preventing donation proceeding until fully investigated. Others tests should be done at the discretion of the investigating clinician and the potential donor made aware that occasionally unrelated medical conditions may be found that warrant further investigation and referral.

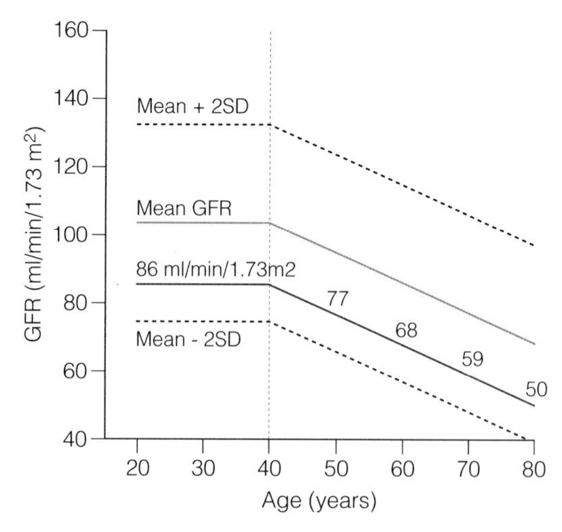

Figure 6.6 • Graphic representation of declining GFR with increasing age of the patient. The solid line demonstrates values for GFR above which donor nephrectomy should be achievable with safety and avoiding the need for dialysis in the donor at a later date assuming a natural decline in kidney function with time. This is based on data from Blake GM, Grewal GS. Reference data for 51Cr-EDTA measurements of GFR derived from live kidney donors. Nucl Med Commun 2005; 26:61–5.[58] It is also included in the BTS/RA Guidelines for Living Donor Kidney Transplantation.[54]

Assessment of donor kidney function

Kidney function should be assessed by both biochemical tests and isotope glomerular filtration studies. Serum creatinine is a poor marker of kidney function although simple to measure in a clinic setting. Its variability is multifactorial and dependent on patient medications, absolute body muscle mass and recent dietary intake. It is also necessary for a significant number of nephrons and functioning kidney mass to have been lost before it starts to numerically deteriorate. Glomerular filtration rate (GFR) is far more accurate and reproducible.[57] Either a DMSA or MAG 3 scan should be performed as calculated GFR values have their limitations for the same reasons that serum creatinine does. The BTS/RA guidelines suggest that donation should not proceed if the predicted GFR will fall to 37 mL/min by the age of 80[58] (**Fig. 6.6**). It is essential that kidney function is split equally between both kidneys so that the donor has sufficient nephron mass remaining following uninephrectomy. If this is not the case then the donor kidney should be the one with slightly less function.

Assessment of kidney anatomy

Firstly, it is important to determine that there are two kidneys that have normal structural anatomy, for example the absence of multiple cysts or potentially malignant lesions. Secondly, it is important to know the vascular anatomy. This enables surgical planning to be done preoperatively. Historically this has been done with a combination of ultrasound, intravenous urography and selective renal angiography. The use of computed tomography (CT) and magnetic resonance imaging (MRI) to assess the donor is now the norm as this allows all aspects of donor kidney anatomy to be assessed in one investigation. There are proponents of either and the choice is dependent on local expertise (**Figs 6.7** and **6.8**).

Multiple renal arteries or veins make retrieval and re-implantation more difficult but do not increase the risk of graft thrombosis.[59] Increased rates of urinary leaks following transplantation do occur, particularly when there are small accessory lower pole arteries,[60] and care needs to be taken not to 'strip' the ureter at the time of retrieval. Duplicate vena caval anatomy and retroaortic left renal veins anecdotally are more difficult to retrieve due to the fragility of the vein wall. Right nephrectomy should be considered if there are multiple arteries to the left kidney and only a single vessel to the right.

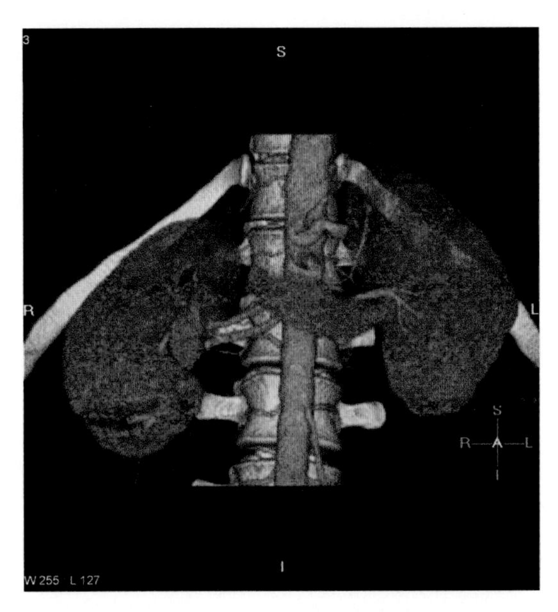

Figure 6.7 • Three-dimensional CT reconstruction of kidney anatomy demonstrating two equal-sized arteries to the right kidney and a main stem and smaller upper pole artery to the left kidney. Courtesy of Dr Richard O'Neill, Consultant Radiologist, Nottingham University Hospitals.

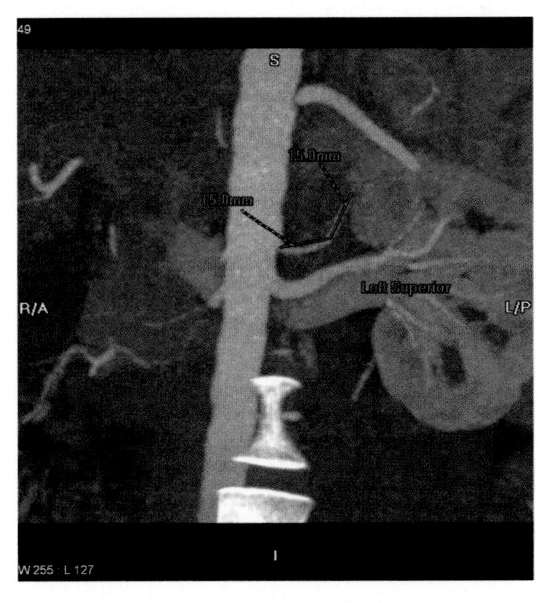

Figure 6.8 • Kidney anatomy mapping using CT angiography to enable preoperative surgical planning. Courtesy of Dr Richard O'Neill, Consultant Radiologist, Nottingham University Hospitals.

Perioperative donor management

All procedures are performed under general anaesthesia with a cuffed endotracheal tube. The decision on insertion of central venous catheters and arterial lines is for the anaesthetist but a urinary catheter should be used as standard. Liberal use of intravenous fluids should be used in the donor to ensure good kidney perfusion throughout the operative procedure. Some units also advocate the use of furosemide and mannitol to force a urinary diuresis immediately prior to removal of the kidney. Mannitol is also a free-radical scavenger, which may help to minimise any subsequent reperfusion injury. Postoperative analgesia is best achieved with either a patient-controlled analgesia (PCA) device or an epidural infusion. Full dietary intake and a return to full mobility can be encouraged from the first postoperative day. The time to discharge is very dependent on the patient and the operative procedure that they have undergone, but should be within 2–5 days.

Operative approaches

Laparoscopic surgery has revolutionised live donor transplantation in that it has made the operation a more acceptable option for the donor, when compared to the traditional flank approach. Improved cosmesis and faster recovery times are the main advantages to the donor patient. Hospital stay can be reduced by several days, with discharge being possible as early as the second postoperative day. However, it is a technically demanding procedure and as with other laparoscopic surgical procedures has a learning curve. Although laparoscopic donor nephrectomy is becoming the procedure of choice (either fully laparoscopic or hand assisted), there are others who favour an open approach through a mini-incision. In deciding which surgical approach to consider it is important to keep the following principles in mind: the risk of life-threatening complications and other morbidity in the donor must be minimised, and the integrity and function of the donated kidney must be preserved.[60]

A single blind, randomised controlled clinical trial of 100 patients comparing laparoscopic with mini-incision open donor nephrectomy showed that the laparoscopic approach resulted in improved quality of life in the first year with equal safety and graft function.[61]

A meta-analysis of laparoscopic vs. open donor nephrectomy showed that although the open technique may be associated with shorter operative and warm ischaemic times, the laparoscopic technique may result in shorter hospital stay and faster return to work, without compromising graft function.[62]

The procedure transplant units offer will very much depend on the surgical expertise available, but patients should be able to make an informed choice as to which procedure they prefer. Realistically, most patients will choose to have their procedure locally by the team they know, but there may be some who choose to travel to another unit for a procedure type that is not offered in their home unit.

The various operative approaches are described below. Unless otherwise stated operative procedures described relate to the left kidney since there is an emphasis on laparoscopic left nephrectomy.

Full laparoscopic donor nephrectomy

Transperitoneal

The donor is placed in a full or semi-lateral position on the operating table with the side of the kidney to be removed upwards. Generally this is performed through three ports with a 5- to 6-cm Pfannansteil incision being used for retrieval of the kidney and insertion of other operative instruments (**Fig. 6.9**). A pneumoperitoneum is maintained with a CO_2 pressure of 8–10 mmHg and a high gas flow rate.

Dissection commences with mobilisation of the left colon, splenic flexure and spleen. This should be done from the upper pole of the spleen to the pelvic brim. Dissection of the mesocolon should be taken back to the right side of the aorta. This exposes Gerota's fascia. The gonadal vein should be identified and cleaned to the level of the left renal vein and the pelvic brim below. This is used as a marker for the ureter, which can then be dissected free from the psoas fascia posteriorly with the gonadal vein attached to it. Once completed, the lower pole of the kidney should be mobilised with this continued superiorly behind the kidney. It is important to leave the lienorenal ligament intact to prevent twisting of the kidney around the hilum. The renal artery is dissected back to the aorta and the left renal vein taken to the right side of the aorta, maximising its length for re-implantation. Branches of these, most notably the gonadal and adrenal veins, are clipped or sealed using one of various methods. Care should be taken if clips are used as these can be easily displaced and can also obstruct the stapling device used later for division of the renal vein. Mobilisation is completed, leaving the kidney attached only by its vascular pedicle and ureter. An endoscopic stapling device or endoscopic metal clips are then used to seal and divide the ureter, renal artery and renal vein sequentially. The kidney is retrieved in a retrieval bag via the Pfannansteil incision and immediately cold perfused on the back operating table by another surgeon.

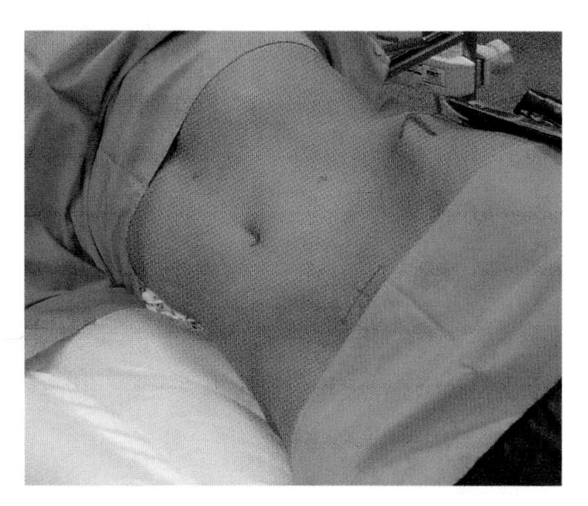

Figure 6.9 • Patient positioning in a semi-lateral (45-degree) set-up with no table break prior to left donor nephrectomy. Surgical incisions have been marked preoperatively with indelible ink. Courtesy of Mr Magdi Shehata, Consultant Surgeon, Nottingham University Hospitals.

The right kidney can be retrieved in a similar manner. It is usually necessary to use an extra 5-mm right subcostal laparoscopic port for retraction of the liver. Mobilisation of the vascular pedicle is more difficult as a consequence of the proximity of the vena cava. In order to maintain right renal vein length a narrow cuff of IVC should be taken when dividing the venous structures with an endovascular stapling device. This will leave a scalloped appearance to the vena cava. The right renal artery should be approached by mobilising the vena cava medially, exposing its posterolateral aspect with care. This again is an attempt to maximise its length for subsequent re-implantation. The right gonadal vein can and should be preserved to prevent future complications.

Retroperitoneal

This approach has been used more commonly for right donor nephrectomy procedures but is practised by very few units. The advantages are related to the avoidance of the peritoneal cavity, hence minimising intraoperative viscus injury and postoperative adhesions. Dissection is performed in a similar manner to open surgery (see later text).

Hand-assisted laparoscopic nephrectomy (HALN)

Proponents of hand-assisted techniques justify use of this technique because of the help that the hand provides with retraction whilst also allowing rapid control of intraoperative bleeding by direct pressure if needed.

The patient is positioned in a full or semi-lateral position with the side of the kidney to be removed upwards. It is not essential to break the table. Surgery is performed with a hand port device placed in either a left upper abdominal transverse incision or a supra-umbilical midline incision. Two further laparoscopic ports are inserted in the left iliac fossa. It is necessary to use an intra-abdominal insufflation pressure of at least 15 mmHg to allow space for the operative hand.

The surgical technique is otherwise as described for full laparoscopic donor nephrectomy. The operative hand, via the hand port wound, retrieves the kidney.

Open donor nephrectomy

Traditional

This is usually performed through a loin incision with the patient in a full lateral position and a break on the table. The 11th rib can be excised to aid access. Retroperitoneal dissection exposes the left kidney and its hilum followed by the aorta.

Mini-incision

A short posterior transcostal or supracostal incision or anterior incision is utilised and the kidney mobilised by either a trans- or a retroperitoneal dissection. The procedure is as already described, with advocates of it claiming the benefits are related to the small incision size and avoidance of muscle tissue destruction, particularly the latissimus dorsi muscle. Operative loupes and a head torch are essential aids in performing the procedure.

Procedure-related morbidity and mortality

As part of the assessment and consent process for live-donor nephrectomy the potential donor has to understand that this is a procedure that offers no direct medical benefit to themselves. In addition to operative morbidity there is a small risk of death which needs to be discussed with the patient openly as part of the consent process, with details being documented accordingly. Registry data from the USA for donors in the 1990s gives a mortality incidence of 2 in 10 828.[63]

UK registry data show no reported donor deaths within the first 30 postoperative days in 2509 donor procedures (601 laparoscopic, 1800 open and 108 not specified) performed between 2000 and 2006. The risk of major morbidity was 4.9% overall (laparoscopic = 4.5%, open = 5.1%) and of any morbidity was 14.3% (laparoscopic = 10.3%, open = 15.7%).[64]

Currently, the BTS/RA guidelines[54] give a documented mortality of 1 in 3000 for donor nephrectomy for all surgical approaches. There is no difference in mortality rates between open and laparoscopic donor procedures.[63,65] Life expectancy for live-donor patients is probably greater[66] than the general population as a whole because of the selection bias when choosing a donor, in that only the fittest patients are likely to become donors.

Live-donor kidney transplantation should be considered a failure if one patient receives a kidney transplant at the expense of another who enters a dialysis programme. The incidence of renal failure leading to transplantation in the US general population is 0.03%, which compares to 0.04% of living donors.[67] What is more difficult to follow is the number of donors who have reached end-stage renal disease but who have not made it to a transplant waiting list. The United Network for Organ Sharing (UNOS) later reported live-donor data[68] listing 148 donors (0.22% of all donors) who had reached end-stage disease and had been listed for kidney transplantation themselves between January 1996 and March 2007. Of these, 5 (0.01% of all donors) had disease that would specifically affect one kidney only, such as malignancy. It is this figure that should be quoted when giving patients a risk of them needing dialysis as a consequence of uninephrectomy; 111 patients had primary renal disease that affected both kidneys and that would potentially have become a problem even without donor nephrectomy having been performed. It can thus be seen that the risk of requiring dialysis following uninephrectomy is extremely low. Twenty-six patients had no documented diagnosis and six patients who had been listed were listed for a second transplant.

A meta-analysis of 44 studies showed an overall complication rate of 13.7% for fully laparoscopic and HALN combined, compared with 16.4% for the open nephrectomy group.[62] Major complications are reported as pneumothorax, vascular injury, intestinal injury and splenic injury. A meta-analysis of 31 studies showed a conversion rate from laparoscopic to open procedures of 1.8%.[62] The commonest reason for conversion is bleeding due to injury to the aorta, vena cava or renal vessels, which may arise as a result of vascular injury from misfiring or malalignment of the endoscopic stapling device. This may reflect the learning curve for the procedure and is an issue that may be resolved with increasing expertise. Minor complications include urinary tract infection, pneumonia, wound infection, port site or wound hernia/loin 'bulge' and slow resolution of bowel function/constipation. The latter is under-reported and in the experience of the authors extremely common, often requiring the use of oral aperients and rectal suppositories.

Following nephrectomy creatinine clearance is reduced due to the reduction in nephron mass. If the donor is chosen correctly then the functioning nephron mass will be halved. As a result the serum creatinine in the donor will rise following surgery by as much as 50% on the first postoperative day. As the single remaining kidney hyperfiltrates over subsequent days this value will fall towards a baseline level by 1 month that should still be in the normal reference range for an individual with two kidneys assuming donor selection has been appropriate. As has already been discussed the impact of the resultant reduction in functioning nephrons on long-term outcomes is less clear but would seem to have very little impact on a patient's potential progression towards kidney failure.

Follow-up of live-donor patients

It is clear that any patient who donates a kidney will have been deemed medically fit to have undergone the procedure. Consequently, long-term morbidity is relatively uncommon, but it is still recommended that these patients are followed up.[54] This need not involve more than an annual nurse-led health check but should include blood pressure monitoring, serum biochemistry and urinalysis for protein and blood as routine. These are merely surrogate markers of kidney function but changes may herald the need for early intervention.

Live donation in other solid-organ systems

Liver

The need for donor livers is rapidly increasing because of the increase in incidence of hepatitis C, alcohol-related liver disease and non-alcoholic steatohepatitis (NASH). At the same time there are fewer deceased-donor livers and an increasing number of patients dying whilst on the waiting list. With improvements in liver surgery and the management of complications, most notably bleeding from the liver parenchyma, live-donor liver transplantation has seen increasing acceptance internationally as a treatment for end-stage liver disease. It is practised widely in the USA and Japan and a programme has just started in the UK.

An Interventional Procedure Guide has been produced by the National Institute for Health and Clinical Excellence,[69] which supports living-donor liver transplantation in suitable patients.

There are no specific guidelines for donor medical assessment, although this should include a full study of the likelihood of a donor progressing to end-stage liver disease themselves, which may be accelerated by partial hepatectomy. Patents should therefore be screened for hepatitis B and C, as well as all the congenital forms of liver disease, including cystic fibrosis, Wilson's disease and haemachromatosis. A liver biopsy may also be performed to assess liver architecture, as steatosis is an increasing problem for the population as a whole. One study showed that 27 of 70 donors were prevented from donating as a consequence of their liver biopsy result for this very reason.[70] Other absolute and relative contraindications are similar to living kidney donation. The aim is to minimise both operative and long-term mortality and morbidity for the patient whilst offering maximum benefit to a potential transplant recipient. Any morbidity and mortality is devastating for all concerned but may also have a potential effect on the transplantation programme, as was demonstrated in 2002 when a live liver donor died following surgery, resulting in a 40% reduction in live-donor liver transplantation rates in the USA.[71] In view of the risks involved a donor advocate team also assesses each donor, this team consisting of a consultant physician, consultant psychiatrist, social worker and transplant coordinator.

Initially in adult-to-adult living liver donation, left liver lobes were resected for transplantation but it soon became apparent that these caused 'small for size syndrome',[72] with inadequate liver reserve available for the size of the recipient even with the hypertrophy that followed in the transplanted liver remnant. Right hepatectomy became the usual procedure performed as a consequence, except in adult to child where left hepatectomy remains sufficient.

There are significant mortality risks, which are summarised as 0.23–0.5% for right lobe donation, 0.05–0.21% for left lobe donation and 0.2% overall. Median donor morbidity is 16%, with bile leaks contributing significantly.[69] By 6 months the majority of donors have returned to normal activities and the liver has regenerated to 89% of its original size.[69]

Liver anatomy is very variable and surgical planning should be thorough to avoid injury to biliary anatomy and vascular structures in either the donor liver segment or remnant hepatic segments. This will minimise morbidity following surgery, particularly bile leaks from direct injury or from ischaemia. Combinations of angiography, high-resolution contrast-enhanced CT and MRI will provide this information preoperatively. Most surgeons also use intraoperative ultrasound imaging to confirm their findings before commencing hepatic transection.

Pancreas

Pancreas transplantation is the best modality of treatment for patients with 'brittle' diabetes where unheralded hypoglycaemia can be disastrous and also for patients with end-stage kidney disease as a consequence of their diabetes. Diabetic kidney transplant recipients have improved graft and patient survival rates when simultaneous pancreas and kidney (SPK) transplants are performed.

The use of pancreatic tissue from live donors is, however, not performed frequently because of the potential morbidity that the donor is exposed to. The advent of laparoscopic and hand-assisted distal pancreatectomy has improved donor rates because of the avoidance of a long, painful bilateral subtotal incision. However, major morbidity is still a problem. In 20 live donors from the University of Minneapolis, four required splenectomy and another two needed re-operation for intra-abdominal abscess and a pseudocyst.[73] The same group also reported that 10 out of 67 donors who they were able to follow up long term had abnormal HbA1C levels, with three of these requiring insulin.[74] With the benefits of SPK transplantation being well documented there is a vogue in a few centres to perform simultaneous laparoscopic live-donor left nephrectomy followed by distal pancreatectomy.

Lung

Living-donor lung transplantation is performed in a limited number of centres in the USA, Japan, some parts of Europe and the UK.

An Interventional Procedure Guide has been produced by the National Institute for Health and Clinical Excellence,[75] which supports living-donor lung transplantation in selected patients who would otherwise die.

In order to transplant sufficient lung parenchyma to allow gaseous exchange, two lung lobes need to be implanted. This clearly has to be done without jeopardising donor morbidity and mortality and will therefore usually necessitate two donors each donating a lobe. Whilst donor morbidity should be no greater than that of a patient undergoing lobectomy for another clinical reason, this will expose two donors to this risk and should therefore not be undertaken lightly. The exception to this is the paediatric recipient who may survive on one lobe. The most common indications for living-donor lung transplantation are cystic fibrosis and pulmonary hypertension.

Because of the rarity of this procedure no mortality has been reported in donor patients. Morbidity occurs in 20%,[75] with the commonest cause being the need for prolonged tube thoracostomy (>14 days) for persistent air leaks or drainage.

Small bowel

Living-donor small-bowel transplantation is still a subspeciality, very much in its infancy. This is the result of the limited number of patients who fulfil criteria for small-bowel transplantation and have failed to be managed on total parenteral nutrition. Only 32 cases had been reported to the Intestinal Transplant Registry by 2003.[76]

The donor procedure is safe and amounts to no more than a small-bowel resection and primary anastomosis. Mortality estimates are difficult due to the limited numbers available. Care needs to be taken to leave the donor with sufficient small bowel such that they are not left with 'short-bowel syndrome' themselves. Estimates of small-bowel length can be made with the aid of preoperative imaging. Return to full function should be rapid, but is occasionally complicated by self-limiting diarrhoea that should resolve without intervention.

Key points

- Equal importance should be given to both the donor and the recipient operation, and this needs to be the case for deceased donation as it has been for living donation. Effective means of delivering training to gain and maintain surgical expertise in the light of reduced working hours need to be developed.
- The number of DBD donors has been decreasing and as a consequence there have been more ECD and DCD donors, which are surgically more challenging. It is hoped that the implementation of the recommendations of the Organ Donor Taskforce in the UK will lead to the predicted 50% increase in deceased-donor kidneys (see Chapter 2).
- The increases in ECD and DCD donors have also prompted developments in preservation techniques and fluids, which will help to minimise organ damage and improve outcomes.
- Living-donor transplantation has expanded, with most of the increase seen in kidneys, although living liver programmes are developing. The expansion in live kidney donation has been accompanied by changes in surgical approaches, most notably laparoscopic donor nephrectomy, which has proven to be safe and effective.

References

1. Human Tissue Act 2004. Available from: http://www.opsi.gov.uk/acts/acts2004/ukpga_20040030_en_1.
2. Human Tissue (Scotland) Act 2006. Available from: http://www.opsi.gov.uk/legislation/scotland/acts2006/asp_20060004_en_1.
3. Human Tissue Authority. Code of Practice 1 – Consent. Available from: http://www.hta.gov.uk/guidance/codes_of_practice.cfm.
4. Human Tissue Authority. Code of Practice 2 – Donation of organs, tissues and cells for transplantation. Available from: http://www.hta.gov.uk/guidance/codes_of_practice.cfm.
5. Intercollegiate Curriculum Surgical Programme. The Intercollegiate Surgical Curriculum – General surgery syllabus. August 2007. Available at: http://www.iscp.ac.uk/Documents/Syllabus_GS.pdf.
6. Organ Donation Taskforce. Organs for transplants: a report from the Organ Donation Taskforce. London: Department of Health, 2008. Available at: http://www.dh.gov.uk/en/Publicationsandstatistics/Publications/PublicationsPolicyAndGuidance/DH_082122.
7. Mayes RD, Waite LR, McKeown D. Effect of 1 year of senior critical care support to an organ retrieval team. Presented as abstract, British Transplantation Society, April 2008.
8. Attia MS, Prasad KR, Bellamy MC et al. Organ retrieval – fluids and techniques. In: Forsythe J (ed.) Transplantation surgery: current dilemmas, 2nd edn. London: WB Saunders, 2001; pp. 25–64.
9. Marson LP. The donor procedure. In: Forsythe J (ed.) Transplantation (Companion to Specialist Surgical Practice), 3rd edn. London: Elsevier Saunders, 2005; pp. 119–38.
10. Maathuis MJ, Leuvenink HGD, Ploeg RJ. Perspectives in organ preservation. Transplantation 2007; 83:1289–98.
11. Bessems M, Doorschodt BM, Albers PM et al. Wash-out of the non-heart-beating donor liver: a matter of flush solution and temperature? Liver Int 2006; 26:880–8.
12. Kay MD, Hosgood SA, Harper SJF et al. Static normothermic preservation of renal allografts using a novel nonphosphate buffered preservation solution. Transplant Int 2007; 20:88–92.
13. Opelz G, Döhler B. Multicentre analysis of kidney preservation. Transplantation 2007; 83:247–53.
   This multicentre analysis studied 91 674 kidney transplants from Europe, North America and Australia on the Collaborative Transplant Database. The effects of kidney preservation methods and ischaemic times were analysed and there a number of key significant results.
14. Wight JP, Chilcott JB, Holmes MW. Pulsatile machine perfusion vs. cold storage of kidneys for transplantation: a rapid and systematic review. Clin Transplant 2003; 17:293–307.
15. Magliocca JF, Magee JC, Rowe SA et al. Extracorporeal support for organ donation after cardiac death effectively expands the donor pool. J Trauma 2005; 58:1095–102.
16. Fondevila C, Hessheimer AJ, Ruiz A et al. Liver transplant using donors after unexpected cardiac death: novel preservation protocol and acceptance criteria. Am J Transplant 2007; 7:1849–55.
17. Amador A, Grande L, Marti J et al. Ischemic preconditioning in deceased donor liver transplantation: a prospective randomized clinical trial. Am J Transplant 2007; 7:2180–9.

18. McNally SJ, Harrison EM, Wigmore SJ. Ethical considerations in the application of preconditioning to solid organ transplantation. J Med Ethics 2005; 31:631–4.
19. European Commission. Organ donation and transplantation: policy actions at EU level. Available at: http://ec.europa.eu/health/ph_threats/human_substance/oc_organs/oc_organs_en.htm.
20. Wigmore SJ, Seeney FM, Pleass HCC et al. Kidney damage during organ retrieval: data from UK National Transplant Database. Lancet 1999; 354:1143–6.
21. Wells A, Watson C, Jamieson N et al. What time is it? A suggestion for unambiguous nomenclature in transplantation. Am J Transplant 2007; 7:1315–16.
22. Brockmann JG, Vaidya A, Reddy S et al. Retrieval of abdominal organs for transplantation. Br J Surg 2006; 93:133–46.
23. Department of Health. A code of practice for the diagnosis of brainstem death. London, 1988. Available at: http://www.dh.gov.uk/en/Publicationsandstatistics/Publications/PublicationsPolicyAndGuidance/DH_4009696.
24. Chiari P, Hadour G, Michel P et al. Biphasic response after brain death induction: prominent part of catecholamines release in this phenomenon. J Heart Lung Transplant 2000; 19:675–82.
25. Gramm HJ, Meinhold H, Bickel U et al. Acute endocrine failure after brain death? Transplantation 1992; 54:851–7.
26. Kendall SW, Bittner HB, Peterseim DS et al. Right ventricular function in the donor heart. Eur J Cardiothorac Surg 1997; 11(4):609–15.
27. Weiss S, Kotsch K, Francuski M et al. Brain death activates donor organs and is associated with a worse I/R injury after liver transplantation. Am J Transplant 2007; 7:1584–93.
28. Sánchez-Fructuoso A, Naranjo Garcia P, Calvo Romero N et al. Effect of the brain-death process on acute rejection in renal transplantation. Transplant Proc 2007; 39:2214–16.
29. Schnuelle P, Lorenz D, Mueller A et al. Donor catecholamine use reduces acute allograft rejection and improves graft survival after cadaveric renal transplantation. Kidney Int 1999; 56:738–46.
30. Nygaard CE, Townsend RN, Diamond DL. Organ donor management and organ outcome: a 6-year review from a level I trauma center. J Trauma 1990; 30:728–32.
31. Pennefather S, Bullock R, Mantle D et al. Use of low dose arginine vasopressin to support brain dead organ donors. Transplantation 1995; 59:58–62.
32. Wheeldon DR, Potter CD, Dunning J et al. Haemodynamic correction in multiorgan donation. Lancet 1992; 339:1175.
33. Salim A, Vassiliu P, Velmahos GC et al. The role of thyroid hormone administration in potential organ donors. Arch Surg 2001; 136:1377–80.
34. Intensive Care Society. Guidelines for adult organ and tissue donation. London, 2005. Available at: http://www.ics.ac.uk/icmprof/downloads/Organ%20%20Tissue%20Donation.pdf.
35. de Ville de Goyet J, Hausleithner V, Malaise J et al. Liver procurement without in situ portal perfusion. A safe procedure for more flexible multiple organ harvesting. Transplantation 1994; 57:1328–32.
36. Gäbel M, Lidén H, Norrby J et al. Early function of liver grafts preserved with or without portal perfusion. Transplant Proc 2001; 33:2527–8.
37. Gruttadauria S, Foglieni CS, Doria C et al. The hepatic artery in liver transplantation and surgery: vascular anomalies in 701 cases. Clin Transplant 2001; 15:359–63.
38. Saidi RF, Elias N, Kawai T et al. Outcome of kidney transplantation using expanded criteria donors and donation after cardiac death kidneys: realities and costs. Am J Transplant 2007; 7:2769–74.
39. Ojo AO, Hanson JA, Meier-Kriesche H et al. Survival in recipients of marginal cadaveric donor kidneys compared with other recipients and wait-listed transplant candidates. J Am Soc Nephrol 2001; 12:589–97.
40. Audard V, Matignon M, Dahan K et al. Renal transplantation from extended criteria cadaveric donors: problems and perspectives overview. Transplant Int 2008; 21:11–17.
41. Devey LR, Friend PJ, Forsythe JL et al. The use of marginal heart beating donor livers for transplantation in the United Kingdom. Transplantation 2007; 84:70–4.
42. Busuttil RW, Shaked A, Millis JM et al. One thousand liver transplants. The lessons learned. Ann Surg 1994; 219:490–9.
43. Irish WD, McCollum DA, Tesi RJ et al. Nomogram for predicting the likelihood of delayed graft function in adult cadaveric renal transplant recipients. J Am Soc Nephrol 2003; 14:2967–74.
44. Department of Health. Guidance on the microbiological safety of human organs, tissues and cells used in transplantation. London, 2000. Available at: http://www.dh.gov.uk/en/Publicationsandstatistics/Publications/PublicationsPolicyAndGuidance/DH_4005526.
45. Sureshkumar KK, Reddy CS, Nghiem DD et al. Superiority of pediatric en bloc renal allografts over living donor kidneys: a long-term functional study. Transplantation 2006; 82:348–53.
46. Dharnidharka VR, Stevens G, Howard RJ. En-bloc kidney transplantation in the United States: an analysis of United Network of Organ Sharing (UNOS) data from 1987 to 2003. Am J Transplant 2005; 5:1513–17.

47. Kootstra G, Wijnen R, van Hooff JP et al. Twenty percent more kidneys through a non-heart beating program. Transplant Proc 1991; 23:910–11.
48. Koostra G, Daemoin J, Oomen A. Categories of non-heart beating donors. Transplant Proc 1995; 27:2893.
49. Muiesan P, Girlanda R, Jassem W et al. Single-center experience with liver transplantation from controlled non-heartbeating donors: a viable source of grafts. Ann Surg 2005; 242:732–8.
50. Brook NR, Waller JR, Richardson AC et al. A report on the activity and clinical outcomes of renal non-heart beating donor transplantation in the United Kingdom. Clin Transplant 2004; 18:627–33.
51. Reddy S, Zilvetti M, Brockmann J et al. Liver transplantation from non-heart-beating donors: current status and future prospects. Liver Transplant 2004; 10:1223–32.
52. British Transplantation Society. Guidelines relating to solid organ transplants from non-heart beating donors. London, 2004. Available at: http://www.bts.org.uk/Forms/Guidelines_Nov_2004.pdf.
53. Gok MA, Shenton BK, Buckley PE et al. How to improve the quality of kidneys from non-heart-beating donors: a randomised controlled trial of thrombolysis in non-heart-beating donors. Transplantation 2003; 76:1714–19.
54. British Transplantation Society/Renal Association. United Kingdom guidelines for living kidney donation, 2nd ed, 2005. Available at: http://www.bts.org.uk/Forms/Guidelines_complete_Oct05.pdf.

    Comprehensive evidence-based guidelines and best-practice guide for all aspects of living kidney donation.
55. The Organ Donor Procurement and Transplantation Network. Living donor evaluation guidelines, 2005. Available at: http://www.unos.org/ContentDocuments/Living_Kidney_Donor_Evaluation_Guidelines2(1).pdf.
56. Gall MA, Hougaard P, Borsch-Johnsen K et al. Risk factors for development of incipient and overt diabetic nephropathy in patients with non-insulin dependent diabetes mellitus: prospective, observational study. BMJ 1997; 314:783–8.
57. Delmonico F. A report of the Amsterdam Forum on the Care of the Living Kidney Donor: data and medical guidelines. Transplantation 2005; 79:S53–66.
58. Blake GM, Grewal GS. Reference data for 51Cr-EDTA measurements of GFR derived from live kidney donors. Nucl Med Commun 2005; 26: 61–5.
59. Benedetti E, Troppmann C, Gillingham K et al. Short- and long-term outcomes of kidney transplants with multiple renal arteries. Ann Surg 1995; 221:406–14.
60. Carter JT, Freise CE, McTaggart RA et al. Laparoscopic procurement of kidneys with multiple renal arteries is associated with increased ureteral complications in the recipient. Am J Transplant 2005; 5:1312–18.
61. Matas AJ, Bartlett ST, Leichtman AB et al. Morbidity and mortality after living kidney donation, 1999–2001: a survey of United States transplant centers. Am J Transplant 2003; 3:830–4.
62. Kok NFM, Lind MY, Hansson BME et al. Comparison of laparoscopic and mini incision open donor nephrectomy: single blind, randomised controlled clinical trial. BMJ 2006; 333:221–6.
63. Nanidis TG, Antcliffe D, Kokkinos C et al. Laparoscopic versus open live donor nephrectomy in renal transplantation: a meta-analysis. Ann Surg 2008; 247:58–70.

    A meta-analysis of 6594 patients from 73 comparative studies between 1997 and 2006 which gives a comprehensive comparison of outcomes between operative approaches.
64. Hadjianastassiou VG, Johnson RJ, Rudge CJ et al. 2509 living donor nephrectomies, morbidity and mortality, including the UK introduction of laparoscopic donor surgery. Am J Transplant 2007; 7:2532–7.

    A good contemporary review of UK practice which looks at type of procedure and associated morbidity and mortality.
65. Fabrizio MD, Ratner LE, Kavoussi LR. Laparoscopic live donor nephrectomy: pro. Urology 1999; 53:668–70.
66. Fehrman-Ekholm I, Elinder CG, Stenbeck M et al. Kidney donors live longer. Transplantation 1997; 64:976–8.
67. Ellison MD, McBride MA, Taranto SE et al. Living kidney donors in need of kidney transplants: a report from the Organ Procurement and Transplantation Network. Transplantation 2002; 74:1349–51.
68. Cherikh W, Taranto S. Final report: update on prior living donors who were subsequently placed on the waiting list. OPTN/UNOS Minority Affairs Committee Descriptive Data Request for Minority Affairs Committee Meeting, 13 July 2007.
69. National Institute for Health and Clinical Excellence. IPG194 Living-donor liver transplantation – guidance, 2006. Available at: http://www.nice.org.uk/nicemedia/pdf/IPG194guidance.pdf.
70. Tran TM, Changsri C, Shackleton CR et al. Living donor liver transplantation: histological abnormalities found on liver biopsies of apparently healthy potential donors. J Gastrohepatol 2005; 11:1440–76.
71. Miller C, Florman S, Kim-Schluger L et al. Fulminant and fatal gas gangrene of the stomach in a healthy live liver donor. Liver Transplant 2004; 10:1315–19.

72. Emond JC, Renz JF, Ferrell LD et al. Functional analysis of grafts from living donors. Implications for the treatment of older recipients. Ann Surg 1996; 224:544–52; discussion 552–4.

73. Gruessner RW, Kendall DM, Drangstveit MB et al. Simultaneous pancreas–kidney transplantation from live donors. Ann Surg 1997; 226:471–80; discussion 480–2.

74. Gruessner R, Sutherland D, Drangstveit M et al. Pancreas transplants from living donors: short- and long-term outcome. Transplant Proc 2001; 33:819.

75. National Institute for Health and Clinical Excellence. IPG170 Living donor lung transplantation for end-stage lung disease – guidance, 2006. Available at: http://www.nice.org.uk/nicemedia/pdf/IPG170guidance.pdf.

76. Grant D, Abu-Elmagd K, Reyes J et al. on behalf of the Intestine Transplant Registry. 2003 report of the intestine transplant registry: a new era has dawned. Ann Surg 2005; 241:607–13.

7

Recent trends in kidney transplantation: a review

Benjamin E. Hippen
Robert S. Gaston

Introduction: the wages of success

As kidney transplantation enters its sixth decade, the challenges facing the field are largely the consequence of its success as a therapy. It is difficult to remember (and for those under a certain age, an experience not available for recollection) the era when the physiological burden of systemic immune suppression was so taxing as to make kidney transplantation a dubious alternative to dialysis. For some time, this has not been the case. At least since the advent of ciclosporin A in 1983, the outcomes for kidney transplantation have been shown to be vastly superior to dialysis for virtually every category of patient with end-stage renal disease (ESRD).[1]

Indeed, while survival among dialysis-dependent patients has not changed appreciably in almost two decades (with only 35% alive after 5 years), survival of kidney transplant recipients is now in excess of 90% at 3 years, with a documented advantage over dialysis even among elderly recipients.[2,3]

Current challenges in kidney transplantation, then, might best be characterised as a consequence of technological progress.[4] While clinical hurdles remain in the attempt to deliver the very best outcomes for recipients, the most pressing issues are now driven by the limiting factor of organ availability. The question is no longer who might benefit from transplantation (answer: nearly everyone), but who shall benefit, when not everyone can. Indeed, it is this seemingly straightforward, pedestrian problem that underlies much of the recent public policy debate in renal transplantation, and that fuels unsavoury practices such as transplant tourism.[5] General approaches to the issue can be loosely classified into:

1. **Efforts to expand the supply of organs.** These range from aggressive, systematic efforts to identify and promulgate best practices in organ recovery, scrutiny of heretofore underexploited sources of organs such as 'expanded criteria' donors, the reintroduction of policies to guide the retrieval of organs from donors after cardiac death, paired donation and desensitisation programmes, non-directed donation, all the way to controversial efforts such as proposals to exchange valuable consideration for organs.[6] The timeliness of this issue was emphasised in the recent pronouncements of the Brown government in the UK supporting presumed consent,[7] a concept often discussed but rarely implemented in full in Western democracies.
2. **Efforts to reduce (or redefine) the overall demand for organs.** This is a relatively new concept in policy debates. Attention is now being paid to the question of whether all individual demands for a transplantable

organ should be judged of equal merit, in part due to the recognition that neither current nor optimistic expectations for the yield of orthodox approaches to organ procurement will meaningfully address the growing demand, and in part due to dissatisfaction with the current system of allocation on a number of levels. One example of this arises from the suggestion that not all candidates currently on the waiting list for a transplant are truly appropriate candidates, and instead represent an artificial inflation of the waiting list that exaggerates the scope of demand.[8]

In contrast, a recent analysis by Schold et al. found that, by applying straightforward demographic criteria defining transplantability, 16807 Americans might indeed be inappropriately listed. The quandary, however, is that these same criteria identified over 134000 others on dialysis who would benefit from transplantation, but are not currently on the waiting list (**Fig. 7.1**).[9]

3. **Efforts to better pair donor and recipient to optimise outcomes.** A radical restructuring of the way in which organs from deceased donors are allocated is increasingly advocated among transplant professionals in the USA.

Using retrospective data collected on deceased donors and recipients, this new system would allocate organs based on a scoring system designed to maximise 'life years following transplant' or 'LYFT'. In so doing, the proponents suggest, scarce organs would be put to better use, and the opportunity to maximise LYFT would offer a clearer metric (in the way time spent on a waiting list does not) to adjudicate competing claims for the same organ.[10]

Though the access issue may seem paramount, and advances in immunosuppression and overall care have transformed the act of transplanting an organ from audacious to routine, rejection is still a problem, and further improvement is most assuredly desirable. Even so, state-of-the-art therapies have made it increasingly difficult to test new approaches that, in

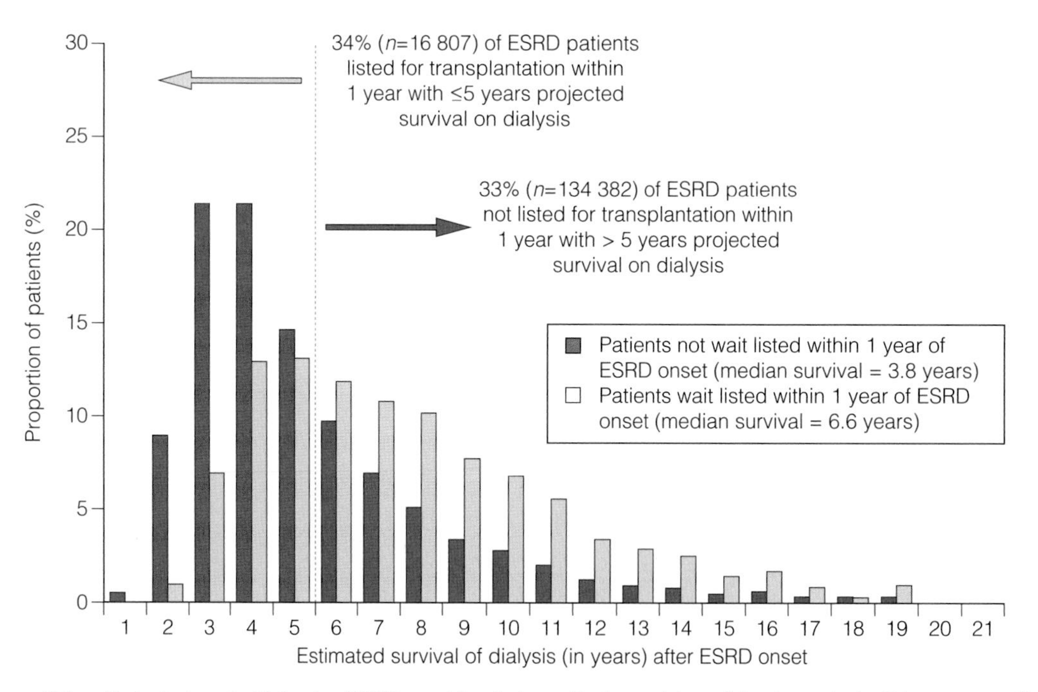

Figure 7.1 • Projected survival following ESRD onset (excludes patients receiving a living transplant within one year after ESRD onset). Reproduced from Schold JD, Srinivas TR, Kayler LK et al. The overlapping risk profile between dialysis patients listed and not listed for renal transplantation. Am J Transplant 2008; 8(1):58–68. With permission from Blackwell Publishing.

the short term at least, may demand deviation from accepted orthodoxy. At the same time, rather than the traditional problem of cellular rejection, it now appears that antibodies and B cells are increasingly the source of clinical conundrums, and the highly sensitised recipient continues to pose daunting clinical challenges.[11] These trends have revitalised the science of histocompatibility, rendered briefly moribund by the use of potent induction and maintenance immunosuppression. In this chapter, we will summarise these emerging trends, focusing on efforts to improve both access to and outcomes from kidney transplantation.

Kidney transplantation: the treatment of choice for end-stage renal disease

As recently as two decades ago, the best treatment option for patients with end-stage renal disease (ESRD) was a debatable question. Innovations such as recombinant erythropoietin and biocompatible membranes dramatically reduced the comorbidity associated with chronic haemodialysis. Peritoneal dialysis offered relative freedom from dietary restrictions and thrice-weekly visits to dialysis units. Although the overall mortality rates for those on dialysis appeared stable, given a substantially older patient population, outcomes were clearly improving.[2,12,13] Transplantation, despite its life-enhancing qualities, was haunted by the complications associated with toxic or inadequate immunosuppression, including rejection, osteonecrosis, opportunistic infections and cancer.[14] Almost half of deceased donor kidneys were lost within 3 years after engraftment. Rennie's 1978 contention that a kidney transplant was merely a 'temporary respite' from dialysis remained an accurate, if depressing, assessment.[15]

In the last decade, advances in transplantation have increasingly shifted the debate over optimal ESRD treatment modality from an open question to an established conclusion. After a decade of stagnant graft survival rates, the statistical half-life of allografts began improving in 1988, and by 1996 had almost doubled.[16] While still associated with substantial morbidity, new options in maintenance immunosuppression lessened rejection rates and reliance on corticosteroids, reducing complications after transplantation.[17] In the USA, where the Social Security Amendments of 1972 had made payment for ESRD therapy an entitlement for most of the population, transplantation proved substantially less expensive than maintenance dialysis.[18] Finally, a new statistical approach made it possible to better compare outcomes with dialysis and transplantation, a comparison previously difficult because of a bias in modality selection.[19] These recent analyses have defined a survival benefit relative to maintenance dialysis for virtually all ages and categories of patients with ESRD. Confirming the widely held notion that ESRD is a systemic disease conferring significant physiological deterioration over time, longevity on dialysis is now considered a major risk factor for graft loss: those patients transplanted earliest in the course of ESRD experience the best outcomes.[20] Just *how* early a recipient might be transplanted and still derive survival benefit, compared to accruing the cardiovascular vicissitudes of chronic kidney disease, remains an open question.

Why such significant changes in favour of renal transplantation? Better allograft survival is the result of several advances. More sensitive crossmatch techniques, in particular single-antigen flow-bead assays, have made donor–recipient compatibility easier to determine, even while perhaps reducing the rates at which sensitised recipients are transplanted.[21] Twenty-five years ago, the immunosuppressive armamentarium included only polyclonal antilymphocyte globulins (useful for only the short term), azathioprine (relatively ineffective) and large doses of corticosteroids. An improved understanding of the immunobiology of rejection has produced agents with greater specificity and less toxicity,[22] though intriguing recent data from the MYSS trial raise the question of the relative benefit of the newer agents.[23] The latter point is of considerable interest, given the cost differences between mycophenolate mofetil and azathioprine.

Under current immunosuppressive protocols, fewer than 25% of patients now experience acute rejection, and immunological graft losses are uncommon early after transplantation.[17,22] Recent data indicate that quality and stability of renal function (as reflected in the serum creatinine) in the early post-transplant period is a strong predictor of long-term graft survival.[24] Current immunosuppression is associated with unprecedented preservation of glomerular filtration rate.[25] Immunosuppressive complications have become both less common and more manageable. Improved antiviral, antibacterial and antifungal agents have reduced morbidity and mortality due to infections. The adoption of prophylactic and pre-emptive approaches to antimicrobial therapies has even reduced the incidence of serious infections among transplanted patients.

Dialysis patients, on average, have 20–25% the life expectancy of healthy, age-matched controls.[2] The most common causes of death are cardiovascular and infectious. While transplant recipients remain at greater risk of cardiovascular and infectious deaths than the general population, transplantation confers a substantial risk reduction relative to remaining on dialysis, and manoeuvres to reduce cardiovascular risk, including treatment of hyperlipidaemia, are proving to be efficacious after transplantation.[26,27] Recent reports indicate risk of death from common malignancies (lung, breast, colon, etc.) is nearly identical among those on dialysis and those with functioning transplants.[28] While the risk of dying in the perioperative period exceeds that of a waiting-list patient on dialysis, mortality risks decline over time, equalising after about 3 months. After that time, the transplanted patient is at substantially less risk of death. Life expectancy with a deceased-donor transplant ultimately is about twice that of remaining on dialysis.[19] In the USA, 70% of patients with ESRD are currently undergoing dialysis. However, of those patients living at least 10 years with ESRD, 70% have functioning allografts.[2] Kidney transplantation saves lives, a newly delineated fact that is increasingly appreciated by patients with ESRD.

Supply and demand

The number of identified dialysis-dependent ESRD patients in the USA has grown substantially over the last two decades. In 2005, 341 000 patients in the USA suffered from ESRD. Current estimates vary, but that number is expected to grow to between 400 000 and 520 000 by 2010, and approach 525 000–700 000 by 2020.[2,29,30] Today, in the USA, there are over 73 000 people waiting for a kidney transplant from a deceased donor, and by 2010 the waiting list is expected to grow to nearly 100 000 candidates. In 2006, some 6570 recipients, or 8% of the total listed, died waiting and 2020 patients were removed after being deemed 'too sick to transplant'.[31] Furthermore, the extended time waiting on dialysis exacts a toll. While receiving a renal transplant, even one of marginal quality, confers a significant mortality benefit compared to remaining on dialysis, the progression of cardiovascular morbidity while on dialysis translates into inferior allograft and patient outcomes.[19,20,32] Utilising a multivariate analysis, Meier-Kriesche et al. demonstrated that any amount of time on dialysis results in inferior graft outcomes when compared to pre-emptive transplantation.[20]

Annual mortality on the waiting list is 10.8% for diabetics and 6.3% for non-diabetics.[33] While this is better than for other non-transplanted patients with ESRD, it also means that a substantial number of patients will die awaiting transplantation.

Currently, a 'low-risk' patient without diabetes who is less than 65 years of age at the time of listing has a greater than 30% chance of dying before being offered a transplant.[34]

Even as some observers report a levelling off in the number of incident patients with ESRD, recent reports suggest that the rate of growth of chronic kidney disease (CKD) continues unabated.[35] Since the vast majority of patients with CKD die from accumulated cardiovascular complications prior to reaching ESRD, at least one explanation for the current and future growth in the number of patients with ESRD is success in treating hypertension, diabetes and other risk factors that may slow progression of heart disease. The attendant irony of this success is that a small but growing fraction of patients are now living long enough to progress to stage V CKD and thus require dialysis or transplantation.[36]

The financial costs of these trends command attention. In the USA, federal expenditures on ESRD alone amount to US \$21 billion in 2005, accounting for 6.4% of the Medicare budget for that year, spent on 0.6% of eligible Medicare beneficiaries.[2] Of that \$21 billion, only \$0.59 billion (2.8%) was spent on organ acquisition and transplant surgical costs. Recent estimates have suggested that the break-even point between the cost of dialysis and transplantation may now be less than 1.5 years (\$101 259 for 1.5 years of dialysis versus \$85 507 after 2 years of transplantation and full reimbursement for immunosuppressant medications.[37] Although current US statutes mandate transplant evaluation for all patients with ESRD, the previously cited study of Schold et al. indicates that there are as many dialysis patients not listed as listed.[9]

Increasing the supply of organs

Expanded criteria organs

In the USA, numerous efforts have been made in the last several years to maximise organ procurement from deceased donors. These efforts have occurred under the aegis of the Organ Donor Collaborative, which has sought to identify best practices by the most successful organ recovery agencies and promulgate them to others. As a consequence of these efforts, the number of deceased-donor organs has increased by about 10%. The largest growth, by category, has been in kidneys from extended criteria donors. As has often been the case in organ recovery from deceased donors, Spain appears to have been leading the way.[38] A conference convened to address the shortage of donated organs found that the principal difference between Spain and the USA was utilisation of donors older than 45 years of age.[39] Subsequently, a new category of donors was defined as 'expanded criteria' donors (ECDs). Previously, a high percentage of these organs were either discarded (18%) or never recovered. Donor factors utilised to define ECD kidneys were known to be independently associated with an increased risk of graft loss, and included donor age greater than 60 years, cerebrovascular accident as cause of death, hypertension and baseline renal insufficiency (serum creatinine greater than 1.5 mg/dL).[40,41] Despite a graft failure rate 70% greater than kidneys from standard donors, Ojo and colleagues demonstrated a substantial survival benefit from these kidneys relative to remaining on dialysis.[33]

Recent analysis of the ECD programme (initiated in the USA in 2002) documented increased access to transplantation for those candidates willing to accept ECD kidneys, with acceptable outcomes.[32]

Retrospective study of outcomes from organs turned down at one institution and used at another have helped to refine what is known about donor quality, and thereby salvage organs that might otherwise never have been used. Still, a recent examination of the data suggests that the category of 'extended criteria' encompasses a wide range of donor quality, so the hopeful outcomes derived from some ECD kidneys suggest that at least some

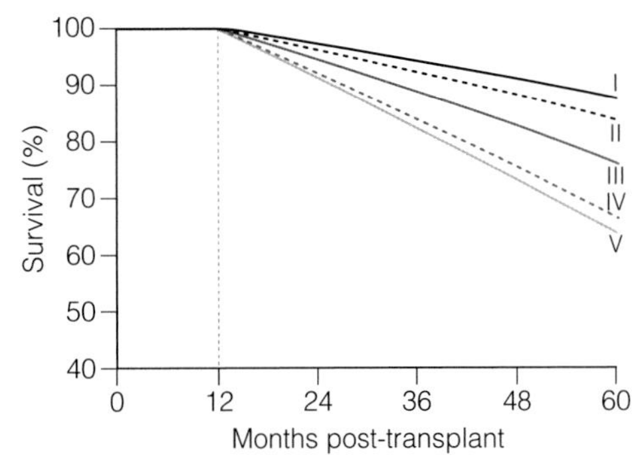

Figure 7.2 • Adjusted long-term overall graft survival by donor grade (I–V, best to worst; adjusted Cox model results for patients with a minimum of 1-year of graft survival). Donor grade was determined by calculation of donor risk scores utilising data from the Scientific Registry of Transplant Recipients (SRTR) incorporating donor–recipient cytomegalovirus match, donor race, donor age, cause of death, HLA mismatch, cold ischaemia time and donor history of hypertension and/or diabetes. Reproduced from Schold JD, Kaplan B, Baliga RS et al. The broad spectrum of quality in deceased donor kidneys. Am J Transplant 2005; 5(4, Pt 1):757–65. With permission from Blackwell Publishing.

organs so classified tend to offset considerably less sanguine outcomes from the majority of 'expanded criteria' donors[42] (**Fig. 7.2**). As we predicted in the previous version of this chapter,[40] the increased use of ECD organs has accounted for most of the 10% increase in the number of organs available in recent years. Which cohorts of recipients would most benefit from these kidneys remains a topic of intense interest, and will attract more attention if the allocation system undergoes revision that emphasises matching donor–recipient pairs to maximise life years gained following transplantation.[10,43–45]

Donation after cardiac death

Still another, though substantially smaller, source of organs has come with the rehabilitation of donation after cardiac death (or DCD), previously known as non-heart-beating donation. DCD can be either 'controlled', in which the donor's cardiopulmonary arrest spontaneously occurs in an operating room after cessation of life-sustaining measures in a critically ill patient, or 'uncontrolled', in which a patient sustaining an out-of-hospital cardiopulmonary arrest, and in which cardiopulmonary resuscitation is not successful, is placed on cardiopulmonary bypass until consent is obtained (or not) for organ procurement. Use of DCD in the USA, and most of the Western world, is essentially all in the controlled situation, and outcomes from these kidney donors appear comparable to those from standard criteria donors.[46]

Uncontrolled donation after cardiac death currently occurs with any regularity only in Spain.[47] In its recent effort to address the growing disparity between the demand for and supply of organs for transplantation, the Institute of Medicine settled on uncontrolled donation after cardiac death as a signature solution, with the promise of yielding as many as 22 000 kidneys per year.[48] Left out of the discussion were the numerous significant logistic and ethical difficulties associated with a widespread programme of recovering organs from uncontrolled DCD situations. Such a programme would necessitate:

1. Correctly identifying a candidate for uncontrolled DCD under circumstances of significant stress (e.g. a witnessed cardiopulmonary arrest in the field, with near-immediate initiation of CPR, hopefully by someone with formal training).
2. Transfer to an emergency room with available personnel to correctly identify an uncontrolled DCD candidate as such, contact the next of kin and secure permission for organ harvesting, all within the scope of approximately 90–120 minutes after an arrest, as it is generally agreed that the maximum time window for harvesting organs after cessation of CPR is 1 hour. An organ recovery professional would have a fairly limited amount of time to confirm prior consent from the decedent, or secure consent from the family, and would likely have to do this *while resuscitation efforts are in progress*, begging the issue of conflict of interest.
3. Already resource-strained emergency departments providing personnel appropriate for DCD management on a 24/7 basis, with round-the-clock availability of organ procurement teams, including transplant surgeons, operating room techs and nurses, and operating rooms, all ready to be mobilised for activity within an hour of cessation of circulation. In essence, this would require a continuous presence of such personnel in the hospital at all times, a state of affairs likely to strain, fiscally and otherwise, all but the largest transplant centres. For outlying emergency departments, this would have to entail a rapid transfer of an uncontrolled DCD candidate almost immediately after being identified as such.
4. Alternatively, uncontrolled DCD candidates would have to be put on cardiopulmonary bypass after cessation of CPR within the allotted time frame, adding substantially to the standard acquisition charge and manpower required to recover the organs. Consideration would also have to be given to the appearance of the donor candidate to the family, arriving to the emergency department to see their *deceased* loved one on cardiopulmonary bypass.
5. By the Institute of Medicine's own estimate, only 7.6% of all out-of-hospital arrests meet the criteria for uncontrolled DCD.[48] Apart from correctly identifying this small fraction of patients with out-of-hospital cardiac arrest, the results of serological testing (e.g. HIV, hepatitis) are unlikely to be available prior to the warm ischaemia time deadline of 1 hour. Undiagnosed conditions such as cancer, in the absence of a readily available medical record or reliable historian, would also likely go undiagnosed. Consequently, an untold number of donors will require either organ retrieval or cardiopulmonary bypass while awaiting the results of these necessary tests and medical information. The financial loss of prospectively procuring or preserving any organs lost due to subsequently ascertained contraindications would be significant.
6. Recipients and physicians would have to be prepared to accept substantially more complications in the post-transplant period. A recent study from Los Angeles showed that the incidences of primary non-function and delayed graft function were essentially twice the rates of those from donors after brain death.[49] Since it is likely these kidneys would be matched with older recipients (both because of a move within the Organ Procurement and Transplantation Network to adjust the allocation scheme to provide better quality kidneys to younger donors, and because the largest cohort of waiting-list patients are increasingly older), the result will be kidney transplantation with higher complication rates for those least physiologically equipped to handle such complications.

Donation after cardiac death has also generated ethical controversy, highlighted by a recent series of events at a hospital in California, where a transplant surgeon stood accused of hastening a patient's death by prescribing supratherapeutic doses of narcotics and anxiolytics in an effort to procure the patient's organs. At the time of writing, the surgeon in question still faces criminal and likely civil charges.[50] Since maintaining a protocol for implementing donation after cardiac death is now a Medicare requirement for hospitals in the USA, renewed efforts have been undertaken by the United Network for Organ Sharing (UNOS) to reiterate crucial facets of model DCD policies. First and foremost, there must be clear divisions of responsibility between the primary care team and transplant team. The decision to withdraw care from a potential candidate for DCD must be the consequence of discussions between the primary physicians and the patient's authorised decision-maker (with no involvement of the transplant team) up to and including the process of withdrawing life-sustaining measures from the patient.

The recovery procedure must not proceed until 2–5 minutes after cardiopulmonary arrest has been documented.[46] Increased interest in controlled donation after cardiac death has generated criticism from other quarters. Long-dormant doubts about the temporal relationship between the onset of cardiopulmonary arrest and the onset of death by neurological criteria have been brought to the forefront of the debate.[51] While the neocortex is quite susceptible to anoxic injury, animal models (it should be emphasised) suggest that the brainstem is thought to be considerably more resilient,[52] begging the question of whether or not a donor with cessation of cardiopulmonary circulation always (or often) meets death by whole-brain criteria (which includes loss of brainstem function) after 2 minutes or 5 minutes, and how this might be known with any reasonable certainty in the absence of a properly constructed and adequately powered study in humans. From an organ recovery standpoint, this is not a trivial issue, given the importance of warm ischaemia time on graft outcomes from DCD donors.[46] On the former point, both defenders and critics of DCD might agree that dispositive data regarding the time to brain death is wanting, though defenders of DCD frequently counter that this may be less ethically important than supposed by critics. As one group puts it, 'If donors were seen by the public as persons who are seriously and irreversibly brain injured, and whose families desperately want to donate organs to give meaning to the death and to respect the patient's final wishes, the number of minutes needed to declare death will be ignored as a theoretical issue of little magnitude.'[53] Not unreasonably, critics of DCD suggest that if this is the case, it would be more intellectually honest to simply abandon the dead donor rule (which requires that donors be certified as 'dead' by whole-brain or by cardiopulmonary criteria), seek a societal consensus for the moral obligations of society and the medical profession to the critically ill who are morally committed to organ donation, and refashion the laws governing organ procurement accordingly.[54] It is likely that this issue will continue to be debated as donation after cardiac death becomes more common.

Allocation of deceased donor kidneys

The increased number of kidneys available for transplantation as a result of the implementation of ECD and DCD programmes remains woefully inadequate to address the demand for kidneys in the USA. Thus, the transplant community has attempted to more optimally match donor and recipient in order to maximise survival after transplantation. Current organ allocation policies, grounded on data accumulated decades ago, are based on a combination of major histocompatibility complex (MHC) similarity (to promote better outcomes, or efficiency) and time spent on the waiting list (equity). Recent data indicate only a minimal impact of human leucocyte antigen (HLA) matching on long-term graft survival, and time on dialysis, as noted above, is now widely recognised as predictive of graft loss. Numerous alternatives have been proposed, some simple (promoting similarity in age like Meier-Kriesche et al.[55]) and some extremely complicated (LYFT; based on equations that include multiple pertinent variables to define relative benefit in projected longevity of transplant versus dialysis[10]). Alternatives are the subject of intense debate in the USA[56] and around the world.[57]

Living donor transplantation and the highly sensitised candidate

After a near tripling in the number of living donors between 1994 and 2004, further growth in living-donor transplantation has not occurred. Typically, the growth of the 1990s was attributed to increased availability of minimally invasive donor nephrectomy and liberalisation of criteria defining appropriate living-donor relationships. At the University of Alabama, however, the increase in number of living donors seemed to merely parallel growth in demand for transplantation; indeed, national data support this finding, with the number of living donors reflecting the increase in referral for transplantation and of the waiting list (**Fig. 7.3**).[58] In the USA, the number of living donors has been essentially unchanged since 2005. Further growth seems dependent not only on identifying additional donors, but on finding ways, via paired donation and desensitisation, to utilise those potential donors previously willing but unable to participate due to ABO or MHC incompatibility with their identified recipient.

The successful transplantation of the highly sensitised candidate continues to be a vexing challenge. In the USA, patients exhibiting any degree of sensitisation comprise 25% of the waiting list, and 12.5% of listed patients have a panel-reactive antibody (PRA) > 80%.[11] In an effort to better characterise the sensitisation 'fingerprint' of individual recipients, Bray et al. reported on their experience using a single-antigen flow-bead assay to generate a highly specific profile of unacceptable HLA antigens, a 'virtual' crossmatch, which was then used to quickly screen donor organs for potential compatibility[21] (**Fig. 7.4**). These investigators found that the actual flow cytometry crossmatch at the time of transplantation corresponded to the 'virtual' more than 80% of the time, and retrospective analysis suggested that the 5-year graft outcomes for highly sensitised recipients were not significantly different from unsensitised recipients. Application of these, or similar, techniques may facilitate sharing of kidneys for sensitised candidates across broader geographic areas.

The principal advance in desensitisation during the past 2–3 years has been better characterisation of which candidates will, and will not, benefit from treatment. It is increasingly clear that ABO incompatibility, especially when low titres are present, is more easily overcome.

Even though most transplanted subjects develop recurrent anti-ABO antibodies and may display C4d on protocol biopsy after transplantation, accommodation occurs in the majority of recipients, and antibody-mediated rejection is very uncommon.[59,60]

Indeed, desensitisation and early transplantation across ABO barriers, though more expensive than ABO-compatible transplantation, has now been shown to be cost-effective compared to remaining on dialysis whilst awaiting a suitable deceased

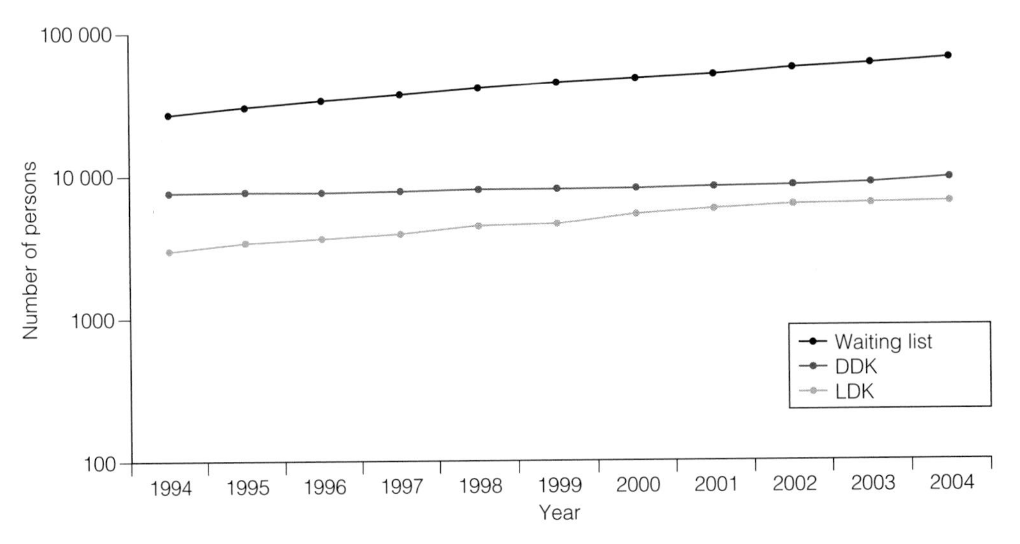

Figure 7.3 • Number of persons awaiting deceased donor kidneys in the USA (waiting list), plotted on a logarithmic scale with the number of available kidneys from deceased donors (DDK) and living donors (LDK), 1999–2004, by year. Note that the increase in LDK parallels, but does not exceed, growth in transplant candidates.
Reproduced from Gaston RS, Danovitch GM, Epstein RA et al. Limiting financial disincentives in live organ donation: a rational solution to the kidney shortage. Am J Transplant 2006; 6:2548–55. With permission from Blackwell Publishing.

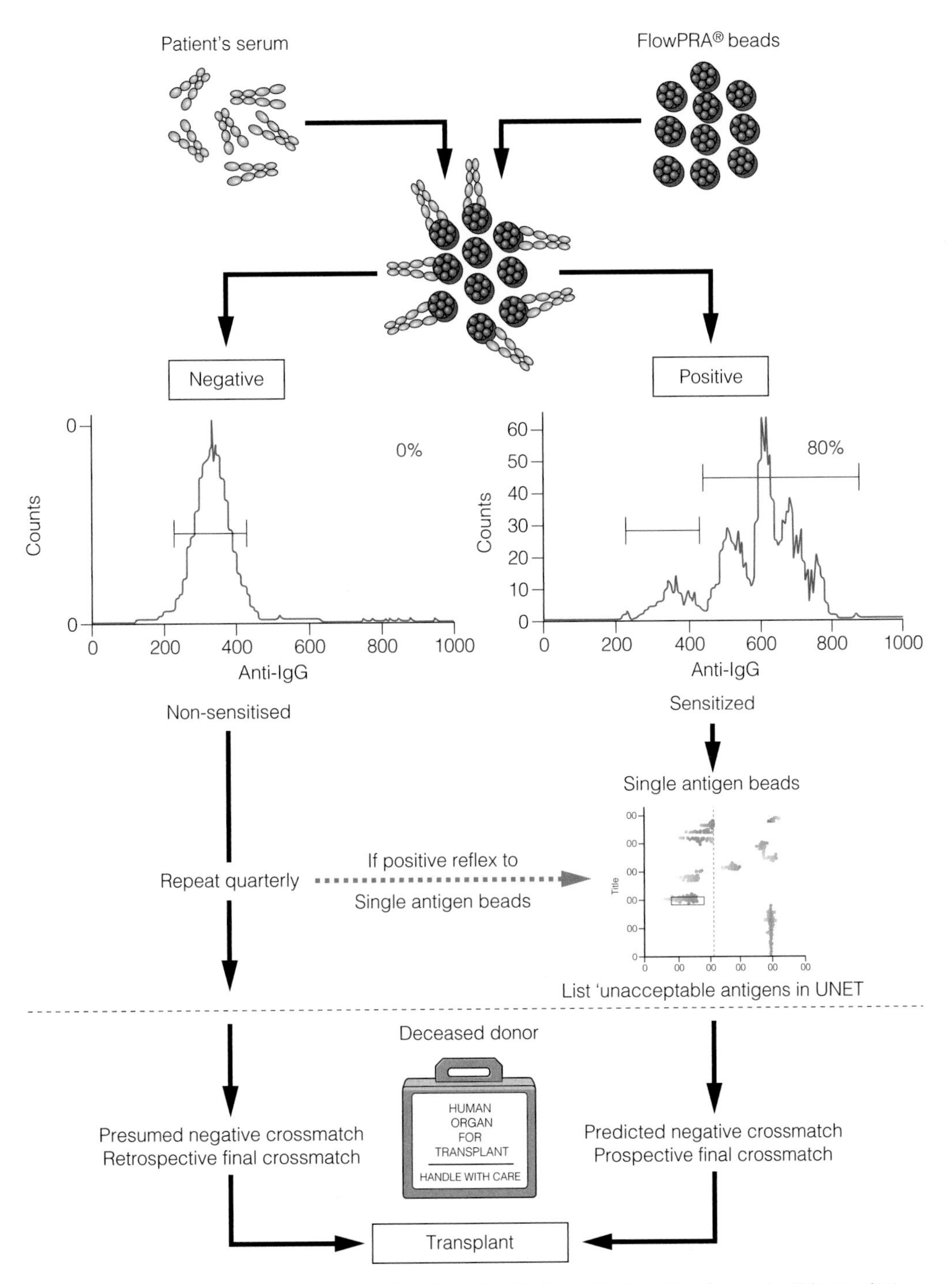

Figure 7.4 • Emory algorithm for classifying patients based on HLA sensitisation. Sera from potential transplant recipients are initially screened for the presence/absence of HLA antibodies. Patient's sera that are 'positive' for HLA antibody are subjected to detailed antibody specificity analysis using HLA-specific microparticle technology. Well-defined HLA specificities are then listed in UNET as 'unacceptable' antigens. When a donor organ is available, patients' unacceptable antigens are used to select potential recipients with the highest probability of a negative crossmatch. Sensitised patients who are listed on the UNOS match run are selected for final crossmatching by flow cytometry. Only flow cytometric crossmatch-negative patients were transplanted. Reproduced from Bray RA, Nolen JD, Larsen C et al. Transplanting the highly sensitized patient: The Emory algorithm. Am J Transplant 2006; 6(10):2307–15. With permission from Blackwell Publishing.

donor kidney. Likewise, the ability of pretransplant testing to identify high titres of anti-MHC antibody is increasingly defining a cohort of sensitised transplant candidates that will not respond satisfactorily to manoeuvres to deplete antibody.[61] Indeed, the recent finding by Perry and colleagues that even when antibody specificity disappears from sera, plasma cell clones remain in the spleen and bone marrow, indicates the limitations of relying on desensitisation to diminish donor–recipient incompatibility.[62]

In a series of articles, the Johns Hopkins group has outlined proposals whereby utilisation of currently identified donors via a kidney paired donation programme (KPD) would increase organ availability by as much as 40%.[63]

This would require implementation of KPD on a nationwide basis, an approach recently authorised by Congress. Indeed, the often frustrating experiences of the Hopkins group with desensitisation on a broad scale has led to advocacy of a combined approach utilising desensitisation of one or two pairs to facilitate KPD among broader numbers of donors and recipients. Going forward, it seems likely that a highly organised multicentre programme incorporating all these techniques offers the best hope of expanding the donor supply for patients in need of transplants.

Evolving immunosuppression

Highly effective immunosuppression became available to the transplant community in the mid-1990s, providing consistent control of cellular rejection, thought at the time to be the most important proximate cause of allograft failure.

However, the precipitous decline in acute rejection episodes in the late 1990s was not accompanied by a commensurate increase in allograft survival, the first time such a dissociation was noted in outcome data.[64,65]

It is now apparent that merely controlling cellular rejection, while important, is not the only variable influencing outcome.[66,67] Half of grafts are lost to the death of the recipient, focusing efforts on early transplantation, effective management of CKD-related morbidity and less toxic immunosuppression.[68,69] The other 50% of graft losses reflect a hodgepodge of events, including antibody-mediated immunological injury, immunosuppressant nephrotoxicity, recurrent disease and non-adherence to treatment regimens.[70–72] While dealing with antibody-mediated injury, especially in the setting of late allograft dysfunction, remains most challenging, minimising the impact of each of these variables has now become the motivating force driving new drug development and more informed use of immunosuppressants in clinical transplantation.[62,71,72]

The immunosuppressant most recently approved in the USA for use in kidney transplantation was sirolimus, in 1998. This followed the rapid-succession approval of tacrolimus, mycophenolate, rabbit antithymocyte globulin (ATG; thymoglobulin), daclizumab and basiliximab in the period 1994–97. Currently undergoing phase II and III testing are belatacept, alefacept, CP 690,550 (a JAK3 kinase inhibitor), AEB071 (a protein kinase C inhibitor), among other agents, none of which is likely to receive FDA approval for widespread use in the near future. Thus, it seems the most important recent clinical trials, rather than forging new frontiers, have served to refine our usage of currently available immunosuppressants.

Calcineurin inhibitors (CNIs) remain the cornerstone of most immunosuppressant regimens, and tacrolimus has become the agent of choice in most centres in the USA.[73] Several studies have documented its advantage over ciclosporin in preventing acute rejection, promoting allograft function and reducing cardiovascular risk factors (hypertension and hyperlipidaemia).

While clearly exerting more adverse impact on glucose tolerance than ciclosporin, the preponderance of evidence is that new-onset diabetes mellitus after kidney transplantation is a common complication with both CNIs.[74]

Thus far, for most patients, there appears to be no widely accepted alternative to CNI-based immunosuppression, although conversion to sirolimus or everolimus remains an attractive option for some patients after the early post-transplant period.[75]

Mycophenolate mofetil (MMF) remains the antiproliferative agent of choice; indeed, MMF may offer the greatest benefit with the fewest adverse effects of any maintenance immunosuppressant, with the previously cited small European study questioning benefit standing in stark contrast to several randomised, controlled trials and ever-enlarging clinical experience.[23] In combination with CNI, sirolimus appears to exacerbate nephrotoxicity.

In combination with MMF, sirolimus has proven relatively ineffective in preventing acute rejection, and with substantial additive toxic effects[76,77] (**Fig. 7.5**).

These factors have thus far precluded widespread usage of sirolimus, at least in the USA, and the tacrolimus/mycophenolate combination remains the most widely utilised for maintenance therapy.

Early corticosteroid withdrawal is becoming increasingly popular in kidney transplant immunosuppression. The experience of a number of single centres appears to document excellent intermediate-term outcomes for patients after antibody induction, and weaned off steroids in the first week after transplantation; as many as 30% of new transplant recipients are now in steroid-free protocols.[78,79] Concerns regarding the long-term impact of steroid avoidance remain.[80,81] In an ongoing blinded, randomised trial of early corticosteroid withdrawal, data after 4 years follow-up are available.[82] With 190 subjects in each arm, and maintenance tacrolimus and MMF, the steroid-free and steroid-maintenance cohorts have equivalent graft and patient survival, rejection rates and renal function. There is less diabetes in the steroid-free group, but very little difference in other metabolic parameters. Finally, in patients undergoing for-cause biopsies, there is more interstitial fibrosis/tubular atrophy in the steroid-free patients. Thus, a final conclusion regarding safety and efficacy of steroid-free immunosuppression for the long term remains indeterminate.

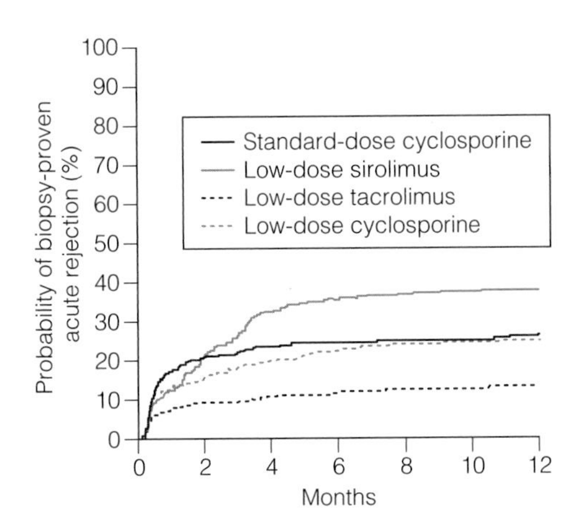

Figure 7.5 • The ELITE-Symphony Study. Cumulative probability of biopsy-proven acute rejection among 1589 patients randomised prospectively to one of four treatment groups. All patients also received mycophenolate mofetil and corticosteroids, with or without daclizumab induction. Reproduced from Ekberg H, Tedesco-Silva H, Demirbas A et al. Reduced exposure to calcineurin inhibitors in renal transplantation. N Engl J Med 2007; 357(25):2562–75. With permission from the Massachusetts Medical Society.

Antibody induction, utilised in only one-third of USA kidney recipients as recently as 1998, is now administered to almost all undergoing transplantation.[10] Traditionally, the underlying rationale for induction was to provide more potent immunosuppression along with protection from CNI nephrotoxicity in the early post-transplant period. Now, antibody induction is increasingly viewed as a requisite component of drug minimisation protocols, facilitating low-dose CNI administration and/or early corticosteroid withdrawal. In achieving these latter goals, lymphocyte depletion is considered by many to be an essential attribute of induction, focusing usage on rabbit ATG and alemtuzumab. In an ongoing, multicentre, steroid avoidance trial testing alemtuzumab against basiliximab in low-risk patients, and against rabbit ATG in high-risk patients (mostly African-Americans), the depleting agents were more effective in preventing rejection overall.[83] Within the high-risk group, alemtuzumab and rabbit ATG were equally effective, although ATG was associated with slightly more infectious complications. This latter finding is in keeping with a recently published report from the UNOS database documenting twice the risk of post-transplant lymphoproliferative disorder (PTLD; thought to be one of the sequelae of lymphocyte depletion) in patients receiving rabbit ATG compared to alemtuzumab.[84] It now appears that which cells are depleted, and which cells repopulate fastest, may be the primary determinant of both efficacy and adverse events rather than depletion per se.

Conclusion

Kidney transplantation remains a rapidly evolving field, with the challenges of current clinical practice reflecting the legacy of enormous success during past decades; nothing less than our best efforts will be necessary if these progressive trends are to continue. The unprecedented demand for transplantation requires inventive approaches to increase organ availability. Short-term successes in outcomes dictate greater focus on long-term survival and more effective, less toxic immunosuppression. It remains too early to judge the outcomes of current approaches to each of these issues. We recall the spirit of a different era in transplantation, a time when what seemed to be required, as Thomas E. Starzl remarked, was 'the courage to fail'.[85] Looking back, the failures seem almost wholly eclipsed by the successes. Looking forward, we hope the field is judged by the same standard, with successful solutions to current problems resulting in better outcomes for more patients, recognising that in so doing we are likely to uncover an entirely new set of challenges for future physicians and surgeons to address.

Key points

- The demand for kidney transplantation will continue to grow.
- There remain some ethical challenges in DCD donation – especially in the 'uncontrolled situation'.
- Interest in kidney allocation will grow as the shortage of organs grows and questions about the suitability of all patients on the list are asked.
- Living-donation levels have probably peaked, though proper organisation of programmes combining desensitisation and paired/pooled donation are needed.
- Interest in antibody induction has grown as regimens are designed to allow drug minimalisation.

References

1. Meier-Kriesche H-U, Schold JD, Gaston RS et al. Kidneys from deceased donors: maximizing the value of a scarce resource. Am J Transplant 2005; 5:1725–30.

2. Collins AJ, Foley R, Herzog C et al. Excerpts from the United States Renal Data System 2007 annual data report. Am J Kidney Dis 2008; 51(1, Suppl 1):S1–320.

   The USRDS annual data report offers a snapshot of CKD and renal replacement therapy in the USA. The full report is available at www.usrds.com, and is mandatory reading for anyone interested in the demographics of kidney transplantation.

3. Oniscu GC, Brown H, Forsythe JL. How old is old for transplantation? Am J Transplant 2004; 4(12):2067–74.
4. Hippen B. Innovation and the persistent challenge of collapsing goods. J Value Inquiry 2007; 40(2–3):227–33.
5. Naqvi SA, Ali B, Mazhar F et al. A socioeconomic survey of kidney vendors in Pakistan. Transplant Int 2007; 20(11):934–9.
6. Hippen BE. In defence of a regulated market in kidneys from living vendors. J Med Phil 2005; 30(6):593–626.
7. Hennesey P, Donneley L. Organs to be taken without consent. Telegraph, 2008.
8. Delmonico FL. What is the system failure? Kidney Int 2006; 69(6):954–5.
9. Schold JD, Srinivas TR, Kayler LK et al. The overlapping risk profile between dialysis patients listed and not listed for renal transplantation. Am J Transplant 2008; 8(1):58–68.
10. Andreoni KA, Brayman KL, Guidinger MK et al. Kidney and pancreas transplantation in the United States, 1996–2005. Am J Transplant 2007; 7(5, Pt 2):1359–75.
11. Hippen B. The sensitized recipient: What is to be done? Am J Transplant 2006; 6(10):2230–31.
12. Eknoyan G, Beck GJ, Cheung AK et al. Effect of dialysis dose and membrane flux in maintenance hemodialysis. N Engl J Med. 2002; 347(25):2010–19.
13. Meier-Kriesche HU, Ojo AO, Port FK et al. Survival improvement among patients with end-stage renal disease: trends over time for transplant recipients and wait-listed patients. J Am Soc Nephrol 2001; 12(6):1293–6.
14. Cameron JL, Whiteside C, Katz J et al. Differences in quality of life across renal replacement therapies:

a meta-analytic comparison. Am J Kidney Dis 2000; 35:629–37.

15. Rennie D. Home dialysis and the costs of uremia. N Engl J Med 1978; 298(7):399–400.
16. Hariharan S, Johnson CP, Bresnahan BA et al. Improved graft survival after renal transplantation in the United States, 1988 to 1996. N Engl J Med 2000; 342(9):605–12.
17. Gaston RS. Maintenance immunosuppression in the renal transplant recipient: an overview. Am J Kidney Dis 2001; 38(Suppl 6):S25–35.
18. Eggers PW. A quarter century of medicare expenditures for ESRD. Semin Nephrol 2000; 20(6):516–22.
19. Wolfe RA, Ashby VB, Milford EL et al. Comparison of mortality in all patients on dialysis, patients on dialysis awaiting transplantation, and recipients of a first cadaveric transplant. N Engl J Med 1999; 341(23):1725–30.
20. Meier-Kriesche HU, Kaplan B. Waiting time on dialysis as the strongest modifiable risk factor for renal transplant outcomes: a paired donor kidney analysis. Transplantation 2002; 74(10):1377–81.
21. Bray RA, Nolen JD, Larsen C et al. Transplanting the highly sensitized patient: the Emory algorithm. Am J Transplant 2006; 6(10):2307–15.
22. Halloran PF. Immunosuppressive drugs for kidney transplantation. N Engl J Med 2004; 351(26):2715–29.
23. Remuzzi G, Cravedi P, Costantini M et al. Mycophenolate mofetil versus azathioprine for prevention of chronic allograft dysfunction in renal transplantation: the MYSS follow-up randomized, controlled clinical trial. J Am Soc Nephrol 2007; 18(6):1973–85.
24. Hariharan S, McBride MA, Cherikh WS et al. Post-transplant renal function in the first year predicts long-term kidney transplant survival. Kidney Int 2002; 62(1):311–18.
25. Gourishankar S, Hunsicker LG, Jhangri GS et al. The stability of the glomerular filtration rate after renal transplantation is improving. J Am Soc Nephrol 2003; 14(9):2387–94.
26. Foley RN, Parfrey PS, Sarnak MJ. Clinical epidemiology of cardiovascular disease in chronic renal disease. Am J Kidney Dis 1998; 32(5, Suppl 3):S112–19.
27. Holdaas H, Fellstrom B, Cole E et al. Long-term cardiac outcomes in renal transplant recipients receiving fluvastatin: the ALERT extension study. Am J Transplant 2005; 5(12):2929–36.
28. Vajdic CM, McDonald SP, McCredie MR et al. Cancer incidence before and after kidney transplantation. JAMA 2006; 296(23):2823–31.
29. Gilbertson DT, Liu J, Xue JL et al. Projecting the number of patients with end-stage renal disease in the United States to the year 2015. J Am Soc Nephrol 2005; 16(12):3736–41.
30. Xue JL, Ma JZ, Louis TA et al. Forecast of the number of patients with end-stage renal disease in the United States to the year 2010. J Am Soc Nephrol 2001; 12(12):2753–8.
31. United Network for Organ Sharing, 2007 (available from: www.optn.org, accessed 13 February 2008).
32. Sung RS, Guidinger MK, Leichtman AB et al. Impact of the expanded criteria donor allocation system on candidates for and recipients of expanded criteria donor kidneys. Transplantation 2007; 84(9):1138–44.

    An important update on the clinical impact of the ECD kidney allocation programme instituted in the USA in 2002.

33. Ojo AO, Hanson JA, Meier-Kriesche H et al. Survival in recipients of marginal cadaveric donor kidneys compared with other recipients and wait-listed transplant candidates. J Am Soc Nephrol 2001; 12(3):589–97.

    Death on the kidney waiting list is a relatively new, controversial topic. Should patients who die on the list have been there in the first place? Data are contradictory in this regard.

34. Danovitch GM, Hariharan S, Pirsch JD et al. Management of the waiting list for cadaveric kidney transplants: report of a survey and recommendations by the Clinical Practice Guidelines Committee of the American Society of Transplantation. J Am Soc Nephrol 2002; 13(2):528–35.
35. Coresh J, Selvin E, Stevens LA et al. Prevalence of chronic kidney disease in the United States. JAMA 2007; 298(17):2038–47.
36. Hippen B. Preventive measures may not reduce the demand for kidney transplantation. There is reason to suppose this is not the case. Kidney Int 2006; 70(3):606–7.
37. Field MJ, Lawrence RL, Zwanziger L (eds). Extending Medicare coverage for preventive and other services. Washington, DC: National Academy Press, 2000.
38. Matesanz R. A decade of continuous improvement in cadaveric organ donation: the Spanish model. Nefrologia 2001; 21(Suppl 5):59–67.
39. Rosengard BR, Feng S, Alfrey EJ et al. Report of the Crystal City meeting to maximize the use of organs recovered from the cadaver donor. Am J Transplant 2002; 2(8):701–11.
40. Hippen B, Gaston R. Renal transplantation in the third millennium. In: Forsythe JLR (ed.) Transplantation (Companion to Specialist Surgical Practice), 3rd edn. Philadelphia: Elsevier Saunders, 2005.
41. Port FK, Bragg-Gresham JL, Metzger RA et al. Donor characteristics associated with reduced graft survival: an approach to expanding the pool of kidney donors. Transplantation 2002; 74(9):1281–6.
42. Schold JD, Kaplan B, Baliga RS et al. The broad spectrum of quality in deceased donor kidneys. Am J Transplant 2005; 5(4, Pt 1):757–65.

43. Danovitch GM, Cecka JM. Allocation of deceased donor kidneys: past, present, and future. Am J Kidney Dis 2003; 42(5):882–90.
44. Gaston RS, Danovitch GM, Adams PL et al. The report of a national conference on the wait list for kidney transplantation. Am J Transplant 2003; 3(7):775–85.
45. Merion RM, Ashby VB, Wolfe RA et al. Deceased-donor characteristics and the survival benefit of kidney transplantation. JAMA 2005; 294:2726–33.
46. Bernat JL, D'Alessandro AM, Port FK et al. Report of a National Conference on Donation after cardiac death. Am J Transplant 2006; 6(2):281–91.
47. Sanchez-Fructuoso AI, Marques M, Prats D et al. Victims of cardiac arrest occurring outside the hospital: a source of transplantable kidneys. Ann Intern Med 2006; 145(3):157–64.
48. Institute of Medicine, Committee on Increasing Rates of Organation, Childress JF, Liverman CT (eds). Organ donation: opportunities for action. Washington, DC: National Academies Press, 2006.
49. Gagandeep S, Matsuoka L, Mateo R et al. Expanding the donor kidney pool: utility of renal allografts procured in a setting of uncontrolled cardiac death. Am J Transplant 2006; 6(7):1682–8.
50. Arnquist S. Civil suit in organ harvesting case will go on. San Luis Obispo Tribune, 2007.
51. Menikoff J. Doubts about death: the silence of the Institute of Medicine. J Law Med Ethics 1998; 26(2):157–65.
52. Safar P, Xiao F, Radovsky A et al. Improved cerebral resuscitation from cardiac arrest in dogs with mild hypothermia plus blood flow promotion. Stroke 1996; 27(1):105–13.
53. Youngner SJ, Arnold RM, DeVita MA. When is "dead"? Hastings Cent Rep 1999; 29(6):14–21.
54. Verheijde JL, Rady MY, McGregor J. Recovery of transplantable organs after cardiac or circulatory death: transforming the paradigm for the ethics of organ donation. Phil Ethics Humanit Med 2007; 2:8.
55. Meier-Kriesche HU, Schold JD, Gaston RS et al. Kidneys from deceased donors: maximizing the value of a scarce resource. Am J Transplant 2005; 5(7):1725–30.
56. Stegall MD. The development of kidney allocation policy. Am J Kidney Dis 2005; 46:974–5.
57. Oniscu GC, Forsythe JLR. Allocation of kidneys for transplantation: practical aspects and ethical dilemmas. In: Weimar W, Bos MA, Busschbach JJ (eds). Organ transplantation: ethical, legal and psychosocial aspects. Leigerich: Pabst Science Publishers, 2008.
58. Gaston RS, Danovitch GM, Epstein RA et al. Limiting financial disincentives in live organ donation: a rational solution to the kidney shortage. Am J Transplant 2006; 6:2548–55.
59. Setoguchi K, Ishida H, Shimmura H et al. Analysis of renal transplant protocol biopsies in ABO-incompatible kidney transplantation. Am J Transplant 2008; 8(1):86–94.
60. Schwartz J, Stegall MD, Kremers WK et al. Complications, resource utilization, and cost of ABO-incompatible living donor kidney transplantation. Transplantation 2006; 82(2):155–63.

Two very important papers that document the outcomes possible with ABO-incompatible transplantation from clinical and financial perspectives.

61. Zachary AA, Montgomery RA, Leffell MS. Factors associated with and predictive of persistence of donor-specific antibody after treatment with plasmapheresis and intravenous immunoglobulin. Hum Immunol 2005; 66(4):364–70.
62. Perry DK, Pollinger HS, Burns JM et al. Two novel assays of alloantibody-secreting cells demonstrating resistance to desensitization with IVIG and rATG. Am J Transplant 2008; 8(1):133–43.
63. Segev DL, Gentry SE, Warren DS et al. Kidney paired donation and optimizing the use of live donor organs. JAMA 2005; 293(15):1883–90.
64. Meier-Kriesche HU, Schold JD, Srinivas TR et al. Lack of improvement in renal allograft survival despite a marked decrease in acute rejection rates over the most recent era. Am J Transplant 2004; 4(3):378–83.
65. Pascual M, Theruvath T, Kawai T et al. Strategies to improve long-term outcomes after renal transplantation. N Engl J Med 2002; 346(8):580–90.
66. Hippen BE, DeMattos A, Cook WJ et al. Association of CD20+ infiltrates with poorer clinical outcomes in acute cellular rejection of renal allografts. Am J Transplant 2005; 5(9):2248–52.
67. Sarwal M, Chua MS, Kambham N et al. Molecular heterogeneity in acute renal allograft rejection identified by DNA microarray profiling. N Engl J Med 2003; 349(2):125–38.
68. Abbud-Filho M, Adams PL, Alberu J et al. A report of the Lisbon Conference on the care of the kidney transplant recipient. Transplantation 2007; 83(8, Suppl):S1–22.
69. Abecassis M, Bartlett ST, Collins AJ et al. Kidney transplantation as primary therapy for end-stage renal disease: a National Kidney Foundation/ Kidney Disease Outcomes Quality Initiative (NKF/ KDOQITM) Conference. Clin J Am Soc Nephrol 2008; 3(2):471–80.
70. Nankivell BJ, Borrows RJ, Fung CL et al. The natural history of chronic allograft nephropathy. N Engl J Med 2003; 349(24):2326–33.
71. Nankivell BJ, Wavamunno MD, Borrows RJ et al. Mycophenolate mofetil is associated with altered expression of chronic renal transplant histology. Am J Transplant 2007; 7(2):366–76.
72. Sis B, Campbell PM, Mueller T et al. Transplant glomerulopathy, late antibody-mediated rejection

and the ABCD tetrad in kidney allograft biopsies for cause. Am J Transplant 2007; 7(7):1743–52.

73. Meier-Kriesche HU, Li S, Gruessner RW et al. Immunosuppression: evolution in practice and trends, 1994–2004. Am J Transplant 2006; 6(5, Pt 2):1111–31.

74. Vincenti F, Friman S, Scheuermann E et al. Results of an international, randomized trial comparing glucose metabolism disorders and outcome with ciclosporin versus tacrolimus. Am J Transplant 2007; 7(6):1506–14.

A prospective, controlled study documenting the true incidence of abnormalities in glucose metabolism occurring after kidney transplantation.

75. Campistol JM, Eris J, Oberbauer R et al. Sirolimus therapy after early ciclosporin withdrawal reduces the risk for cancer in adult renal transplantation. J Am Soc Nephrol 2006; 17(2):581–9.

76. Ekberg H, Tedesco-Silva H, Demirbas A et al. Reduced exposure to calcineurin inhibitors in renal transplantation. N Engl J Med 2007; 357(25):2562–75.

The largest (n = 1600) randomised, controlled trial of immunosuppression after kidney transplantation ever performed, documenting superiority of a tacrolimus/MMF-based regimen in preventing rejection, preserving allograft function and promoting allograft survival.

77. Oberbauer R, Segoloni G, Campistol JM et al. Early ciclosporin withdrawal from a sirolimus-based regimen results in better renal allograft survival and renal function at 48 months after transplantation. Transplant Int 2005; 18(1):22–8.

78. Gallon LG, Winoto J, Leventhal JR et al. Effect of prednisone versus no prednisone as part of maintenance immunosuppression on long-term renal transplant function. Clin J Am Soc Nephrol. 2006; 1(5):1029–38.

79. Matas AJ, Kandaswamy R, Gillingham KJ et al. Prednisone-free maintenance immunosuppression – a 5-year experience. Am J Transplant 2005; 5(10):2473–8.

80. Meier-Kriesche HU, Magee JC, Kaplan B. Trials and tribulations of steroid withdrawal after kidney transplantation. Am J Transplant 2008; 8(2):265–6.

81. Vincenti F, Schena FP, Paraskevas S et al. A randomized, multicenter study of steroid avoidance, early steroid withdrawal or standard steroid therapy in kidney transplant recipients. Am J Transplant 2008; 8(2):307–16.

82. Woodle ES. A randomized double blind, placebo-controlled trial of early corticosteroid cessation versus chronic corticosteroids: four year results (abstract). Am Transplant Congr, San Francisco, CA, 2007.

83. Hanaway M, Woodle ES, Mulgaonkar S et al. Results of a multicenter, randomized trial comparing three induction agents (alemtuzumab, Thymoglobulin, and basiliximab) with tacrolimus, MMF, and rapid steroid withdrawal in renal transplantation (abstract). Am Transplant Congr, San Francisco, CA, 2007.

84. Kirk AD, Cherikh WS, Ring M et al. Dissociation of depletional induction and post transplant lymphoproliferative disease in kidney recipients treated with alemtuzumab. Am J Transplant 2007; 7(11):2619–25.

85. Fox RC, Swazey JP. The courage to fail: a social view of organ transplants and dialysis. Chicago: University of Chicago Press, 1973.

8

Liver transplantation

Steve A. White
Derek M. Manas

Introduction

The single most effective therapy for end-stage liver failure (ESLF) is liver transplantation (LT). Data from the European Liver Transplant Registry show that over 70 000 LT transplants have been performed in 137 participating centres around Europe.[1] Currently 680 liver transplants are performed yearly in the UK for end-stage liver disease (ESLD) and to date more than 6000 patients have been transplanted. Nationally, in 2007, over 1000 patients were listed for LT, with a mean waiting time of 60 days. Unfortunately the supply cannot meet demand. A recent audit of the seven LT centres in the UK suggested that up to 14% of patients listed will die waiting,[2,3] while in the USA more than 1500 patients per annum (from a waiting list of over 17 000 patients) will die waiting for a transplant.[4]

Indications

In the UK in 1999, following the death of a young woman with liver failure who was not listed for LT, a colloquium was set up to discuss ethical issues and guidelines for appropriate patient selection. It was agreed that donated livers should be considered a national resource and that patients should be considered for LT if they had an anticipated length of life (in the absence of LT) of less than 1 year or an unacceptable quality of life, providing patients had a 50% chance of being alive at 5 years after transplant. For many conditions there are guidelines and algorithms to help the clinician decide whether a patient meets certain acceptable criteria. Guidelines for LT also reflect the reality of the donor pool, aiming to maximise benefit from this scarce resource. It is imperative that these guidelines are not set in stone and are undergoing a constant process of refinement to meet the changing need.[5,6]

Patients may present with minimal symptoms or at the extreme have an accumulation of complications such as varices, ascites, infection Spontaneous Bacterial Peritonitis (SBP or cholangitis) combined with poor synthetic liver function, e.g. hypoalbuminaemia (<30 g/L), hyperbilirubinaemia (>50 μmol/L) and prolonged clotting times (prothrombin time prolonged over control). Once infection ensues patients with ESLF have a 50% chance of dying within 6 months. The common indications for LT differ between adults (Box 8.1) and children (Box 8.2), although some do overlap. To date, over 60 conditions have been treated by LT.

Alcohol

The most controversial indication for adults in the UK is alcoholic liver disease (ALD). There have been recent concerns that the results of LT are poor for patients who return to alcohol. Patients suffering from ALD have to demonstrate a period of

Box 8.1 • Indications for LT (adults)

Common

Alcoholic liver disease (ALD)
Cryptogenic cirrhosis
Primary biliary cirrhosis
Primary sclerosing cholangitis (PSC)
Hepatitis (B, C, non-A, non-B)
Hepatocellular cancer
Autoimmune hepatitis

Rare

Haemochromatosis
Wilson's disease
α_1-Antitrypsin deficiency
Budd–Chiari
Polycystic disease
Hyperoxaluria, familial hypercholesterolaemia, porphyrias, amyloidosis, neuroendocrine tumours (e.g. carcinoid)

Box 8.2 • Commoner indications for LT in children

Biliary atresia
Familial cholestasis syndromes
Metabolic disorders:

- Cystic fibrosis
- α_1-Antitrypsin deficiency
- Crigler–Najjar type 1
- Wilson's disease

Unresectable tumours, e.g. hepatoblastomas
Acute liver failure – viral, drugs (e.g. paracetamol toxicity, autoimmune)

6 months abstinence. This will allow the treating clinician time to assess the need for LT as well as assess the potential for recidivism. Unfortunately up to 25–50% will return to alcohol after LT, but only 10–15% of these patients will relapse to harmful drinking.[7,8] Current graft survival rates are 72% at 5 years and compare well with other chronic liver diseases after LT.[9] A working party has recently published guidelines for assessing patients with ALD and listing them for LT (Box 8.3).[10,11]

Box 8.3 • Recommendations for alcohol-related liver disease

Assessment

Alcohol specialist

Contraindications

Alcoholic hepatitis
More than two episodes of medical non-adherence
Return to drinking after full professional assessment
Current illicit drug misuse

Listing

Sign agreement indicating intention of abstinence

Follow-up

Alcohol specialist and transplant clinician

From Bathgate AJ on behalf of the UK Liver Transplant Units. Recommendations for alcohol-related liver disease. Lancet 2006; 367:2045–6. With permission from Elsevier.

Hepatitis C (HCV)

The commonest indication for LT in the USA and Europe is HCV and it is predicted that HCV will place a major strain on future waiting lists. One of the problems in this cohort is recurrence of disease in almost all recipients. This can either occur in an accelerated form or insidiously, resulting in gradual graft fibrosis and dysfunction. There is anecdotal evidence to suggest that recurrence can be worse after live-donor LT (LDLT)[12] but with reduced ischaemic time in LDLT, theoretically, these donor grafts should actually do better. The universal recurrence of HCV in the graft leads to chronic hepatitis in most patients, progressing to cirrhosis in 20–50% of patients after 5 years.[13,14]

Not all patients with HCV require LT; most with compensated cirrhosis can be managed with antiviral therapy. Pegalated interferon with ribavirin is the most common regimen but these change according to response rates.[15] A recent study reports a survival benefit (74% vs. 63% after 7 years) in patients receiving antiviral therapy compared with those who do not. Most centres should expect to achieve 5-year survival rates in excess of 70%.[16]

After hepatectomy HCV RNA levels fall rapidly and then, after the first few days, start to climb, often exceeding pretransplant levels; in fact after 4 weeks the level may be 20 times increased. Even antiviral therapy that depletes the virus pretransplant cannot prevent recurrence in the early postoperative period[17,18] and few patients can tolerate this therapy immediately prior to transplantation. What remains controversial are those factors associated with progression of fibrosis and how the disease can be modified, thus prolonging graft survival. No single factor but rather a combination of factors are likely to be involved.

The main factors associated with accelerated disease recurrence include high viral load pretransplant (>10^6 IU/mL) and/or early after transplantation (>10^7 IU/mL), older donors 40–50 years, prolonged ischaemia, cytomegalovirus (CMV) coinfection, over-immunosuppression, genotype 1b and human immunodeficiency virus (HIV) coinfection.[19]

It has been suggested that treatments should therefore focus more on minimising viral replication in the early postoperative period but no regimen has proven superiority and regimes can be too toxic so soon after LT.[20,21] Better outcomes may be developed by tailoring protocols with specific immunosuppression. HCV genotype and HCV RNA do not appear to correlate with survival, but in the presence of alcohol abuse outcome is likely to be worse than with HCV alone.

Hepatitis B (HBV)

The situation for HBV is very different. Historically 80% of patients would develop recurrence but now antiviral therapy with nucleoside and nucleotide analogues (such as lamivudine) and HbIg provide excellent survival rates with low rates of reinfection.[22,23] Adefovir is an alternative if patients develop resistance to lamivudine.[24] Ten-year survival rates for HBV can be expected to be 70%.[25] Unlike HCV, HBV may not be detectable even up to 2 years post-transplant in 90% of patients.[26] HBV is also a common cause of fulminant hepatic failure (FHF), occurring in 4% of patients presenting with acute HBV infection.

Primary sclerosing cholangitis (PSC)

Patients with PSC are often managed expectantly. Some patients may require drainage procedures, either surgical or by stenting and some by medical means (e.g. antihistamines, cholestyramine, ursodeoxycholic acid or rifampacin). Nevertheless, for those progressing to ESLD, LT remains the only option; 70% will also have inflammatory bowel disease and this is predominantly ulcerative colitis. These patients are often a challenge to manage. For example, it would make sense to perform a colectomy before LT in anyone at high risk of developing a colonic malignancy.[27,28]

Hilar cholangiocarcinoma, and its treatment with LT, has been revisited in recent years. Aggressive surgical resection remains the treatment of choice whenever possible, but for a select group of patients with irresectable disease LT may be the only hope. There is no accurate method of differentiating a benign from a malignant stricture either biochemically or radiologically. A CA19.9 >100 U/mL has a sensitivity and specificity of over 80%; this can be improved to >90% if considered with a carcinoembryonic antigen (CEA). Peripheral intrahepatic cholangiocarcinoma or extrahepatic distal tumours do not cause so much of a diagnostic problem, but the ‘Klatskin’ or hilar tumours are more problematic as they often present late and involve major vascular structures at the hilum, making them unresectable. In the presence of chronic liver disease (e.g. PSC, fibrocystic disease) it would therefore appear logical to treat by LT, except the patient would then need to be immunosuppressed, which goes against current teaching. Early reports showed 5-year survival rates of 26%, with median survival generally less than 12 months.[29] Many centres have reported their experiences over the last 5 years using various combinations of neoadjuvant and adjuvant regimens. A recent analysis from the United Network for Organ Sharing (UNOS) database of over 280 patients over 18 years indicate 1- and 5-year survival rates of 74% and 38% respectively.[30] Newer regimens with better survival have now been advocated. The Mayo Clinic have the largest experience using staging laparoscopy to exclude nodal disease combined with external beam irradiation, brachytherapy and treatment with fluorouracil (FU) and/or capcitabine. The most recent report (n = 64) suggests a 5-year patient and disease-free survival of 76% and 60% respectively.[31] A CEA >100, older age, previous cholecystectomy, a mass on computed tomography (CT) scan, residual tumour >2 cm and tumour grade were independent risk factors for recurrence.[31] A non-randomised retrospective comparison between resection and LT demonstrated superior survival for LT, 82% vs. 21%, with 13% vs. 27% for recurrence.[32] The jury is still out as to whether cholangiocarcinoma should be treated by LT and only a randomised clinical trial (RCT) will give the answer, but this will be difficult to do given that there is a wide variation between surgeons as to what is deemed resectable.

Hepatocellular carcinoma (HCC)

In 1967 one of the world's first LTs was performed for HCC but the initial results for this indication were poor and in 1989 LT for HCC was considered a contraindication due to 5-year survival rates <40%.[33] It is now well recognised that LT for HCC does have an important role in carefully selected patients, particularly those with less advanced disease.

Staging of patients is very important and a number of staging systems exist. The ideal staging classification for HCC and LT should consider two factors: the extent of liver disease and tumour staging. The conventional TNM[34] classification does not consider liver disease. Okuda[35] staging includes tumour size, serum albumin, bilirubin and ascites. The Cancer of the Liver Italian Program[36] (CLIP) is based on the Child–Pugh class,[37,38] tumour progression, α-fetoprotein and presence of portal venous thrombosis. The CLIP was thought to be the most accurate in terms of prognosis,[39] but now other systems are in place such as the Barcelona Clinic Liver Cancer (BCLC) Staging System.[40] This classification uses variables such as tumour stage, liver function, physical status and cancer-related symptoms, and links these in a widely used treatment algorithm (**Fig. 8.1**). Patients with stage A disease are usually suitable candidates for LT.

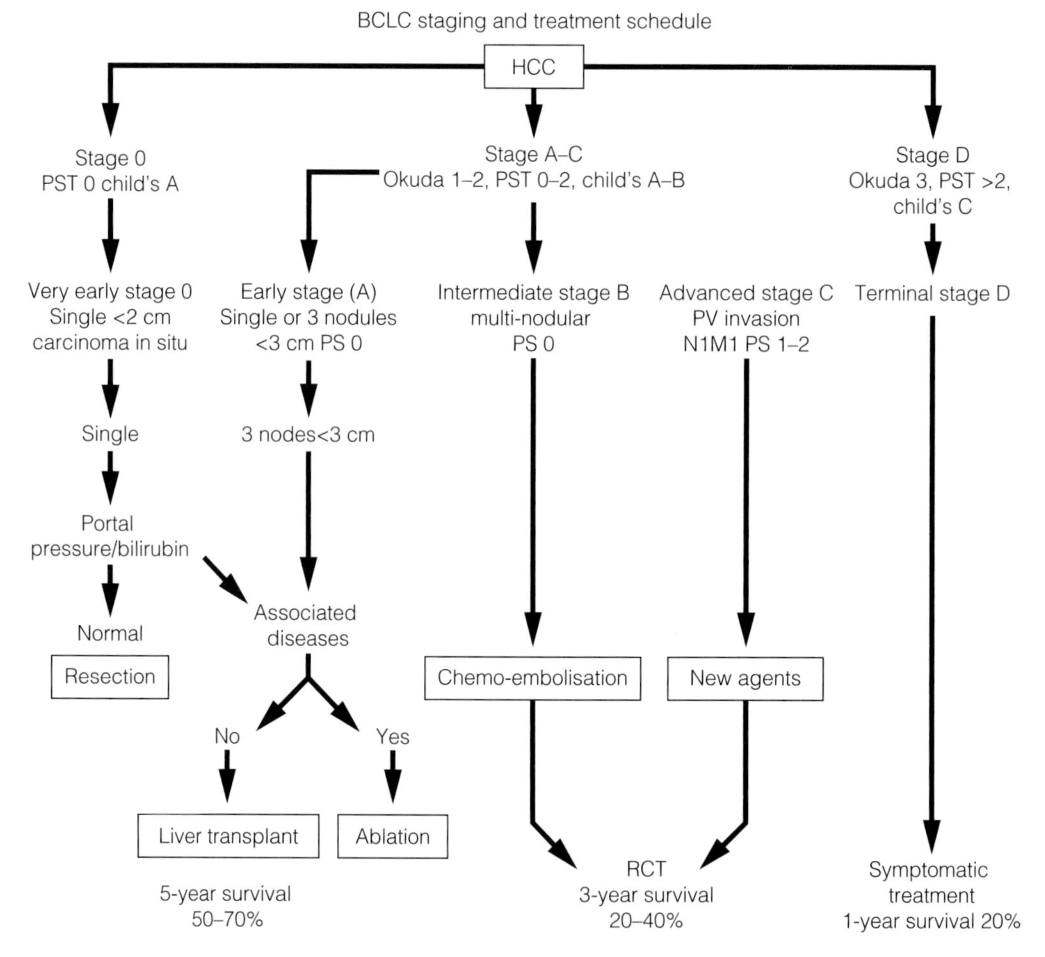

Figure 8.1 • Barcelona Clinic Liver Cancer (BCLC) staging classification and treatment schedule.[40] Stage 0: very early HCC suitable for resection. Stage A: early HCC suitable for more radical treatment such as resection, liver transplant or ablation. Stage B: patients with intermediate HCC may benefit from chemoembolisation. Stage C: patients with advanced HCC may receive new agents in the setting of an RCT. Stage D: advanced disease suitable for only palliative/symptomatic treatment. With kind permission from Springer Science & Business Media: Journal of Gastroenterology, Llovet JM. Updated treatment approach to hepatocellular carcinoma. J Gastroenterol 2005; 40:225-35.

Pioneering work from Milan, Italy revisited the idea of LT for HCC.[41] Those patients with cirrhosis (Child's B or C) with a single tumour <5 cm or up to three tumours <3 cm, in the absence of macrovascular portal vein invasion, showed 5-year survival rates of 70% with only a 15% recurrence rate. Although these 'Milan criteria' were used widely for definition of acceptability onto the liver transplant list, it is now argued that the Milan criteria do not consider patients who have larger tumours with more favourable histology (such as well-differentiated tumours with no vascular invasion). At least 30% of patients who meet Milan criteria will have poorly differentiated tumours with vascular invasion and a poorer survival. Others suggest that expanding criteria does not adversely affect outcome.[42] The most widely quoted expanded criteria are from University of California San Francisco (UCSF)[42] (single tumour <6.5 cm, maximum of three total tumours with none greater than >4.5 cm and cumulative tumour size <8 cm). When applied to a large series of patients the results are marginally worse than those reported for Milan criteria (64% vs. 79% survival at 5 years).[43]

Another difficulty is that radiological staging does not equal histological staging. Consequently, patients meeting histological UCSF criteria have better 5-year survival than those meeting radiological UCSF criteria.[44] A recent report of 168 patients by Yao et al. suggests that a 5-year recurrence-free probability of 59% can be achieved in those with explants exceeding UCSF criteria. This compares to 96.7% for those within UCSF criteria.[45]. In the UK, HCC patients have to fulfil category 2 listing criteria. A lesion has to be present in the same location on at least two radiological imaging modalities (either ultrasound, triple phase CT, magnetic resonance imaging (MRI) or hepatic angiogram) and has to fulfil Milan criteria. The widest diameter on CT is the maximum recorded diameter.[5,46]

Since the UK HCC consensus meeting in March 2008, there has been agreement that solitary lesions between 5 and 7cm with <20% progression over 6 months will be elligible for consideration of LT. The outcome of LT for HCC after bridging treatments such as resection, transarterial chemoembolisation (TACE) or radiofrequency ablation has also been reviewed recently.[47,48] Some studies suggest that a good response to TACE can predict survival at 5 years when compared to those tumours that do not respond at all (85% vs. 51%). This has not been translated into showing a benefit in reducing dropout rates for those exceeding listing criteria.[49] Indeed, bridging therapies on the whole have shown no benefit in reducing dropout rates for those that exceed listing criteria. Controversy is likely to continue to surround LT for HCC, particularly in the setting of LDLT, when potential recipients may exceed conventional listing criteria by virtue of supplying their own donor.

Human immunodeficiency virus (HIV)

The survival of patients infected with HIV has improved dramatically in recent years because of advances in highly active antiretroviral therapy (HAART). Until recently, HIV-positive patients were almost universally considered to be an absolute contraindication to transplantation. Early reports of LT in HIV-infected individuals showed poor results. Pittsburgh were the first to report their experience with 25 patients.[50] Failure in the early 1990s was due to poorer patient selection, lack of potent immunosuppression and availability of good antiviral regimens. It also appears that HCV coinfection with HIV has a significant impact on survival, with poor outcome compared to HBV coinfection.[51] This is especially important given that rates of coinfection of HIV and hepatitis B or C is 5–10% in the UK.

Guidelines[52] and recent reviews[53] of LT in HIV-infected individuals have recently been published. Current practice targets those with a low HIV viral load with limited or treated opportunistic infections and a $CD4^+$ T-cell count >200/mL or >100/μL in the presence of portal hypertension.[54]

A recent analysis from the THEVIC study group showed that with use of newer antiviral agents 5-year survival is 51% in patients with HIV/HCV coinfection compared with 81% in those with HCV alone.[55] A National Institutes of Health (NIH)-funded multicentre trial has been initiated to evaluate further treatment outcomes.

Hepatopulmonary syndrome (HPS)

HPS is a pulmonary complication observed in patients who have chronic liver disease and/or portal hypertension. This is due to intrapulmonary vascular dilatation causing severe hypoxaemia. HPS is usually detected when patients are being preassessed for listing. A diagnosis is made by blood gas analysis, transthoracic contrast-enhanced echo or a body scan with ^{99m}Tc-labelled macroaggregated albumin perfusion. When the P_aO_2 is ≥80 mmHg it is unlikely that the patient has HPS. When it is less than this imaging should be tailored to demonstrate pulmonary vascular dilatation. Knowing the degree of hypoxaemia is crucial to optimising patient management. Severely hypoxaemic patients (e.g. P_aO_2 < 60 mmHg) should be listed for LT; less than 50 mmHg may preclude LT because mortality is greatly increased. Between 60 and 80 mmHg, patients should be carefully followed to note any deterioration.[56]

Children

In children the commonest indication for LT is biliary atresia. This is a congenital fibro-inflammatory disease that destroys the biliary tree. Many such infants benefit initially from a porto-enterostomy (Kasai procedure) but the majority progress to secondary biliary cirrhosis and the need for a subsequent LT. The timing of transplantation in biliary atresia is largely determined by the onset of growth failure, portal hypertension, ascites and poor synthetic liver function. Children presenting with acute liver failure also make up a significant proportion requiring LT. The commonest cause is viral hepatitis (A, B and non-A–G). Metabolic syndromes also present in children with acute liver failure (ALF), including neonatal haemochromatosis and Wilson's disease.

Patient selection, risk assessment and allocation

Survival and outcome depends on the patient's clinical status and whether the LT is performed semi-electively or as an emergency for those with ALF. Traditionally cirrhotic patients are assessed in terms of their Child–Pugh score (CTP)[37,38] (Table 8.1). This score was devised to predict outcome in patients with portal hypertension and not those awaiting LT. Nevertheless, patients who are Child's A can remain stable for a considerable length of time without decompensation or the need for LT. Of those classified as Child's B/C, 75% will eventually meet criteria for LT. The limitations of CTP relate to sick Child's C patients. In particular it fails to discriminate those with FHF and does not take into account renal

Table 8.1 • Child–Pugh scoring

	Score		
	1	2	3
Bilirubin (μmol/L)	<34	<34–51	>51
Albumin (g/L)	>35	28–35	<28
Prothrombin time	<3	3–10	>10
Ascites	None	Slight	Moderate/severe
Encephalopathy	None	Slight	Moderate/severe

Grade A: 5–7 points; grade B: 8–9 points; grade C: 10–15 pts.
Pugh RN, Murray-Lyon IM, Dawson JL et al. Transection of the oesophagus for bleeding oesophageal varices. Br J Surg 1973; 60:646–9.

Box 8.4 • King's criteria for LT in fulminant hepatic failure

Paracetamol toxicity

1. pH <7.3
2. Prothrombin time >100 s and creatinine >300 μmol/L in patients with grade III and IV encephalopathy

Other patients

1. Prothrombin time >100 s (irrespective of grade of encephalopathy)
2. Or any three of the following:
   Age <10 or >40 years
   Non-A, non-B hepatitis, drug reaction or halothane hepatitis
   Jaundice for >7 days prior to encephalopathy
   Prothrombin time >50 s
   Serum bilirubin >300 μmol/L

From O'Grady JG, Alexander GJ, Hayllar KM et al. Early indicators of prognosis in fulminant hepatic failure. Gastroenterology 1989; 97:439–45. With permission from Elsevier/AGA Institute.

function, which is an important predictor of outcome. As a result the King's College Hospital criteria were widely adopted in the UK to define poor prognosis groups requiring LT with FHF[57] (Box 8.4).

More recently, because of concerns over priority for organ allocation, the Model for End-stage Liver Disease (MELD; Box 8.5) was developed in the USA.[58] The MELD score does have limitations but it is not within the scope of this chapter to discuss all its pitfalls, which have been reviewed in detail elsewhere.[59,60] It is estimated that it does not accurately predict mortality in up to 20% of patients based on concordance (c-statistic).[61] For example, the MELD score is limited by not considering other important factors which influence outcome such as the use of marginal grafts (e.g. older donors, presence of steatosis, tumour histology, etc.).

Box 8.5 • The Model for End-stage Liver Disease

MELD score = $3.8\log_e$ (bilirubin mg/dL) + $11.2\log_e$ (international normalised ratio) + $9.6\log_e$ (creatinine mg/dL) + 6.4(aetiology: 0 if cholestatic or alcoholic, 1 otherwise)

From Kamath PS, Wiesner RH, Malinchoc M et al. A model to predict survival in patients with end-stage liver disease. Hepatology 2001; 33:464–70. With permission from John Wiley & Sons/American Society for the Study of Liver Diseases.

MELD was originally developed to assess the prognosis of cirrhotic patients' prognosis who underwent a transjugular intrahepatic portosystemic shunt procedure (TIPS) and had been shown to predict short-term outcome in patients awaiting LT. Patients are given a score of severity, the higher the score the higher the risk of death.

MELD is good at predicting mortality within the first 3 months of LT for patients with many different aetiologies awaiting LT.[62] More importantly, in the USA, the use of MELD to aid allocation of livers has reduced waiting time and waiting-list deaths by allowing the sicker patients to be transplanted.[63–65]

To date MELD has undergone several revisions by the UNOS to allow better equity of access to donor livers – for example, capping of serum creatinine so those patients with renal failure do not have the advantage of a higher overall score. Another problem that was encountered was for patients with HCC. Those with long waiting times had a 50% chance of not meeting criteria at the time of LT because of tumour progression. Therefore UNOS modified the MELD score for HCC patients by loading the final score with additional points; this was originally 29. However, this score was deemed too high and the current UNOS policy is for patients with stage 2 disease (single tumour 2–5 cm or 2–3 nodules <3 cm) to receive 24 points and stage 1 patients no longer have any loading of points.[66] The score is also capped to limit futile transplants for a score higher than 40.[60]

Other specific recipient risk factors for LT include advancing age (>65 years), but there is no upper age limit so long as the patient has a predicted survival of more than 50% at 5 years. Other adverse factors include obesity (body mass index (BMI) >35 kg/m^2) and coronary artery disease, which should be treated before transplant if possible. Patients with advanced cardiac disease, poor ventricular function or cardiomyopathy are not candidates for LT. Many patients have transient and reversible renal failure (e.g. hepatorenal syndrome), which is not a contraindication for LT. Some patients will have reversible renal dysfunction and therefore may have combined liver and kidney transplants prematurely. Indeed, the number of combined transplants has increased during the MELD era. A general view is that those who have renal dysfunction but who are not dialysis dependent should not have a combined transplant, as suggested in recent guidelines.[67]

As the MELD score has limitations, the UK has developed its own scoring system to guide allocation (Box 8.6) and predict waiting-list mortality.[68] This is called the UK Model (UK Model for End-stage Liver Disease, or UKELD). This is a modified MELD system incorporating serum sodium, which has been shown to be an independent predictor of death after LT.[69] A UKELD score of 49 predicts a greater than 9% mortality at 1 year and is the minimum entry criterion to the waiting list.[5]

In the UK there are seven designated centres for LT and a recent working party has published new criteria for selection and de-listing for adult LT.[5] The principles of selection and listing are summarised in Box 8.7. Listing and allocation for those patients with ALF is very different and these patients can be offered a donor liver from any region in the UK and take national priority. The waiting list is prioritised based on time spent on the list and exclusions are made by blood group compatibility and donor/recipient size. Criteria for registration as a super-urgent LT in the UK are summarised in Box 8.8.

Box 8.6 • UKELD (www.uktransplant.org)

UKELD = 5{1.5 ln (international normalised ratio) + 0.3 ln (creatinine) + 0.6 ln (bilirubin) – 3 ln (Na^+) + 70}

Box 8.7 • The principles of patient selection

- Selection is based primarily on the risk of death without transplantation
- Selection is secondarily assessed on the ability of transplant to improve quality of life
- All cases are regularly reviewed to ensure that they continue to meet criteria and have neither improved nor become too sick to benefit from transplantation
- Criteria for removal from the list to be agreed and applied to all patients

From Neuberger J, Gimson A, Davies M et al. Selection of patients for liver transplantation and allocation of donated livers in the UK. Gut 2008; 57:252–7. With permission from BMJ Publishing Group Ltd.

Box 8.8 • Current UK criteria for registration as super-urgent

Category 1

Aetiology: paracetamol poisoning – pH <7.25 more than 24 h after overdose and fluid resuscitation

Category 2

Aetiology: paracetamol poisoning – coexisting prothrombin time >100 s or INR >6.5, and serum creatinine >300 μmol/L or anuria, and grade 3–4 encephalopathy

Category 3

Aetiology: paracetamol poisoning – serum lactate more than 24 h after overdose >3.5 mmol/L on admission or >3.0 mmol/L after fluid resuscitation

Category 4

Aetiology: paracetamol poisoning – two of the three criteria from category 2 with clinical evidence of deterioration (e.g. increased ICP, F_iO_2 >50%, increasing inotrope requirements) in the absence of clinical sepsis

Category 5

Aetiology: seronegative hepatitis, hepatitis A, hepatitis B or an idiosyncratic drug reaction, prothrombin time >100 s or INR >6.5 and any grade of encephalopathy

Category 6

Aetiology: seronegative hepatitis, hepatitis A, hepatitis B or an idiosyncratic drug reaction and any grade of encephalopathy, and any three from the following: unfavourable aetiology (idiosyncratic drug reaction, seronegative hepatitis), age >40 years, jaundice to encephalopathy time >7 days, serum bilirubin >300 mmol/L, prothrombin time >50 s or INR >3.5

Category 7

Acute presentation of Wilson's disease, or Budd–Chiari syndrome. A combination of coagulopathy and any grade of encephalopathy

Category 8

Hepatic artery thrombosis on days 0–21 after liver transplantation

Category 9

Early graft dysfunction on days 0–7 after liver transplantation with at least two of the following: AST >10 000 IU/L, INR >3.0, serum lactate >3 mmol/L, absence of bile production

Category 10

Any patient who has been a live liver donor who develops severe liver failure within 4 weeks of the donor operation

AST, aspartate aminotransferase; ICP, intracranial pressure; INR, international normalised ratio.

From Neuberger J, Gimson A, Davies M et al. Selection of patients for liver transplantation and allocation of donated livers in the UK. Gut 2008; 57:252–7. With permission from BMJ Publishing Group Ltd.

For elective registration all patients must fit into one of three categories:

- Category 1 – all patients must have an expected mortality exceeding 9% without LT.
- Category 2 – patients include those patients with HCC ('Milan criteria'; see above).
- Category 3 – comprises listing of patients based on variant syndromes (Box 8.9).

With the increasing number of waiting-list deaths and reduction in organ donors a number of

Box 8.9 • Variant syndromes and definitions for elective LT listing

Diuretic-resistant ascites

Ascites unresponsive to or intolerant of maximum diuretic dosage and non-responsive to TIPS or where TIPS deemed impossible or contraindicated and in whom the UKELD score at registration is ≤49

Hepatopulmonary syndrome

Arterial PO_2 <7.8 kPa. Alveolar–arterial oxgen gradient <20 mmHg. Calculated shunt fraction >8% (brain uptake following technetium macroaggregate albumin)

Chronic hepatic encephalopathy

Confirmed by EEG or trail-making tests with at least two admissions in 1 year due to exacerbations of encephalopathy that have not been manageable by standard therapy. Structural or neurological disease must be excluded by appropriate imaging and if necessary psychometric testing

Persistent and intractable pruritus

Pruritis consequent on cholestatic liver disease which is intractable after therapeutic trials which might include cholestyramine, ursodeoxycholic acid, rifampicin, odansetron, naltrexone and after exclusion of psychiatric comorbidity that might contribute to the itch

Familial amyloidosis

Confirmed transthyretin mutation in the absence of significant debilitating cardiac involvement or autonomic neuropathy

Primary hyperlipidaemias

Homozygous familial hypercholesterolaemia with absent LDL receptor expression and LDL gene mutation

Polycystic liver disease

Intractable symptoms due to the mass of liver or pain unresponsive to cystectomy or severe complications secondary to portal hypertension

LDL, low-density lipoprotein; TIPS, transjugular intrahepatic portosystemic shunt.

From Neuberger J, Gimson A, Davies M et al. Selection of patients for liver transplantation and allocation of donated livers in the UK. Gut 2008; 57:252–7. With permission from BMJ Publishing Group Ltd.

innovative strategies have been developed to overcome this shortfall. The main developments have been the use of marginal donors (older age, steatotic livers and non-heart-beating donors), reduction hepatectomy or split-liver transplantation.[70] Split-liver transplantation has had such a dramatic effect that it has reduced paediatric waiting-list mortality to almost zero,[71] but regrettably this has not advanced the adult situation despite some use of the residual right lobe into adult recipients. In recent years this has provided a compelling argument to develop the ultimate application of split-liver transplantation, living-related liver transplantation (LRLT).[72] In certain situations patients should be assessed for LT not just for aetiology but also for types of transplant, particularly if a living donor is available. LDLT has several advantages over cadaver liver transplantation. It reduces waiting-list mortality[73] and increases the rate of transplantation, but can also offer the potential for pre-emptive transplantation before patients develop the serious complications associated with cirrhosis and portal hypertension, thus making liver transplantation more cost-effective.

Work-up for liver transplantation

Assessment for conventional deceased donor

All patients will have had a conventional liver screen as part of their diagnosis. Important additional tests include tumour markers such as α-fetoprotein, CA19.9, CEA and CMV serology. Any patient with significant cardiopulmonary disease will need further investigation by either angiography, myocardial perfusion scanning and/or bubble contrast echo or cardiopulmonary exercise testing. Hepatopulmonary syndrome also has to be excluded by arterial blood gas analysis (see above). Specific investigations relating to LT include an assessment of the portal vein by Doppler ultrasound and/or magnetic resonance angiography to exclude any thrombosis. For recipients having had malignancy it is important to exclude metastatic disease by CT, bone scan and ascitic tap. Optimising pretransplant nutrition is also important, especially in children; for adults there is no good randomised evidence strong enough to exclude patients from the LT waiting list based on nutritional status.

Assessment for living donation

Although the first successful LDLT was done nearly 20 years ago, this still remains a controversial and complex area. The ethics of LDLT do not differ from those relating to renal transplantation. Donors must accept the risks of surgery and should agree to donation voluntarily without coercion. However, the stakes are higher for both donor and recipient. The only benefit to the donor is psychological. It is imperative to leave sufficient remnant liver to avoid hepatic insufficiency in the donor and provide a graft of sufficient size for restoration of health in the recipient. The ethical issues surrounding donation have been rigorously reviewed in detail elsewhere and it is well recognised that the risk is more appreciable than for renal transplantation (0.5% vs. 0.03%).[74–76] There have been possibly 33 reported donor deaths,[77,78] although the precise number is speculative as most probably go unreported.[77–79] A recent population study suggests that donors would accept LDLT even with a 10% mortality risk.[80]

The majority of donor/recipient combinations have been ABO compatible, but not exclusively.[81] ABO-incompatible recipients generally have worse outcomes.[82] Unfortunately, at least 50% of potential recipients are unable to find a suitable donor. Furthermore, 10% of donors discover they have significant medical comorbidity that needs investigation.[83] In total, less than 40% make suitable donors,[84] making the cost of donor work-up significant.

In the USA, obesity is becoming an increasingly frequent contraindication to LDLT, as there is an appreciable risk of the donor liver having significant steatosis. For example, patients with a BMI >28 cm/kg^2 have a 78% chance of more than 10% steatosis,[85] although this has been disputed by others.[86] It is well documented that steatosis increases the probability of compromised graft function and reduces hepatic regenerative capacity.[87] Some centres therefore assess the degree of steatosis with a predonation biopsy despite the risks for the donor (0.4–1%). In a recent survey of LDLT practice only 15% of centres biopsied routinely, whereas 25% did not biopsy at all.[88] A less invasive approach is to quantify steatosis with imaging, either CT or MRI (see below), but both these modalities have poor specificity.[89]

Hepatitis B surface antigen-positive donors are usually excluded from donation. Hepatitis B core antibody-positive donors can be used on the proviso they are also HepBsAg negative. In Taiwan up to 80% of the population can be hepatitis B core antibody positive. These donors should have a biopsy in order to exclude cirrhosis and fibrosis. Fortunately, donor hepatectomy in these patients has been shown to have no additional risk.[90] The use of lamivudine and hepatitis B Ig can significantly minimise the acquisition or reactivation of hepatitis B in the recipient, even in living donation.[91] LDLT in patients with HCV has been reviewed elsewhere.[92,93] In summary, anti-HCV-positive donors are refused donation in most European countries, but in the USA HCV donors can donate to HCV recipients. Preliminary concerns that HCV recurrence is more severe compared to cadaver liver transplantation have not been substantiated.[94,95]

It is also important to exclude occult thromboembolic disorders in living donors since thrombotic complications such as deep vein thrombosis or pulmonary embolism are a potential source of major morbidity and early postoperative death (see below). Donors should be carefully screened for abnormalities with respect to prothrombin, protein C, protein S, antithrombin III, factor V Leiden, factor VIII, cardiolipin and antiphospholipin.

Imaging for LDLT and graft selection

LDLT anamolous hepatic anatomy has to be fully appreciated and is vital for successful living donation. Arterial anomalies are common and can be a contraindication to donation (see below). Preoperative imaging is important for donor hepatectomy to ensure that remnant segments are left with their inflow (arterial and portal) and drainage systems intact (venous and biliary).

New generation multislice helical CT and 3-D reconstructions are used for volumetry analysis and to identify important anatomical variants.[96–98] Computerised analysis has been shown to correlate well with functional hepatic mass.[99–100]

MRI also provides acceptable high-resolution imaging without the added risk of radiation and contrast exposure when compared to CT.[101–103]

Determining the correct size of the graft for LDLT is vital. For adults donor/recipients have to be well matched in terms of BMI. In most cases, when the body weight of the recipient is 30% more than the donor, it is unlikely that the donor operation will yield a graft of sufficient size and leave an adequate remnant. Different formulae have been devised for guidance when assessing graft size. To predict a safe remnant volume the graft is expressed as a percentage of the standard liver volume (SLV).[104] It is generally agreed that between 30% and 40% SLV is sufficient to be safe.[105] Another important issue which can be overlooked is that most formulae do not assess functional volume as it is very difficult to assess, particularly if the donor has a degree of steatosis or fibrosis. Graft/recipient body weight ratio (GRWR) of more than 0.8–1.0% (i.e. graft weight vs. body weight) is also used.[106] If recipients are graded Child's B or C then a GRWR of more than 0.85% will be required to avoid small-for-size syndrome.[107]

Techniques of liver transplantation

For best results livers should be transplanted in the recipient within 12 hours of removal from the organ donor. University of Wisconsin (UW) solution has been the gold standard cold perfusate and storage solution for many years, but this has been questioned in recent years by several RCTs comparing it with Celsior (histidine–tryptophan–ketoglutarate, HTK) solution.

Although there is still no universal agreement the problems associated with UW are hyperkalaemic cardiac arrest, high viscosity, ischaemic-type biliary complications, microcirculatory disturbances due to undissolved particles and it is expensive. Celsior is cheaper, and has low potassium concentration with no need to flush before re-perfusion, no microparticles and improved biliary protection.[108]

There are many important anaesthetic considerations unique to LT. Special provision must be made for prolonged surgery (4–12 hours), hypothermia, hypoxia and excessive blood loss (10 units). Patients with ESLD have a high cardiac output with low peripheral resistance and are prone to metabolic acidosis and hyperkalaemia (at reperfusion), hypocalcaemia, thrombocytopenia and severe coagulopathy. There are in general four types of liver grafts: (1) whole; (2) reduced size; (3) split; (4) LDLT. The vast majority of grafts done worldwide are whole grafts. The operation has two phases: a hepatectomy phase and implantation phase.

Hepatectomy

The hepatectomy phase focuses on safe removal of the diseased liver with as little blood loss as possible while maintaining the integrity of the three important structures required to implant the new graft: venous drainage with the inferior vena cava (IVC; the retrohepatic IVC may be removed), portal inflow and hepatic artery or alternative arterial inflow. The fourth such structure is the bile duct, but sometimes a Roux-en-Y hepatico-jejunostomy is used as opposed to the conventional duct-to-duct technique and therefore the recipient's native extrahepatic bile duct may not need to be preserved.

The hepatectomy phase can be very demanding because of previous surgery, severe portal hypertension and/or previous subacute bacterial peritonitis. Conventional LT includes removing the recipient liver (anhepatic phase) and the hepatic segment of the IVC. This causes significant haemodynamic instability comprising reduced venous return, hypotension, splanchnic congestion and reduced cardiac output. Some centres overcome this problem by using veno-venous bypass (VVB; **Fig. 8.2**) or more commonly a temporary portocaval shunt if the IVC is preserved (portal vein to IVC). The main disadvantage of VVB is an increased risk of air embolism. It is not always necessary to institute these measures and a trial of IVC clamping may suffice whilst the hepatectomy is being completed when trying to preserve the IVC. If the patient has a good collateral circulation or portosystemic shunts a portocaval shunt is not usually necessary either. The piggyback technique involves preservation of the retrohepatic IVC and usually eliminates the need for VVB, but can be problematic if the caudate lobe completely surrounds the IVC at the time of hepatectomy (e.g. Budd–Chiari syndrome). VVB is rarely used in paediatric LT since children tolerate the haemodynamic consequences of IVC and portal cross-clamping better than adults.

The hepatectomy phase can be more challenging when the IVC is going to be preserved as the liver has to be completely mobilised off the retrohepatic IVC. Huge cholestatic livers and LT for Budd–Chiari can

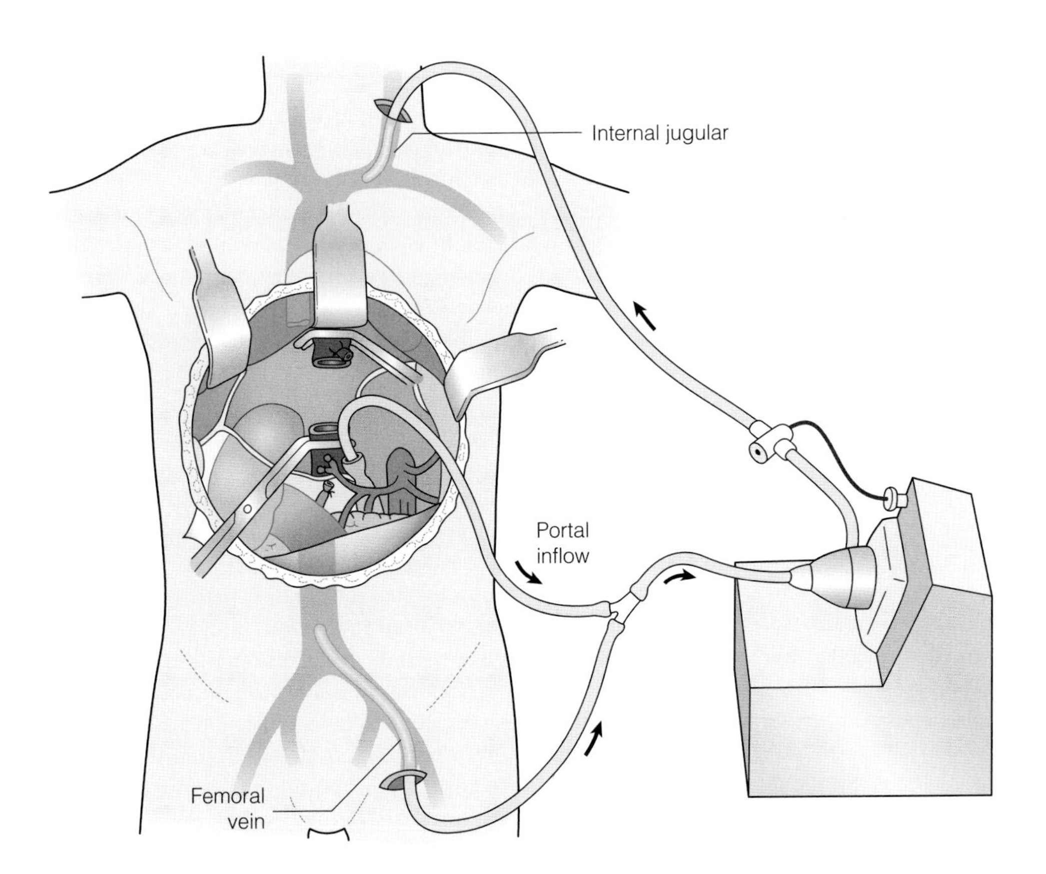

Figure 8.2 • Veno-venous bypass (VVB) circuit.

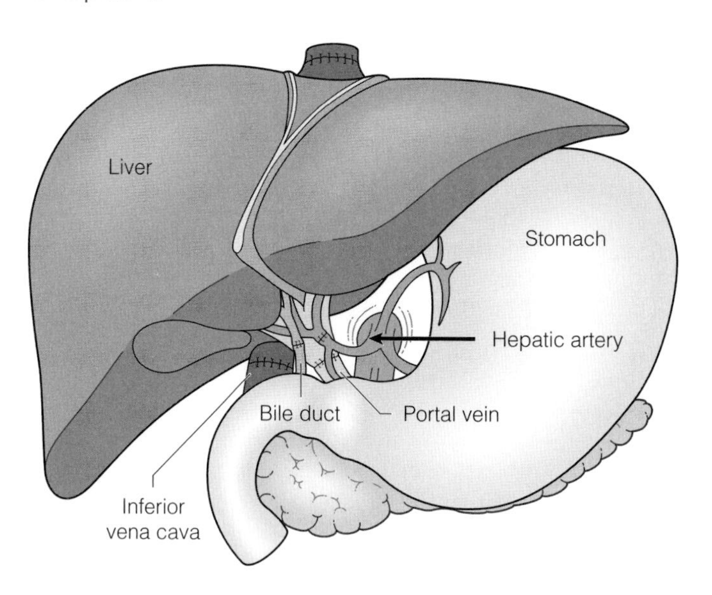

Figure 8.3 • Reconstruction of conventional orthotopic liver transplant.

make caval preservation very demanding indeed and sometimes a caval clamp will be needed for safety, but it is usually only for a short time and shunting is not usually needed. Caval preservation prolongs the hepatectomy phase but is associated with less blood loss. One of its advantages is that it speeds up the implantation phase as a single IVC anastomosis is required as opposed to two with the conventional method. The hepatic artery (HA) and common bile duct are ligated in the liver hilum. Aberrant arteries used to be a contraindication for LT but there are now various methods that can be used to circumvent this problem (see below). At the end of the anhepatic phase it is useful to ensure adequate length of inflow from the recipient HA and portal vein. The HA often needs to be mobilised and if required it is easier to do it when the liver is out before the new liver is put in.

Implantation

Various implantation techniques of LT have been described, and vary between centres and depending on the surgeon's preference. These include conventional (standard; **Fig. 8.3**) or piggyback techniques. If the native liver is removed en bloc with the recipient's retrohepatic IVC, two end-to-end anastomoses are performed using the retrohepatic donor IVC as an interposition IVC graft to restore its continuity. If the three remnant recipient hepatic veins remain these can be refashioned to form a single ostium with the IVC for anastomosis to the donor suprahepatic IVC. The lower anastomosis is performed just above the level of the renal veins.

A major change in this area has been the more widespread use of caval preservation. If the IVC is preserved several options exist. A piggyback technique can be performed where the donor suprahepatic IVC is anastomosed directly to the unified stumps of the recipient's hepatic veins. Modifications have been described where a longitudinal extension is made on both the recipient and donor IVC for anastomosis (**Fig. 8.4**). A latero-lateral (cavo-cavostomy or cavaplasty) is when the reconstruction does not involve the hepatic veins at all (**Fig. 8.5**). The anastomosis can be either anterior or slightly lateral. In most cases a side-biting Satinsky clamp can be used to perform the anastomosis without complete IVC occlusion. If there is a size discrepancy, to make this easier and to avoid venous outflow obstruction or kinking a reduced graft may be used, or an interposition venous graft or using the confluence of the recipient hepatic veins with a longitudinal extension, thus allowing a wider anastomosis. Alternatively, the left and middle hepatic vein can be modified to also form a common trunk and the recipient right hepatic vein is ligated. It is important not to stretch the donor hepatic veins in any way as this will lead to a venous outflow obstruction.

The portal vein anastomosis is performed next with an inverting suture on the back wall. A growth factor suture allows expansion at reperfusion without a concertina effect reducing the likelihood of stenosis. If the portal vein is thrombosed several options exist: thrombectomy (dissection or with a Fogarty catheter), a portal vein jump graft (with donor iliac vein) from either the superior mesenteric vein (SMV) or splenic vein or a major vein tributary

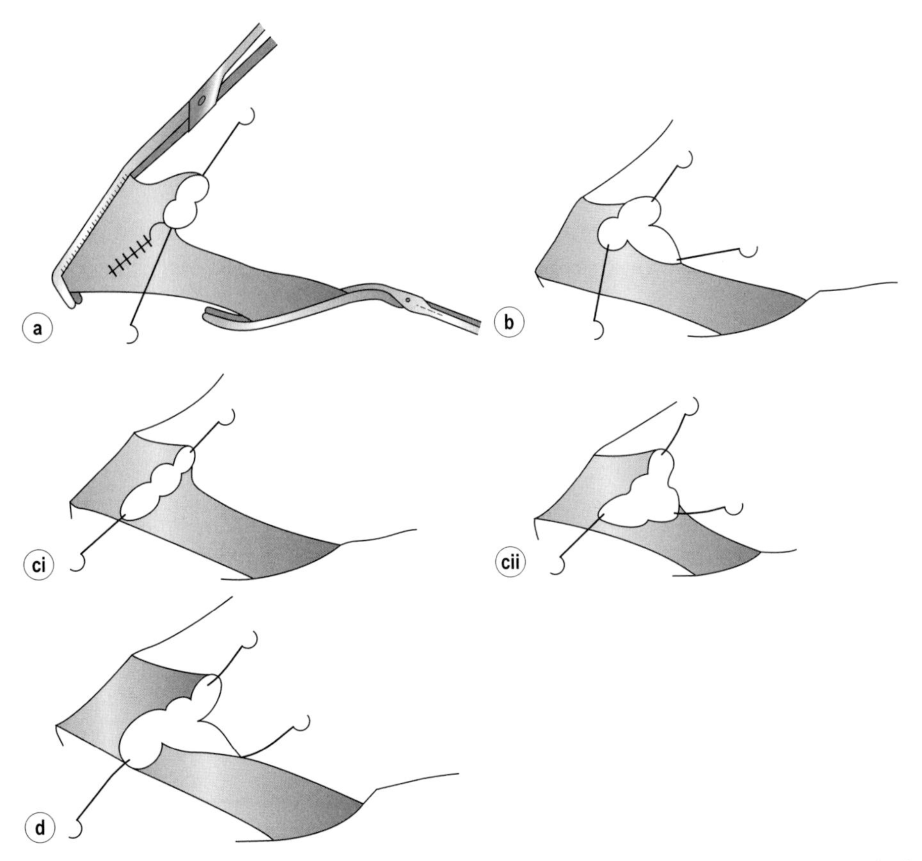

Figure 8.4 • Cavo-caval anastomosis. Caval implantation anastomotic site may vary according to surgeon's choice. In order to decrease the risk of further outflow problems, preference must be given to the largest anastomotic line and to triangulation technique, when possible (cii and d). **(a)** Trunk of middle and left hepatic veins. **(b)** Same as (a) with longitudinal split of the vena cava and triangulation. **(ci)** All hepatic veins ostia. **(cii)** Same as (ci) with triangulation. **(d)** Same as (ci), adding vena cava longitudinal split and triangulation technique.

(**Fig. 8.6**). A last resort would be a portocaval hemitransposition (rarely done; **Fig. 8.7**). A recent 11-year experience of 23 patients having hemitransposition has been reported but 3-year survival was only 38%.[109] The most common cause of death in this cohort is multiorgan failure, sepsis and bleeding. Previous portocaval shunts should be dismantled and a distal splenorenal (Warren) shunt must also be ligated as portal flow can be compromised, leading to higher rates of thrombosis. If a previous TIPS has been placed it is important to realise this before surgery as these can be prone to migration, making IVC preservation difficult. VVB is probably safer in this situation as portal collaterals have been decompressed by the TIPS.

The arterial anastomosis is performed using a patch of donor aorta or donor coelic artery with the recipient common Hepatic artery (HA) (end to end) or via a branch angioplasty at the confluence of the recipient HA and gastroduodenal artery (GDA). If the donor HA is too short an interposition graft can be used (e.g. aberrant recipient right hepatic artery, RHA) or donor iliac conduit either from the recipient iliac artery (**Fig. 8.8**) or from the infrarenal aorta. If there is a large donor accessory RHA then this can be reconstructed on the back table to either the splenic artery, GDA stump or right gastroepiploic artery (**Fig. 8.9**). Sometimes the recipient common HA proper is tiny and an accessory RHA is dominant; in this situation it makes sense to use this as the source of arterial inflow.

The final manoeuvres are removal of the donor gallbladder and creation of a biliary anastomosis. This is usually an end-to-end, duct-to-duct anastomosis. If the bile ducts are too far apart, have a size discrepancy or the recipient bile duct has to be excised (e.g. PSC), a Roux-en-Y bilioenteric reconstruction may be needed. A retrocolic Roux-en-Y is

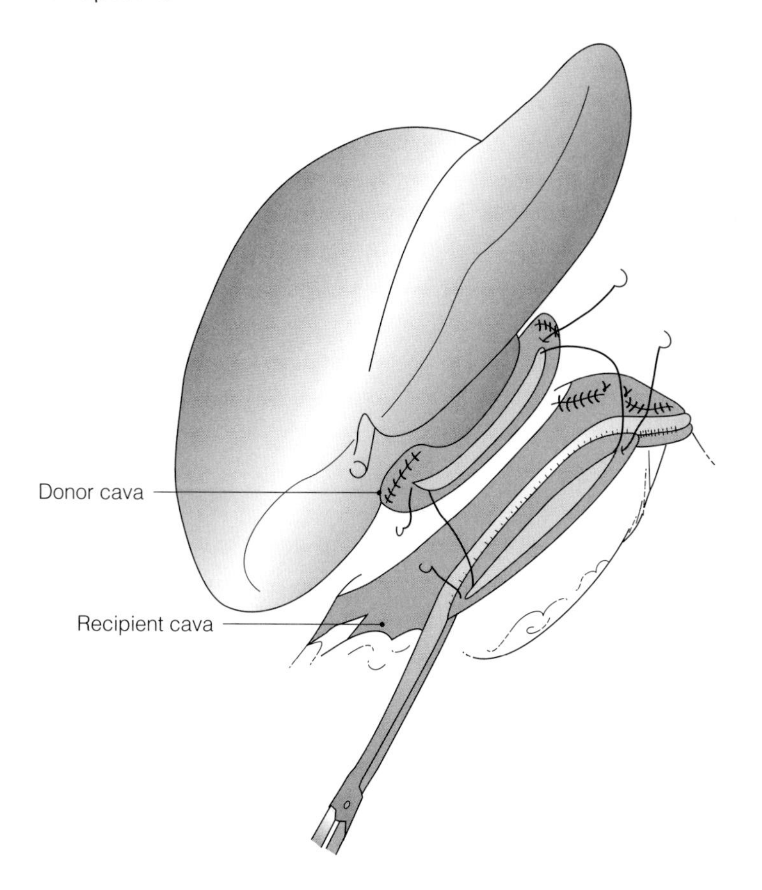

Figure 8.5 • Latero-lateral cavo-cavostomy. The donor and recipient venae cavae are preserved and a large side-to-side anastomosis is performed after suturing both ends of the donor cava. A large clamp is positioned but preserving the recipient caval flow during the procedure.

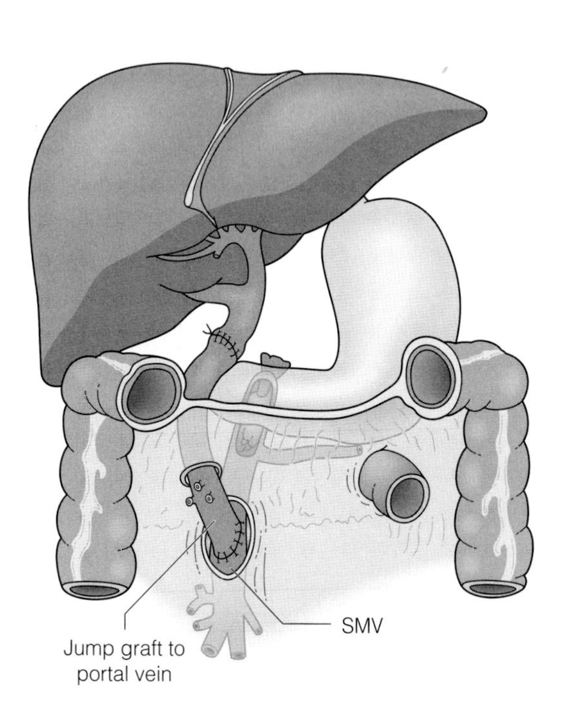

Figure 8.6 • Use of donor iliac vein for a portal vein jump graft from the SMV.

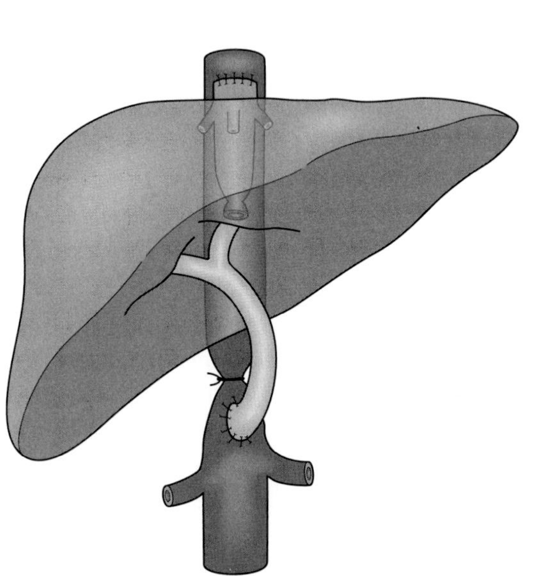

Figure 8.7 • Portocaval hemitransposition for portal venous thrombosis.

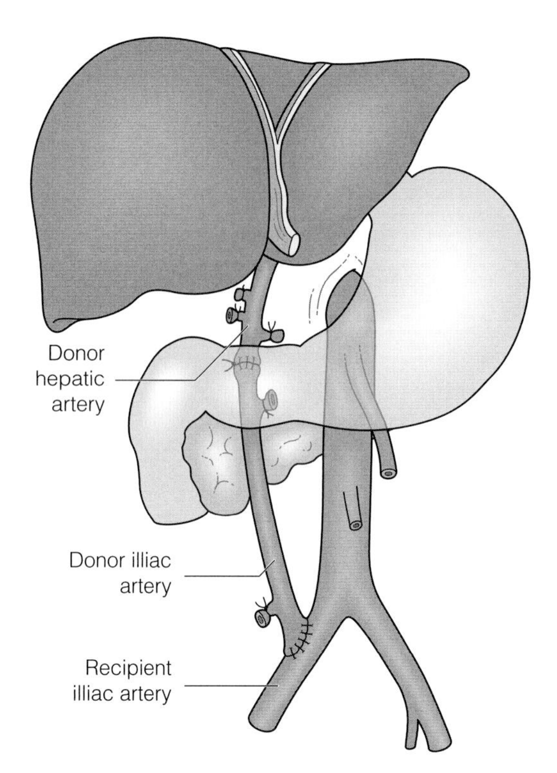

Figure 8.8 • Donor iliac artery jump graft from the recipient iliac artery (infrarenal aorta can be an alternative recipient artery).

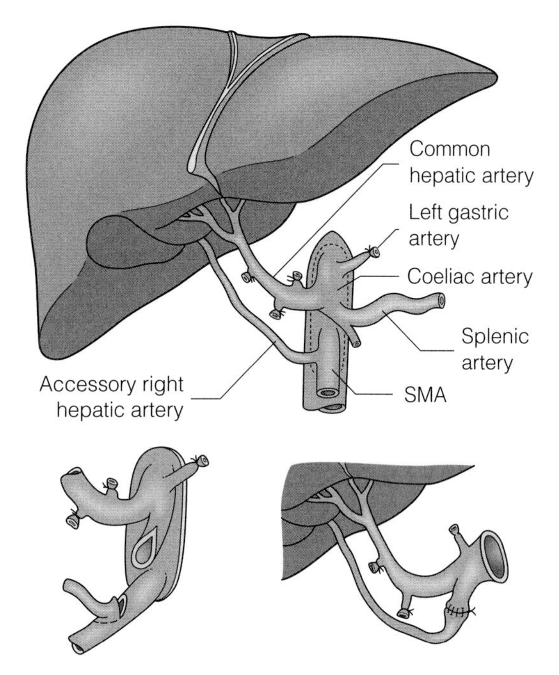

Figure 8.9 • Reconstruction of a donor right accessory hepatic artery onto the donor gastroduodenal stump. SMA, superior mesenteric artery.

best if the colon remains in a PSC patient. A biliary anastomosis over a T tube is now virtually obsolete. For LDLT the biliary anastomosis is principally duct to duct unless there is a specific indication for biloenteric (e.g. PSC). There is no agreement as to whether biliary stents or T tubes have any advantage in reducing biliary complications,[110] but in the authors' centre they are rarely used.

Reduced size grafts

A reduced size LT (RLT) is when a cadaver liver is reduced and the remainder discarded.[111] This is particularly useful for creating a liver graft for a small child where the shortage of small-sized matched donors is most acute. The common bile duct, common HA, the portal vein and IVC are retained with the graft. The most common reduced-sized graft is a left lobe (segments I–IV; **Fig. 8.10**) when the left lateral segment from a formal split will be too small.

RLT was originally used for emergency paediatric recipients and its main advantage was reducing preoperative death by reducing waiting time. Size reduction is generally feasible if the donor is no more than 4–6 times the weight of the recipient. Good results were obtained with 50% survival in emergency situations comparable to that of full-sized grafts at the time.[112] Eventually with the good results reported for RLT it was applied to elective patients, achieving 80% survival. The main complications were related to ischaemic damage and portal venous thrombosis.[112] One disadvantage of RLT is the back-table dissection significantly increases the cold ischaemic time compared to whole grafts. Overall RLT undoubtedly minimises time spent on the waiting list but compared to whole grafts does have increasing morbidity, particularly with biliary complications (30 days: 16% vs. 7% for whole grafts). Compared to whole grafts RLT also has lower rates of patient survival at 4 years (89% vs. 79%).[113] The major disadvantage of RLT is that half the liver has to be discarded to the detriment of potential adult recipients and consequently this technique has been superseded by split-liver transplantation, which maximises the use of donor organs.

Split-liver transplantation

RLT laid the groundwork for the development of split-liver transplantation (SLT). SLT typically yields a left lateral segment (LLS; II and III) with donor

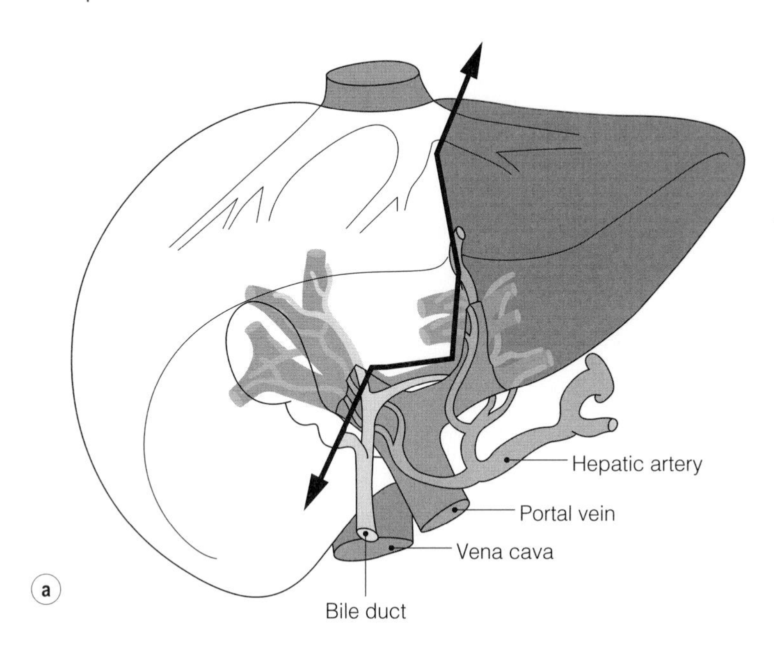

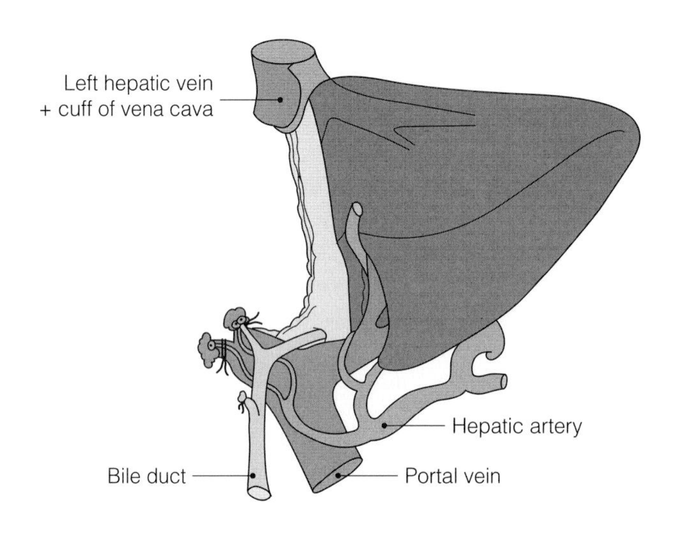

Figure 8.10 • Liver graft reduction technique. **(a)** Section line for preparation of a reduced liver graft using a straight transhilar approach. Right portal vein, hepatic artery and bile duct are electively divided at porta hepatis level. **(b)** Final aspect of the reduced graft minus segment four.

IVC removed and an extended right lobe graft with IVC attached for an adult[114] (**Fig. 8.11**). Size matching between donor grafts and recipients is based on the donor/recipient weight ratio. Thus an LLS can be used to transplant a child whose weight is 10–15 times less than the donor. An LLS can be further reduced to a monosegment to allow small infants weighing less than 5 kg to be transplanted.

An LLS split-liver graft is prepared either in situ or ex situ,[115,116] the latter being the most common in the UK. The liver is divided just to the right of the falciform ligament (across segment IV) and along the line of the ligamentum venosum posteriorly. The parenchyma is divided using either forceps or bipolar diathermy and ligatures. Traditionally the left hepatic artery, left portal vein and segment II/III duct are retained with the graft. The left hepatic vein is divided from the main donor IVC with a cuff. The liver is split. In the UK, in order to encourage adult-only centres to split whole grafts, the Liver Advisory Group's recommendation is that the extended right graft will receive all the vessels in their entirety. Implantation of an LLS requires some minor modifications. The left hepatic vein (LHV) is

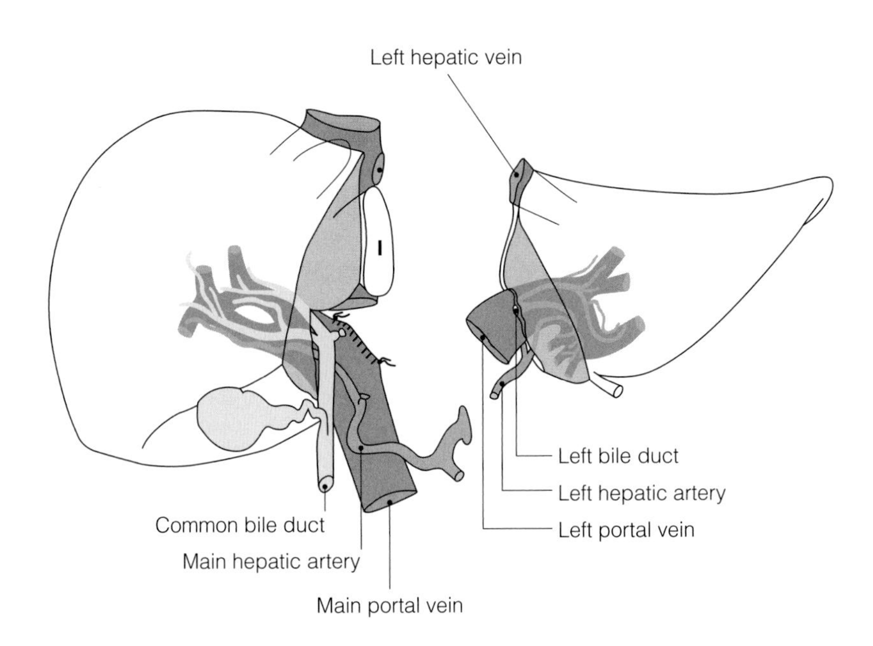

Figure 8.11 • Schematic drawing of two grafts obtained after splitting. Each hemiliver has its own vascular supply and bile duct.

anastomosed directly to the recipient IVC, the left portal vein may require reconstruction as the recipient portal vein is typically small in biliary atresia. A Roux-en-Y biliary anastomosis is routinely used (**Fig. 8.12**).

With respect to extended right lobe grafts it is often debated as to whether segment IV should be resected. This is because it is anatomically part of the left side. Initially this was commonly resected but this appears to be clinically insignificant.[70,116,117] However, resecting it may increase bleeding because of the bigger cut surface area.

Full-right (V–VIII) and full-left (I–IV) splits are less common and are more often retrieved using the in situ splitting technique. Graft weight (at least 0.8% recipient body weight) and quality are very important. One advance has been splitting the IVC down the middle, providing large venous cuffs for an anastomosis, as described by Broering et al.[118] (**Figs 8.13 and 8.14**). This reduces the risk of venous outflow obstruction and crucially preserves valuable functional volume from the caudate lobe.[119] Another technique, described by the Minnesota group, is preserving the IVC with the right lobe graft to maximise venous outflow. Thus all small, short draining veins from the right lobe graft are maintained with the IVC. Previously all

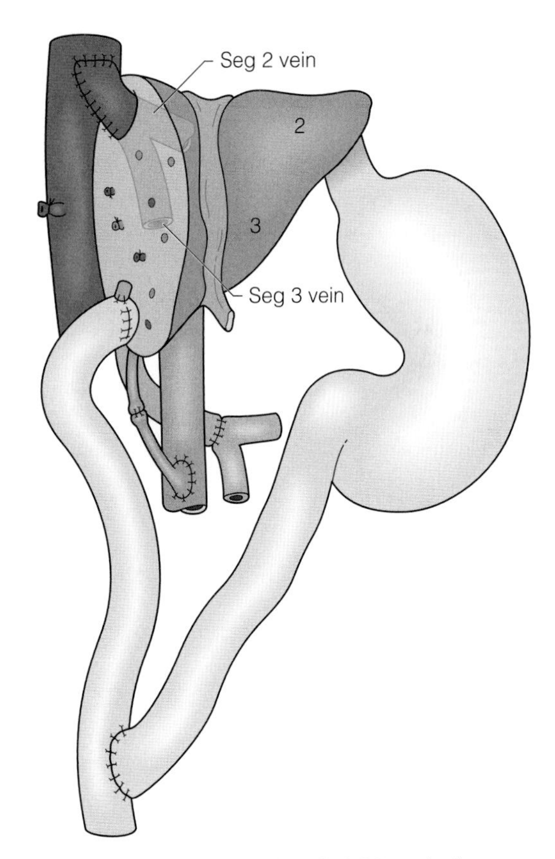

Figure 8.12 • Reconstruction of a left lateral split segment for a small child, demonstrating a Roux-en-Y hepatico-jejunostomy onto the left hepatic duct.

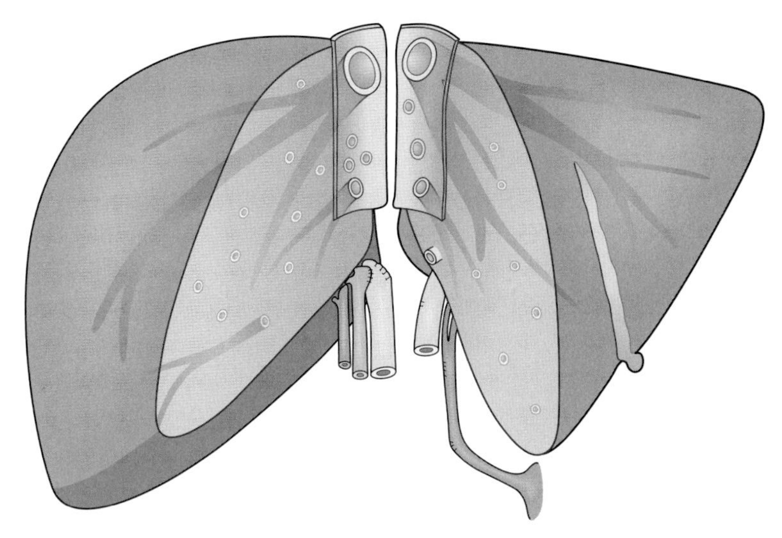

Figure 8.13 • Technique of full-right, full-left split for two adults by splitting the IVC down the middle. Reproduced from Gundlach M, Broering D, Topp S et al. Split-cava technique: liver splitting for two adult recipients. Liver Transplant 2000; 6:703–6. With permission from the American Association for the Study of Liver Diseases.

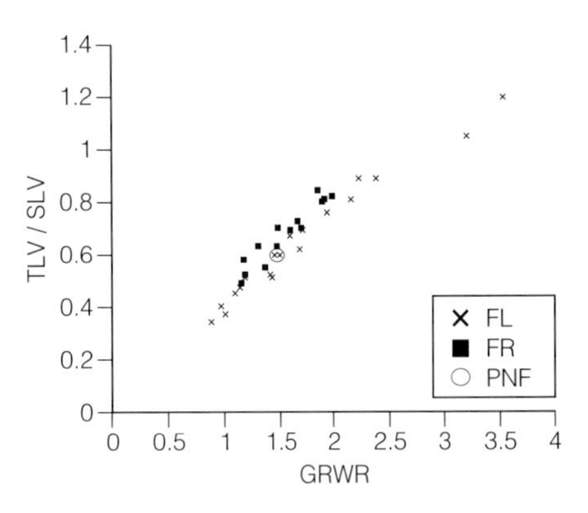

Figure 8.14 • Modified technique of full-right and full-left split by splitting the IVC and MHV. TLV - Total Liver Volume, SLV - Split Liver Volume, GRWR - Graft Recipient Weight Ratio FL - Full Left, FR - Full Right, PNF - Primary Non Function. Reproduced from Broering DC, Wilms C, Lenk C et al. Technical refinements and results in full-right full-left splitting of the deceased donor liver. Ann Surg 2005; 242:802–12. With permission from Lippincott, Williams & Wilkins.

significantly sized small veins (>5 mm) would have been individually re-implanted into the recipient IVC significantly adding to the warm ischaemic time. Major tributaries from the middle hepatic vein (MHV) are reconstructed on the back table by using a reversed segment of saphenous vein for direct implantation onto donor IVC that is retained with the graft. The venous anastomosis of the left lobe graft is usually between the confluence of the donor left and MHV with the recipient IVC.[120]

Results of full-right and full-left splitting are still inferior and should still be regarded as experimental, even though some centres have good results. The results of SLT are marginally better than those of RLT but still not as good as for a whole graft. The main problems relate to higher rates of vascular complications, particularly portal, compared to whole grafts.

A recent report from the SPLIT LT group describes 4-year patient survival of 89% (whole) vs. 85% (split).[113] Similarly, a matched paired analysis of extended right lobe grafts vs. whole grafts shows 5-year patient survival of 82.6% vs. 75.6% respectively.[121]

LDLT

Anatomical variants of the liver are very common and knowledge of these is very important for LDLT. Briefly, in 13–25% of cases the right hepatic artery (RHA) arises from the superior mesenteric artery.[122] Likewise, 25% have a left hepatic artery (LHA) completely replaced by a branch from the left gastric artery. Only a RHA arising from the LHA or a LHA arising from the RHA may preclude donation.[123] The presence of multiple arteries increases the complexity of arterial reconstruction and the risk of arterial thrombosis. Two portal vein branches (≈86%) or a trification (≈6%)[124] with anterior and posterior branches is not a contraindication to donation.[125] Biliary anatomy is more variable than vascular anatomy. Some authors defer delineation until surgery by intraoperative cholangiography after cholecystectomy.[126] The presence of multiple biliary orifices does not prevent donation but will necessitate multiple anastomosis.

The management of the middle hepatic vein for right-sided grafts is most often discussed and has been reviewed in detail elsewhere.[127] As a general rule it is safe to use right lobe grafts either with or without the MHV. In the latter case the use of venous grafts and hepatic venoplasty can give good results. The anterior sector (V and VIII) of a right lobe graft taken without the MHV can be subject to venous congestion and reduced functional capacity. Reconstruction with various interposition vein grafts has been described but does increase the risk of thrombosis. The native portal vein of the explant is most commonly used for grafting.[128] If the MHV is kept with the graft there is a risk of venous congestion in segment IV of the donor.[129–133] If the right hepatic vein (RHV) is small the MHV should be taken with the graft so long as the remnant size is more than 30% of the total donor liver. Often sizeable inferior RHVs are encountered (30%). Those more than 5 mm should be preserved for separate anastomosis[134,135] but increases warm ischaemic time.

Graft selection for LDLT

Grafts can be classified according to whether they are segmental (e.g. III) or contain multiple segments. The most common types of graft are left lateral segmental (II and III, 15–20% of SLV),[136] left lobe (II–IV, 40% SLV) or right lobe (V–VIII, 60% SLV) with or without the MHV. In general, left lateral grafts are for paediatric recipients up to 30 kg, left lobe grafts for 30–60 kg recipients and right lobe grafts for recipients more than 60 kg.[137,138] If the recipient is more than 60 kg and the left lobe does not yield a graft of sufficient size a posterior sectoral graft (VI and VII) should be considered.[139] However, this is technically more difficult to harvest and most centres have limited experience. In this scenario it may be safer to list the recipient for a cadaver liver or perform a dual LRLT where two grafts are harvested from two recipients.[140] An alternative approach would be to take the caudate lobe with a left lobe graft as it increases graft mass by about 9%. This could be critical if the graft size is considered marginal.[141] Techniques to reduce small-for-size syndrome include portal flow reduction by shunting.[142]

Results of LDLT

Important recipient adverse events are thrombosis, either arterial or portal. Initial reports were around 6%,[143] but this has now fallen to 2%.[144] The most common complications are again biliary related (15–20%).[137] Graft survival is up to 89% at 2 years.[141] The largest registry data (n = 796) reports a 75% survival at 3 years.[145] A recent Japanese survey of paediatric (n = 950) and adult recipients (n = 614) described 5-year cumulative survival of 82% and 70% respectively.[146] The commonest causes of death relate to sepsis (53%), haemorrhage (14%) and vascular complications (14%).[147] Patient survival ranges between 70% and 93% up to 3 years[148] and is highly dependent on the recipient's Child's status.

Finally, outcomes after LDLT for HCC need to be considered. Using the Milan criteria recurrence at 4 years is 75%.[41] Time spent on the waiting list may make prognosis worse; therefore LDLT is an attractive alternative as it reduces waiting time sevenfold.[149] Kyoto has one of the largest series.[150] Three-year survival was 55% but 45% did not achieve Milan criteria. Tumour-free survival was 68% at 3 years. Taken together, many patients who would not be accepted for orthotopic LT as judged by the Milan criteria mandate consideration for LDLT.

Laparoscopic live donor hepatectomy

A recent advance concerning paediatric LDLT has been laparoscopic donor hepatectomy (LDH).[151] Cherqui has the largest reported series of 16 cases, with one needing an open conversion. Their technique describes hepatectomy without hand-assisted devices using five ports. Each graft is retrieved through an endocatch device placed through a suprapubic incision. The mean operating time is under 5 hours with a warm ischaemic time of up to 15 minutes. The operation takes longer than an open procedure and there appears to be no difference in the mean hospital stay or analgesia requirements when compared to an open operation. There was a high rate of arterial thrombosis (2/7) leading to the death of one recipient in their first series of seven reported cases.[152]

Non-heart-beating donors (NHBDs)

The justification for utilising NHBDs is not new. Historically there have been serious concerns that organs retrieved from NHBDs are *second class*. Despite enthusiasm for NHBD renal transplantation, the use of NHBDs for liver transplantation has not been reciprocated as early results were considered significantly inferior to those of heart-beating donors (HBDs). However, these results were attributed to the use of poorly selected uncontrolled donors.

With judicious donor selection, a recent analysis from the UNOS demonstrates 3-year survival of 63% compared to 72% for HBDs.[153] At the 2nd European Meeting on Transplantation from NHBDs, the combined UK experience (King's, Leeds, Newcastle) of liver transplantation from NHBDs demonstrated a very acceptable 1-year graft survival of 84%.[154]

The major problems in NHBD liver transplantation continue to be higher rates of primary non-function (12% vs. 6%) and biliary complications when compared to those from HBDs,[155,156] thus more long-term results are eagerly awaited.

Auxiliary liver transplantation (ALT)

The initial results of ALT in patients with chronic liver disease were poor due to lack of liver regeneration. However, with the increasing use of split-liver transplantation the enthusiasm for ALT has increased in recent years. The largest experience is for patients whose native liver is expected to recover (e.g. after paracetamol-induced ALF). ALT allows immunosuppression to be gradually withdrawn as the native liver regenerates and recovers. There are two types of ALT: one is heterotopic, where the graft, usually partial, is placed below the unresected native liver; the second is an auxillary partial orthotopic, where the left or right lobe is resected and replaced by either a partial[157] or a whole graft.[158] In most cases the graft is a right lobe placed orthotopically after a right hepatectomy in the recipient. In this case the cut surfaces of the donor graft and the recipient remnant are positioned face to face. In general most centres report 1-year patient survival greater than 60% and comparable to whole grafts, but the added advantage is the lack of immunosuppression in the ALT group in over 50%.[157,159] However, this is at the expense of a higher rate of postoperative morbidity after ALT.

Domino liver transplantation

In this technique a donor liver is used to salvage a patient with a single metabolic liver-based defect (e.g. familial amyloid polyneuropathy, FAP). The recipient's liver can be used in turn for transplantation into an older patient (60 years) who is unlikely to develop the clinical manifestations of amyloid during their lifetime.[160] This subject has been reviewed in detail elsewhere.[161] A recent report of over 600 patients demonstrates that the results of domino liver transplantation for FAP are as good as orthotopic LT for other chronic liver diseases.[162] Readers are also directed to the world FAP liver transplant registry website at www.fapwtr.org.

Immunosuppression

It is not within the scope of this chapter to discuss all the aspects of immunosuppression relevant to LT and therefore only the outcome of important trials will be briefly discussed. LT is unique in that it is one of the few solid-organ transplants that sometimes does not require immunosuppression in that it can, in a selected few, be withdrawn totally.[163] Many different immunosuppressive regimens have been described but most consist of tacrolimus in combination with either azathioprine or mycophenolate mofetil (MMF). Two major RCTs in the USA and Europe reported superiority of tacrolimus compared to ciclosporin-based regimens.[164,165]

The TMC (neoral (C_0) vs. tacrolimus combined with azathioprine and prednisolone)[166] study has recently reported 3-year results and demonstrated that patients receiving tacrolimus are much more likely to have longer graft survival when compared to neoral (62% vs. 42%), less acute rejection and tacrolimus is better tolerated. A recent Cochrane review comparing neoral versus tacrolimus in LT recipients also supports these findings.[167]

However, the major disadvantage of tacrolimus is an increased risk of developing diabetes (27%). MMF, as an alternative to azathioprine, in combination with either tacrolimus or neoral and steroids appears to show no significant difference in rejection or long-term outcome.[168,169]

Since this trial commenced, a change in neoral monitoring has been advocated to reduce toxicity by measuring levels at 2 hours postingestion (C_{-2} monitoring). The LIS2T study was designed to evaluate any advantage. At 12 months there was no difference between graft survival (only in HCV; see below) and at 3 months no difference in biopsy-proven acute rejection. There was also no difference in renal function, but a higher rate of de novo diabetes was observed in those taking tacrolimus.[170,171]

For LDLT, steroid-free protocols are used in less than 10% at 2 years post-transplantation.[172] This is very important for paediatric recipients as this improves quality of life and does not retard growth.

Some liver disorders require modification of standard regimens. For example, those transplanted with HCV require a rapid steroid taper with a view to discontinuation after 2–3 months. The opposite is true for autoimmune disorders: higher doses of steroids are required to prevent onset of acute cellular rejection and recurrent autoimmunity.

The largest current study relating to HCV is the HCV-3 study.[173] This recently reported an interim 1-year analysis suggesting that a steroid-free regimen in combination with dacluzimab, tacrolimus and MMF reduces acute rejection ($P < 0.05$) and HCV recurrence (although non-significant). Dacluzimab has the added advantage of sparing the use of calcineurin inhibitors (CNIs) and steroids in the early postoperative period, particularly in those with renal impairment. Interestingly the TMC study showed no difference between outcome in HCV between tacrolimus or neoral, whereas in the LIS2T study there were higher rates of death and graft loss in those receiving tacrolimus.[170,171]

The widespread use of CNIs for immunosuppression often leads to renal impairment; it can be as much as 20% at 5 years. Changing immunosuppression in stable grafts does have its challenges as there is always a risk of acute rejection. Azathioprine is a less potent drug and has rarely been tried for conversion; MMF and sirolimus are more commonly tried. Sirolimus has its own disadvantages, including higher risks of thrombosis, but does have some advantages as it is thought to reduce myeloproliferation and carcinogenesis, although most studies are anecdotal.[174] Its use in LT for HCC will be clarified after completion of the proposed SiLVER study (www.silver-study.org). If CNIs are withdrawn sirolimus improves renal function but there are well-recognised side-effects.[175–177]

MMF as an alternative for patients with renal impairment has been more widely documented. If proteinuria is not present and the changes are reversible, up to 80% may have an improvement in creatinine clearance.[178,179] This policy works best for those that have not previously experienced rejection or are at low risk. In those at higher risk it may be safer to just reduce the CNI dose instead and add MMF as maintenance.[179] MMF replacing a CNI as monotherapy can lead to rejection and re-transplantation and is not advised.[180] MMF allows the use of steroid-free regimens but acute rejection appears to be a risk factor for accelerated hepatic fibrosis in HCV recipients and therefore MMF may have some benefits in this cohort compared to azathioprine.[181]

Lastly there is conflicting evidence as to whether tacrolimus is deleterious to recurrence of primary biliary cirrhosis after LT. Up to 40% of LT patients will develop recurrence but both neoral and tacrolimus are safe drugs to use,[182,183] with no obvious superiority between the two.

Complications after LT

Primary non-function (PNF)

The development of PNF is a devastating complication that in its worst form resembles FHF and requires urgent re-transplantation. PNF can result from an unstable donor, pre-existing disease in the donor, inadequate or overly long preservation, an imperfect recipient operation or a perioperative immunological reaction. These factors can occur separately or in combination. PNF rates of 5–10% have been reported.[184,185] A number of donor-specific factors have been identified which predict the development of initial poor function (IPF) or PNF in the immediate postoperative period:

- advanced age (>40 years);
- elevated bilirubin >20 mmol/L;
- ITU stay >3 days;
- hepatic steatosis >50%.

In most cases of PNF, the liver produces little or no bile after reperfusion; the pre-existing coagulopathy worsens (or occurs de novo) and the lactate level either fails to decrease or even increases. Occasionally, liver function is good or fair during the first 24 hours or so, only to deteriorate rapidly thereafter. Postoperatively, the patient is either comatose or extremely agitated. The urine output usually decreases, with a concomitant increase in blood urea and creatinine. The coagulation parameters are abnormal, the liver enzymes are very high and bilirubin increases rapidly. If the situation does not improve within 24–36 hours, the patient's only chance for survival is emergency re-transplantation. Recently, repeated sessions of plasmapheresis have been used with some success in buying time to allow the liver to return to normal (unpublished material).

The morbidity and mortality of PNF complication is high. Survival following re-transplantation for PNF is only half of that seen in the general liver transplant population. Re-transplantation for PNF has rates of 60% at 5 years compared to 67% for re-transplantation from other causes of initial graft failure. However, re-transplantation for another PNF is best avoided as long-term results are less than 50% at 5 years.[186] The incidence of adult graft loss from PNF within the first week following LT in our centre is less than 2%. The reasons for the reduction in IPF and PNF may include improved donor management, a policy of keeping the cold ischaemic time short and caution in the use of steatotic livers.

Hepatic artery thrombosis (HAT)

HAT is one of the most common and most disastrous arterial complications. It occurs in 4–10% of adult LT patients and in 9–42% of paediatric LT patients.[187] Causes include allograft rejection, hepatic artery kinking due to vascular redundancy, underlying hepatic artery stenosis (HAS) and technical problems at the anastomosis. Doppler ultrasonography (US) has a sensitivity of approximately 90% in detecting HAT.

Doppler findings include absent flow at the porta hepatis with absent intrahepatic arterial flow and absent flow in the donor hepatic artery with abnormal intrahepatic flow. Collateralised intrahepatic flow can be differentiated from normal flow by its tardus parvus Doppler waveform morphology. Tardus parvus can be diagnosed when the intrahepatic arterial resistive index (RI) is less than 0.5 and the systolic acceleration time (SAT) is 0.08 seconds or longer.[188]

Without aggressive diagnosis and treatment, HAT is associated with a mortality rate of greater than 80%. Unfortunately, percutaneous thrombolytic therapy of HAT is problematic because it often occurs soon after LT. Thrombolysis and angioplasty of underlying HAS have been successful in a few patients; however, HAT usually requires urgent re-transplantation for patient survival.

Hepatic artery stenosis (HAS)

HAS occurs in approximately 11–13% of all LTs. HAS can be caused by clamp injury, intimal injury due to perfusion catheters, anastomotic ischaemia due to a disrupted vasa vasorum or acute rejection. The consequences of haemodynamically significant HAS (>50% stenosis) are essentially identical to those of HAT. Although HAS can be corrected by balloon angioplasty, it usually requires operative vascular reconstruction or re-transplantation. Doppler US is used to screen for HAS. If the surgical anastomosis can be interrogated directly, stenosis is diagnosed if peak systolic velocity exceeds 200–300 cm/s.[188]

In most patients, direct Doppler evaluation of the HA anastomosis is not possible because the donor–recipient arterial anastomosis is tortuous, and because it is usually obscured by overlying bowel gas. However, similar to HAT, haemodynamically significant HAS results in intrahepatic arterial tardus parvus waveform morphology. By using an RI of less than 0.5 in conjunction with an SAT 0.08 seconds or longer yields a sensitivity of 45% for the detection of HAS. The sensitivity increases to 70% if two of the following three criteria are abnormal: (1) RI <0.5; (2) SAT 0.08 seconds or longer; and (3) peak systolic velocity >200 cm/s at the anastomotic site.[188,189]

Because the specificity of Doppler is only 64% in detecting marked arterial disease (i.e. HAT or haemodynamically significant HAS), angiography is usually required to confirm the diagnosis. At the time of diagnostic angiography, a decision can be made regarding the effectiveness of percutaneous correction of documented HAS.[190]

Pseudoaneurysms

Extrahepatic pseudoaneurysms occur after LT in approximately 2% of patients. The most common location of pseudoaneurysm formation is at the donor–recipient arterial anastomosis. A less common site is found at the ligation of the donor gastroduodenal artery. Pseudoaneurysms can be caused by infection or technical failure. Because of the potential for rupture and life-threatening haemorrhage, pseudoaneurysms need to be treated promptly. To ensure maximum arterial flow, most extrahepatic lesions are repaired surgically.

Intrahepatic pseudoaneurysms usually result from percutaneous biopsy, biliary procedures or infection. Most intrahepatic lesions can be treated by transcatheter embolisation or stent placement.[191]

Biliary complications

Biliary complications occur in up to 20% of adults following LT. With adult LT, a primary, end-to-end choledochocholedochostomy is performed after donor cholecystectomy. In most cases of LDLT, a biliary–enteric hepatojejunostomy is used to reconstitute biliary drainage. T tubes are no longer used for either type of biliary anastomosis; therefore, T-tube complications are no longer a problem.

Bile duct strictures can be anastomotic or non-anastomotic. Some anastomotic strictures result from technical difficulties or are caused by fibrosis and scarring that can be associated with bile leaks. Anastomotic stenosis can be related to marginal blood supply of the cut ends of the donor and recipient ducts; stricturing results from anastomotic ischaemia.

Non-anastomotic strictures can be caused by HAS, HAT, prolonged cold ischaemia time, rejection, CMV infection, intraductal sludge and stone formation, and recurrent primary sclerosing cholangitis in the allograft.

Non-ischaemic strictures can develop over time and present with deterioration of liver function tests and bouts of cholangitis. Dominant biliary strictures can be detected using US, CT, MRI, magnetic resonance cholangiopancreatography, percutaneous transhepatic cholangiography (PTC) and endoscopic retrograde cholangiopancreatography (ERCP).

The choices for mechanical revision of a bile duct stricture include PTC, ERCP or repeat operation. Ischaemic bile duct strictures can present with rapid onset of allograft dysfunction and sepsis. Unlike the native biliary tree that is supplied by collaterals from the gastroduodenal artery, the harvested donor bile duct and the anastomotic end of the recipient duct are solely dependent on the hepatic artery. Compromise of the hepatic artery by HAS or HAT results in abrupt ischaemia and necrosis of the biliary epithelium. This causes strictures, disruption of the ducts, bile leaks, bilomas, infected bilomas and abscesses.

Timely angiographic correction of HAS or HAT can occasionally result in graft salvage. Percutaneous drainage of associated fluid collections is often only a temporising measure because severe underlying biliary tree damage has usually occurred. Patients with sepsis who have severe biliary and hepatic damage from HAS or HAT require re-transplantation for ultimate survival.

Portal vein complications

Complications involving the portal vein occur in approximately 1–13% of LT recipients. Portal vein stenosis (PVS) usually occurs at the donor–recipient anastomosis. Portal vein thrombosis (PVT) often involves the main extrahepatic segment. Causes of PVS and PVT include surgical difficulties, thrombus formation from the portal venous bypass cannula, excessive vessel redundancy and hypercoagulability. PVS can lead to PVT. PVT and haemodynamically significant PVS cause graft dysfunction and portal hypertension.

US, CT, MRI and magnetic resonance venography can be used to detect both PVS and PVT. An abrupt three- to fourfold increase in velocity occurs within the portal vein on spectral Doppler US and aliasing on colour Doppler US reflects turbulent flow associated with PVS (PVS can be successfully corrected in many patients by using percutaneous transluminal balloon angioplasty).[192]

Haemodynamically significant PVS must be distinguished from portal vein pseudostenosis, which results when the recipient portal vein is somewhat larger than the donor portal vein. The difference in calibre causes increased velocity and turbulence at the anastomosis that are not physiologically significant.

If portal vein narrowing is associated with a velocity increase of less than three- to fourfold on spectral Doppler analysis, the narrowing is likely to be a pseudostenosis. PVT also can be diagnosed non-invasively (US, CT, MRI).

- Percutaneous intervention can be successful in restoring venous patency. Techniques include infusion of a thrombolytic (e.g. rTPA), angioplasty of associated PVS (if present), mechanical thrombectomy, and catheter embolisation of competing collaterals.[193]
- Surgical options in PVT include thrombectomy, segmental portal vein resection, placement of a jump graft or creation of a portosystemic shunt. PVT can necessitate re-transplantation.
- Occasionally PVT is detected in patients with normal allograft function and without portal hypertension. In these patients, sufficient hepatopetal collateralisation has developed to maintain adequate venous inflow.

Complications involving the IVC and associated hepatic vein

Post-transplantation IVC stenosis or thrombosis is an uncommon complication that occurs in approximately 1–4% of cases. IVC stenosis and thrombosis tend to occur at the superior and inferior caval anastomoses. Causes include:

- technical problems;
- IVC compression due to a fluid collection or haematoma; and
- mass effect from hepatic regeneration.

Patients with suprahepatic lesions tend to present with signs and symptoms of Budd–Chiari syndrome (pleural effusion, hepatic enlargement and ascites). IVC obstruction can cause lower-extremity oedema, which can be a prominent feature of infrahepatic stenosis. Clinically, leg swelling helps distinguish a suprahepatic caval complication from a portal vein problem, both of which are associated with features of portal hypertension.

US is the imaging study of choice for screening for IVC complications. A stenosis is detected as a grey-scale narrowing with a three- to fourfold increase in velocity on spectral Doppler analysis and associated colour Doppler aliasing. Indirect findings of suprahepatic IVC stenosis or thrombosis include distension of the hepatic veins with dampening of the hepatic venous spectral Doppler waveform and loss of its usual phasicity. Haemodynamically significant IVC stenosis can be differentiated from pseudostenosis by the presence of features of Budd–Chiari syndrome.

Stenosis and thrombosis of the IVC can also be detected by using CT scans. Inferior venacavographic results can confirm stenosis and thrombosis. Pressure gradient measurements can distinguish physiologically significant lesions from pseudostenoses. Balloon angioplasty and stent placement can be used to correct an IVC stenosis.

Isolated complications of the hepatic vein are rare. Strictures at the suprahepatic caval anastomosis can involve the confluence of the hepatic vein. Like caval strictures, lesions of this vein are amenable to percutaneous transjugular angioplasty and stent placement.

Acute rejection

Acute (cellular) hepatic allograft rejection can occur in as many as 40% of patients during the first 3 months post-transplantation. Acute rejection normally occurs 7–14 days after operation but can occur earlier or much later.

- Hyperacute rejection of the liver, comparable to that observed in kidney transplantation, is controversial and difficult to diagnose.
- Early accelerated rejection certainly occurs.

Liver biopsy may be required to distinguish between rejection and viral infection.

Rejection is most commonly manifested by malaise, fever, graft enlargement and diminished graft function, with a rise in bilirubin and transaminase levels. Graft biopsy should be performed, if safe, to document rejection. Rejection episodes are managed sequentially by pulse steroids, antithymocyte globulin, and/or the use of CellCept/tacrolimus switch (if patient was on ciclosporin), or the addition of rapamycin. Re-transplantation is the last resort when therapy fails and the patient develops hepatic failure.

Survival

Most large single-centre studies and registry data report that adult 30-day mortality after LT is in the region of 10–15%.[194–196] Most centres now report greater than 90% 1-year survival for elective LT in adults. Paediatric liver transplant centres now achieve 90% survival at 1 year and >80% at 5–10 years for ESLD, less after ALF. The European Liver Transplant Registry reports 5-year survival at 66% for adults. In comparison, those patients transplanted for ALF fare less well, with approximately 70% surviving at 1 year.[197] During the first 90 days after LT the risk-adjusted mortality is higher in the UK than in the USA, but after 1 year the results are reversed.[197] Most recipients (80%) report a good quality of life but there can be limitations because of nephrotoxicity, hypertension and other side-effects from immunosuppressive medication.[198]

Key points

- Liver transplantation has achieved excellent results – greater than 90% 1-year survival for elective LT in adults.
- Modifications of technique – even for liver transplants from deceased donors – continue to be made.
- LDLT has undergone major development in recent years – though ethical concerns still remain.
- The trend to use donors after cardiac death (or NHBDs) as a source of livers is likely to increase.
- Allocation systems for liver transplantation have undergone significant development in recent years.

References

1. www.eltr.org. Transplant statistics. Accessed 2008.
2. www.uktransplant.org. Transplant statistics. Accessed 2008.
3. van der Meulen JH, Lewsey JD, Dawwas MF et al. Adult orthotopic liver transplantation in the United Kingdom and Ireland between 1994 and 2005. Transplantation 2007; 84:572–9.
4. www.UNOS.org. Transplant statistics. Accessed 2008.

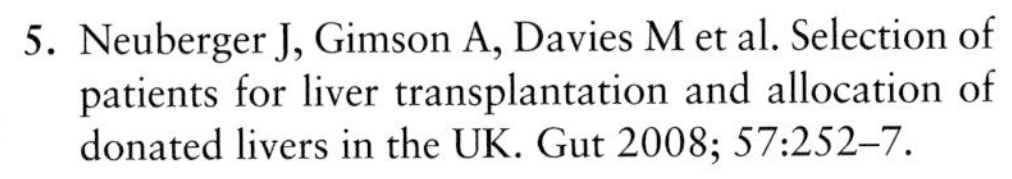

5. Neuberger J, Gimson A, Davies M et al. Selection of patients for liver transplantation and allocation of donated livers in the UK. Gut 2008; 57:252–7.
6. Neuberger J, James O. Guidelines for selection of patients for liver transplantation in the era of donor-organ shortage. Lancet 1999; 354:1636–9.
7. Tome S, Lucey MR. Timing of liver transplantation in alcoholic cirrhosis. J Hepatol 2003; 39:302–7.
8. Mackie J, Groves K, Hoyle A et al. Orthotopic liver transplantation for alcoholic liver disease: a retrospective analysis of survival, recidivism, and risk factors predisposing to recidivism. Liver Transplant 2001; 7:418–27.
9. Young TA, Neuberger J, Longworth L et al. Survival gain after liver transplantation for patients with alcoholic liver disease: a comparison across models and centers. Transplantation 2003; 76:1479–86.
10. Bathgate AJ on behalf of the UK Liver Transplant Units. Recommendations for alcohol-related liver disease. Lancet 2006; 367:2045–6.
11. O'Grady JG. Liver transplantation alcohol related liver disease: (deliberately) stirring a hornet's nest! Gut 2006; 55:1529–31.
12. Russo MW, Shrestha R. Is severe recurrent hepatitis C more common after adult living donor liver transplantation? Hepatology 2004; 40:524–6.
13. Gane EJ, Portmann BC, Naoumov NV et al. Long-term outcome of hepatitis C infection after liver transplantation. N Engl J Med 1996; 334:815–20.
14. Prieto M, Berenguer M, Rayon JM et al. High incidence of allograft cirrhosis in hepatitis C virus genotype 1b infection following transplantation: relationship with rejection episodes. Hepatology 1999; 29:250–6.
15. Booth JC, O'Grady J, Neuberger J. Clinical guidelines on the management of hepatitis C. Gut 2001; 49(Suppl 1):I1–21.
16. Berenguer M, Palau A, Aguilera V et al. Clinical benefits of antiviral therapy in patients with recurrent hepatitis C following liver transplantation. Am J Transplant 2008; 8:679–87.
17. Shergill AK, Khalili M, Straley S et al. Applicability, tolerability and efficacy of preemptive antiviral

therapy in hepatitis C-infected patients undergoing liver transplantation. Am J Transplant 2005; 5:118–24.

18. Crippin JS, McCashland T, Terrault N et al. A pilot study of the tolerability and efficacy of antiviral therapy in hepatitis C virus-infected patients awaiting liver transplantation. Liver Transplant 2002; 8:350–5.
19. Roche B, Samuel D. Risk factors for hepatitis C recurrence after liver transplantation. J Viral Hepat 2007; 14(Suppl 1):89–96.
20. Gane EJ, Lo SK, Riordan SM et al. A randomized study comparing ribavirin and interferon alfa monotherapy for hepatitis C recurrence after liver transplantation. Hepatology 1998; 27:1403–7.
21. Samuel D, Bizollon T, Feray C et al. Interferon-alpha 2b plus ribavirin in patients with chronic hepatitis C after liver transplantation: a randomized study. Gastroenterology 2003; 124:642–50.
22. Samuel D, Muller R, Alexander G et al. Liver transplantation in European patients with the hepatitis B surface antigen. N Engl J Med 1993; 329:1842–7.
23. Grellier L, Mutimer D, Ahmed M et al. Lamivudine prophylaxis against reinfection in liver transplantation for hepatitis B cirrhosis. Lancet 1996; 348:1212–15.
24. Schiff E, Lai CL, Hadziyannis S et al. Adefovir dipivoxil for wait-listed and post-liver transplantation patients with lamivudine-resistant hepatitis B: final long-term results. Liver Transplant 2007; 13:349–60.
25. Roche B, Feray C, Gigou M et al. HBV DNA persistence 10 years after liver transplantation despite successful anti-HBS passive immunoprophylaxis. Hepatology 2003; 38:86–95.
26. Angus PW, McCaughan GW, Gane EJ et al. Combination low-dose hepatitis B immune globulin and lamivudine therapy provides effective prophylaxis against posttransplantation hepatitis B. Liver Transplant 2000; 6:429–33.
27. Vera A, Gunson BK, Ussatoff V et al. Colorectal cancer in patients with inflammatory bowel disease after liver transplantation for primary sclerosing cholangitis. Transplantation 2003; 75:1983–8.
28. Papatheodoridis GV, Hamilton M, Mistry PK et al. Ulcerative colitis has an aggressive course after orthotopic liver transplantation for primary sclerosing cholangitis. Gut 1998; 43:639–44.
29. Jenkins RL, Pinson CW, Stone MD. Experience with transplantation in the treatment of liver cancer. Cancer Chemother Pharmacol 1989; 23(Suppl):S104–9.
30. Becker NS, Rodriguez JA, Barshes NR et al. Outcomes analysis for 280 patients with cholangiocarcinoma treated with liver transplantation over an 18-year period. J Gastrointest Surg 2008; 12:117–22.
31. Heimbach JK, Gores GJ, Haddock MG et al. Predictors of disease recurrence following neo-adjuvant chemoradiotherapy and liver transplantation for unresectable perihilar cholangiocarcinoma. Transplantation 2006; 82:1703–7.
32. Rea DJ, Heimbach JK, Rosen CB et al. Liver transplantation with neoadjuvant chemoradiation is more effective than resection for hilar cholangiocarcinoma. Ann Surg 2005; 242:451–8.
33. Iwatsuki S, Gordon RD, Shaw BW Jr et al. Role of liver transplantation in cancer therapy. Ann Surg 1985; 202:401–7.
34. Vauthey JN, Lauwers GY, Esnaola NF et al. Simplified staging for hepatocellular carcinoma. J Clin Oncol 2002; 20:1527–36.
35. Okuda K, Ohtsuki T, Obata H et al. Natural history of hepatocellular carcinoma and prognosis in relation to treatment. Study of 850 patients. Cancer 1985; 56:918–28.
36. The Cancer of the Liver Italian Program (CLIP) Investigators. Prospective validation of the CLIP score: a new prognostic system for patients with cirrhosis and hepatocellular carcinoma. Hepatology 2000; 31:840–5.
37. Turcotte JG, Child CG III. Portal hypertension. Pathogenesis, management and prognosis. Postgrad Med 1967; 41:93–102.
38. Pugh RN, Murray-Lyon IM, Dawson JL et al. Transection of the oesophagus for bleeding oesophageal varices. Br J Surg 1973; 60:646–9.
39. Levy I, Sherman M. Staging of hepatocellular carcinoma: assessment of the CLIP, Okuda, and Child–Pugh staging systems in a cohort of 257 patients in Toronto. Gut 2002; 50:881–5.
40. Llovet JM. Updated treatment approach to hepatocellular carcinoma. J Gastroenterol 2005; 40:225–35.
41. Mazzaferro V, Regalia E, Doci R et al. Liver transplantation for the treatment of small hepatocellular carcinomas in patients with cirrhosis. N Engl J Med 1996; 334:693–9.
42. Yao FY, Ferrell L, Bass NM et al. Liver transplantation for hepatocellular carcinoma: expansion of the tumor size limits does not adversely impact survival. Hepatology 2001; 33:1394–403.
43. Duffy JP, Vardanian A, Benjamin E et al. Liver transplantation criteria for hepatocellular carcinoma should be expanded: a 22-year experience with 467 patients at UCLA. Ann Surg 2007; 246:502–9.
44. Cillo U, Vitale A, Bassanello M et al. Liver transplantation for the treatment of moderately or well-differentiated hepatocellular carcinoma. Ann Surg 2004; 239:150–9.
45. Yao FY, Xiao L, Bass NM et al. Liver transplantation for hepatocellular carcinoma: validation of the UCSF-expanded criteria based on preoperative imaging. Am J Transplant 2007; 7:2587–96.

46. Sutcliffe R, Maguire D, Portmann B et al. Selection of patients with hepatocellular carcinoma for liver transplantation. Br J Surg 2006; 93:11–18.

47. Belghiti J, Carr BI, Greig PD et al. Treatment before liver transplantation for HCC. Ann Surg Oncol 2008; 15:993–1000.

48. Schwartz M, Roayaie S, Uva P. Treatment of HCC in patients awaiting liver transplantation. Am J Transplant 2007; 7:1875–81.

49. Martin P. Can response to TACE predict survival after liver transplantation in patients with hepatocellular carcinoma? Nat Clin Pract Gastroenterol Hepatol 2007; 4:540–1.

50. Tzakis AG, Cooper MH, Dummer JS et al. Transplantation in HIV+ patients. Transplantation 1990; 49:354–8.

51. Prachalias AA, Pozniak A, Taylor C et al. Liver transplantation in adults coinfected with HIV. Transplantation 2001; 72:1684–8.

52. O'Grady J, Taylor C, Brook G. Guidelines for liver transplantation in patients with HIV infection (2005). HIV Med 2005; 6(Suppl 2):149–53.

53. Samuel D, Weber R, Stock P et al. Are HIV-infected patients candidates for liver transplantation? J Hepatol 2008; 48:697–707.

54. Neff GW, Sherman KE, Eghtesad B et al. Review article: current status of liver transplantation in HIV-infected patients. Aliment Pharmacol Ther 2004; 20:993–1000.

55. Duclos-Vallee JC, Feray C, Sebagh M et al. Survival and recurrence of hepatitis C after liver transplantation in patients coinfected with human immunodeficiency virus and hepatitis C virus. Hepatology 2008; 47:407–17.

56. Pastor CM, Schiffer E. Therapy insight: hepatopulmonary syndrome and orthotopic liver transplantation. Nat Clin Pract Gastroenterol Hepatol 2007; 4:614–21.

57. O'Grady JG, Alexander GJ, Hayllar KM et al. Early indicators of prognosis in fulminant hepatic failure. Gastroenterology 1989; 97:439–45.

58. Kamath PS, Wiesner RH, Malinchoc M et al. A model to predict survival in patients with end-stage liver disease. Hepatology 2001; 33:464–70.

59. Cholongitas E, Senzolo M, Triantos C et al. MELD is not enough – enough of MELD? J Hepatol 2005; 42:475–7.

60. Kamath PS, Kim WR. The model for end-stage liver disease (MELD). Hepatology 2007; 45:797–805.

61. Wiesner RH, McDiarmid SV, Kamath PS et al. MELD and PELD: application of survival models to liver allocation. Liver Transplant 2001; 7:567–80.

62. Yoo HY, Thuluvath PJ. Short-term postliver transplant survival after the introduction of MELD scores for organ allocation in the United States. Liver Int 2005; 25:536–41.

63. Wiesner R, Edwards E, Freeman R et al. Model for end-stage liver disease (MELD) and allocation of donor livers. Gastroenterology 2003; 124:91–6.

64. Freeman RB, Wiesner RH, Edwards E et al. Results of the first year of the new liver allocation plan. Liver Transplant 2004; 10:7–15.

65. Brown RS Jr, Lake JR. The survival impact of liver transplantation in the MELD era, and the future for organ allocation and distribution. Am J Transplant 2005; 5:203–4.

Reviews of the application of MELD to liver allocation.

66. Wiesner RH, Freeman RB, Mulligan DC. Liver transplantation for hepatocellular cancer: the impact of the MELD allocation policy. Gastroenterology 2004; 127:S261–7.

67. Eason JD, Gonwa TA, Davis CL et al. Proceedings of Consensus Conference on Simultaneous Liver Kidney Transplantation (SLK). Am J Transplant 2008; 8:2243–51.

68. Barber KM, Pioli S, Blackwell JE et al. Development of a UK score for patients with end stage liver disease. Hepatology 2007; 46.

69. Dawwas MF, Lewsey JD, Neuberger JM et al. The impact of serum sodium concentration on mortality after liver transplantation: a cohort multicenter study. Liver Transplant 2007; 13:1115–24.

70. Azoulay D, Astarcioglu I, Bismuth H et al. Split-liver transplantation. The Paul Brousse policy. Ann Surg 1996; 224:737–46.

71. de Ville DG, Reding R, Sokal E et al. Related living donor for liver transplantation in children: results and impact [in French]. Chirurgie 1997; 122:83–7.

72. Strong RW, Lynch SV, Ong TH et al. Successful liver transplantation from a living donor to her son. N Engl J Med 1990; 322:1505–7.

73. Liu CL, Lam B, Lo CM et al. Impact of right-lobe livedonor liver transplantation on patients waiting for liver transplantation. Liver Transplant 2003; 9:863–9.

74. Shapiro RS, Adams M. Ethical issues surrounding adult-to-adult living donor liver transplantation. Liver Transplant 2000; 6:S77–80.

75. Matas AJ, Bartlett ST, Leichtman AB et al. Morbidity and mortality after living kidney donation, 1999–2001: survey of United States transplant centers. Am J Transplant 2003; 3:830–4.

76. Surman OS. The ethics of partial-liver donation. N Engl J Med 2002; 346:1038.

77. Akabayashi A, Slingsby BT, Fujita M. The first donor death after living-related liver transplantation in Japan. Transplantation 2004; 77:634.

78. White SA, Pollard SG. Living donor liver transplantation. Br J Surg 2005; 92:262–3.

79. Ringe B, Strong RW. The dilemma of living donor death: to report or not to report. Transplantation 2008;85:790–3.

80. Neuberger J, Farber L, Corrado M et al. Living liver donation: a survey of the attitudes of the public in Great Britain. Transplantation 2003; 76:1260–4.

81. Tanabe M, Shimazu M, Wakabayashi G et al. Intraportal infusion therapy as a novel approach to adult, ABO incompatible liver transplantation. Transplantation 2004; 73:1959–61.

82. Todo S, Furukawa H, Jin MB et al. Living donor liver transplantation in adults: outcome in Japan. Liver Transplant 2000; 6:S66–72.

83. Trotter JF, Wachs M, Trouillot T et al. Evaluation of 100 patients for living donor liver transplantation. Liver Transplant 2000; 6:290–5.

84. Ghobrial RM, Freise CE, Trotter JF et al. Donor morbidity after living donation for liver transplantation. Gastroenterology 2008;135:468–76.

85. Rinella ME, Alonso E, Rao S et al. Body mass index as a predictor of hepatic steatosis in living liver donors. Liver Transplant 2001; 7:409–14.

86. Ryan CK, Johnson LA, Germin BI et al. One hundred consecutive hepatic biopsies in the workup of living donors for right lobe liver transplantation. Liver Transplant 2002; 8:1114–22.

87. Selzner M, Clavien PA. Failure of regeneration of the steatotic rat liver: disruption at two different levels in the regenerative pathway. Hepatology 2000; 31:35–42.

88. Brown RS Jr, Russo MW, Lai M et al. A survey of liver transplantation from living adult donors in the United States. N Engl J Med 2003; 348:818–25.

89. Rinella ME, McCarthy R, Thakrar K et al. Dual-echo, chemical shift gradient-echo magnetic resonance imaging to quantify hepatic steatosis: implications for living liver donation. Liver Transplant 2003; 9:851–6.

90. Hwang S, Moon DB, Lee SG et al. Safety of anti-hepatitis B core antibody-positive donors for living-donor liver transplantation. Transplantation 2003; 75:S45–8.

91. Chen YS, Wang CC, de Villa VH et al. Prevention of de novo hepatitis B virus infection in living donor liver transplantation using hepatitis B core antibody positive donors. Clin Transplant 2002; 16:405–9.

92. Everson GT, Trotter J. Role of adult living donor liver transplantation in patients with hepatitis C. Liver Transplant 2003; 9:S64–8.

93. Zimmerman MA, Trotter JF. Living donor liver transplantation in patients with hepatitis C. Liver Transplant 2003; 9:S52–7.

94. Troppmann C, Rossaro L, Perez RV et al. Early, rapidly progressive cholestatic hepatitis C reinfection and graft loss after adult living donor liver transplantation. Am J Transplant 2003; 3:239–40.

95. Van Vlierberghe H, Troisi R, Colle I et al. Hepatitis C infection-related liver disease: patterns of recurrence and outcome in cadaveric and living-donor liver transplantation in adults. Transplantation 2004; 77:210–14.

96. Schroeder T, Nadalin S, Stattaus J et al. Potential living liver donors: evaluation with an all-in-one protocol with multi-detector row CT. Radiology 2002; 224:586–91.

97. Leelaudomlipi S, Sugawara Y, Kaneko J et al. Volumetric analysis of liver segments in 155 living donors. Liver Transplant 2002; 8:612–14.

98. Kamel IR, Kruskal JB, Warmbrand G et al. Accuracy of volumetric measurements after virtual right hepatectomy in potential donors undergoing living adult liver transplantation. Am J Roentgenol 2001; 176:483–7.

99. Kamel IR, Kruskal JB, Raptopoulos V. Imaging for right lobe living donor liver transplantation. Semin Liver Dis 2001; 21:271–82.

100. Kamel IR, Kruskal JB, Pomfret EA et al. Impact of multidetector CT on donor selection and surgical planning before living adult right lobe liver transplantation. Am J Roentgenol 2001; 176:193–200.

101. Cheng YF, Chen CL, Huang TL et al. Single imaging modality evaluation of living donors in liver transplantation: magnetic resonance imaging. Transplantation 2001; 72:1527–33.

102. Fulcher AS, Szucs RA, Bassignani MJ et al. Right lobe living donor liver transplantation: preoperative evaluation of the donor with MR imaging. Am J Roentgenol 2001; 176:1483–91.

103. Lee VS, Morgan GR, Teperman LW et al. MR imaging as the sole preoperative imaging modality for right hepatectomy: a prospective study of living adult-to-adult liver donor candidates. Am J Roentgenol 2001; 176:1475–82.

MRI imaging is good for LDLT assessment.

104. Urata K, Hashikura Y, Ikegami T et al. Standard liver volume in adults. Transplant Proc 2000; 32:2093–4.

105. Lo CM, Fan ST, Liu CL et al. Minimum graft size for successful living donor liver transplantation. Transplantation 1999; 68:1112–16.

106. Kiuchi T, Kasahara M, Uryuhara K et al. Impact of graft size mismatching on graft prognosis in liver transplantation from living donors. Transplantation 1999; 67:321–7.

107. Ben Haim M, Emre S, Fishbein TM et al. Critical graft size in adult-to-adult living donor liver transplantation: impact of the recipient's disease. Liver Transplant 2001; 7:948–53.

108. Feng L, Zhao N, Yao X et al. Histidine–tryptophan–ketoglutarate solution vs. University of Wisconsin solution for liver transplantation: a systematic review. Liver Transplant 2007; 13:1125–36.

Comparison between UW and other solutions, notably HTK.

109. Selvaggi G, Weppler D, Nishida S et al. Ten-year experience in porto-caval hemitransposition for liver transplantation in the presence of portal vein thrombosis. Am J Transplant 2007; 7:454–60.

110. Liu CL, Lo CM, Chan SC et al. Safety of duct-to-duct biliary reconstruction in right-lobe live-donor liver transplantation without biliary drainage. Transplantation 2004; 77:726–32.

111. Bismuth H, Houssin D. Reduced-sized orthotopic liver graft in hepatic transplantation in children. Surgery 1984; 95:367–70.

112. Broelsch CE, Emond JC, Thistlethwaite JR et al. Liver transplantation, including the concept of reduced-size liver transplants in children. Ann Surg 1988; 208:410–20.

113. Diamond IR, Fecteau A, Millis JM et al. Impact of graft type on outcome in pediatric liver transplantation: a report from Studies of Pediatric Liver Transplantation (SPLIT). Ann Surg 2007; 246:301–10.

114. Bismuth H, Morino M, Castaing D et al. Emergency orthotopic liver transplantation in two patients using one donor liver. Br J Surg 1989; 76:722–4.

115. Noujaim HM, Gunson B, Mayer DA et al. Worth continuing doing ex situ liver graft splitting? A single-center analysis. Am J Transplant 2003; 3:318–23.

116. Reyes J, Gerber D, Mazariegos GV et al. Split-liver transplantation: a comparison of ex vivo and in situ techniques. J Pediatr Surg 2000; 35:283–9.

117. Rela M, Vougas V, Muiesan P et al. Split liver transplantation: King's College Hospital experience. Ann Surg 1998; 227:282–8.

118. Broering DC, Wilms C, Lenk C et al. Technical refinements and results in full-right full-left splitting of the deceased donor liver. Ann Surg 2005; 242:802–12.

119. Gundlach M, Broering D, Topp S et al. Split-cava technique: liver splitting for two adult recipients. Liver Transplant 2000; 6:703–6.

120. Humar A, Khwaja K, Sielaff TD et al. Split-liver transplants for two adult recipients: technique of preservation of the vena cava with the right lobe graft. Liver Transplant 2004; 10:153–5.

121. Wilms C, Walter J, Kaptein M et al. Long-term outcome of split liver transplantation using right extended grafts in adulthood: a matched pair analysis. Ann Surg 2006; 244:865–72.

122. Erbay N, Raptopoulos V, Pomfret EA et al. Living donor liver transplantation in adults: vascular variants important in surgical planning for donors and recipients. Am J Roentgenol 2003; 181:109–14.

123. Brandhagen D, Fidler J, Rosen C. Evaluation of the donor liver for living donor liver transplantation. Liver Transplant 2003; 9:S16–28.

124. Varotti G, Gondolesi GE, Goldman J et al. Anatomic variations in right liver living donors. J Am Coll Surg 2004; 198:577–82.

125. Lee SG, Hwang S, Kim KH et al. Approach to anatomic variations of the graft portal vein in right lobe living-donor liver transplantation. Transplantation 2003; 75:S28–32.

126. Tanaka K, Inomata Y, Kaihara S. Evaluation of vascular and biliary anatomy. In: Tanaka K, Inomata Y, Kaihara S (eds) Living donor liver transplantation surgical techniques and innovations. Barcelona: Prous Science, 2004; pp. 9–12.

127. Yu PF, Wu J, Zheng SS. Management of the middle hepatic vein and its tributaries in right lobe living donor liver transplantation. Hepatobil Pancreat Dis Int 2007; 6:358–63.

128. Cattral MS, Greig PD, Muradali D et al. Reconstruction of middle hepatic vein of a living-donor right lobe liver graft with recipient left portal vein. Transplantation 2001; 71:1864–6.

129. Cheng YF, Chen CL, Huang TL et al. Magnetic resonance of the hepatic veins with angular reconstruction: application in living-related liver transplantation. Transplantation 1999; 68:267–71.

130. Fan ST, de Villa VH, Kiuchi T et al. Right anterior sector drainage in right-lobe live-donor liver transplantation. Transplantation 2003; 75:S25–7.

131. Fan ST, Lo CM, Liu CL et al. Safety and necessity of including the middle hepatic vein in the right lobe graft in adult-to-adult live donor liver transplantation. Ann Surg 2003; 238:137–48.

132. Ghobrial RM, Hsieh CB, Lerner S et al. Technical challenges of hepatic venous outflow reconstruction in right lobe adult living donor liver transplantation. Liver Transplant 2001; 7:551–5.

133. Sugawara Y, Makuuchi M, Sano K et al. Vein reconstruction in modified right liver graft for living donor liver transplantation. Ann Surg 2003; 237:180–5.

134. Egawa H, Inomata Y, Uemoto S et al. Hepatic vein reconstruction in 152 living-related donor liver transplantation patients. Surgery 1997; 121:250–7.

135. Kido M, Ku Y, Fukumoto T et al. Significant role of middle hepatic vein in remnant liver regeneration of right-lobe living donors. Transplantation 2003; 75:1598–600.

136. Kapoor V, Peterson MS, Baron RL et al. Intrahepatic biliary anatomy of living adult liver donors: correlation of mangafodipir trisodium-enhanced MR cholangiography and intraoperative cholangiography. Am J Roentgenol 2002; 179:1281–6.

137. Trotter JF, Wachs M, Everson GT et al. Adult-to-adult transplantation of the right hepatic lobe from a living donor. N Engl J Med 2002; 346:1074–82.

138. Lo CM, Fan ST, Liu CL et al. Adult-to-adult living donor liver transplantation using extended right lobe grafts. Ann Surg 1997; 226:261–9.

139. Sugawara Y, Makuuchi M, Takayama T et al. Liver transplantation using a right lateral sector

graft from a living donor to her granddaughter. Hepatogastroenterology 2001; 48:261–3.

140. Kaihara S, Ogura Y, Kasahara M et al. A case of adult-to-adult living donor liver transplantation using right and left lateral lobe grafts from 2 donors. Surgery 2002; 131:682–4.

141. Sugawara Y, Makuuchi M, Imamura H et al. Living donor liver transplantation in adults: recent advances and results. Surgery 2002; 132:348–52.

142. Boillot O, Delafosse B, Mechet I et al. Small for size partial liver graft in an adult recipient: a new technique. Lancet 2002; 359:406–7.

143. Hayashi M, Cao S, Concepcion W et al. Current status of living-related liver transplantation. Pediatr Transplant 1998; 2:16–25.

144. Hatano E, Terajima H, Yabe S et al. Hepatic artery thrombosis in living related liver transplantation. Transplantation 1997; 64:1443–6.

145. Adam R. Living Donor Registry European Liver Transplant Registry Report, 2002.

146. Sugawara Y, Makuuchi M. Small-for-size graft problems in adult-to-adult living-donor liver transplantation. Transplantation 2003; 75:S20–2.

147. Todo S, Furukawa H, Jin MB et al. Living donor liver transplantation in adults: outcome in Japan. Liver Transplant 2000; 6:S66–72.

148. Broering DC, Sterneck M, Rogiers X. Living donor liver transplantation. J Hepatol 2003; 38(Suppl 1): S119–35.

149. Gondolesi GE, Roayaie S, Munoz L et al. Adult living donor liver transplantation for patients with hepatocellular carcinoma: extending UNOS priority criteria. Ann Surg 2004; 239:142–9.

150. Kaihara S, Kiuchi T, Ueda M et al. Living-donor liver transplantation for hepatocellular carcinoma. Transplantation 2003; 75:S37–40.

151. Cherqui D, Soubrane O, Husson E et al. Laparoscopic living donor hepatectomy for liver transplantation in children. Lancet 2002; 359:392–6.

152. Soubrane O, Cherqui D, Scatton O et al. Laparoscopicleft lateral sectionectomy in living donors: safety and reproducibility of the technique in a single centre. Am J Transplant 2006; 244:815–20.

153. Abt PL, Desai NM, Crawford MD et al. Survival following liver transplantation from non-heart-beating donors. Ann Surg 2004; 239:87–92.

154. White SA, Prasad KR. Liver transplantation from non-heart beating donors. BMJ 2006; 332:376–7.

155. Abt P, Crawford M, Desai N et al. Liver transplantation from controlled non-heart-beating donors: an increased incidence of biliary complications. Transplantation 2003; 75:1659–63.

156. Muiesan P, Girlanda R, Jassem W et al. Single-center experience with liver transplantation from controlled non-heartbeating donors: a viable source of grafts. Ann Surg 2005; 242:732–8.

157. van Hoek B, de Boer J, Boudjema K et al. Auxiliary versus orthotopic liver transplantation for acute liver failure. EURALT Study Group. European Auxiliary Liver Transplant Registry. J Hepatol 1999; 30:699–705.

158. Lodge JP, Dasgupta D, Prasad KR et al. Emergency subtotal hepatectomy: a new concept for acetaminophen-induced acute liver failure: temporary hepatic support by auxiliary orthotopic liver transplantation enables long-term success. Ann Surg 2008; 247:238–49.

159. Azoulay D, Samuel D, Ichai P et al. Auxiliary partial orthotopic versus standard orthotopic whole liver transplantation for acute liver failure: a reappraisal from a single center by a case–control study. Ann Surg 2001; 234:723–31.

160. Azoulay D, Samuel D, Castaing D et al. Domino liver transplants for metabolic disorders: experience with familial amyloidotic polyneuropathy. J Am Coll Surg 1999; 189:584–93.

161. Monteiro E, Freire A, Barroso E. Familial amyloid polyneuropathy and liver transplantation. J Hepatol 2004; 41:188–94.

162. Ericzon BG, Larsson M, Herlenius G et al. Report from the Familial Amyloidotic Polyneuropathy World Transplant Registry (FAPWTR) and the Domino Liver Transplant Registry (DLTR). Amyloid 2003; 10(Suppl 1):67–76.

163. Starzl TE, Demetris AJ, Trucco M et al. Systemic chimerism in human female recipients of male livers. Lancet 1992; 340:876–7.

164. European FK506 Multicentre Liver Study Group. Randomised trial comparing tacrolimus (FK506) and cyclosporin in prevention of liver allograft rejection. Lancet 1994; 344:423–8.

165. The U.S. Multicenter FK506 Liver Study Group. A comparison of tacrolimus (FK 506) and cyclosporine for immunosuppression in liver transplantation. N Engl J Med 1994; 331:1110–15.

166. O'Grady JG, Hardy P, Burroughs AK et al. Randomized controlled trial of tacrolimus versus microemulsified cyclosporin (TMC) in liver transplantation: poststudy surveillance to 3 years. Am J Transplant 2007; 7:137–41.

167. Haddad EM, McAlister VC, Renouf E et al. Cyclosporin versus tacrolimus for liver transplanted patients. Cochrane Database Syst Rev 2006; CD005161.

CNI usage in liver transplantation.

168. Fisher RA, Ham JM, Marcos A et al. A prospective randomized trial of mycophenolate mofetil with neoral or tacrolimus after orthotopic liver transplantation. Transplantation 1998; 66:1616–21.

169. Sterneck M, Fischer L, Gahlemann C et al. Mycophenolate mofetil for prevention of liver allograft rejection: initial results of a controlled clinical trial. Ann Transplant 2000; 5:43–6.

170. Levy G, Villamil F, Samuel D et al. Results of lis2t, a multicenter, randomized study comparing

cyclosporine microemulsion with C2 monitoring and tacrolimus with C0 monitoring in de novo liver transplantation. Transplantation 2004; 77:1632–8.

171. Levy G, Grazi GL, Sanjuan F et al. 12-month follow-up analysis of a multicenter, randomized, prospective trial in de novo liver transplant recipients (LIS2T) comparing cyclosporine microemulsion (C2 monitoring) and tacrolimus. Liver Transplant 2006; 12:1464–72.

172. Asonuma K, Inomata Y, Uemoto S et al. Growth and quality of life after living-related liver transplantation in children. Pediatr Transplant 1998; 2:64–9.

173. Klintmalm GB, Washburn WK, Rudich SM et al. Corticosteroid-free immunosuppression with daclizumab in HCV(+) liver transplant recipients: 1-year interim results of the HCV-3 study. Liver Transplant 2007; 13:1521–31.

174. Guba M, von Breitenbuch P, Steinbauer M et al. Rapamycin inhibits primary and metastatic tumor growth by antiangiogenesis: involvement of vascular endothelial growth factor. Nat Med 2002; 8:128–35.

175. Watson CJ, Gimson AE, Alexander GJ et al. A randomized controlled trial of late conversion from calcineurin inhibitor (CNI)-based to sirolimus-based immunosuppression in liver transplant recipients with impaired renal function. Liver Transplant 2007; 13:1694–702.

176. Morard I, Dumortier J, Spahr L et al. Conversion to sirolimus-based immunosuppression in maintenance liver transplantation patients. Liver Transplant 2007; 13:658–64.

177. Shenoy S, Hardinger KL, Crippin J et al. Sirolimus conversion in liver transplant recipients with renal dysfunction: a prospective, randomized, single-center trial. Transplantation 2007; 83:1389–92.

178. Creput C, Blandin F, Deroure B et al. Long-term effects of calcineurin inhibitor conversion to mycophenolate mofetil on renal function after liver transplantation. Liver Transplant 2007; 13:1004–10.

179. Pageaux GP, Rostaing L, Calmus Y et al. Mycophenolate mofetil in combination with reduction of calcineurin inhibitors for chronic renal dysfunction after liver transplantation. Liver Transplant 2006; 12:1755–60.

180. Stewart SF, Hudson M, Talbot D et al. Mycophenolate mofetil monotherapy in liver transplantation. Lancet 2001; 357:609–10.

181. Kato T, Gaynor JJ, Yoshida H et al. Randomized trial of steroid-free induction versus corticosteroid maintenance among orthotopic liver transplant recipients with hepatitis C virus: impact on hepatic fibrosis progression at one year. Transplantation 2007; 84:829–35.

182. Schreibman I, Regev A. Recurrent primary biliary cirrhosis after liver transplantation – the disease and its management. MedGenMed 2006; 8:30.

183. Neuberger J, Gunson B, Hubscher S et al. Immunosuppression affects the rate of recurrent primary biliary cirrhosis after liver transplantation. Liver Transplant 2004; 10:488–91.

184. Anselmo DM, Baquerizo A, Geevarghese S et al. Liver transplantation at Dumont-UCLA Transplant Center: an experience with over 3,000 cases. Clin Transplant 2001; 179–86.

185. Strasberg SM, Howard TK, Molmenti EP et al. Selecting the donor liver: risk factors for poor function after orthotopic liver transplantation. Hepatology 1994; 20:829–38.

186. Uemura T, Randall HB, Sanchez EQ et al. Liver retransplantation for primary nonfunction: analysis of a 20-year single-center experience. Liver Transplant 2007; 13:227–33.

187. Silva MA, Jambulingam PS, Gunson BK et al. Hepatic artery thrombosis following orthotopic liver transplantation: a 10-year experience from a single centre in the United Kingdom. Liver Transplant 2006; 12:146–51.

188. Vaidya S, Dighe M, Kolokythas O et al. Liver transplantation: vascular complications. Ultrasound Q 2007; 23:239–53.

189. Saad WE. Management of hepatic artery steno-occlusive complications after liver transplantation. Tech Vasc Interv Radiol 2007; 10:207–20.

190. Ueno T, Jones G, Martin A et al. Clinical outcomes from hepatic artery stenting in liver transplantation. Liver Transplant 2006; 12:422–7.

191. Banga NR, Kessel DO, Patel JV et al. Endovascular management of arterial conduit pseudoaneurysm after liver transplantation: a report of two cases. Transplantation 2005; 79:1763–5.

192. Jia YP, Lu Q, Gong S et al. Postoperative complications in patients with portal vein thrombosis after liver transplantation: evaluation with Doppler ultrasonography. World J Gastroenterol 2007; 13:4636–40.

193. Woo DH, Laberge JM, Gordon RL et al. Management of portal venous complications after liver transplantation. Tech Vasc Interv Radiol 2007; 10:233–9.

194. Busuttil RW, Farmer DG, Yersiz H et al. Analysis of long-term outcomes of 3200 liver transplantations over two decades: a single-center experience. Ann Surg 2005; 241:905–16.

195. Jain A, Reyes J, Kashyap R et al. Long-term survival after liver transplantation in 4,000 consecutive patients at a single center. Ann Surg 2000; 232:490–500.

196. Barber K, Blackwell J, Collett D et al. Life expectancy of adult liver allograft recipients in the UK. Gut 2007; 56:279–82.

197. Dawwas MF, Gimson AE, Lewsey JD et al. Survival after liver transplantation in the United Kingdom and Ireland compared with the United States. Gut 2007; 56:1606–13.

198. Adam R, McMaster P, O'Grady JG et al. Evolution of liver transplantation in Europe: report of the European Liver Transplant Registry. Liver Transplant 2003; 9:1231–43.

9

Pancreas transplantation

Murat Akyol

Introduction

Transplantation of the pancreas is the only treatment currently available that reliably offers insulin independence and normal glucose metabolism for patients with type I diabetes mellitus. Islet transplantation may ultimately supersede solid-organ pancreas transplantation in regularly providing insulin independence for diabetic patients. There are formidable obstacles before this can be achieved. Unless there is a significant breakthrough altering the pace of development in islet transplantation, solid-organ transplantation is here to stay for a number of decades rather than years.

The first pancreas transplant was performed in the University of Minnesota in 1966.[1] Early experience with pancreatic transplantation was disappointing and this situation remained for many years. Difficulties were related to the management of the exocrine secretions and septic complications, a high incidence of thrombosis, acute rejection and pancreatitis. For the first 22 years of its 42-year history less than 1200 pancreas transplants were performed worldwide. Even after the introduction of ciclosporin in 1983, 1-year patient and graft survival rates were only 75% and 37% respectively. Understandably in the 1970s and 1980s enthusiasm for pancreas transplantation was scarce; the predominant sentiment was scepticism.

Throughout the 1990s significant changes occurred. These came about as a consequence of improvements in organ retrieval and preservation methods, refinements in surgical techniques, advances in immunosuppression, advances in the prophylaxis and the treatment of infection, and the experience gained in donor and recipient selection. Success rates following pancreas transplantation are now comparable with other forms of organ transplantation. More than 30 000 pancreas transplants have been performed worldwide. Pancreas transplantation has never been compared with insulin therapy in a prospective controlled trial. It is very unlikely that such a trial will ever be performed. However, considerable experience and a substantial body of evidence have accumulated, which now favour the viewpoint of the enthusiasts rather than the sceptics.

Indications for pancreas transplantation

Pancreas transplantation aims to provide patients who have type I diabetes with an alternative source of insulin. In 1998 Sasaki et al. reported a small number of patients with insulin-requiring type II diabetes who had received pancreas transplants with short-term success.[2] The long-term outlook for such patients remains unknown.

There are a number of theoretical concerns in considering pancreatic transplantation for patients with type II diabetes.[3] The pathogenesis of type II diabetes is fundamentally different from that of type I diabetes. Patients with type II diabetes have insulin resistance as part of their clinical syndrome and the influence of this on islet function in the long term after transplantation remains unclear. Furthermore, type II diabetes is much more prevalent than type I diabetes worldwide and carries a different, often much poorer, prognosis.[4] The allocation of a scarce resource to such patients at the expense of those with type I diabetes clearly poses ethical difficulty.

Box 9.1 • Potential risks and benefits of simultaneous pancreas–kidney (SPK) transplantation

Risks

Perioperative morbidity and mortality
Potential for pancreas transplant to adversely affect kidney transplant outcome
Consequences of higher immunosuppression

Benefits

Improved quality of life, insulin independence
Potential benefits on diabetic complications
Improved life expectancy

Pancreas transplantation from living donors

The pancreas transplant database of all US transplants between 1988 and 2003 records 66 living-donor transplants out of approximately 15 000 pancreas transplants.[5] The experience regarding perioperative morbidity and the long-term risks for the donor is inadequate to allow comment. The fact that it remains a fringe activity confined to very few centres presumably indicates that living-donor pancreas transplantation is not endorsed by the transplantation community in general. This is certainly the view that prevails in the UK and the rest of Europe. The discussion in this chapter is confined to allogeneic pancreas transplantation from deceased organ donors.

Patient selection for pancreas transplantation

Pancreas transplantation is performed in three distinct clinical settings, as presented below.

Simultaneous pancreas–kidney transplantation (SPK)

There is little debate that diabetic patients with renal failure should be offered kidney transplantation and there is good evidence that their prognosis is poor on dialysis.[6] Such patients will already be obligated to immunosuppression on account of kidney transplantation. Combined kidney and pancreas transplantation offers these patients the opportunity to become insulin independent as well as dialysis independent. The risks and potential benefits of SPK transplantation compared with kidney transplantation alone for diabetic patients are summarised in Box 9.1. The evidence for the risks and benefits alluded to in Box 9.1 is reviewed in the final part of the chapter.

Experience has taught us the crucial importance of recipient selection to the success of pancreas transplantation. Cardiovascular comorbidity is the most important factor leading to patient death in the early postoperative period after transplantation in diabetic patients. Patient selection needs to include a comprehensive medical evaluation, ideally performed by a multidisciplinary team within each transplant unit. Asymptomatic ischaemic heart disease is not uncommon in diabetics. Therefore, as a minimum the cardiac assessment should include a 12-lead echocardiography, exercise tolerance test and a non-invasive test of myocardial perfusion (a radioisotope scan or dobutamine stress echo). There is insufficient evidence to comment on the value of routine angiography.[7] Any abnormality detected in non-invasive tests or those patients with symptoms of ischaemic heart disease should be investigated further, including angiography. Any correctable coronary artery disease should be dealt with prior to transplantation.[8]

Most pancreas transplant units will have an arbitrary and flexible upper age limit in determining suitability for SPK transplantation. With increasing experience, criteria for suitability of individuals for SPK transplantation have become more liberal. In the USA 19% and 15% of patients receiving SPK transplantation were aged 50 or older in 2004 and 2005 respectively.[9]

Previous contraindications to transplantation have become relative contraindications or risk factors (Box 9.2).[9]

Strong evidence to differentiate between contraindications and relative contraindications does not exist. Neither blindness nor previous amputation are regarded as contraindications. Hepatitis B, hepatitis C or human immunodefiency virus (HIV) infection

Box 9.2 • Contraindications and risk factors for pancreas transplantation

- Inability to give informed consent
- Active drug abuse
- Major psychiatric illness or non-compliant behaviour
- Recent history of malignancy
- Active infection
- Recent myocardial infarction
- Evidence of significant uncorrectable ischaemic heart disease
- Insufficient cardiac reserve with poor ejection fraction
- Any other illness that significantly restricts life expectancy
- Age greater than 60
- Significant obesity (BMI >30)
- Severe aortoiliac atherosclerosis

in the potential recipient are not contraindications to pancreas transplantation either. Box 9.2 is intended as a guide only. No attempt has been made to separate absolute and relative contraindications. The use of imprecise definitions such as 'significant', 'severe' or 'recent' is also intentional. Appropriate patient selection requires a balanced assessment by experienced clinicians of the risk factors versus potential benefits.

Pancreas after kidney transplantation (PAK)

Historically the large majority of pancreas transplants performed have been SPKs. This was largely because of the relatively poor outcomes following solitary pancreas transplantation. Improved success of pancreas transplantation in the last 10 years has encouraged many transplant units to offer pancreas transplantation for diabetic patients who have previously undergone successful kidney transplantation. Until the end of 2000, PAK transplants constituted 11% of all pancreas transplants performed in the USA and 5% of pancreas transplants performed outside the USA.[5] Around a quarter of US pancreas transplants between 2000 and 2005 have been solitary pancreas transplants.[9] More detailed analysis of activity and outcome in different forms of pancreas transplantation is given in the next section.

Contraindications to PAK transplantation are the same as those for SPK (Box 9.2). Within the confines of these contraindications, all diabetics who have previously undergone successful kidney transplantation are potentially suitable candidates for PAK. In practice the procedure is most useful for those diabetics who have a potential living donor for kidney transplantation. With PAK outcomes now approaching those of SPK, an elective and early living-donor kidney transplantation has obvious benefits to the potential recipient and has additional benefits to the overall pool of patients awaiting transplantation by releasing another cadaveric kidney. Clearly some time should elapse after kidney transplantation to allow for recovery from surgery and stabilisation of allograft function and immunosuppression. The optimal timing has not been determined. Available data comparing early (within the first few months) with late (more than 4 months after kidney transplantation) PAK do not reveal any significant difference in the incidence of morbidity or outcome.[10]

There are also unique immunological considerations for patients who are being considered for PAK. Previous kidney transplantation may have influenced sensitisation profiles for such patients. In the presence of good renal allograft function and no history of acute rejection following kidney transplantation, shared human leucocyte antigens (HLAs) between the pancreas allograft and the previous renal allograft may have a favourable impact on the outcome, an assertion that remains untested. Whether acute rejection following previous kidney transplantation predisposes PAK recipients to acute rejection of the pancreatic allograft is not known either.

Finally, an important consideration, in particular for patients who are being assessed some considerable time after kidney transplantation, is the adequacy of kidney function. Criteria based on strong evidence do not exist but a creatinine clearance of greater than 40 mL/min is commonly quoted as a minimum requirement.[11] For recipients of kidney transplants who are not receiving calcineurin inhibitors, a higher creatinine clearance of not less than 55 mL/min is recommended. A renal allograft biopsy prior to PAK is useful for documentation of the baseline renal reserve and for the monitoring of the progression of allograft nephropathy (diabetic or otherwise).

Pancreas transplantation alone (PTA)

Universally agreed criteria that constitute indications for PTA are more difficult to find compared with those for SPK or PAK. Risks of transplantation for non-uraemic diabetics needs to take account of not only perioperative morbidity but also complications associated with immunosuppression. At present the majority of patients developing type I diabetes will be better served by insulin injections at the onset of disease. Clear indications for PTA are life-threatening complications of diabetes in patients managed with insulin, namely hypoglycaemic unawareness and cardiac autonomic neuropathy. In these two subpopulations of diabetic patients pancreas transplantation is truly life saving. Other indications for PTA are more controversial. The presence of two or more diabetic complications has been advocated as an indication.[12] In the absence of conclusive data on some of the diabetic complications, as reviewed in the final section of this chapter, it is difficult to defend the logic behind the assertion that two or more complications constitute an indication for PTA. Disabling and intractable symptoms of diabetic neuropathy or early nephropathy with preserved renal function may be considered as indications. In diabetics with overt nephropathy and impaired renal function, the nephrotoxicity of the immunosuppressive drugs will need to be considered. Available evidence does not permit precise guidelines but patients with creatinine clearance less than 40 mL/min may be better served by SPK transplantation.

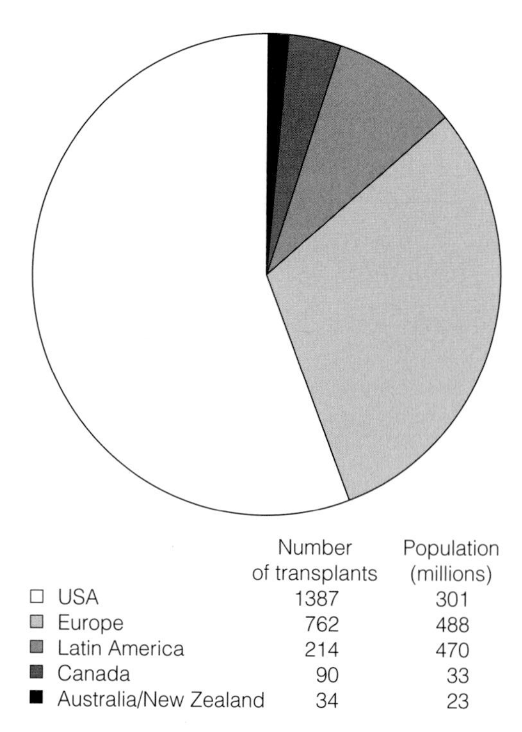

Figure 9.1 • Pancreas transplants performed worldwide in 2006. Data from International Figures on Organ Donation and Transplantation 2006. Transplant Newsletter 12(1), September 2007. Council of Europe.

Pancreas transplantation activity worldwide

At the end of 2007 the total number of pancreas transplants performed worldwide exceeded 30 000. Nearly two-thirds of these have been performed in the USA. The International Pancreas Transplantation Registry (IPTR), based at the University of Minnesota, shares US transplant data with the United Network for Organ Sharing (UNOS). Reporting of data to the UNOS is compulsory, hence the IPTR records regarding US pancreas transplantation activity are accurate and complete. Non-US pancreas transplants are reported to the IPTR on a voluntary basis. Some national organisations, such as Eurotransplant, share data with the international registry. The IPTR estimates that >90% of non-US transplants are reported. Worldwide pancreas transplantation activity reported in a recent Council of Europe document suggests[13] that a much greater proportion of non-US transplants are missing from the IPTR database (**Fig. 9.1**). Despite this probable under-representation of non-US transplants, there seems to be a genuine difference in the utilisation of pancreas transplantation between the USA and the remainder of the world. **Figure 9.2** shows worldwide pancreas transplantation activity as recorded by the IPTR at the beginning of 2005.[14]

In the USA, SPK transplantation activity reached a peak of 972 transplants in 1998. Thereafter there has been a small decline and the activity has remained stable at a plateau of around 900 SPK transplants per annum. In contrast solitary pancreas transplantation activity (PAK + PTA) continued to increase from 64 in 1992 to 554 in 2002 (**Fig. 9.3**).

The total numbers of pancreas transplants performed in the UK and in the Eurotransplant zone are shown in Table 9.1 for comparison.[15,16] In Europe, within the Eurotransplant zone pancreas transplantation activity has declined by about one-third since the peak of 2003. Conversely transplant activity has been steadily increasing in the UK. In 2006 and 2007 more pancreas transplants per million population were performed in the UK compared with the Eurotransplant zone.

The age profile of pancreas transplant recipients has also changed significantly over the last 15 years. In the USA between 1987 and 1992, 9% of pancreas transplant recipients were older than 45; in the period 2000–2004 this proportion had risen to 34%.[9]

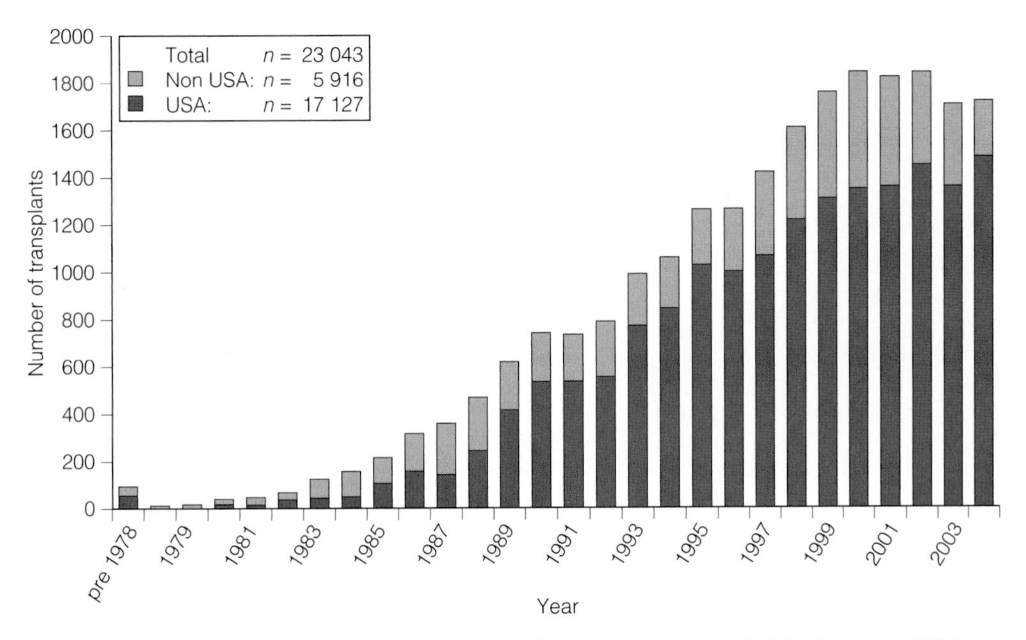

Figure 9.2 • Pancreas transplants reported to the International Pancreas Transplant Registry (1 January 2005).

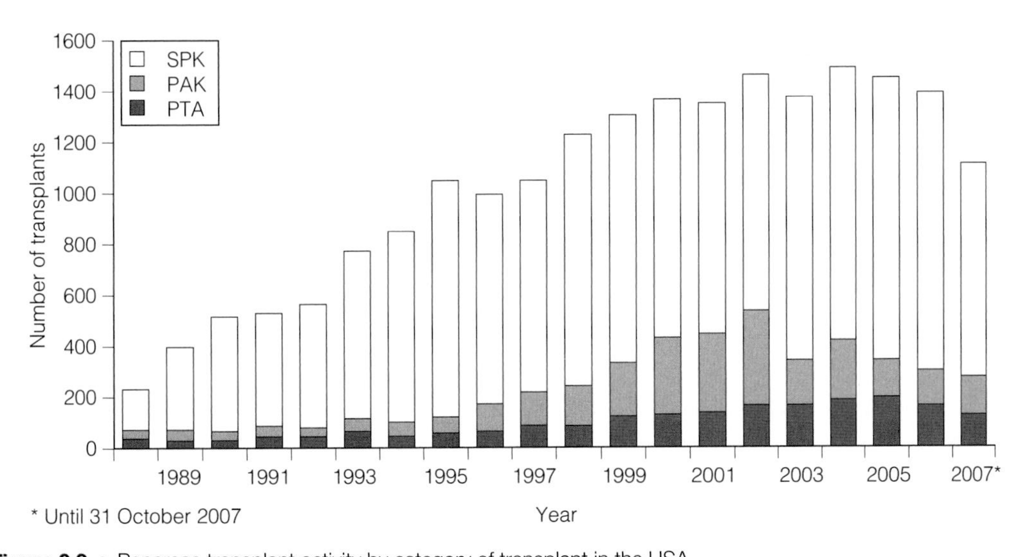

Figure 9.3 • Pancreas transplant activity by category of transplant in the USA.

Interestingly 5% of patients receiving pancreas transplants in the USA between 1994 and 2004 were classified as having type II diabetes at the time of transplant.[9] There are no universally agreed criteria defining type I versus type II diabetes for the purposes of registration. Detectable or high C-peptide levels at the time of diagnosis of diabetes is often used to designate patients as type II diabetics. This may be misleading since the detection of C-peptide may simply reflect impaired elimination as a consequence of renal failure.[3]

Management of exocrine secretions has also seen significant change (**Fig. 9.4**). Enteric drainage has been the predominant method in Europe but in the USA until the mid-1990s less than 10% of pancreas transplants were enterically drained. Enteric drainage has since surged in popularity in the USA also. Between 1996 and 2002 about half of US solitary pancreas transplants and two-thirds of US SPK transplants were enterically drained. In the most recent IPTR analysis,[14] 75% of all pancreas transplants (81% of SPK transplants) in the USA were enterically drained.

Table 9.1 • Pancreas transplantation activity in Eurotransplant and the UK

	Eurotransplant*		United Kingdom	
Year	No. of transplants	pmp†	No. of transplants	pmp†
2000	309	2.58	33	0.55
2001	267	2.23	47	0.78
2002	337	2.81	60	1.0
2003	385	3.21	54	0.9
2004	334	2.78	79	1.31
2005	302	2.52	118	1.96
2006	245	2.04	162	2.69
2007	249	2.08	199	3.31

*Eurotransplant zone includes Austria, Belgium, Germany, Luxembourg, The Netherlands and Slovenia. Population: 120 million.
†pmp = per million population.

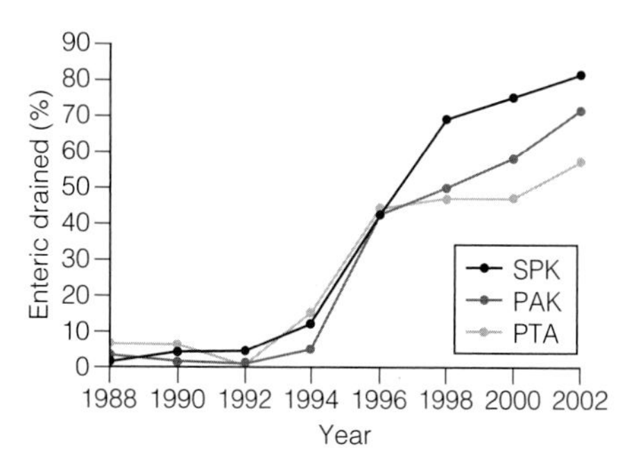

Figure 9.4 • Evolution of exocrine management technique for pancreas transplantation. US pancreas transplants 1 January 1988–31 December 2003. Data from the IPTR.

Enteric drainage seems to have emerged as the preferred method of exocrine drainage worldwide in all but a small proportion of pancreas transplants.[14]

Portal venous drainage gained popularity in the mid-1990s. Utilisation of portal venous drainage has remained relatively constant during the late 1990s and has increased slightly since 1999. In the USA between 1996 and 2002, 1091 of the 4394 (25%) enterically drained primary SPK cases have been performed with the portal venous drainage technique.[17] Data from the IPTR 2000–2004 cohort reveal that this rate has since remained stable at around 25%.[14]

Other demographic data for patients transplanted in the USA between 1996 and 2002 are summarised in Table 9.2.

Table 9.2 • Demographics of pancreas transplant recipients for US cadaveric primary pancreas transplants, 1 January 2000–6 June 2004

	SPK	PAK	PTA
No. of transplants	3947	1148	453
Recipient age (years)	40.6 ± 8.3	41.4 ± 7.7	40.4 ± 9.4
Male recipients (%)	59.6	57.4	41.9
Minorities (%)	20	13	5
Diabetes duration (years)	26 ± 8	28 ± 8	25 ± 10
Enterically drained (%)	81	67	56
Preservation time (h)	13 ± 6	15 ± 6	16 ± 6
Donor age (years)	27 ± 11	27 ± 11	27 ± 11
No. of HLA-A, -B-DR mismatches	4.6 ± 1.3	4.7 ± 1.3	3.9 ± 1.5
Waiting list time (median, months)	9.5	5.1	2.9
Interquartile range	4–17	2–11	1–8

SPK, simultaneous pancreas–kidney transplants; PAK, pancreas after kidney transplants; PTA, pancreas transplantation alone.

The pancreas donor and the organ retrieval procedure

Criteria for eligibility for pancreas donors

Absolute contraindications to organ donation are HIV infection, diagnosis of Creutzfeldt–Jakob disease (CJD), history of malignancy (except non-melanoma skin cancer and certain primary central nervous system tumours) and active systemic sepsis. Additional specific contraindications to pancreas donation are the presence of diabetes mellitus or gross pancreatic disease in the donor (including trauma, acute or chronic pancreatitis, excessive fatty infiltration). Neither hyperglycaemia nor hyperamylasaemia should preclude pancreas donation providing the pancreas appears normal on inspection. Increasing donor age, cerebrovascular/cardiovascular cause of death and donor obesity are associated with poorer pancreas transplant outcomes.[18–20] A precise upper age limit for pancreas donors is difficult to define because of many confounding variables, the retrospective nature of the studies and small sample sizes of individual reports. IPTR analyses use 45 as the age cut-off distinguishing 'young' and 'old' donors. Historically no more than 3% of the US pancreas donors were older than 49.[5] In the most recent cohort this has remained unchanged, donors older than 49 accounting for 3.4% of all deceased pancreas donors in the USA.[9] In practice an upper age limit of around 55 is used by most transplant units and the general consensus is that the ideal pancreas donor will be younger than 45.

It is similarly difficult to quote an acceptable weight limit for donors. Body mass index (BMI) >30 should be regarded as a contraindication. Pancreas grafts from paediatric donors (age >4 years) and selected non-heart-beating donors can be used with excellent results.[21,22] The influence of donor-related factors on pancreas transplant outcome is discussed on p. 189.

Pancreas retrieval operation

The pancreas is a close neighbour of the liver and shares important vascular structures with it. During multiorgan retrieval, priority clearly needs to be given to the liver. The key to successful retrieval of both the liver and the pancreas is good cooperation between the two retrieval teams. The optimum scenario is for both organs to be retrieved and transplanted by the same team. Specific anomalies in the arterial blood supply to the liver that preclude successful liver and successful pancreas transplantation are very rare.

It is not intended to give a detailed description of the surgical procedure for pancreas retrieval in this section, but several pertinent points about pancreatic retrieval from multiorgan donors are highlighted below.

University of Wisconsin (UW) solution was first developed as a pancreatic preservation solution[23] and remains the benchmark for pancreas preservation. Other preservation solutions have been used but experience with these is limited.[24]
The cold ischaemia tolerance of the pancreas is somewhere between that of the liver and the kidney. In pancreas allografts perfused with UW solution, 20–24 hours appears to be the limit for successful preservation, beyond which a time-dependent deterioration in outcome is demonstrable.[25]
Although there are no data that demonstrate a clear benefit from a preservation time cut-off below 20 hours, most surgeons intuitively aim for shorter preservation times. The mean (± SE) preservation time for the 2000–2004 cohort of pancreas transplants in the USA was 13.3 (± 5.9) hours for SPK, 15.3 (± 5.5) hours for PAK and 16.2 (± 5.8) hours for the PTA group.[14]

The pressure gradient between mean arterial pressure and portal venous pressure that maintains blood flow through the pancreas can be significantly diminished during the perfusion of the abdominal organs in retrieval operations. Particular attention is required to maintain an adequate gradient if a cannula for perfusion is placed in the portal venous system as well as the aorta. Many transplant units perfuse abdominal organs with an aortic cannula only. Some evidence supports the view that additional portal perfusion is unnecessary.[26] For the interests of the pancreatic allograft, aortic perfusion alone is the most 'physiological' state that allows satisfactory perfusion and adequate drainage of the effluent.

It is common practice to flush the donor duodenum through a nasogastric tube with an antiseptic or antibiotic solution during the retrieval operation. No evidence exists to demonstrate the superiority of any solution used for duodenal decontamination.

Povidone–iodine during cold storage may be toxic to duodenal mucosa.[27] If povidone–iodine is used for flushing the allograft duodenal segment during organ retrieval, further flushing of the duodenal segment with preservation solution on the back table should be considered.

Donor duodenal contents should be submitted for bacterial and fungal culture. The results may be important in guiding the management of infection in pancreas transplant recipients.[28]

Careful and minimal handling of the pancreas during retrieval is important. Removal of the spleen and the pancreatico-duodenal graft en bloc with the liver is the quickest and the safest method for both organs. The organs are then easily and quickly separated on the back table at the retrieval centre.

Further back-table preparation of the pancreas, which takes place in the recipient centre, is a crucial part of the procedure and takes a minimum of 2 hours. The short stumps of the proximal superior mesenteric artery (SMA) and splenic artery should be marked with fine polypropylene sutures at the time of retrieval. Demonstration of good collateral circulation between these two arteries by flushing them individually at the back table is reassuring. An iliac artery Y graft of donor origin anastomosed to the SMA and the splenic artery is the most common method of reconstruction for the graft arterial inflow (**Fig. 9.5**).

Meticulous dissection and ligation of the lymphatic tissue and small vessels around the pancreas is important to prevent haemorrhage upon reperfusion of the graft in the recipient. Particular attention should be paid to secure the duodenal segment staple lines by inversion with further sutures.

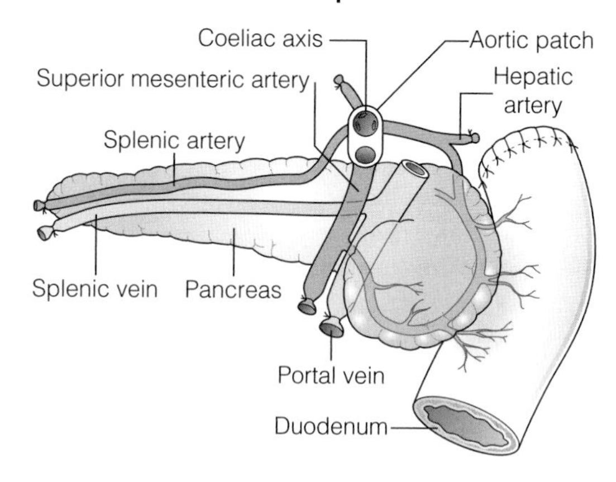

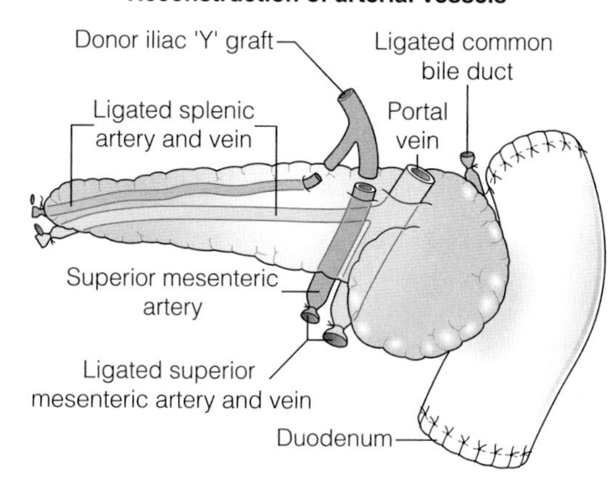

Figure 9.5 • In the absence of an aortic patch containing hepatic artery and the gastroduodenal artery, the pancreatic graft reconstruction requires a donor iliac 'Y' graft.

The pancreas transplant operation

General considerations

In SPK transplantation, pancreatic implantation is usually performed first because of the lower ischaemia tolerance of the pancreas. It is easier to implant the pancreatic graft on the right side. The renal allograft can also be placed intra-abdominally with anastomoses to the left iliac vessels. Alternatively, and perhaps more easily, an extraperitoneal renal transplant on the left side can be performed using the same incision or through a separate left iliac fossa incision.

In PAK transplantation, even in the presence of a right-sided renal allograft, the preferred site for the pancreas would be on the right-hand side cranial to the renal allograft. The arterial inflow to the pancreatic allograft usually comes from the right common iliac artery. Unless the lower aorta is clamped the renal graft does not suffer any ischaemia during pancreatic implantation.

Severely atherosclerotic and calcified vessels in some diabetic recipients can be a challenge during pancreatic implantation. Iliac Y grafts used for reconstruction offer greater flexibility in choosing a suitable arterial anastomotic site in the recipient vessels.

In the last 10 years the most common technique used for pancreas transplantation has been intra-abdominal implantation of the whole pancreas together with a donor duodenal segment. A list of previously employed techniques is given below. These are all associated with poorer outcomes and are rarely performed nowadays.

- Segmental pancreas transplantation using only the tail and the body of the pancreas.
- Pancreatic duct occlusion or ligation.
- Free drainage of exocrine secretions into the peritoneal cavity.
- Direct anastomosis of the transected pancreatic head into the bladder.
- Utilisation of a small duodenal button.

Currently the choices available to the surgeon are related to the management of the exocrine secretions and the management of the venous drainage, as discussed below.

Management of exocrine secretions

Drainage of the exocrine secretions of pancreatic grafts into the recipient's bladder was the most common technique, accounting for 90% of US pancreas transplants during the 1980s and early 1990s.[29] The popularity of this technique was due to its perceived safety, primarily less serious consequences of anastomotic leaks (compared with enteric drainage) in the days of higher corticosteroids and unrefined immunosuppression. The ability to monitor amylase levels in the urine had been considered an additional advantage of bladder drainage. The latter point is contentious. The evidence with respect to the usefulness of urinary amylase monitoring is reviewed later in this chapter. The unphysiological diversion of pancreatic exocrine secretions into the urinary bladder causes frequent complications, often leading to chronic and disabling symptoms. As a consequence, conversion of the urinary diversion to enteric drainage is required in many patients. Complications of pancreatic transplantation, including specific problems associated with bladder drainage, are discussed in detail later in the chapter.

As discussed earlier, enteric drainage has replaced bladder drainage as the method of choice for the management of exocrine secretions in the USA. Data regarding the prevalence of enteric versus bladder drainage for pancreas transplantation in Europe are not easy to find. Historically enteric drainage has been the preferred method in Europe and it is likely that the majority of European pancreas transplant units employ this technique.

Any part of the recipient's small bowel can be used for the anastomosis to the allograft duodenum. No data exist to demonstrate the superiority of one particular site over others. Roux-en-Y loops, which were commonly used, are becoming rare and a simple side-to-side entero-enterostomy is preferred.[14] The correct alignment of the vascular anastomoses is also easier with enteric drainage.

Delayed endocrine function from the transplanted graft is rare and insulin infusion should be discontinued at the time of reperfusion. Achieving insulin independence for the first time in many years in the recipient is a gratifying part of the operation for the surgeon. Patients can become hypoglycaemic at this stage. Blood sugar levels should be checked frequently and a low rate of dextrose infusion is often required.

Management of the venous drainage

Drainage of the venous outflow from pancreas grafts into the portal circulation was first described by Calne in 1984.[30] This complex surgical technique using a segmental graft and gastric exocrine diversion in a paratopic position has never gained popularity. Drainage of the venous outflow into the systemic circulation at the level of the lower inferior vena cava (IVC) has been the norm in pancreatic transplantation. Since the mid-1990s there has been a resurgence of interest in portal venous (PV) drainage. The technique, described by Shokouh-Amiri et al.,[31] involves placement of the graft slightly more cranially in the abdominal cavity with the utilisation of the superior mesenteric vein for the venous drainage. Several studies, including prospective randomised comparisons, have shown that this offers at least equivalent outcome to that of systemic venous (SV) drainage with no compromise in safety.[32–34] The impetus for PV drainage was to achieve a more physiological insulin delivery. A theoretical benefit was considered to be avoidance of hyperinsulinaemia, which has been linked with atherogenesis.[35] Systemic drainage does not always cause hyperinsulinaemia. Nor may it be the only factor, since hyperinsulinaemia occurs in non-diabetic recipients of kidney transplants receiving steroids.[36,37] None of the studies of metabolic function after PV drainage have shown a clear benefit in terms of glucose metabolism, lipid profiles or atherogenesis. There is, however, evidence suggestive of an immunological advantage to PV drainage in the form of a reduction in the incidence of acute rejection. The mode of antigen delivery is known to modulate the immune response, which has been proposed as the mechanism responsible for the observed reduction in acute rejection rates in PV-drained grafts.[12,32–34]

Technically PV drainage may be attractive for retransplants or for patients who have had previous lower abdominal surgery. It requires the use of a long donor iliac Y graft to reach a suitable arterial anastomotic site on the recipient vessels. Suturing the graft portal vein to the recipient superior mesenteric vein (SMV) or its main feeding tributary requires delicate handling of these fragile vessels. In obese patients and in those with thickened or foreshortened bowel mesentery the SMV may not be easily accessible and PV drainage may be difficult to perform.

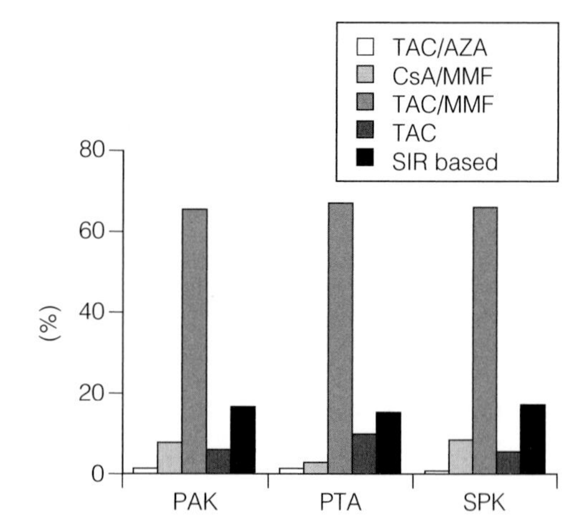

Figure 9.6 • Immunosuppressive protocols used for maintenance in pancreas transplant recipients in the USA. US primary deceased donor pancreas transplants 1 January 2000–6 June 2004. AZA, azathioprene; CsA, ciclosporin; MMF, mycophenolate mofetil; SRL, sirolimus; TAC, tacrolimus. Data from the IPTR.

Immunosuppression in pancreas transplantation

Historically there is ample evidence that the incidence of acute rejection is higher after pancreas transplantation compared with kidney transplantation.[12,25,29] The reasons for this difference are not clear. Nevertheless there has been general acknowledgement of the higher immunological risk of pancreas transplantation. This has resulted in the evolution of strategies that employ more intense immunosuppressive protocols for pancreas transplantation compared with kidney transplantation.

In Europe immunosuppressive protocols in solid-organ transplantation in general have been less aggressive compared with the USA.

In the evolution of immunosuppression for pancreas transplantation tacrolimus has largely replaced ciclosporin starting from the mid-1990s. Similarly mycophenolate mofetil (MMF) has replaced azathioprine in most immunosuppressive protocols (**Fig. 9.6**). There is a sound evidence base for this evolution. This evidence, reviewed below, has been provided not only by single-centre reports or registry analyses but also by prospective randomised trials.

Studies of induction therapy with antithymocyte globulin (ATG) or OKT3 were conducted in the early 1990s, in the ciclosporin era. Two prospective multicentre randomised trials[38,39] show that these agents delay the onset and lessen the severity of rejection episodes at the expense of increased cytomegalovirus (CMV) disease but with no demonstrable influence on patient survival. Pancreas graft survival in the medium term, in this Sandimmun era, was better in patients given ATG or OKT3 for induction.

Two prospective randomised multicentre studies were published in 2003 investigating the role of induction therapy combined with tacrolimus- and MMF-based immunosuppression.

Stratta et al. reported a significant reduction in the incidence of kidney and pancreas rejection with daclizumab induction.[40] Adverse events, including infectious complications, were not different with or without induction therapy. A modified two-dose daclizumab regime gave overall better results than the standard five-dose regime. Another prospective randomised trial at 18 US centres compared induction using any one of the biological agents (OKT3, ATG, basiliximab or daclizumab) with no induction.[41] A trend towards reduction in the incidence of acute rejection by induction therapy was seen. The 1-year incidence of acute rejection (kidney or pancreas) was 24.6% and 31.2% in induction therapy and control groups respectively ($P = 0.28$). The incidence of biopsy-confirmed acute kidney allograft rejection was 13.1% and 23% ($P = 0.08$).

MMF has been shown to reduce the incidence of acute rejection compared with azathioprine in two large prospective randomised trials in the USA and Europe in kidney transplant recipients[42,43] and one prospective randomised trial in pancreas transplant recipients.[44]

Two other prospective controlled studies and many single-centre reports and registry analyses demonstrate improved outcomes with tacrolimus in pancreas transplant recipients compared with ciclosporin.[45–47] Other recent studies have compared rapamycin versus MMF in tacrolimus-based protocols. A multicentre study of 167 patients found no significant differences in any of the outcome measures comparing the two agents.[48] A small single-centre prospective randomised trial of tacrolimus–MMF versus tacrolimus–rapamycin and steroid withdrawal at 6 months in PAK transplant recipients showed similar outcome for primary end-points (patient survival, graft survival, graft loss from rejection). Wound infections, intra-abdominal infections and the need for lipid-lowering agents at 1 year was higher with rapamycin.[49] Another small single-centre prospective randomised trial compared rapamycin with MMF, in the setting of antibody induction and tacrolimus–steroid maintenance. A significantly higher incidence of acute rejection was noted in the MMF group and the tolerability of MMF was poorer.[50]

Steroid withdrawal or avoidance has been a focus of study in the last 5 years. There is as yet no evidence demonstrating significant benefit from steroid avoidance or withdrawal but experience suggests that it is feasible without adversely affecting outcome in pancreas transplant patients.[49,51]

Figure 9.7 shows the use of antibody therapy as part of the induction immunosuppression in pancreas transplant recipients. Induction therapy with biological agents is used more often for pancreas transplant recipients compared with recipients of any other solid-organ transplant. In the early years of pancreas transplantation (1987–1993), 90% of pancreas transplant recipients were given antibody induction. This gradually decreased to 83% in 1994–1997 and to 76% for those transplanted in 2001.[14] There has since been a gradual increase again in the use of antibodies for induction. In 2005 88% of SPK and 85% of PAK transplant recipients were given antibodies as part of their induction immunosuppression.[52] No clear and consistent pattern emerges to demonstrate the superiority of any

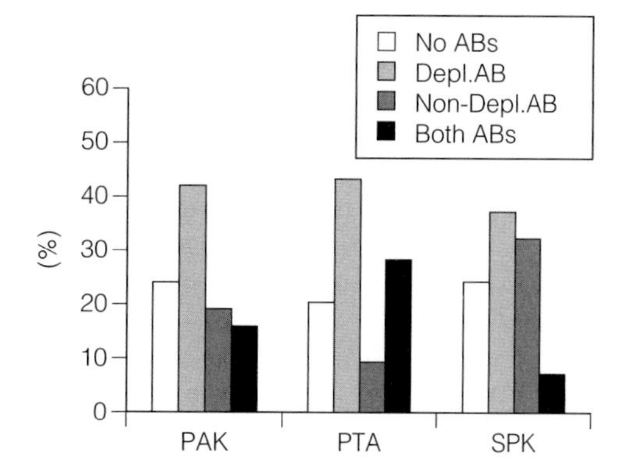

Figure 9.7 • The use of biological agents as part of induction in pancreas transplantation in the USA. US primary deceased donor pancreas transplants 1 January 2000–6 June 2004. Depl AB, depleting antibodies; Non-Depl AB, non-depleting antibodies. Data from the IPTR.

one of the biological agents when used in conjunction with tacrolimus-based immunosuppression.

Individualised immunosuppressive therapy is required for the greater immunological risk associated with young patients, black patients and recipients of re-transplants. SPK transplant recipients with delayed graft function can be maintained on antibodies for longer periods and commence calcineurin inhibitors late, after the kidney function starts to improve. Older recipients require less potent immunosuppression.

Acute rejection following pancreas transplantation

Diagnosis of acute rejection

One of the notable features about pancreas transplantation over the last 15 years has been the considerable reduction in the incidence of acute rejection. In 1992, 74% of SPK transplant recipients and 50% of solitary pancreas transplant recipients (this probably underestimates the true incidence) were reported to have received antirejection therapy.[12,25,29] This had reduced to 19% and 17% respectively by 2000.[14,17]

An important feature of pancreatic graft rejection for the purposes of patient management is the lack of a reliable early marker. In SPK transplants diagnosis of acute rejection almost completely relies on monitoring of the renal allograft by serum creatinine levels and further assessment when indicated by renal biopsy. Discordant rejection of allografts is thought to occur rarely following SPK transplantation. Isolated rejection of the pancreas is said to represent no more than 5–10% of acute rejection episodes.[12,53] Direct evidence from series with simultaneous kidney and pancreas biopsies is very limited. Experimental work in dogs by the Westmead team in Sydney showed isolated pancreas rejection to occur with an incidence of no more than 2% following combined kidney and pancreas transplantation.[54] The most convincing evidence for the security of the implied diagnosis of rejection in the pancreas by diagnosing kidney rejection comes from consistent clinical observations of higher pancreatic graft survival rates following SPK transplantation compared with solitary pancreas transplantation. Monitoring for acute rejection and patient management in the early postoperative period is therefore a particular challenge in solitary pancreas transplants.

Acute rejection of the pancreas affects the exocrine pancreas first. The inflammation may cause pain and a low-grade fever associated with a rise in serum amylase. These symptoms and signs are non-specific, can be subtle and do not distinguish between acute rejection and other causes of graft inflammation (such as ischaemia–reperfusion injury or allograft pancreatitis). Islets of Langerhans are scattered sparsely throughout the exocrine pancreas and beta cells have considerable functional reserve. Therefore dysfunction of the majority of islets resulting in hyperglycaemia as a consequence of rejection occurs only very late in the course of pancreatic rejection. Imaging modalities such as computed tomography (CT) or magnetic resonance imaging (MRI) visualise the pancreas and are helpful to exclude other pathology (such as lack of perfusion which may be segmental or intra-abdominal collections). There are no specific signs of acute rejection on any radiological investigation. Unlike the kidney, the pancreas lacks a firm capsule. Therefore vascular resistance is not reliably altered as a consequence of inflammation and duplex scan monitoring is not useful. Detection of urinary amylase in bladder-drained grafts is a sensitive indicator of exocrine function. However, detection of hypoamylasuria lacks specificity. Factors other than acute rejection such as diuresis or fasting also cause hypoamylasuria.[55] Greater than 25% reduction in urinary amylase correlates with acute rejection in no more than half of the cases when assessed by biopsy.[56] A stable urinary amylase may therefore be helpful in excluding acute rejection but detection of hypoamylasuria is unhelpful. Other tests have been shown to be better correlated with early endocrine dysfunction of pancreas grafts such as glucose disappearance rate or insulin secretion dynamics following an intravenous glucose load.[57,58] Owing to their complexity and lack of availability, such tests are not practical for use as routine daily monitoring tools in patient management.

Pancreas allograft biopsy has recently become established as a reliable and safe technique and is the gold standard in the diagnosis of acute rejection in solitary pancreas transplants. Percutaneous biopsy under ultrasound or CT guidance is the most common method. Histological criteria for the diagnosis and grading of rejection have been standardised.[59]

Management of acute rejection

Recipients of pancreas transplants have more to lose than recipients of other organ transplants from unnecessary treatment for presumed acute rejection. Diagnosis of acute rejection of the pancreas should always be confirmed histologically prior to the institution of treatment.

Early or mild acute rejection of the pancreas allograft concurrent with kidney rejection can be treated with high-dose corticosteroids. Recurrent acute rejection or moderate to severe rejection episodes should be treated with anti-T-cell agents. IPTR data show that steroids were used in 85% of SPK and 80% of solitary pancreas transplant recipients diagnosed as having acute rejection.[14] However, 48% of SPK recipients and 80% of solitary pancreas recipients with acute rejection were given anti-T-cell agents also, suggesting that many patients were treated with both.

Acute rejection in pancreas allografts is not life threatening and caution is advised against overimmunosuppression. If diagnosed before the onset of hyperglycaemia most rejection episodes are reversible.

Impact of acute rejection on outcome

The UNOS data for 4251 patients who received SPK transplants between 1988 and 1997 were analysed by Reddy et al. in order to determine the influence of acute rejection on long-term outcome.[60]

Acute rejection of either graft increased the relative risk of pancreas and kidney graft failure at 5 years. The relative risks, adjusted for other risk factors, were 1.32 and 1.53 for pancreas and kidney respectively if acute rejection occurred. In this analysis 45% of the cohort had no acute rejection. The worst outcome was in patients who had both kidney and pancreas rejection.

Complications of pancreas transplantation

Introduction

Pancreas transplantation is associated with a higher incidence and a greater range of complications than kidney transplantation, and postoperative patient management constitutes a greater challenge (Box 9.3). Between one-fifth and one-quarter of patients require relaparotomy following pancreas transplantation to deal with complications. Part of the reason for the increased incidence of complications is the higher level of immunosuppression in a high-risk diabetic population who already exhibit impaired infection resistance, poor healing and a high prevalence of comorbidity. Other factors relate to the allograft, which unlike kidney or liver allografts is not sterile and uniquely possesses rich proteolytic enzymes, making it susceptible to specific complications such as pancreatitis, leaks and fistula formation. The blood flow to the pancreas is much lower compared with the kidney, and this is a further risk factor specifically for thrombotic complications. Finally, bladder drainage of the exocrine secretions is associated with a high incidence of complications unique to this unphysiological diversion.

Box 9.3 • Complications of pancreas transplantation

Infective complications

- Systemic infection (opportunistic infections associated with immunosuppression)
- Local infections (peritonitis, localised collections, enteric or pancreatic fistulas)

Vascular complications

- Haemorrhage: early haemorrhage from allograft vessels and late haemorrhage (rupture of pseudoaneurysms)
- Thrombosis: allograft arterial or venous thrombosis

Allograft pancreatitis

- Ischaemia–reperfusion injury or reflux pancreatitis (especially after bladder drainage)

Complications specific to bladder drainage

- Chronic dehydration, acidosis, recurrent urinary tract infections, haematuria, chemical cystitis, urethral strictures or urethral disruption

Increasing donor age, prolonged preservation time, recipient obesity and donor obesity are factors associated with increased probability of complications and early graft loss.

Infective complications

CMV disease is more common after pancreas transplantation compared with kidney or liver transplantation. Antiviral prophylaxis in CMV-mismatched donor/recipient pairs is mandatory. Unique to pancreatic transplantation are intra-abdominal septic complications that occur as a consequence of bacteria or fungi transmitted from the donor via the allograft or those that occur as a consequence of anastomotic leaks. Patients on peritoneal dialysis at the time of transplantation may have a higher rate of intra-abdominal infection compared with those on haemodialysis.[25]

It is not known whether duodenal decontamination during organ retrieval has any influence on recipient intra-abdominal or wound infections. A bacteriology specimen of the donor duodenal contents should be used to guide antimicrobial therapy in the event of intra-abdominal sepsis. Abdominal lavage with warm saline or antibiotic/antifungal solutions is also common practice after implantation of pancreatic allografts. This practice, intuitively useful, is supported by indirect evidence from other clinical settings but there are no controlled trials demonstrating efficacy for pancreas transplantation.

Vascular complications

Thrombosis

Allograft venous or arterial thrombosis occurs more commonly following pancreatic transplantation compared with kidney transplantation. Retrospective analysis of data reported to registries regarding the causes of graft loss is understandably prone to error and difficult to interpret. Nevertheless graft thrombosis appears to be one of the most common causes of early graft loss following pancreas transplantation; 5.1% of US pancreas transplants

Table 9.3 • Reasons for early technical graft loss in US cadaveric primary pancreas transplants, 1 January 2000–6 June 2004

	SPK (%) n = 3947	PAK (%) n = 1149	PTA (%) n = 453	Total (%) n = 5549
Thrombosis	4.9	5.3	7.3	5.1
Infection	1.2	1.4	1.7	1.3
Pancreatitis	0.3	0.1	0.6	0.3
Anastomotic leak	1.1	1.3	1.1	1.1
Bleeding	0.4	0.3	1.5	0.5
Total graft loss	7.9	8.4	12.2	8.3

SPK, simultaneous pancreas–kidney transplants; PAK, pancreas after kidney transplants; PTA, pancreas transplantation alone.

performed between 2000 and 2004 failed due to graft thrombosis[9,14] (Table 9.3). A predisposing factor could be the use of venous extension grafts for the portal vein anastomosis, which should be only very rarely required. Concern about a potentially higher incidence of thrombosis following PV drainage has not been borne out by clinical experience. There is no difference in the incidence of technical failure rate with PV drainage compared with SV drainage. Routine use of heparin for prophylaxis against allograft vascular thrombosis is associated with increased haemorrhage. Most transplant centres do not use heparin; however, authors from the largest pancreas transplant unit in the world have reported a small reduction in the incidence of graft thrombosis with heparin.[25]

Table 9.3 illustrates the relative prevalence of early complications of pancreas transplantation leading to graft loss from the IPTR database.

Haemorrhage

Release of the vascular clamps and reperfusion of the pancreatic allograft during the recipient operation can be a tricky moment, with potential for bleeding from multiple points on the allograft. The key to avoiding this is meticulous preparation of the allograft on the back table prior to implantation.

Haemorrhage in the early postoperative hours is often a result of the proteolytic and fibrinolytic activity of the pancreatic exocrine secretions which may come into contact with vascular anastomoses as a consequence of a leak from the surface of the pancreas, a complication unique to pancreas transplantation.

Late haemorrhage following pancreas transplantation is an uncommon but catastrophic complication, often due to the rupture of a pseudoaneurysm or direct erosion of one of the anastomoses secondary to a leak. Any unexplained fever, tachycardia, leucocytosis or abdominal pain in recipients of pancreas transplants should lead to investigations to look for a leak or an intra-abdominal collection.

Allograft pancreatitis

Cold storage and ischaemia–reperfusion injury inevitably result in a degree of oedema of the pancreatic allograft. This is a commonly encountered finding if a relaparotomy becomes necessary in the first few postoperative days and it is not always associated with an elevation in serum amylase. There is no universally agreed definition of allograft pancreatitis. The condition has a different clinical course to native pancreatitis. It is rarely severe or life threatening. Ischaemia–reperfusion injury may be the cause or a predisposing factor. It is not known (although likely) whether drugs associated with native pancreatitis can also cause allograft pancreatitis. Bladder drainage (especially with autonomic neuropathy affecting the bladder and causing high intravesical pressures) can be associated with recurrent episodes of allograft pancreatitis due to reflux. Catheter drainage of the bladder for at least 7–10 days is usually adequate for the management of the acute episode but ultimately enteric conversion may be required.

During the pancreas transplant operation, the allograft exocrine function starts very promptly upon revascularisation and the duodenal segment fills with the pancreatic juice quickly. Excessive distension of the stapled duodenal segment and the consequent reflux could cause postoperative pancreatitis. Even if the exocrine diversion is not going to be performed straight away, the duodenal segment should be decompressed and excessive distension of the graft duodenum should be avoided during the recipient operation.

The distinction between allograft pancreatitis and acute rejection in the presence of an oedematous pancreas, abdominal pain and a slightly raised serum amylase is a difficult clinical diagnosis.

Complications specific to bladder drainage

The most common consequence of the diversion of the exocrine pancreatic secretions into the bladder is a chemical cystitis, which predisposes patients to infection, persistent haematuria and troublesome dysuria. Dysuria is more troublesome in men, with urethritis that can progress to urethral disruption. Failure of reabsorption of the exocrine secretions results in chronic dehydration and acidosis. Urinary tract infections are much more common compared with intestinal drainage. Persistent haematuria can require repeated blood transfusions. In the presence of autonomic neuropathy affecting the bladder, repeated episodes of reflux allograft pancreatitis is another potential complication. As a consequence of one or more of these complications enteric conversion of the exocrine drainage may become necessary. The enteric conversion rate in bladder-drained pancreas transplants increases with increasing follow-up and could be as high as 40% at 5 years.[12]

Outcome following pancreas transplantation

Introduction

There is little doubt that patient and graft survival rates following pancreas transplantation continued to improve throughout the 1990s. Detailed analyses of outcomes reported to the IPTR have been available as annual reports form the registry. They remain as valuable sources of data. The IPTR reports, however, refer to short- or medium-term outcome and should be interpreted within the context of their limitations. For instance, the analysis of 1996–2000 US pancreas transplants refers to 5276 transplants performed in this period. The outcome figures were based on data from 4073 of these patients for whom complete information was available. It is not mentioned what information was missing in the remaining 1203 patients.[61] Table 9.4 summarises the improvements observed in the success rate of pancreatic transplantation in the USA between 1987 and 2000.

Another comprehensive database that is readily accessible on the internet and regularly publishes and updates pancreas transplantation activity and outcome from the USA is the OPTN/SRTR (Organ Procurement and Transplantation Network/ Scientific Registry of Transplant Recipients) database.[5,9] Table 9.5 summarises patient survival and graft survival data from the database as of May 2006.

Short-term patient survival and graft survival rates remain as the standard primary outcome measures in organ transplantation. As illustrated in Table 9.5 these rates have improved considerably in pancreas transplantation, to the extent that demonstration of any further significant improvement as a consequence of any intervention requires prospective studies with very large patient groups. In order to circumvent this difficulty in kidney transplantation, surrogate end-points (or secondary outcome measures) such as the incidence of acute rejection or the quality of graft function have been used. Similar surrogate outcome measures applicable to pancreas transplantation are more difficult to define. As an example of difficulties with the interpretation of

Table 9.4 • Improvements observed in 1-year survival rates of pancreatic transplantation in the USA between 1987 and 2000

	SPK		PAK		PTA	
	1987–1990	1998–2000	1987–1990	1998–2000	1987–1990	1989–2000
Patient survival (%)	89	95	91	94	93	100
Pancreas graft survival (%)	72	82	52	74	47	76
Kidney graft survival (%)	84	92	–	–	–	–

SPK, simultaneous pancreas–kidney transplants; PAK, pancreas after kidney transplants; PTA, pancreas transplantation alone.

Table 9.5 • Patient and graft survival rates for US pancreas transplants

	Patient survival (%)			Graft survival (%)		
	1-year	3-year	5-year	1-year	3-year	5-year
SPK	95.1 *n* =1728	90.8 *n* = 3497	85.8 *n* = 5317	85.2 n = 1746	79.3 *n* = 3534	71.2 *n* = 5372
PAK	95.5 *n* = 600	89.9 n = 1123	83.6 *n* = 1483	78.7 *n* = 763	67.3 *n* = 1439	56.4 *n* = 1958
PTA	94.9 *n* = 195	91.6 *n* = 411	90.2 *n* = 590	72.8 *n* = 225	58.4 *n* = 462	53.4 *n* = 660

SPK, simultaneous pancreas–kidney transplants; PAK, pancreas after kidney transplants; PTA, pancreas transplantation alone. One-year data in this analysis refer to the outcome (actual survival rates) for the cohort transplanted in 2003–2004. Three-year data give the actuarial survival rates for transplants performed in 2001–2004. Five-year data refer to actuarial survival rates for transplants performed in 1999–2004. The number of patients included in each of the analyses is shown.
Data from the OPTN/SRTR database, May 2006.

data, one may recall several single-centre reports referring to the immunological benefits of PV drainage. The fact that this is not reflected in improved graft survival rates may indeed be a genuine finding. However, it could also be attributed to already excellent graft survival rates with SV drainage and the relatively infrequent application of PV drainage in possibly selected patients. Distorted reporting with less incomplete data in the registry from the portal-drained subgroup of patients may also be a confounding factor.

Any enquiry into the outcome of pancreas transplantation worldwide suffers from the lack of a truly representative and complete international database. The IPTR and OPTN/SRTR databases have given us a remarkably useful insight into the picture in the USA. Information with respect to the outcome of pancreas transplantation outside the USA is more sketchy. The responsible attitude for the pancreas transplantation community worldwide should be to regard complete and accurate reporting of pancreas transplantation activity and outcome as an indispensable priority.

Factors influencing pancreas transplantation outcome

Recipient age

Increasing recipient age is a small but significant risk factor in the outcome of pancreas transplantation. Historically patient and graft survival rates have been higher in younger recipients. In recent years more careful patient selection has influenced the outcome in older recipients favourably to the extent that the short-term outcome following pancreas transplantation is no different for patients older than 45 at the time of transplant compared with those who are younger than 45.[9] The number of patients in solitary pancreas transplant categories is smaller and most patients receiving solitary transplants tend to be in the younger age group. As a consequence the influence of recipient age is more readily demonstrable in SPK transplantation and becomes more pronounced with longer follow-up.

Five years after SPK transplantation patient survival is 86.0% for recipients aged 35–49 at the time of transplantation compared with 81.7% for those aged 50–64.[9]

Re-transplantation

Re-transplantation appears as a consistent and significant risk factor for graft survival in all categories. One-year pancreas graft survival rate after re-transplantation in the SPK category is 70.8 % (± 9.3%) compared with 85.4% (± 0.9%) in primary SPK transplants (**Fig. 9.8**).[17]

HLA matching

HLA matching has a small influence on the outcome of pancreas transplants depending on the category. In SPK transplantation the outcome appears to be independent of HLA. Similarly in PAK transplantation no influence of HLA matching on outcome is demonstrable. However, in pancreas transplants alone there is an incremental increase of borderline significance in the rate of immunological graft loss with increasing mismatches for class I HLA antigens. In a multivariate model, two HLA-B mismatches were associated with a relative risk of 2.17

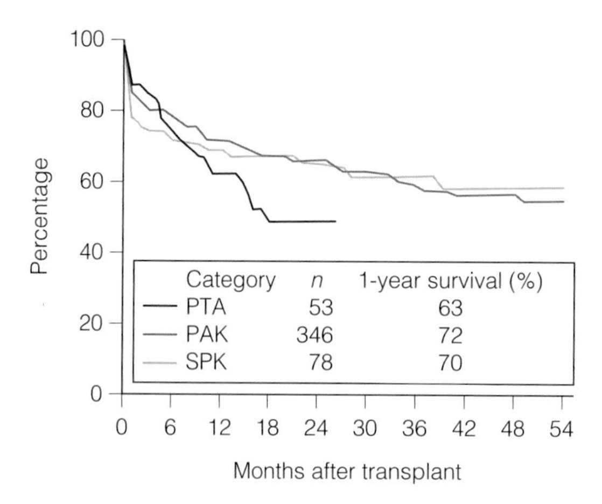

Figure 9.8 • Pancreas *re-transplant* graft function. These are US deceased donor re-transplants from January 1996 to July 2003.

for immunological graft loss (*P* = 0.09). HLA-DR matching has no influence in any group.[17]

Management of exocrine secretions: management of venous drainage

The surgical technique employed for the exocrine diversion has no influence on outcome in pancreas transplantation. Pancreas graft survival rates at 1 year following SPK transplantation for the 2000–2004 cohort in the USA were 87% and 85% for bladder-drained and enteric-drained grafts respectively. Kidney graft survival was 91% and 93% for bladder-drained versus enteric-drained grafts. Minor advantages in some short-term outcome measures in favour of selected subgroups need to be interpreted with caution since there is no consistent pattern that emerges.[14,17]

PV drainage was used in 23% of SPK, 27% of PAK and 44% of PTA transplants in the USA during 2000–2004. SV or PV drainage made no difference to the outcome (immunological failure rate, technical failure rate, graft survival).[9,17]

Immunosuppression

Immunosuppressive therapy has a major influence on the outcome of pancreas transplantation. In the last decade, tacrolimus and MMF have been the basis of maintenance immunosuppression in the large majority of patients (Fig. 9.6). Multivariate analyses reveal the tacrolimus–MMF combination to be associated with highly significant reductions in pancreas graft loss (relative risk (RR) = 0.74, *P* = 0.08 for SPK; RR = 0.51, *P* = 0.001 for PAK; RR = 0.46, *P* = 0.014 for PTA). In multivariate analyses, sirolimus use considered as a variable is also associated with a significant reduction in pancreas graft loss (RR = 0.81, *P* = 0.17 for SPK; RR = 0.54, *P* = 0.01 for PAK; RR = 0.16, *P* = 0.001 for PTA).[14]

Donor factors

Donor age, donor obesity and donor cause of death have been linked with the outcome of pancreas transplantation. Whilst all of these variables may be independently associated with outcome, they are likely to be inter-related, for instance younger donors are more likely to have died of trauma whilst older and obese donors are more likely to have died of causes other than trauma.

An analysis of all SPK transplants performed in the USA during a 12-year period between 1994 and 2005 (*n* = 8850) has conclusively shown increasing donor age to be a risk factor in pancreas transplantation.[18] During this period 776 transplants (8.8%) were performed using organs from deceased donors aged 45 or older (mean age ± SD = 49.2 ± 3.7). A large majority of patients (8074, 91.2%) received organs from donors younger than 45 (mean age ± SD = 24.6 ± 9.1). Pancreas and kidney graft survival and patient survival were significantly inferior in the 'old donor' group (77% vs. 85%, 89% vs. 92% and 95% vs. 93% respectively at 1 year). A larger decrement in early graft survival after old versus young donor transplants was observed for pancreas compared to kidney transplants. Graft survival for young and old donor transplants continued to diverge over time for both grafts. Five-year pancreas graft survival was 72% versus 60% for young versus old donor transplants respectively.

Humar et al.[19] reported the outcome of 711 deceased donor pancreas transplants in Minneapolis performed between 1994 and 2001. The outcomes were analysed for three groups based on donor BMI (BMI <25, *n* = 434; BMI 25–30, *n* = 196; BMI >30, *n* = 81). Patients who received grafts from obese donors (BMI >30) had higher incidence of complications and inferior graft survival. The incidence of technical failure was 9.7% in the BMI <25 group, 16.3% in the BMI 25–30 group and 21% in the BMI >30 group (*P* = 0.04). More recently the same group of authors published an analysis in a slightly larger cohort of pancreas transplant patients from Minneapolis[20] in order to determine risk factors for technical failure following pancreas transplantation. Technical failure, defined as thrombosis, bleeding, leaks, infections or pancreatitis, was responsible for the loss of 13.1% of transplants (131 of 973). By multivariate analysis the following were significant risk factors: recipient BMI >30 (RR = 2.42, P = 0.0003), preservation time >24 hours (RR = 1.87, P = 0.04) and cause of donor death other than trauma (RR = 1.58, P = 0.04). Donor obesity had borderline significance.

Pancreas grafts from paediatric donors (aged >4 years) and non-heart-beating donors (NHBDs; or donation after cardiac death, DCD) can be used with excellent results, not inferior to those obtained from optimum deceased donor grafts.[21,22]

Donor selection has a major influence on the outcome of pancreas transplantation. Therefore other risk factors such as preservation time and recipient BMI need to be considered when using suboptimal grafts for pancreas transplantation.

Long-term outlook following pancreas transplantation

Pancreas transplantation and life expectancy

Clearly one of the most important factors for patients considering pancreas transplantation is whether their life expectancy will be influenced by the transplant. Numerous studies and analyses of databases have consistently shown that successful pancreas transplantation is associated with improved survival prospects in diabetic patients. No prospective controlled study has ever been carried out that compares pancreas transplantation with insulin therapy. Hence all available evidence is subject to selection bias. It would be reasonable to expect that in clinical practice younger, fitter patients with lower risk would have been chosen for pancreas transplantation, creating bias in favour of this group.

Nevertheless a strong body of indirect evidence suggests that successful pancreas transplantation may truly increase life expectancy as well as improving quality of life.

Several studies have addressed the question of the impact of pancreas transplantation on long-term mortality in a number of different ways. The University of Wisconsin experience in 500 SPK transplant recipients published in 1998[62] simply quotes a 10-year patient survival rate of 70%. This is matched in their experience only by recipients of living-donor kidney transplants. Unsurpassed as they are, these results were obtained in a highly selective group of young patients with a relatively short duration of diabetes and kidney failure and strict eligibility criteria excluding those with ischaemic heart disease.

A large registry analysis from the USA published in 2001 by Ojo et al. looked at the outcome in 13 467 adults with type I diabetes registered in kidney and SPK transplant waiting lists between 1988 and 1997.[63] Adjusted 10-year patient survival was 67% for SPK transplant recipients, 65% for living-donor kidney (LKD) transplant recipients and 46% for cadaveric kidney (CAD) transplant recipients. Taking the mortality of patients who remained on dialysis as reference, the adjusted relative risk of 5-year mortality was 0.40, 0.45 and 0.74 respectively for SPK, LKD and CAD recipients. Another large review of the UNOS database published by Reddy et al.[64] in 2003 analysed long-term survival in 18 549 type I diabetic patients transplanted between 1987 and 1996. There was a long-term survival advantage in favour of pancreas transplant recipients (8-year crude survival rates: 72% for SPK, 72% for LKD and 55% for CAD). This diminished but persisted after adjusting for donor and recipient variables and kidney graft function. Tyden and colleagues from Sweden demonstrated a striking difference in long-term survival in a small group of diabetic patients transplanted between 1982 and 1986 and followed up for at least 10 years.[65] Fourteen patients received successful SPK transplants. The control group consisted of 15 patients who had undergone the same assessment, had been accepted for SPK transplants but either declined pancreas transplantation or received SPK but lost the pancreas graft within the first year because of a technical complication. Ten years later three of the SPK transplant recipients had died (20%) compared with 12 of the 15 kidney transplant recipients (80%). Another interesting study addressing the long-term survival after transplantation in type I diabetics came from The Netherlands. Smets et al. studied all 415 adults with type I diabetes who started renal replacement therapy in The Netherlands between 1985 and 1996. Patients were divided into two groups depending on where they lived. The basis for this was the fact that in the Leiden area the treatment of choice for such patients was SPK transplantation (73% of transplants), whereas in the remainder of the country kidney transplantation alone was the predominant type of therapy and SPK was performed uncommonly (37%). The relative risk for death for patients who lived outside the Leiden area was 1.9. When the transplanted patients only were analysed the relative risk for death was 2.5 for those outside the Leiden area.[66]

Influence of pancreas transplantation on diabetic complications

Nephropathy

There is convincing evidence that successful pancreas transplantation can stop the progression of diabetic nephropathy and reverse associated histological changes. This evidence largely comes from studies that have assessed the course of diabetic nephropathy in kidney allografts in SPK or PAK transplant recipients.[67,68]

Fioretto et al. have shown that established lesions of diabetic nephropathy in native kidneys can be reversed with successful pancreas transplantation (PTA).[69] In the setting of clinical transplantation the beneficial effect of the pancreas graft is counterbalanced by the nephrotoxicity of the immunosuppressive drugs.

Retinopathy

Patients with type I diabetes often quote preservation of eyesight as one of the main reasons for considering pancreas transplantation. There is good evidence that better blood glucose control reduces the risk of progression of retinopathy.[70]

However, in practice the large majority of patients undergoing pancreas transplantation will have advanced proliferative retinopathy. This would have been stabilised in most patients with laser photocoagulation, which is an effective treatment for proliferative retinopathy. Hence prevention of retinopathy should not be a major factor in the consideration of the risks and benefits of pancreas transplantation.[71]

For the minority of patients who present with non-proliferative retinopathy or only recently underwent treatment for proliferative retinopathy there is a risk of rapid progression of the retinopathy following transplantation. Such patients require close ophthalmic follow-up within the first 3 years of transplantation. Stabilisation of retinopathy after pancreas transplantation takes 3 years.[71–73] During this time any patient who has an indication or develops an indication for laser treatment should undergo treatment.

Improvement in macular oedema after transplantation can result in early improvement of vision. It is unclear whether this is a consequence of euglycaemia or the kidney transplantation resulting in improved fluid balance. Euglycaemia offered by pancreas transplantation results in elimination of osmotic swelling of the lens, hence improving fluctuations in vision that diabetic patients experience.[71]

Neuropathy

Patients with end-stage renal failure and type I diabetes almost universally exhibit an autonomic and peripheral (somatic) diabetic polyneuropathy as well as uraemic neuropathy. Improvement in neuropathy following SPK transplantation using objective measures of nerve function has been demonstrated by several transplant centres.[74,75] For individuals with intractable and distressing symptoms of neuropathy the clinical benefit may be considerable. Obesity, presence of advanced neuropathy and poor renal allograft function may be predictors of poor recovery in nerve function after SPK transplantation.[76]

Cardiovascular disease

Pancreas transplantation has demonstrable benefits on microangiopathy in diabetics.[77] Some of its effects on retinopathy, nephropathy and neuropathy may be mediated through this mechanism. It has been more difficult to demonstrate any improvement in macroangiopathy. The enhanced survival prospects after pancreas transplantation ought to be at least in part due to the improvement in the cardiovascular risk profile. Evidence to support this is accumulating. Fiorina and colleagues in Milan demonstrated favourable influences of pancreas[78] and islet[79] transplantation on atherosclerotic risk factors, including plasma lipid profile, blood pressure, left ventricular function and endothelial function. This translates into reduced cardiovascular death rate.[80,81] Similar improvements occur in early non-uraemic diabetics after PTA.[82]

Key points

- By the end of 2007, over 30000 pancreas transplants had been performed worldwide. In the USA alone there are nearly 100000 patients with functioning transplants.
- The outcome following pancreas transplantation has improved considerably in the last 10–15 years. It is now comparable to the outcome for other solid-organ transplants.
- The number of SPK transplants has remained stable since 2000. Solitary pancreas transplantation activity continues to increase and accounts for nearly 40% of all pancreas transplants.
- Induction immunosuppression with biological agents is used in pancreas transplantation more often than any other solid-organ transplant. The tacrolimus–MMF combination is the basis of the most commonly used maintenance immunosuppression protocols. Steroid withdrawal or avoidance is gaining momentum.
- Over the last 10 years enteric drainage has gradually replaced bladder drainage as the preferred technique for the management of exocrine secretions in pancreas transplantation.
- Portal venous drainage, introduced in the mid-1990s, is currently used in about one-quarter of the pancreas transplants.
- Evidence regarding the influence of pancreas transplantation on diabetic complications and life expectancy is not available from prospective controlled trials. Nevertheless accumulating evidence from many studies strongly suggests that successful pancreas transplantation has a favourable influence on diabetic complications and survival prospects for patients.

References

1. Kelly KD, Lillehei KC, Merkle FK et al. Allotransplantation of pancreas and duodenum along with kidney in diabetic nephropathy. Surgery 1967; 61:827–37.
2. Sasaki TM, Gray RS, Ratner RE et al. Successful long term kidney-pancreas transplants in diabetic patients with high C-peptide levels. Transplantation 1998; 65:1510–12.
3. Nath DS, Gruessner AC, Kandaswamy R et al. Outcomes of pancreas transplants for patients with Type 2 diabetes mellitus. Clin Transplant 2005; 19:792–7.
4. Rogers J, Stratta RJ. The metabolic syndrome as a risk factor in simultaneous pancreas kidney transplantation. Nat Clin Proc Endocr Metab 2007; 3(7):498–9.
5. URREA, UNOS. 2003 Annual Report of the US Organ Procurement and Transplantation Network and the Scientific Registry of Transplant Recipients; http://www.OPTN.org
6. McMillan MA, Briggs JD, Junor BJ. Outcome of renal replacement treatment in patients with diabetes mellitus. BMJ 1990; 301:540–4.
7. Rabbat CG, Treleaven DR, Russell DJ et al. Prognostic value of myocardial perfusion studies in patients with end-stage renal disease assessed for kidney or kidney–pancreas transplantation: a meta-analysis. J Am Soc Nephrol 2003; 14:431–9.
8. Lin K, Stewart D, Cooper S et al. Pre-transplant cardiac testing for kidney–pancreas transplant candidates and association with cardiac outcomes. Clin Transplant 2001; 15:269–75.
9. US Organ Procurement and Transplantation Network and the Scientific Registry of Transplant Recipients. 2006 Annual Report: Transplant Data 1996–2005. Rockville, MD: Health Resources and Services Administration, Healthcare Systems Bureau, Division of Transplantation (2006 OPTN/SRTR Annual Report 1996–2005. HHS/HRSA/HSB/DOT); http://www.optn.org/

   Comprehensive report of US organ donation and transplantation activity. Includes outcome data, tables and detailed explanation of methodology – free access online.
10. Humar A, Sutherland DE, Ramcharan T et al. Optimal timing for a pancreas transplant after successful kidney transplant. Transplantation 2000; 70:1247–50.
11. Hariharan S, Pirsch JD, Lu CY et al. Pancreas after kidney transplantation. J Am Soc Nephrol 2002; 13:1109–18.
12. Odorico JS, Sollinger HW. Technical and immunological advances in transplantation for insulin dependent diabetes mellitus. World J Surg 2002; 26:194–211.
13. International Figures on Organ Donation and Transplantation 2006. Transplant Newsletter 12(1), September 2007. Council of Europe.
14. International Pancreas Transplant Registry. 2004 Annual Report, August 2006; http://www.iptr.umn.edu/IPTR/annual_reports/2004_annual_report.html

15. http://www.uktransplant.org.uk/statistics/statistics.htm
16. http://www.eurotransplant.org/index.php?id=statistics
17. International Pancreas Transplant Registry. 2002 Annual Report, 15(1), August 2003; http://www.iptr.umn.edu
18. Salvalaggio PR, Schnitzler MA, Abbott KC et al. Patient and graft survival implications of simultaneous pancreas kidney transplantation from old donors. Am J Transplant 2007; 7:1561–71.
19. Humar A, Thigarajan R, Kandaswamy R et al. The impact of donor obesity on outcomes after cadaver pancreas transplants. Am J Transplant 2004; 4:605–10.
20. Humar A, Thigarajan R, Kandaswamy R et al. Technical failures after pancreas transplants: why grafts fail and the risk factors – a multivariate analysis. Transplantation 2004; 78(8):1188–92.
21. Krieger NR, Odorico JS, Heisey DM et al. Underutilization of pancreas donors. Transplantation 2003; 75:1271–6.
22. Salvalaggio PR, Davies DB, Fernandez LA et al. Outcomes of pancreas transplantation in the United States using cardiac death donors. Am J Transplant 2006; 6:1059–65.
23. D'Allessandro AM, Stratta JR, Sollinger HW et al. Use of UW solution in pancreas transplantation. Diabetes 1989; 38(Suppl 1):7–9.
24. Potdor S, Eghtesad B, Jain A et al. Comparison of early graft function and complications of pancreas transplant recipients in Histidine–Tryptophan–Ketoglutarate (HTK) solution and University of Wisconsin (UW) solutions. Transplantation 2003; 76:S288.
25. Sutherland DER, Gruessner RW, Dunn DL et al. Lessons learned from more than 1000 pancreas transplants at a single institution. Ann Surg 2001; 233:463–501.
26. DeVilleDeGoyet J, Hausleithner V, Malaise J et al. Liver procurement without in-situ portal perfusion. Transplantation 1994; 57:1328–32.
27. Olson DW, Kadota S, Cornish A et al. Intestinal decontamination using povidone iodine compromises small bowel storage quality. Transplantation 2003; 75:1460–2.
28. Woeste G, Wallstein C, Vogt J et al. Value of donor swabs for intra-abdominal infection in simultaneous pancreas kidney transplantation. Transplantation 2003; 76:1073–8.
29. Pancreas transplants for United States (US) and non-US cases as reported to the International Pancreas Transplant Registry (IPTR) and to the United Network for Organ Sharing (UNOS). Clinical Transplants 1997; 45–59.
30. Calne RY. Para-topic segmental pancreas grafting: a technique with portal venous drainage. Lancet 1984; 1:595–7.
31. Shokouh-Amiri MH, Gaber AO, Gaber LW et al. Pancreas transplantation with portal venous drainage and enteric exocrine diversion: a new technique. Transplant Proc 1992; 24:776–7.
32. Petruzzo PA, Palmina A, DaSilva MA et al. Simultaneous pancreas–kidney transplantation: portal versus systemic venous drainage of the pancreas allografts. Clin Transplant 2000; 14:287–91.
33. Stratta RJ, Shokouh-Amiri MH, Egidi MF et al. A prospective comparison of simultaneous kidney–pancreas transplantation with systemic–enteric versus portal–enteric drainage. Ann Surg 2001; 233:740–51.
34. Stratta RJ, Lo A, Shokouh-Amiri MH et al. Improving results in solitary pancreas transplantation with portal enteric drainage, thymoglobulin induction and tacrolimus/mycophenolate mofetil based immunosuppression. Transplant Int 2003; 16:154–60.
35. Despres JP, Lamarche B, Mauriege P et al. Hyperinsulinemia as an independent risk factor for ischaemic heart disease. N Engl J Med 1996; 334:952–7.
36. Ost LD, Tyden G, Fehrman I. Impaired glucose tolerance in ciclosporine–prednisolone treated renal allograft recipients. Transplantation 1988; 46:370–2.
37. Christiansen E, Vestergaard H, Tibell A et al. Impaired insulin-stimulated non-oxidative glucose metabolism in pancreas–kidney transplant recipients: dose–response effects of insulin on glucose turnover. Diabetes 1996: 45:1267–75.
38. Wadstrom J, Brekke B, Wrammer L et al. Triple versus quadruple induction immunosuppression in pancreas transplantation. Transplant Proc 1995; 27:1317–18.
39. Cantarovich D, Karam G, Giral-Classe M et al. Randomized comparison of triple therapy and anti-thymocyte globulin induction treatment after simultaneous pancreas kidney transplantation. Kidney Int 1998; 54:1351–6.
40. Stratta RJ, Alloway RR, Lo A et al. Two dose daclizumab regimen in simultaneous kidney pancreas transplant recipients: primary endpoint analysis of a multi-center randomised study. Transplantation 2003; 75:1260–6.
41. Kaufman DB, Burke GW, Bruce DS et al. Prospective randomised multi-center trial of antibody induction therapy in simultaneous kidney-pancreas transplantation. Am J Transplant 2003; 3:855–64.
42. European Mycophenolate Mofetil Co-operative Study Group. Placebo controlled study of mycophenolate mofetil combined with ciclosporine and corticosteroids for prevention of acute rejection. Lancet 1995; 345:1321–5.
43. Sollinger HW for the US Renal Transplant Mycophenolate Mofetil Study Group. Mycophenolate mofetil for the prevention of acute rejection in primary

cadaveric renal allograft recipients. Transplantation 1995; 60:225–32.

44. Merion RM, Henry ML, Melzer JS et al. Randomized prospective trial of mycophenolate mofetil versus azathioprine for prevention of acute renal allograft rejection after simultaneous kidney–pancreas transplantation. Transplantation 2000; 70:105–11.

45. Stratta RJ. Review of immunosuppressive usage in pancreas transplantation. Clin Transplant 1999; 13:1–12.

46. Gruessner RWG. Tacrolimus in pancreas transplantation: a multi-center analysis. Clin Transplant 1997; 11:299–12.

47. Bartlett ST, Schweitzer EJ, Johnson LB et al. Equivalent success of simultaneous pancreas–kidney and solitary pancreas transplantation. A prospective trial of tacrolimus immunosuppression with percutaneous biopsy. Ann Surg 1996; 224:440–9.

48. Garcia VD, Keitel E, Santos AF et al. Immunosuppression in pancreas transplantation: mycophenolate mofetil versus sirolimus. Transplant Proc 2004; 36(4):975–7.

49. Kandaswamy R, Khwaja K, Gruessner A et al. A prospective randomized trial of steroid withdrawal with mycophenolate mofetil (MMF) versus sirolimus (SRL) in pancreas after kidney transplants. Am J Transplant 2003; 3(Suppl 5):292.

50. Burke GW, Ciancio G, Mattiazzi A et al. Lower rate of acute rejection with rapamycin than with mycophenolate mofetil in kidney pancreas transplantation. A randomized prospective study with Thymoglobulin/Zenapax induction, tacrolimus and steroid maintenance: comparing rapamycin with mycophenolate mofetil. Am J Transplant 2003; 3(Suppl 5):322.

51. Gruessner RWG, Kandaswamy R, Humar A et al. Calcineurin inhibitor and steroid free immunosuppression in pancreas kidney and solitary pancreas transplantation. Transplantation 2005; 79(9):1184–9.

52. Andreoni KA, Brayman KL, Guidinger MK et al. Kidney and pancreas transplantation in the United States, 1996–2005. Am J Transplant 2007; 7(Part 2):1359–75.

53. Allen RDM. Pancreas transplantation. In: Forsythe JLR (ed.) Transplantation surgery, 1st edn. London: WB Saunders, 1997; pp. 167–201.

54. Hawthorne WJ, Allen RDM, Greenberg ML et al. Simultaneous pancreas and kidney transplant rejection: separate or synchronous events. Transplantation 1997; 63:352–8.

55. Munn SR, Engen DE, Barr D et al. Differential diagnosis of hypoamylasuria in pancreas allograft recipients with urinary exocrine drainage. Transplantation 1990; 49:359–62.

56. Benedetti E, Najarian JS, Gruessner A et al. Correlation between cystoscopic biopsy results and hypoamylasuria in bladder drained pancreas transplants. Surgery 1995; 118:864–72.

57. Elmer DS, Hathaway DK, Bashar AA et al. Use of glucose disappearance rates (kG) to monitor endocrine function of pancreas allografts. Clin Transplant 1998; 12:56–64.

58. Osei K, Henry ML, O'Dorioso TM et al. Physiological and pharmacological stimulation of pancreatic islet hormone secretion in Type 1 diabetic pancreas allograft recipients. Diabetes 1990; 39:1235–42.

59. Drachenberg G, Klassen D, Bartlett S et al. Histologic grading of pancreas acute allograft rejection in percutaneous needle biopsies. Transplant Proc 1996; 28:512–13.

60. Reddy KS, Davies D, Ormond D et al. Impact of acute rejection episodes on long term graft survival following simultaneous kidney pancreas transplantation. Am J Transplant 2003; 3:439–44.

61. Gruessner AC, Sutherland DCR. Pancreas transplant outcomes for United States cases reported to the United Network for Organ Sharing (UNOS) and non-US cases reported to the International Pancreas Transplant Registry (IPTR) as of October 2000. Clinical Transplants 2000; 45–72.

62. Sollinger HW, Odorico JS, Knechtle SJ et al. Experience with 500 simultaneous pancreas–kidney transplants. Ann Surg 1998; 228:284–96.

63. Ojo AO, Meier-Kriesche H, Hanson J et al. The impact of simultaneous pancreas kidney transplantation on long-term patient survival. Transplantation 2001; 71:82–9.

64. Reddy KS, Stablein D, Taranto S et al. Long-term survival following simultaneous kidney–pancreas transplantation versus kidney transplantation alone in patients with type 1 diabetes mellitus and renal failure. Am J Kidney Dis 2003; 41:464–70.

65. Tyden G, Bolinder J, Solders G et al. Improved survival in patients with insulin-dependent diabetes mellitus and end-stage diabetic nephropathy 10 years after combined pancreas and kidney transplantation. Transplantation 1999; 67:645–8.

66. Smets YFC, Westendorp RGJ, Van der Pijl JW et al. Effect of simultaneous pancreas–kidney transplantation on mortality of patients with Type 1 diabetes and end stage renal failure. Lancet 1999; 353:1915–20.

67. Wilczek HE, Jaremko G, Tyden G et al. Evolution of diabetic nephropathy in kidney grafts. Transplantation 1995; 59:51–7.

68. El-Gebely S, Hathaway DK, Elmer DS et al. An analysis of renal function in pancreas kidney and diabetic kidney alone recipients at two years following transplantation. Transplantation 1995; 59:1410–15.

69. Fioretto P, Steffes MW, Sutherland DER et al. Reversal of lesions of diabetic nephropathy after pancreas transplantation. N Engl J Med 1998; 339:69–75.

70. The Diabetes Control and Complications Trial Research Group. The effect of intensive treatment of diabetes on the development and progression of long-term complications in insulin dependent diabetes mellitus. N Engl J Med 1993; 329:977–86.

71. Walsh AW. Effects of pancreas transplantation on secondary complications of diabetes – retinopathy. In: Gruessner RWG, Sutherland DER (eds) Transplantation of the pancreas. New York: Springer, 2004; pp. 462–71.

72. Chow VCC, Pai RP, Chapman JR et al. Diabetic retinopathy after combined kidney–pancreas transplantation. Clin Transplant 1999; 13:356–62.

73. Pearce IA, Ilango B, Sells RA et al. Stabilisation of diabetic retinopathy following simultaneous pancreas and kidney transplant. Br J Ophthalmol 2000; 84:736–40.

74. Cashion AK, Hathaway DK, Milstead EJ et al. Changes in patterns of 24 hour heart rate variability after kidney and kidney–pancreas transplant. Transplantation 1999; 68:1426–30.

75. Hathaway DK, Abell T, Cardoso S et al. Improvement in autonomic and gastric function following pancreas–kidney versus kidney alone transplantation and the correlation with quality of life. Transplantation 1994; 57:816–22.

76. Allen RDM, Al-Harbi IS, Morris JG et al. Diabetic neuropathy after pancreas transplantation: determinants of recovery. Transplantation 1997; 63:830–8.

77. Abendroth D, Schmand J, Landgraf R et al. Diabetic microangiopathy in Type 1 (insulin dependent) diabetic patients after successful pancreatic and kidney or solitary kidney transplantation. Diabetology 1991; 34:131–4.

78. Fiorina P, LaRocca E, Venturini M et al. Effects of kidney–pancreas transplantation on atherosclerotic risk factors and endothelial function in patients with uraemia and Type 1 diabetes. Diabetes 2001; 50:496–501.

79. Fiorina P, Folli F, Maffi P et al. Islet transplantation improves vascular diabetic complications in patients with diabetes who underwent kidney transplantation: a comparison between kidney–pancreas and kidney alone transplantation. Transplantation 2003; 75:1296–301.

80. LaRocca E, Fiorina P, DiCarlo V et al. Cardiovascular outcomes after kidney–pancreas and kidney alone transplantation. Kidney Int 2001; 60:1964–71.

81. Jukema JW, Smets YF, van der Pijl JW et al. Impact of simultaneous pancreas and kidney transplantation on progression of coronary atherosclerosis in patients with end-stage renal failure due to type 1 diabetes. Diabetes Care 2002; 25:906–11.

82. Copelli A, Giannarelli R, Mariotti R et al. Pancreas transplant alone determines early improvement of cardiovascular risk factors and cardiac function in type 1 diabetic patients. Transplantation 2003; 76:974–6.

10

Pancreatic islet transplantation

Juan L. Contreras
Devin E. Eckhoff

Introduction

Diabetes mellitus (DM) comprises a group of common metabolic disorders that share the phenotype of hyperglycaemia and is caused by progressive loss of insulin secretion or action. The metabolic dysregulation associated with DM causes secondary pathophysiological changes in multiple organs. Insulin is produced by β cells within the islets of Langerhans. It is estimated that in the adult human there are approximately 1 million islets scattered through the pancreas.[1] DM affects 15.7 million individuals in the USA, is the leading cause of end-stage renal disease, non-traumatic lower extremity amputations and adult blindness, and contributes $92–138 billion directly to healthcare costs and an additional $45 billion in lost productivity.[2] The incidence of DM is increasing in the USA and worldwide, and 30 000 new cases are diagnosed each year in the USA.[3] As such, diabetes represents a target disease that will have tremendous impact on society in terms of raising the quality of life and freeing healthcare resources for other diseases.

The principal determinant of the risk of devastating diabetes complications is the time of exposure to hyperglycaemia. In this regard, the therapeutic use of insulin significantly improved the survival of patients with type I diabetes by preventing acute lethal complications like diabetic ketoacidosis.[4] Improved patient survival, however, allowed the development of secondary complications. Several studies demonstrated that the incidence of these complications can be reduced by tight glycaemic control but the risk of severe hypoglycaemic reactions is then increased.[5,6] The only approach at present to restore sustained normoglycaemia is endocrine replacement therapy. In this regard, transplantation of the whole pancreas (normally simultaneously with a kidney transplant) is now an established treatment option for some type I diabetics. But this is only an option for a small group of diabetic patients with advanced disease, and even in the most experienced hands it has a high morbidity and modest mortality (40% and 3–5% respectively).[7] The pancreatic islets only account for around 1% of the volume of the pancreas and so transplantation of the exocrine pancreatic tissue is unnecessary and carries added risk.

Transplantation of islets alone offers an attractive alternative to whole pancreas transplantation and may be associated with a lower incidence of serious complications.[8]

The first islet transplant was attempted in 1893 in Bristol – 28 years before the discovery of insulin – whereby fragments of a sheep's pancreas were implanted in the subcutaneous tissues of a 15-year-old boy dying from uncontrolled ketoacidosis.[9] This xenograft was destined to fail without immunosuppression. The era of experimental islet research began in 1911, when Bensley stained islets within the guinea-pig pancreas using a number of dyes and was able to pick free the occasional islet for morphological study.[10] Mass isolation of large numbers of viable islets from the human pancreas has proven to be a challenge ever since. The techniques used today for the mass isolation of human islets evolved from earlier techniques used to prepare islets in rodent models.[11]

Islet transplantation for type I diabetes has been attempted in many units over the last 10–15 years with poor long-term success. Of over 400 islet transplants reported to the International Transplant Registry between 1990 and 2000, less than 10% resulted in 1-year insulin independence.[12] These transplants were carried out on a heterogeneous group of diabetic patients and, in the majority of cases, included glucocorticoids in the immunosuppression protocol.

A report published by the Edmonton group of seven consecutive type I diabetic patients who all attained long-term insulin independence after islet transplantation renewed interest in islet transplantation as a feasible option for a select group of diabetic patients.[13] These patients were all transplanted using a glucocorticoid-free immunosuppressive regimen, immunogenic xenoproteins were eliminated from islet preparations and high islet numbers, mostly from two pancreases, were used to achieve insulin independence.

Islet transplant activities

Islet transplantation has been conducted mostly in non-uraemic type I diabetic recipients as islet transplants alone. However, patients can be considered for islet transplantation while immunosuppressed for another organ transplant, usually kidney. The results of combined islet and kidney transplantation now match those of islet alone[14] and recent data suggest that islets transplanted with a kidney may prolong the patient and kidney graft survival and protect against diabetic vascular complications.[15,16] In addition, islets can be infused in combination with other abdominal organs (islets and liver or islets in combination with liver and small bowel).[17] Autologous islet transplantation has been successfully executed in patients subjected to total pancreatectomy for chronic pancreatitits or trauma.[18] Few cases of islet xenotransplantion (pig to human) have been reported.[19]

Patient selection for islet transplantation

Inclusion criteria for islet transplantation are outlined in Box 10.1. The major indications for islet transplantation are hypoglycaemic unawareness and metabolic lability. Assessment of these complications is subjective and a number of scoring systems have been devised to allow quantification of these problems. The mean amplitude of glycaemic excursion (MAGE) using 14 blood glucose values over a 2-day period has previously been used in the assessment of potential recipients;[20,21] however, it has been superseded recently by the combination of a lability index (LI) and composite HYPO score based on 4 weeks of glucose values.[22] The HYPO scoring system takes into account the frequency, severity and degree of unawareness of hypoglycaemia, and has not been shown to be significantly higher in islet transplant patients pretransplant. It becomes normal post-islet transplant. Ryan et al. suggested that the HYPO score provides a more objective assessment of the metabolic instability of an individual patient and allows pre- and post-transplant comparison. Patients must be assessed by an endocrinologist and to benefit from transplant they should have continuing problems despite an optimum insulin regimen. Ultimately, the decision whether to offer islet transplantation depends on the individual patient and should be arrived at by balancing the risks of the islet transplant procedure itself and potentially lifelong immunosuppression against the daily risks taken by a patient with type I diabetes. Current recipient exclusion criteria for solitary islet transplantation are listed in Box 10.2.

Box 10.1 • Indications for islet-alone transplantation

- Age 18–65 years
- Type I diabetes mellitus for more than 5 years
- C-peptide negative (<0.2 ng/mL)
- Evidence of good compliance
- Type I diabetes mellitus complicated by:
  - hypoglycaemic unawareness (absence of adequate autonomic symptoms at blood glucose levels <50 mg/dL)
  - metabolic lability (lability index, HYPO score)
  - progressive secondary complications

(Despite 'state-of-the-art' insulin therapy defined by monitoring of blood glucose values no less than four times each day and by the administration of three or more insulin injections each day)

Box 10.2 • Exclusion criteria for islet-alone transplantation

- Age <18 years
- BMI >26 (female) and >27 (male) kg/m^2
- Insulin requirement >0.7 U/kg/day
- C-peptide >0.2 ng/mL following i.v. arginine stimulation (5 g)
- Untreated proliferative retinopathy
- Creatinine clearance <80 mL/min/1.73 m^2
- Serum creatinine >1.3 mg/dL (female) or >1.5 mg/dL (male)
- History of panel-reactive anti-HLA antibodies >20%
- Positive pregnancy test, breast-feeding or failure to follow effective contraception
- Active infection
- History of malignancy (except adequately treated squamous or basal cell carcinoma of the skin)
- Active alcohol or substance abuse, smoking (abstinence for 6 months is required)
- Non-adherence to prescribed medical regimens
- Psychiatric disorders
- Baseline haemoglobin <11.7 g/dL (females) or <13 g/dL (male); lymphopenia (<1000/μL); leucopenia (<3000 total leucocytes/μL); and absolute CD4$^+$ count <500/μL; platelets <150 000/μL
- History of coagulopathy or long-term anticoagulant therapy (e.g. warfarin)
- Severe cardiovascular disease, characterised by any one of these conditions: myocardial infarction within the past 6 months; angiographic evidence of non-correctable coronary artery disease; ischaemia on functional testing; left ventricular ejection fraction <30%
- Evidence of liver disease, gallstones or haemangiomas
- Active peptic ulcer
- Hyperlipidaemia (fasting low-density lipoprotein cholesterol >130 mg/dL, treated or untreated; and/or fasting triglycerides >200 mg/dL
- Addison's disease
- Current medical condition requiring chonic steroid administration
- Severe diarrhoea potentially interfering with the absortion of medications

Human islet processing

Regulatory aspects

Despite considerable clinical success, islet transplantation is still considered an investigational procedure in most countries. Therefore, it is regulated to safeguard public health and monitor the development of new products.[23] In the USA, islet grafts are regulated by the Food and Drug Administration (FDA) as biological products, and an Investigational New Drug (IND) application must be submitted and approved by the FDA before the initiation of any studies in humans subjected to islet transplantation. Pancreas processing for islet isolation must be performed in purpose-built Clean Room Facilities with strict adherence to current Good Manufacturing Practices (cGMP).[24]

Pancreas procurement for islet isolation

An ideal donor for human islet recovery will have a favourable medical, sexual and social history, pass the physical examination requirements, and clear all standard laboratory tests used in multiorgan donor work-ups to show low risk of disease transmission. Pancreases are generally obtained from deceased, heart-beating organ donors (DBD), but successful isolations have been reported from non-heart-beating donors and living hemipancreas donors.[25,26] The surgical technique for removal of the pancreas for islets should be the same as that for solid pancreas transplantation, taking care to avoid direct handling of the pancreas and keeping the capsule intact. The pancreas can be removed before or after the liver, or en bloc with the liver and separated on the back table.[27] Care should be taken to keep the pancreas cold after cross-clamping and during back-table preparation/packaging of the organ, as this has been shown to double the yield of viable islets.[28] Ryan et al. have shown that the ischaemia index, which is the cold ischaemic time for a given transplanted islet mass, correlates with insulin secretion.[29] Ideally, the cold ischaemia time should not exceed 8–12 hours. Donor factors associated with successful isolations include body mass index (BMI) >25 kg/m^2, age (lower islet recovery in donors <18 years old), adequate donor glycaemic control before organ procurement, and local procurement team.[30,31]

Islet isolation

The currently recommended, semi-automated process for human islet isolation was described by Ricordi et al. and is based on combined collagenase digestion and mechanical dissociation of the donor pancreas.[32]

First, the pancreas is briefly exposed to an antibiotic and antifungal solution. Extrapancreatic fat and non-pancreatic tissue is removed and the pancreatic duct is cannulated. The pancreas is then perfused under controlled conditions with collagenase

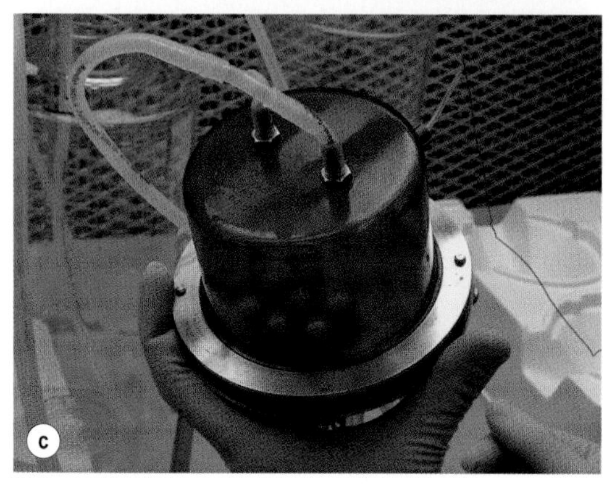

Figure 10.1 • Human islet recovery. **(a)** Pressure-controlled pancreas distension with digestive enzymes. **(b,c)** Ricordi's chamber loaded with human pancreas before (b) and after (c) pancreas dissociation.

(**Fig. 10.1a**). For several years, the most consistent results were obtained using a purified enzyme blend (Liberase HI, Roche), which has more consistent collagenase activity combined with neutral protease activity and low endotoxin levels;[33] however, newer enzyme blends such as Collagenase NB1 (Serva) are now the only available product for clinical islet recovery for transplantation. Following distension, the pancreas is transferred to the Ricordi digestion chamber and circuit (**Fig. 10.1b,c**). Here the temperature of the collagenase is raised to 37°C to allow activation of the enzyme and digestion of the pancreas. Samples are taken at regular intervals to monitor, via inverted microscope, the breakdown of the pancreas. The digestion can be stopped when the intact islets are free from the surrounding exocrine tissue. This part of the isolation process requires skills and experience and is critical to the whole process. The pancreatic digest is purified to separate the islets from other products of digestion such as ductal tissue, lymphoid tissue and necrotic cells with purification gradients (ficoll or iodixanol) and a Cobe 2991 cell separator.[34] Transplantation of unpurified digests can result in insulin independence; however, this approach increases the risk of portal vein thrombosis, portal hypertension and disseminated intravascular coagulation (DIC).[35]

The resulting purified islet preparation is then washed, counted and assayed to make sure that it meets the product release criteria of the isolation facility. These criteria are essential to ensure that a safe, pure and effective islet preparation is consistently produced by the isolation facility (Box 10.3). Safety is measured by negative Gram stain and endotoxin assay and islet purity (ideally >50%) determined by eye, although computerised counting software is currently under evaluation. The likely

Box 10.3 • Current human islet product release criteria

- Sufficient islet mass: 5000 IE/kg recipient body weight
- ABO blood group matched
- Gram stain negative
- Endotoxin load <5 endotoxin units (EU/kg)
- Islet packed cell volume <10 mL (normally 2–3 mL)
- Purity >30% (usually >50%)
- Viability >70%
- Glucose-stimulated insulin release in vitro: stimulation index >1.0

potency of an islet preparation is difficult to determine prior to transplantation and several investigations are under way to develop reliable assays of islet potency and function. Current practice is to use a combination of islet counts and cell viability stains such as fluorescein diacetate/propidium iodide and SytoGreen/ethidium bromide to determine the viable β-cell mass.[13] Islet function is quantified by glucose-stimulated insulin release assays.[13] Newer techniques, such as islet oxygen consumption rate, fractional β-cell viability, ADP/ATP content, and quantification of basal and stimulated reactive oxygen species (ROS) levels showed good correlation between product testing and in vivo islet function in animal studies and may be useful in the near future.[36–40] Islet graft characterisation by histological analysis has also shown good correlation with post-transplant outcomes.[41]

Patients transplanted under the initial Edmonton protocol received islet infusions immediately after purification. Recent evidence, however, suggests that a period of culture may enhance the purity of the islet preparation without loss of potency. Data from the Miami programme suggest that a period of culture of 24–48 hours increases islet purity and reduces the infused volume but can result in insulin independence in the majority of recipients.[42] This ability to culture islets has implications for both immunomodulation of the islet prep and the graft recipient, and for effective product release testing.

Islet infusion

The intraportal site for islet infusion was recognised to be the most efficient location for islet implantation in the rodent, with the benefit of high vascularity, proximity to islet-specific nutrient factors and physiological first-pass insulin delivery to the liver.[43] While many different sites have been tried for islet implantation, the optimal site appears to be through portal venous embolisation. Attempts to embolise the spleen have led to significant life-threatening complications of splenic infarction, rupture and even gastric perforation.[44]

The main portal vein is cannulated under ultrasound and fluoroscopic control using a 4-Fr catheter in the radiology department. The Edmonton group have reported 100% success using this route and have found that avoiding aspirin and completely occluding the catheter track with fibrin glue markedly reduces the risk of bleeding following percutaneous implantation.[45] Alternatively, the portal vein can be cannulated at laparotomy via a mesenteric or omental vein. Most groups are now using a closed-bag system for islet infusion with gravity drainage rather than a syringe to infuse the cells. This minimises trauma to the islets during implantation and allows continuous monitoring of portal pressure. The closed-bag system also ensures that the islet preparation remains in a sterile environment and complies with cGMP guidelines.[46] Prior to infusion unfractionated heparin is added to the islet preparation (35 U/kg if the packed cell volume is less than 5 mL and 70 U/kg if greater than 5 mL) to minimise the risk of portal vein thrombosis.[47]

Islet engraftment

Although well documented in small animals, controversy exists about the liver as an implantation site in clinical islet transplantation.[48,49] Portal vein pressures increase during the islet infusion and this is particularly marked in patients receiving their second and third islet transplants. Portal venous pressure should therefore be monitored throughout the islet infusion and if excessive changes in portal pressure are detected the infusion can be slowed down or stopped. Once implanted in the liver, the islets are incorporated in the liver parenchyma and revascularised from branches of the hepatic artery, a process which, in animals, is complete within 14 days.[50] It has been well recognised that a major limitation in islet transplantation is the limited capacity of islet to successfully engraft after the transplant[51] and therefore a large number of islets (mostly from two organs) is required to achieve insulin independence. In this regard, islets and ductal cells constitutively express tissue factor, a major activator of the coagulation cascade. Thus, when islets are exposed to blood in the portal vein, they trigger a non-specific activation of the coagulation cascade, referred to as instant blood-mediated inflammatory reaction (IBMIR),[52] which is associated with significant loss of islet cells within minutes post-transplant.[53] Strategies to improve islet engraftment are under investigation and most likely will increase the efficacy of islet transplantation.[51]

Immunosuppression

Long-term immunosuppression helps to prevent graft loss, but it can often be associated with worsening of pre-existing conditions or new complications, especially in diabetic patients.[54] Proper selection of immunosuppressive agents and close monitoring of immunosuppressive through levels is a key factor to prevent complications.

Current immunosuppressive protocols for islet transplantation alone are based on a glucocorticoid-free regimen that includes induction therapy with monoclonal antibody against interleukin-2 receptor (Daclizumab (Zenapax®)) and maintenance with sirolimus (Rapamune®) and tacrolimus (Prograf®). This regimen is known as the Edmonton protocol[13] and its efficacy has been reproduced by other groups.[55] Daclizumab is administered at 1 mg/kg pretransplant, then repeated at 2-weekly intervals post-transplant for a total of five doses.[13] In cases where a second islet graft was given beyond the 10-week induction window, the induction course of daclizumab was repeated. Sirolimus is given as a loading dose of 0.2 mg/kg orally immediately pretransplant, with maintenance initially at 0.1 mg/kg/day adjusted to 24-hour target serum trough levels of 12–15 ng/mL for 3 months, then reduced to 7–10 µg/L thereafter. Low-dose tacrolimus is initiated at 2 mg orally given twice daily, but adjusted to target 12-hour trough levels of 3–6 µg/L – representing between one-quarter and one-half of the usual standard dose for other transplants. This regimen appears to provide adequate immunosuppression while avoiding the diabetogenic side-effects of glucocorticoids and tacrolimus.

On some occasions, due to drug toxicity, mycophenolate mofetil (CellCept®) may be introduced instead of tacrolimus or sirolimus.[56] Newer and more potent alternatives to the standard induction therapies have arisen in the past few years.

Induction therapy with rabbit antithymocyte globulin (ATG; Thymoglobulin®) or humanised anti-CD3 antibody (hOKT3γl (Ala-Ala)) showed remarkable results on a series of recipients of single-donor islet transplantation.[57,58]

Lifelong immunosuppression remains a major limitation to the broad application of islet transplantation. However, promising approaches to induce immunological tolerance without chronic immunosuppressive therapy has been recently reported in humans.[59,60]

Outcomes of islet transplantation

A landmark in the history of islet transplantation came with the introduction of the Edmonton protocol in 2000.[13] More than 600 islet transplants have been performed in the last 8 years in comparison to 237 performed in the decade.[61] Although the medical community and the public place a great emphasis on insulin independence as a mark of success, good glycaemic control should be the main goal of islet transplantation.

Insulin independence

The success of the initial Edmonton series (100% insulin independence at 1-year post-transplant)[13] has been replicated by other centres.[55]

Combined data from Edmonton, Miami and Minnesota reveal an initial 90% insulin independence rate, although at newly developed transplant centres the insulin independence rate averages only 23%.[61] However, despite the high percentage rates of insulin independence early after the transplant, this rate quickly starts to decline thereafter, with 40–50% remaining insulin free at 3 years and only 10% at 5 years.[62] However, being insulin dependent does not necessarily equate with complete graft failure as 80% of patients remain C-peptide positive at 5 years.[61] Reasons for this progressive loss of islet graft function remain unknown. A significant number of patients become sensitised to human leucocyte antigens (HLAs) after transplantation, testing positive for panel-reactive antibodies (PRAs). Around 50% of those with high PRAs may lose graft function.[61] Therefore, alloimmunity could be an important factor. Autoimmunity is another potential factor whereby patients who develop autoantibodies may exhibit a decrease in graft function.[61] The liver as an implantation site could be another explanation of progressive islet graft dysfunction, as noted above.[48,49]

Metabolic control

It has been demonstrated that patients with functioning islet grafts (detectable C-peptide levels), insulin dependent or independent, achieve significant reduction in HbA_{1c} to near-normal levels in comparison with both pretransplant levels and patients who lost graft function.[62]

In addition, hypoglycaemia and glycaemic liability (the main indications for islet transplantation) are significantly reduced in all C-peptide-positive patients, insulin free or not, and hypoglycaemia is rarely encountered.[61]

Based on the limited islet engraftment,[51] the majority of insulin-independent patients will demonstrate impaired glucose tolerance test, which could be attributed to reduced functional islet mass compared with the normal pancreas.[61]

Prevention of long-term diabetes complications

In islet after kidney transplant recipients, C-peptide-positive patients had a lower rate of cardiovascular mortality and decreased signs of endothelial injury in comparison to kidney-alone transplant recipients.[63–65] An improvement in cardiac function demonstrated by increased ejection fraction and peak-filling rate in end-diastolic volume has been also noted.[64] Moreover, improved kidney graft survival and function have been demonstrated in islet-after-kidney transplant recipients compared with kidney transplant alone.[16]

Quality of life

Health-related quality-of-life data indicate that fear of hypoglycaemic reactions is significantly reduced in islet recipients.[66,67] Despite secondary effects of chronic immunosuppressive therapy, islet transplantation is associated with overall improvement in patient quality of life.[68]

Post-transplant complications

The overall procedural complication rate is low for percutaneous islet transplantation.[45,69] The more serious procedure-related complications of segmental portal vein thrombosis and bleeding have been reported at 4% and 10% respectively.[70,71] The risk of portal vein thrombosis can be minimised by heparinisation of the recipient and by using only low-volume, high-purity preparations. Bleeding from the liver puncture can be avoided by using a fine-bore (4-Fr) cannula and by ablating the track in the liver using coils, thrombostatic agents or a coagulative laser.[45] The short-term risks of the immunosuppression protocol appear low, with no reports of malignancy or post-transplant lymphoproliferative disease (PTLD), although one would anticipate that the long-term risk of malignancy is at least similar to those transplant patients on conventional immunosuppression.[54] No mortality has been reported to date. Significant elevation of liver enzymes is seen in over 50% of patients receiving intraportal islet grafts.[72] This is more pronounced after the first graft and resolves spontaneously in the majority of patients by 4 weeks. The long-term effect of islets transplanted into the liver remains to be seen, but fatty infiltration of the liver has been reported in 20–30% of the patients who undergo magnetic resonance imaging (MRI) post-islet transplant.[73]

To achieve insulin independence, multiple islet infusions (usually two) are required. Hence, islet transplant recipients are at risk of sensitisation after transplantation. Recent studies demonstrated appearance of HLA antibodies while islet transplant recipients are immunosuppressed and the incidence rises abruptly in subjects weaned completely from immunosuppression. The significance of these findings are currently unknown but may negatively impact on the ability of these patients to undergo further islet, pancreas or kidney transplantation and should be discussed upfront during evaluation in candidates for islet transplantation.[74,75]

Future developments

Optimal use of the donor pool

The worldwide shortage of cadaveric donors and the expansion in potential islet recipients means that the current need for two or even three donors to achieve insulin independence is not sustainable.

Substantial steps have been taken towards single-donor islet transplantation by using a combination of careful donor and recipient selection and meticulous islet preparation.[57,58,76]

Improved laboratory preparation of islets coupled with more reliable and effective collagenases, better assays to evaluate islet potency and predict functionality post-transplant, and implementation of strategies to promote islet engraftment and less toxic immunotherapy are also required to enhance the efficacy of islet transplantation.

New sources of islets

The shortage of organ donors coupled with the increased demand for islets has led to much research into alternative sources of insulin-producing cells, which would be renewable and not depend solely on the availability of human cadaveric donors. The use of porcine islets for human xenotransplants has been explored[77] and remarkable progress has been achieved in pig-to-primate xenotransplantation.[78,79] Stem cells are capable of both self-renewal and multilineage differentiation. They have the potential to proliferate and differentiate into any type of cell and to be genetically modified in vitro, thus providing a renewable source of cells for transplantation. Pancreatic islets have been produced in animal models by in vitro manipulation of both embryonic stem cells and adult pancreatic ductal stem cells. Islets produced in these experiments have been shown to produce endocrine hormones and islet differentiation markers and release insulin in response to glucose stimulation in vitro.[80] In vivo, they can reverse experimentally induced diabetes in mice and maintain vascularised islet-like clusters.[81,82] The production of functional β cells for transplantation is the goal of many research laboratories; however, it is unclear whether transplanted β cells will function adequately or maintain hypoglycaemic counter-regulation outside the islet cluster.

Better immunosuppression and tolerance induction

Sirolimus-based steroid-free immunosuppression has been central to the success of the Edmonton protocol. However, chemical immunosuppression and the associated long-term risks of infection and malignancy are major barriers to the wider application of islet transplantation. Calcineurin inhibitor-free regimens including agents such as hOKT3γ1 (Ala-Ala) and Campath 1H are showing some promise and may be effective in reducing recurrence of autoimmunity as well as suppressing the rejection response. Non-diabetogenic immunosuppressive drugs such as inhibitors of co-stimulation (Belatacept®) are currently under investigation and may represent unique immunotherapy in the settings of islet transplantation.[83] However, the ideal situation will be long-term allograft function without the necessity of immunosuppressive medications. In this regard, recent studies demonstrated the feasibility and efficacy of immunological tolerance induction in clinical transplantation.[59,60] Giving these opportunities, it will be possible to expand the application of islet transplantation beyond the most unstable forms of type I diabetes to more patients, including children, in the future.

Key points

- Islet transplantation is now a viable option in the treatment of selected patients with type I diabetes.
- The Edmonton protocol has been successfully reproduced in several centres worldwide: short-term outcomes are now comparable with solid pancreas transplantation.
- Protocols are emerging that allow single-donor islet transplants. However, better immunosuppression and alternative sources of islets are required to allow the application of islet transplantation to a wider population of diabetic patients.

References

1. Stefan Y, Orci L, Malaisse-Lagae F et al. Quantitation of endocrine cell content in the pancreas of nondiabetic and diabetic humans. Diabetes 1982; 31:694–700.
2. Hogan P, Dall T, Nikolov P. Economic costs of diabetes in the US in 2002. Diabetes Care 2003; 26:917–32.
3. Zimmet P, Alberti KG, Shaw J. Global and societal implications of the diabetes epidemic. Nature 2001; 414:782–7.
4. Banting FG, Best CH, Collip JB et al. Pancreatic extracts in the treatment of diabetes mellitus: preliminary report 1922. CMAJ 1991; 145:1281–6.
5. The Diabetes Control and Complications Trial Research Group. The effect of intensive treatment of diabetes on the development and progression of long-term complications in insulin-dependent diabetes mellitus. N Engl J Med 1993; 329:977–86.
6. UK Prospective Diabetes Study Group. Intensive blood-glucose control with sulphonylureas or insulin compared with conventional treatment and risk complications in patients with type 2 diabetes (UKPDS 33). Lancet 1998; 352:837.
7. Gruessner AC, Sutherland DE. Pancreas transplant outcomes for United States (US) and non-US cases as reported to the United Network for Organ Sharing (UNOS) and the International Pancreas Transplant Registry (IPTR) as of May 2003. Clin Transplant 2003; 21–51.
8. Robertson RP. Islet transplantation as a treatment for diabetes – a work in progress. N Engl J Med 2004; 350:694–705.
9. Williams P. Notes on diabetes treated with extract and by grafts of sheep's pancreas. Br Med J 1894; 2:1303–4.
10. Bensley RR. Studies on the pancreas of the guinea pig. Am J Anat 1911; 12:297–388.
11. Lacy PE, Kostianovsky M. Method for the isolation of intact islets of Langerhans from the rat pancreas. Diabetes 1967; 16:35–9.
12. Brendel MD, Hering BJ, Schultz AO et al. International Islet Transplant Registry. University of Giessen, Germany, 1999; pp. 1–20.
13. Shapiro AM, Lakey JR, Ryan EA et al. Islet transplantation in seven patients with type 1 diabetes mellitus using a glucocorticoid-free immunosuppressive regimen. N Engl J Med 2000; 343:230–8.

   Seven consecutive patients rendered insulin independent for > 1 year post-islet transplantation. This paper demonstrated that islet transplantation is a viable treatment option for type I diabetes and served as the catalyst for the increase in islet transplant activity over the past 4 years.
14. Kaufman DB, Baker MS, Chen X et al. Sequential kidney/islet transplantation using prednisone-free immunosuppression. Am J Transplant 2002; 2:674–7.
15. Fiorina P, Gremizzi C, Maffi P et al. Islet transplantation is associated with an improvement of cardiovascular function in type 1 diabetic kidney transplant patients. Diabetes Care 2005; 28:1358–65.
16. Fiorina P, Folli F, Zerbini G et al. Islet transplantation is associated with improvement of renal function among uremic patients with type I diabetes mellitus and kidney transplants. J Am Soc Nephrol 2003; 14:2150–8.
17. Tzakis AG, Ricordi C, Alejandro R et al. Pancreatic islet transplantation after upper abdominal exenteration and liver replacement. Lancet 1990; 336:402–5.
18. Ahmed SA, Wray C, Rilo HL et al. Chronic pancreatitis: recent advances and ongoing challenges. Curr Probl Surg 2006; 43:127–238.
19. Groth CG, Korsgren O, Tibell A et al. Transplantation of porcine fetal pancreas to diabetic patients. Lancet 1994; 344:1402–4.
20. Service FJ, Molnar GD, Rosevear JW et al. Mean amplitude of glycemic excursions, a measure of diabetic instability. Diabetes 1970; 19:644–55.
21. Service FJ, O'Brien PC, Rizza RA. Measurements of glucose control. Diabetes Care 1987; 10:225–37.
22. Ryan EA, Shandro T, Green K et al. Assessment of the severity of hypoglycemia and glycemic lability in type 1 diabetic subjects undergoing islet transplantation. Diabetes 2004; 53:955–62.
23. Wonnacott K. Update on regulatory issues in pancreatic islet transplantation. Am J Ther 2005; 12:600–4.
24. Johnson PR. Challenges in setting up a new islet transplant program. In: Shapiro AM, Shaw JA (eds) Islet transplantation and beta cell replacement therapy, 1st edn. New York: Informa Healthcare USA, 2007; pp. 203–14.
25. Markmann JF, Deng S, Desai NM et al. The use of non-heart-beating donors for isolated pancreatic islet transplantation. Transplantation 2003; 75:1423–9.
26. Matsumoto S, Okitsu T, Iwanaga Y et al. Insulin independence of unstable diabetic patient after single living donor islet transplantation. Transplant Proc 2005; 37:3427–9.
27. Barshes NR, Lee TC, Udell LI et al. The surgical aspects of pancreas procurement for pancreatic islet transplantation. In: Shapiro AMJ, Shaw JAM (eds) Islet transplantation and beta cell replacement therapy, 1st edn. New York: Informa Healthcare USA, 2007; pp. 81–97.
28. Lakey JR, Kneteman NM, Rajotte RV et al. Effect of core pancreas temperature during cadaveric procurement on human islet isolation and functional viability. Transplantation 2002; 73:1106–10.

29. Ryan EA, Lakey JR, Rajotte RV et al. Clinical outcomes and insulin secretion after islet transplantation with the Edmonton protocol. Diabetes 2001; 50:710–19.

30. Lakey JR, Warnock GL, Rajotte RV et al. Variables in organ donors that affect the recovery of human islets of Langerhans. Transplantation 1996; 61:1047–53.

31. Ponte GM, Pileggi A, Messinger S et al. Toward maximizing the success rates of human islet isolation: influence of donor and isolation factors. Cell Transplant 2007; 16:595–607.

32. Ricordi C, Lacy PE, Scharp DW. Automated islet isolation from human pancreas. Diabetes 1989; 38(Suppl 1):140–2.

The method described here for the first time for human islet isolation is now adopted in every major islet isolation facility worldwide and is the 'gold standard' for comparison of newer isolation techniques.

33. Linetsky E, Bottino R, Lehmann R et al. Improved human islet isolation using a new enzyme blend, liberase. Diabetes 1997; 46:1120–3.

34. Lake SP, Bassett PD, Larkins A et al. Large-scale purification of human islets utilizing discontinuous albumin gradient on IBM 2991 cell separator. Diabetes 1989; 38(Suppl 1):143–5.

35. Walsh TJ, Eggleston JC, Cameron JL. Portal hypertension, hepatic infarction, and liver failure complicating pancreatic islet autotransplantation. Surgery 1982; 91:485–7.

36. Armann B, Hanson MS, Hatch E et al. Quantification of basal and stimulated ROS levels as predictors of islet potency and function. Am J Transplant 2007; 7:38–47.

37. Goto M, Holgersson J, Kumagai-Braesch M et al. The ADP/ATP ratio: a novel predictive assay for quality assessment of isolated pancreatic islets. Am J Transplant 2006; 6:2483–7.

38. Ichii H, Inverardi L, Pileggi A et al. A novel method for the assessment of cellular composition and beta-cell viability in human islet preparations. Am J Transplant 2005; 5:1635–45.

39. Papas KK, Colton CK, Nelson RA et al. Human islet oxygen consumption rate and DNA measurements predict diabetes reversal in nude mice. Am J Transplant 2007; 7:707–13.

40. Sweet IR, Gilbert M, Scott S et al. Glucose-stimulated increment in oxygen consumption rate as a standardized test of human islet quality. Am J Transplant 2008; 8:183–92.

41. Street CN, Lakey JR, Shapiro AM et al. Islet graft assessment in the Edmonton Protocol: implications for predicting long-term clinical outcome. Diabetes 2004; 53:3107–14.

42. Pileggi A, Ricordi C, Kenyon NS et al. Twenty years of clinical islet transplantation at the Diabetes Research Institute – University of Miami. Clin Transplant 2004; 177–204.

43. Kemp CB, Knight MJ, Scharp DW et al. Effect of transplantation site on the results of pancreatic islet isografts in diabetic rats. Diabetologia 1973; 9:486–91.

44. White SA, London NJ, Johnson PR et al. The risks of total pancreatectomy and splenic islet autotransplantation. Cell Transplant 2000; 9:19–24.

45. Owen RJ, Ryan EA, O'Kelly K et al. Percutaneous transhepatic pancreatic islet cell transplantation in type 1 diabetes mellitus: radiologic aspects. Radiology 2003; 229:165–70.

46. Baidal DA, Froud T, Ferreira JV et al. The bag method for islet cell infusion. Cell Transplant 2003; 12:809–13.

47. Casey JJ, Lakey JR, Ryan EA et al. Portal venous pressure changes after sequential clinical islet transplantation. Transplantation 2002; 74:913–15.

48. Robertson RP. Intrahepatically transplanted islets – strangers in a strange land. J Clin Endocrinol Metab 2002; 87:5416–17.

49. Contreras JL. Extrahepatic transplant sites for islet xenotransplantation. Xenotransplantation 2008; 15(2):99–101.

50. Jansson L, Carlsson PO. Graft vascular function after transplantation of pancreatic islets. Diabetologia 2002; 45:749–63.

51. Korsgren O, Lundgren T, Felldin M et al. Optimising islet engraftment is critical for successful clinical islet transplantation. Diabetologia 2008; 51:227–32.

52. Moberg L, Johansson H, Lukinius A et al. Production of tissue factor by pancreatic islet cells as a trigger of detrimental thrombotic reactions in clinical islet transplantation. Lancet 2002; 360:2039–45.

53. Eich T, Eriksson O, Lundgren T. Visualization of early engraftment in clinical islet transplantation by positron-emission tomography. N Engl J Med 2007; 356:2754–5.

54. Faradji RN, Cure P, Ricordi C et al. Care of the islet transplant recipient: immunosuppressive management and complications. In: Shapiro AM, Shaw JAM (eds) Islet transplantation and beta cell replacement therapy, 1st edn. New York: Informa Healthcare USA, 2007; pp. 147–78.

55. Shapiro AM, Ricordi C, Hering BJ et al. International trial of the Edmonton protocol for islet transplantation. N Engl J Med 2006; 355:1318–30.

This is the first multicentre trial that confirmed the efficacy of islet transplantion in type I diabetes patients using the Edmonton protocol.

56. Senior PA, Paty BW, Cockfield SM et al. Proteinuria developing after clinical islet transplantation resolves with sirolimus withdrawal and increased tacrolimus dosing. Am J Transplant 2005; 5:2318–23.

57. Hering BJ, Kandaswamy R, Harmon JV et al. Transplantation of cultured islets from two-layer preserved pancreases in type 1 diabetes with anti-CD3 antibody. Am J Transplant 2004; 4:390–401.

58. Hering BJ, Kandaswamy R, Ansite JD et al. Single-donor, marginal-dose islet transplantation in patients with type 1 diabetes. JAMA 2005; 293:830–5.
59. Kawai T, Cosimi AB, Spitzer TR et al. HLA-mismatched renal transplantation without maintenance immunosuppression. N Engl J Med 2008; 358:353–61.
60. Scandling JD, Busque S, Jbakhsh-Jones S et al. Tolerance and chimerism after renal and hematopoietic-cell transplantation. N Engl J Med 2008; 358:362–8.
61. Al SF, Shapiro AMJ. Clinical outcomes and future directions in islet transplantation. In: Shapiro AMJ, Shaw JAM (eds) Islet transplantation and beta cell replacement therapy, 1st edn. New York: Informa Healthcare USA, 2007; pp. 229–49.
62. Ryan EA, Paty BW, Senior PA et al. Five-year follow-up after clinical islet transplantation. Diabetes 2005; 54:2060–9.

This paper describes the long-term outcomes of the Edmonton group in islet transplantation. Progressive loss of functional islet mass was demonstrated. Only 10% of the islet transplant recipients remained insulin-free 5 years post-transplant. However, most of the recipients remain C-peptide positive (which indicates islet graft function), they sustain excellent metabolic control (despite use of exogenous insulin) and most remain free from hypoglycaemic events.

63. Fiorina P, Folli F, Maffi P et al. Islet transplantation improves vascular diabetic complications in patients with diabetes who underwent kidney transplantation: a comparison between kidney–pancreas and kidney-alone transplantation. Transplantation 2003; 75:1296–1301.
64. Fiorina P, Gremizzi C, Maffi P et al. Islet transplantation is associated with an improvement of cardiovascular function in type 1 diabetic kidney transplant patients. Diabetes Care 2005; 28:1358–65.
65. Fiorina P, Folli F, Bertuzzi F et al. Long-term beneficial effect of islet transplantation on diabetic macro-/microangiopathy in type 1 diabetic kidney-transplanted patients. Diabetes Care 2003; 26:1129–36.
66. Johnson JA, Kotovych M, Ryan EA et al. Reduced fear of hypoglycemia in successful islet transplantation. Diabetes Care 2004; 27: 624–5.
67. Toso C, Shapiro AM, Bowker S et al. Quality of life after islet transplant: impact of the number of islet infusions and metabolic outcome. Transplantation 2007; 84:664–6.
68. Poggioli R, Faradji RN, Ponte G et al. Quality of life after islet transplantation. Am J Transplant 2006; 6:371–8.
69. Goss JA, Soltes G, Goodpastor SE et al. Pancreatic islet transplantation: the radiographic approach. Transplantation 2003; 76:199–203.
70. Casey JJ, Lakey JR, Ryan EA et al. Portal venous pressure changes after sequential clinical islet transplantation. Transplantation 2002; 74:913–15.
71. Ryan EA, Lakey JR, Paty BW et al. Successful islet transplantation: continued insulin reserve provides long-term glycemic control. Diabetes 2002; 51:2148–57.
72. Rafael E, Ryan EA, Paty BW et al. Changes in liver enzymes after clinical islet transplantation. Transplantation 2003; 76:1280–4.
73. Bhargava R, Senior PA, Ackerman TE et al. Prevalence of hepatic steatosis after islet transplantation and its relation to graft function. Diabetes 2004; 53:1311–17.
74. Campbell PM, Senior PA, Salam A et al. High risk of sensitization after failed islet transplantation. Am J Transplant 2007; 7:2311–17.
75. Cardani R, Pileggi A, Ricordi C et al. Allosensitization of islet allograft recipients. Transplantation 2007; 84:1413–27.
76. Markmann JF, Deng S, Huang X et al. Insulin independence following isolated islet transplantation and single islet infusions. Ann Surg 2003; 237:741–9.
77. Groth CG, Korsgren O, Tibell A et al. Transplantation of porcine fetal pancreas to diabetic patients. Lancet 1994; 344:1402–4.
78. Cardona K, Korbutt GS, Milas Z et al. Long-term survival of neonatal porcine islets in nonhuman primates by targeting costimulation pathways. Nat Med 2006; 12:304–6.
79. Hering BJ, Wijkstrom M, Graham ML et al. Prolonged diabetes reversal after intraportal xenotransplantation of wild-type porcine islets in immunosuppressed nonhuman primates. Nat Med 2006; 12:301–3.
80. Bonner-Weir S, Weir GC. New sources of pancreatic beta-cells. Nat Biotechnol 2005; 23:857–61.
81. Lumelsky N, Blondel O, Laeng P et al. Differentiation of embryonic stem cells to insulin-secreting structures similar to pancreatic islets. Science 2001; 292:1389–94.
82. Ramiya VK, Maraist M, Arfors KE et al. Reversal of insulin-dependent diabetes using islets generated in vitro from pancreatic stem cells. Nat Med 2000; 6:278–82.
83. Vincenti F, Larsen C, Durrbach A et al. Costimulation blockade with belatacept in renal transplantation. N Engl J Med 2005; 353:770–81.

11

Cardiothoracic transplantation

Christopher H. Wigfield
Stephen C. Clark
Asif Hasan

Introduction

Within 40 years heart transplantation has evolved from an experimental procedure to an effective therapeutic strategy for end-stage heart disease. In the current era heart transplant programmes around the world are showing medium-term survival in excess of 80–85%.[1] Significant improvements have been achieved, particularly for 30-day and 1-year mortality. Nevertheless, the success of heart transplantation has raised expectations that under present circumstances it cannot fulfil. On one hand, due to improved management of ischaemic heart disease and increased longevity, the number of patients with heart failure is growing.[2] On the other hand, there is a decrease in the number of cardiac transplantations due to donor organ constraints. This disparity between the number of donors and potential recipients has stimulated research to find new alternatives to transplantation. However, these so far had little impact on the current practice of heart transplantation. The advent of novel therapeutics and surgical options for impaired ventricles may in selected patients defer consideration for transplantation, and clinical guidelines have been provided for this purpose.[3] The contemporary practice of heart transplantation with respect to indications, surgical techniques and donor and recipient management will now be reviewed.

Indications for heart transplantation

The reason for undertaking heart transplantation is to prolong life and to improve its quality. The indications for adult heart transplantation have remained essentially unchanged over the last three decades and at present are predominantly coronary-related heart failure (38%) and cardiomyopathies (45%). Valvular (2%) and miscellaneous diagnoses (10%), adult congenital (3%) and re-transplantation (2%) constitute the rest.[1]

The indications for paediatric heart transplantation (<16 years) are different to adults. In our series of 153 paediatric heart transplants (1987–2008), 60% were undertaken for cardiomyopathy and 40% for congenital heart disease.

Aetiology of heart disease

Introduction

End-stage heart failure has become a major medical problem.[4] The increasing prevalence with rising age of the general population in most societies accounts for a large proportion of healthcare spending due to frequent hospital admissions.[5] In the aetiology of congestive heart failure (CHF), we differentiate primarily ischaemic from other cardiomyopathies and congenital heart disease during transplant candidate assessment.

Ischaemic heart disease

This constitutes the largest group requiring heart transplantation. These patients can present in a variety of ways, from being acutely ill after myocardial infarction on mechanical support to being chronically ill with heart failure with or without previous surgical or catheter-based intervention. Unfortunately there are no conclusive prospective studies comparing conventional treatment methods with heart transplantation to provide guidance in risk–benefit assessment. A digest of current thinking would indicate that heart transplantation would definitely be indicated in a patient with severe heart failure with poor ventricular function (ejection fraction <15%), symptoms of heart failure with little or no angina, diffuse coronary artery disease, absence of reversible ischaemia and/or poor right ventricular function (ejection fraction <35%). What is clear is that patients with ischaemic cardiomyopathy who develop heart failure are likely to have a worse prognosis than non-ischaemic patients.[6]

Non-ischaemic cardiomyopathy

This group includes a variety of aetiologies with marked left and/or right ventricular dysfunction. Disease processes that result in changes in heart muscle are classified as: (a) dilated cardiomyopathy; (b) hypertrophic cardiomyopathy; (c) restrictive cardiomyopathy; and (d) arrhythmogenic right ventricular dysplasia. In patients with non-ischaemic cardiomyopathy, transplantation is indicated if there is failure of aggressive medical treatment.

Certain types of cardiomyopathies can show reversibility, and a period of observation with medical treatment should be tried before listing. These include lymphocytic myocarditis, peripartum cardiomyopathy, hypertensive cardiomyopathy and alcoholic cardiomyopathy.[7]

Indications for paediatric patients are similar; however, the risk of death is highest during the first 3 months after presentation, therefore failure of aggressive medical treatment early in the course of the disease should lead to early assessment for transplantation. Acute myocarditis needs a special mention as the finding of acute inflammation on biopsy is a favourable prognostic sign for subsequent recovery.[8]

Congenital heart disease

In the present era the diagnosis of congenital heart disease is as common as cardiomyopathy as an indication for heart transplantation in children. This group of patients has grown due to the success of paediatric cardiac surgery. The positive outcome of palliation for congenital structural cardiac defects has resulted in patients potentially becoming candidates for heart transplantation at a later stage. The indications for transplantation in this group can be for either life-saving reasons or for improvement in quality of life.

Recipient evaluation and selection

Patients are evaluated for transplantation once a referral has been made. We admit the patient for a few days for assessment. During this period there is a systematic evaluation of both the physical and psychological state of the patient; it also gives an opportunity to develop a rapport between the patient, relatives and the multidisciplinary team. The protocol used in our own centre for assessment is summarised in Box 11.1. The assessment process is designed to answer the following questions:

1. Does the patient fulfil the selection criteria for heart transplantation?
2. Are there any contraindications to transplantation?
3. Is there any possibility of any other treatment option?

Selection criteria

The process of selection of patients for transplantation remains an inexact science. In the majority of cases the referral for transplantation is of a patient with chronic heart failure. In these cases there is remarkable divergence of opinions when a patient should be listed for heart transplantation, further compounded by a paucity of evidence to guide day-to-day clinical practice. Avoidance of transplantation when a patient is 'too good' has important prognostic indications, as the 10-year survival after orthotopic heart transplantation remains 50% in most centres.

Mancini et al.[8] and others[9–11] showed that patients with a peak exercise oxygen consumption of <14 mL/kg/min had a significantly higher mortality than patients with a peak exercise oxygen consumption of >14 mL/kg/min.

Box 11.1 • Recipient assessment protocol for heart transplantation

1. Full medical assessment

Full history and physical examination. Investigations include:

- Full blood count, platelets and coagulation screen
- Blood group
- Urea and electrolytes, liver function and thyroid function
- Microbiology – sputum, midstream specimen of urine (MSU), nose/throat/axilla/perineum swabs for culture
- Full viral screen (with patient consent)
- Fasting glucose and lipids
- 12-lead ECG
- Chest X-ray (PA and LAT)
- Spirometry
- Echocardiogram
- Chromium EDTA glomerular filtration rate (GFR) (renal opinion and abdominal ultrasound would be required if GFR <32.5 mL/min)
- Estimation of peak oxygen consumption (VO_{2max})
- Right heart catheter to assess filling pressures and calculate pulmonary vascular resistance, after discussion with the transplant cardiologist (as per protocol)
- Bone density (if >50 years or symptoms)
- Urine flow rate/residual (if male >50 years or symptoms)
- Carotid/peripheral artery Doppler (if symptoms)

2. A structured educational package – provided by the transplant coordinator

Discussion points include:

- Patient's understanding of his or her illness
- Donor compatibility
- Introduction to the concept of transplantation
- Preparation for admission
- Reason for assessment
- Travelling arrangements
- Explain investigations and visits
- Accommodation
- Survival figures
- Outpatient routine
- Surveillence biopsies
- Waiting lists and waiting period
- Adjusting to family life
- Bleeper
- Driving
- Returning to work

3. Social assessment

This looks at both practical and emotional aspects of the transplant process with the patient and carer. Areas covered include:

- Feelings about what is happening to them
- Social security benefits
- Support networks
- Coping strategies

The aim is to evaluate whether the patient understands and whether he or she will cope with having a transplant and to prepare the ground for future involvement throughout the patient's contact with the transplant team

4. Physiotherapy assessment

An assessment and education package from the transplant physiotherapist with regard to exercise pre- and post-transplant, and postoperative chest care

The limitation of this technique is that it can be influenced by body composition, individual motivation or general deconditioning. Some centres have incorporated the heart failure survival score (HFSS) to their preoperative assessment. The score consists of seven variables – resting heart rate, left ventricular ejection fraction, mean arterial blood pressure, interventricular conduction delay, peak exercise oxygen consumption (VO_2), serum sodium and ischaemic cardiomyopathy. Using these variables Aaronson et al. developed a mathematical model to predict outcome with medical management.[9] This score, along with maximal oxygen consumption (VO_{2max}) and clinical assessment, can bring some rigour to the selection process for transplantation.

Deng et al. showed that cardiac transplantation does not benefit patients with medium- and low-mortality risk as assessed by calculation of heart failure survival score.[10]

More recent evidence suggests that for patients with ejection fraction <20% and severely reduced functional capacity, best outcomes are achieved in the absence of comorbidities that raise perioperative mortality risks. The predictors of survival have been refined and patients requiring inotropic or mechanical support to maintain systemic perfusion in advanced heart failure have a good prognosis with heart transplantation.[1]

Contraindications

Contraindications to heart transplantation are summarised in Box 11.2. These can be classed in three groups:

1. Factors that increase perioperative mortality, e.g. elevated pulmonary vascular resistance.
2. Factors affecting long-term prognosis.
3. Factors related to life-threatening non-compliance.

These exclusion criteria have continued to change with improvement in medical treatment and increasing experience with heart transplantation, and now successful outcome can be obtained in cases previously excluded. Some of the contraindications deserve special mention.

Pulmonary vascular resistance (PVR) of more than 6 Wood units has been considered an absolute contraindication to heart transplantation but with the introduction of nitric oxide, use of a bicaval anastomotic technique, early implantation of ventricular assist devices and increasing use of perioperative phosphodiesterase inhibitors and isoprenaline, good results can be obtained in patients who formerly would not have been offered the opportunity of transplantation. Nevertheless, the presence of an elevated PVR should not be taken lightly as the donor right ventricle generally tolerates a systolic pressure of more than 50 mmHg poorly and would acutely fail. In our own practice a PVR >3 Wood units would be considered a relative contraindication to transplantation. The **transpulmonary gradient (TPG)** (see Box 11.3) represents the pressure gradient across the pulmonary vascular bed and is independent of the pulmonary blood flow. Some consider the elevation of this above 14 mmHg as a more useful indication of significantly raised PVR as this is independent of the

Box 11.2 • Contraindications for heart transplant

Factors increasing perioperative mortality

- Irreversible pulmonary hypertension:
  - PVR > 6 Wood units despite standardised reversibility testing protocol
  - TPG >14 mmHg
- Active infection
- Recent peptic ulcer disease
- Severe obesity (>140% ideal body weight)
- Cachexia (<80% ideal body weight)
- Pulmonary infarction within 6–8 weeks

Factors affecting long-term prognosis

- Age >65 years
- Severe renal impairment measured by EDTA GFR and kidney biopsy
- Brittle diabetes
- Active or recent malignancy
- Significant chronic lung disease, FEV_1 <40% predicted, FVC <50% of normal and DL_{CO} <40% of predicted
- Severe peripheral vascular disease
- Significant hepatic impairment

Factors that impair compliance

- Active mental illness
- Drug abuse within last 6 months refractory to treatment
- Chronic illness affecting function

DL_{co}, carbon monoxide diffusing capacity; FEV_1, forced expiratory volume in 1 second; FVC, forced vital capacity; GFR, glomerular filtration rate; PVR, pulmonary vascular resistance (see Box 11.3); TPG, transpulmonary gradient (see Box 11.3).

Box 11.3 • Definitions

Pulmonary vascular resistance

PVR (Wood units) = [PA mean – pulmonary capillary wedge pressure (PCW)/CO]

Pulmonary vascular resistance index

PVRI (Wood units/m^2) = (PA mean – PCW)/CI = PVR BSA

Transpulmonary gradient

TPG (mmHg) = PA mean – PCW

cardiac output, which may be poor in these patients. We rely more on this criterion and generally consider a fixed TPG of 12 mmHg and above as an absolute contraindication. In paediatric patients a higher TPG can be considered as it could be overcome with a larger sized donor heart. Heterotopic transplantation can also be considered in these circumstances to overcome elevated pulmonary vascular resistance.[11]

Renal dysfunction is one of the most common problems encountered in the assessment of these patients. Multiple studies have shown that it is a major risk factor for mortality after heart transplantation. A common dilemma is to distinguish between renal dysfunction due to intrinsic renal disease or severe heart failure and aggressive diuretic therapy. It is essential to measure the glomerular filtration rate (GFR) and a low GFR may occasionally indicate renal biopsy to further elucidate the cause. Others have used measurement of effective renal plasma flow (ERPF) as an investigative modality, and less than 200 mL/min is considered indicative of major intrinsic renal dysfunction and an indication for combined heart and kidney transplantation which can be peformed with good outcomes.[12]

Diabetic candidates have been shown to have good outcomes in the absence of significant end-organ damage (retinopathy, nephropathy, autonomic dysfunction and neuropathy or advanced peripheral vascular disease). Previously often excluded from heart transplantation, the 1- and 3-year survival, as well as rejection rates and infection prevalence achieved after transplantation, are comparable to non-diabetic recipients.

Compliance is the neurobehavioural capacity to adhere to a complex lifelong medical regimen. Non-compliance following heart transplantation can lead to major morbidity or death. Unfortunately there are no proven psychological or sociological factors to predict poor compliance or adverse outcome after transplantation. Adherence to medical treatment and ability to keep appointments can provide some pointers towards compliance. Psychiatric disorders that impair compliance, such as severe depression or untreated schizophrenia, are contraindications to heart transplantation.[13]

Other options

It is not unusual to find patients who have been referred for transplantation to be suitable for alternative treatments. In addition there are newer methods of treatment of heart failure in both medical and surgical disciplines being developed, and some of these patients could derive benefit from them.[14] Some developments are worth mentioning.

Biventricular pacing

In 20–30% of patients with symptomatic heart failure there is a prolonged PR interval, wide QRS complexes and intraventricular conduction disorders leading to a discoordinate contraction pattern. The result is earlier atrial contraction causing mitral regurgitation. This is further compromised by paradoxical septal motion due to wide QRS and conduction abnormalities. Biventricular pacing has been shown to synchronise contractility and improve the functional class of patients.[15–17] We have used biventricular pacing in several of our patients with improvement in functional class and subsequent delisting from transplantation.

Implantable cardio defibrillators

Implantable defibrillators have had benefical impact on survival of patients with implantation including previous ventricular arrhythmias and poor ejection fraction status.[18]

Novel end-stage heart disease therapeutics

New agents to treat end-stage disease have become available and the benefit of antagonists to stabilise symptoms and delay patients entering waiting lists for transplantation has been evident. A recent landmark publication by the American College of Cardiology has confirmed the advanced administration of medical therapeutics in a large cohort of patients and specific guidelines have been widely implemented.[19,20] Other novel therapeutics are in various stages of clinical trials.

Ventricular assist devices

Over approximately 15 years ventricular assist device support has developed into a realistic option for selected patients with refractory congestive heart failure of various aetiologies. This has been established in the REMATCH trial, where medical management of New York Heart Association (NYHA) class IV heart failure patients was inferior to mechanical assistance when comparing 1- and 2-year survival.[21,22] The limiting factors for this approach, either as a bridge to transplantation or as destination therapy, are availability and device selection. More recently the role of assist devices for right heart failure indications is evolving.

Donor selection and matching

Specific guidelines for optimal donor selection have been published by the International Society for Heart and Lung Transplantation.[23]

Management of the potential organ donor has evolved and requires a multidisciplinary approach.[24] Donor allocation for hearts in the UK is run by UK Transplant (UKT). The hearts are allocated on a pro rata basis; however, a category of 'urgent' was created in 1999 to deal with acutely ill patients. Once a donor is identified certain criteria apply before acceptance.

Donor age

An upper limit of 60 years is generally advocated and used by our own unit but there is variation in other centres. The current mean age including paediatric donors at our programme currently is 44 years. It is important that donor age should not be viewed in absolute terms but should be considered along with other factors such as cardiac function, recipient urgency and projected ischaemic times. However, older donors are more likely to have coronary artery disease and there is increased mortality for the recipient if the heart has come from a donor over 40 years of age.[1] The presence of coronary artery disease should not be considered an absolute exclusion criterion as satisfactory outcomes can be achieved with concomitant coronary revascularisation.[25] United Network for Organ Sharing (UNOS) data from the USA show that in 1982 2.1% of donors were aged 50 years or greater but by 1994 this percentage had increased to 8.9% and has remained the same over the last 10 years.[26] It remains difficult to evaluate pre-existing donor coronary artery disease at the time of organ procurement. Some centres advocate a single plane coronary angiography (on table), but the availability and interpretation remain problematic.

Cardiac function

Brain death leads to myocardial changes with abnormalities seen on ECG of ST segment elevation, T-wave inversion and Q waves, often signifying subendocardial ischaemia. Events following brain death, namely prolonged hypotension, cardiopulmonary resuscitation and high-dose inotropic support also contribute to cardiac dysfunction, particularly acute right ventricular impairment. The assessment of cardiac function is undertaken by echocardiogram, Swan–Ganz catheter and finally by the surgeon procuring the organ. Troponin I may be useful in detecting donor myocardial injury and elevated levels are associated with impaired cardiac function.[27]

There is no consensus on what degree of inotropic support correlates with structural and functional damage sufficient to compromise graft function. It has been recommended that hearts should not be used if the inotropic requirements exceed 20 μg/kg/min of dopamine. Often, inotropes are used in conjunction with fluid infusions to fill a vasodilated circulation and bolster perfusion pressure. We utilise arginine vasopressin under these conditions and wean the inotropes. Failure to wean the inotropes under these conditions is a bad prognostic sign and suggests cardiac dysfunction. Donor hearts developing arrhythmias are not considered suitable.

Donor disease

Donors with an active infective focus are usually turned down. However, donors with a history of meningitis that has been adequately treated are considered for donation. Hepatitis C patients are not considered unless the recipient is positive for hepatitis C or is acutely ill on the urgent list. Hepatitis B donors with positive surface antigen are avoided, but core antibody-positive donors (surface antigen-negative) can be considered.

Donor hearts from donors with primary brain tumours are considered for transplantation. Glioblastoma patients and those with shunts are not considered.

A history of intravenous drug abuse would disqualify the donor, but an exception can be made in the very ill recipient and a normal echocardiogram of the donor heart in the presence of negative serological viral testing. High-risk donors require careful evaluation as the routine enzyme-linked immunosorbent assay (ELISA) testing serology used does not achieve the same specificity as DNA-based tests. Chronic cocaine use causes cardiomyopathic changes and caution should be used in accepting hearts from such donors.

Size matching

As a general rule for routine adult heart transplantation with a normal PVR, 30% undersizing is acceptable, although much smaller donors have been reported with satisfactory outcome.[28,29] In patients with a raised PVR deliberate oversizing is routinely undertaken to overcome pulmonary vascular resistance. In the paediatric group oversizing is often done to utilise all available hearts. In our last 30 consecutive paediatric transplants the average size discrepancy between donor and recipient was 150%, and in a cohort of patients who had a failing Fontan circulation as an indication for transplantation, the oversizing was 250%. The adverse consequences of oversizing are delayed sternal closure, collapse of the left lower lobe and systemic hypertension. However, all these factors can resolve with time and appropriate treatment.

ABO compatibility

ABO compatibility is required to avoid hyperacute or accelerated acute rejection. Rhesus incompatibility is acceptable. The only ABO exception would be the A_2 subgroup as donors with this subgroup may be less prone to producing hyperacute rejection, because A_2 antigen is not readily displayed on the endothelial surface of the heart. However, in the paediatric group successful heart transplantation has been undertaken in the presence of ABO incompatibility.[18] This is possible as the immune system in infants is immature and their anti-A and anti-B titres remain low until 12–14 months of age. We have successfully undertaken 10 transplants in children with ABO incompatibility.[30] The oldest was 18 months old when heart transplantation was performed.

Immunological matching

The rationale for undertaking immunological testing is to identify potential recipients with circulating anti-HLA antibodies to avoid mismatch between donor and recipient that could lead to hyperacute or accelerated acute rejection (see Chapter 4).

Common causes of sensitisation are pregnancy, prior blood transfusion or insertion of a ventricular assist device. Rarely, a patient may be sensitised for unknown reasons. The test is considered positive if a 10% threshold is reached on testing the donor serum with the control group. When the recipient has detectable panel-reactive antibodies (PRAs), a prospective crossmatch is undertaken, which has implications regarding the timing of transplantation and in our experience does disadvantage the recipient. The long-term results of recipients having more than 25% PRAs show that they are more prone to rejection.[31]

Donor heart procurement

It is important to optimise the haemodynamic, metabolic and respiratory condition of the donor to maximise the yield of donor organs. This may entail using a multidisciplinary team to manage and optimise the donor before retrieval. Some poorly functioning hearts could be resuscitated by careful manipulation of inotropes and loading conditions of the heart. Using this strategy up to 30% of such hearts can be successfully 'resuscitated' and used for transplantation.[32]

The thoracic organs are accessed by midline sternotomy; this might have already been performed by the liver retrieval team. It is important to secure haemostasis carefully due to coagulopathy and volume replacement should continue actively. This requires careful consideration of other organs procured, especially the lungs.

Whilst the abdominal dissection is being undertaken the pericardium is opened and the heart is inspected for its functional state as well as the presence of any congenital abnormality. The heart is palpated to feel any thrill for valvular heart disease or any coronary plaques. When the mobilisation of abdominal organs is completed, heparin at a dose of 300 units/kg is administered. If a central line is in place, it is withdrawn. The superior vena cava is ligated and the inferior vena cava is completely divided. This allows the heart to exsanguinate into the right pleural cavity. The aorta is now clamped and cardioplegic solution is infused via the aortic root. We use 1 litre of St Thomas's cold crystalloid cardioplegic solution; this is augmented with cold topical saline. The dose for paediatric donors is 30 mL/kg. During the administration of cardioplegia the right superior pulmonary vein is incised to decompress the left side of the heart.

Once the cardioplegia has been given the cardiectomy can proceed further. The superior vena cava is incised above the previous ligature. The aorta is now divided below the innominate artery; this exposes the pulmonary artery, which is divided on the left side where the left pulmonary artery is attached to the pericardial reflection and the right pulmonary artery is divided behind the aorta. The left atrium is now incised at the level of the pericardial reflection. Due consideration to leave sufficient left atrial tissue along each pulmonary vein is essential when lungs are procured for transplantation.

The heart is now inspected for the presence of a patent foramen ovale and if found is oversewn. The heart is now packed inside three bags with cold saline in the intervening bags. The heart is then placed in a transport cooler packed in ice to be transported.

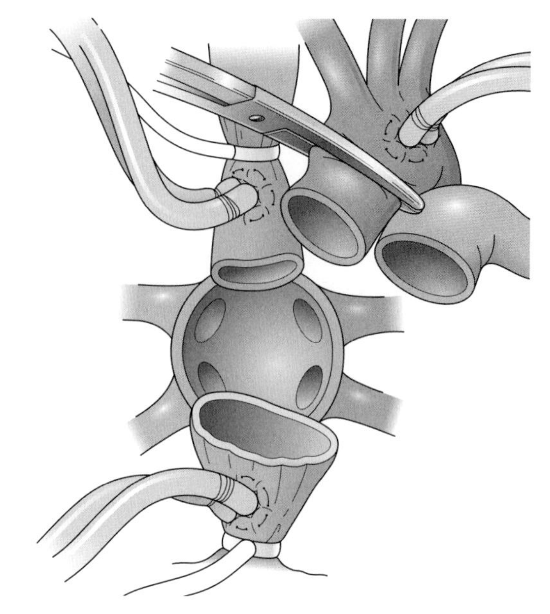

Figure 11.1 • Division of the right atrium to create superior and inferior vena caval cuffs for bicaval technique. The great vessels are divided as in the standard orthotopic method.

Heart transplantation (Figs 11.1 and 11.2)

The classical technique of orthotopic heart transplantation as described by Lower et al. has remained the standard operation for 30 years.[33]

The most significant modification to heart transplant procedure has been the use of a bicaval technique. This results in less tricuspid regurgitation and better haemodynamic performance of the implanted heart.[34]

The operation is undertaken with a midline sternotomy. Cardiopulmonary bypass (CPB) is established with right atrial venous cannulation to allow decompression of the heart; this allows for easier cannulation of superior and inferior venae cavae. The patient is then cooled to 32 °C. The cavae are now snared and the aorta is clamped. To facilitate bicaval anastomosis it is recommended that at this stage the interatrial groove is dissected to develop a cuff of left atrium. The cardiectomy proceeds with a right atrial incision, which runs parallel to the atrioventricular groove; care is taken at the inferior caval end of this incision to preserve as much tissue as possible to facilitate inferior caval anastomosis. An incision is then made in the roof of the left atrium to further decompress the heart before dividing the aorta just above the aortic valve. Retracting the heart downwards now exposes the pulmonary artery, which is again divided above the pulmonary valve. The superior vena cava is now divided just at its right atrial junction. The heart is now only attached to the pulmonary veins and via a small bridge of tissue to the inferior vena cava. The inferior caval attachment is divided, again being mindful of the inferior vena cava (IVC) cuff; the incision

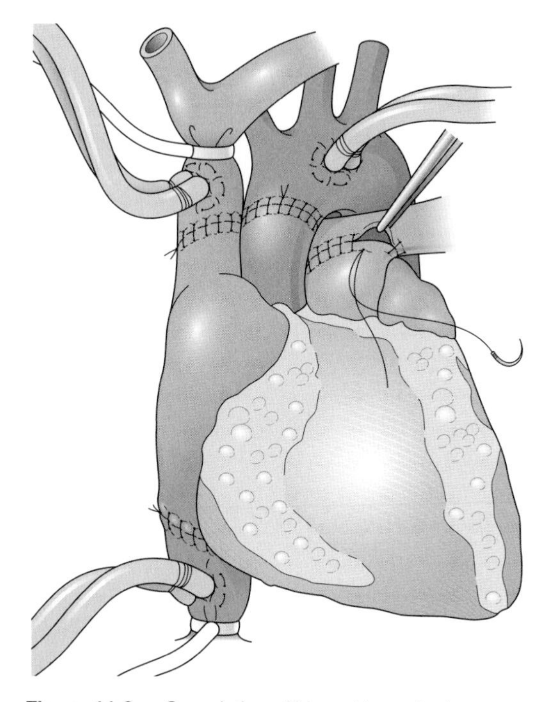

Figure 11.2 • Completion of bicaval transplant technique, showing the inferior vena caval, aortic and pulmonary artery anastomoses. Reproduced from Kirklin JK, Young JB, McGriffin DC. Heart transplantation. Edinburgh: Churchill Livingstone, 2002; Ch. 10.8, p. 343. With permission from Churchill Livingstone.

in the left atrium is now extended to encircle the pulmonary veins, leaving behind two pulmonary veins on each side attached with a bridge of tissue.

The donor heart is now prepared for implantation. The pulmonary veins are joined together by incisions removing any excess tissue, the pulmonary artery is cut back to its bifurcation, and a dose of blood cardioplegia is given in the aortic root. The donor and recipient atria are anastomosed with 3/0 polypropylene; care is taken not to leave excess tissue behind as it could be thrombogenic or can obstruct the pulmonary venous orifices. The suture line is not completed as a vent is left in the left atrium to take away the warm blood. The aortic anastomosis is now undertaken with a 4/0 polypropylene suture. At this stage the aortic cross-clamp can be removed to reduce the donor ischaemic time. De-airing is undertaken through the aortic root and a dose of steroids is given before the clamp is removed. This is a critical period as the heart has been reperfused after a prolonged period of ischaemia. The heart usually starts to beat, but if ventricular fibrillation occurs the heart is promptly defibrillated. Careful attention is paid to the perfusion pressures and the ventricle is kept decompressed; the atrial vent is left in until satisfactory contractility is resumed. The pulmonary artery anastomosis is next undertaken, care being taken in trimming of the pulmonary arterial cuff to avoid redundancy. The IVC and then the superior vena caval anastomoses are completed.

Once the implantation is complete the body temperature is brought back to normothermia. We would reperfuse the heart for at least 10 minutes of every hour of ischaemic time before making an attempt at weaning CPB. Weaning from CPB is undertaken carefully avoiding distension of the right ventricle. Before closure of the chest, ventricular and atrial temporary pacing wires are attached, and a left atrial monitoring line is left in situ. Isoprenaline is frequently used for rate control and initial right ventricular afterload reduction.

Special situations

Heart transplantation for congenital heart disease

This group of patients presents special technical challenges due to unusual anatomy, previous operations and raised pulmonary vascular resistance. Heart transplantation can be undertaken to overcome most structural abnormalities.[35]

Heterotopic heart transplantation

This describes the placement of the donor heart in parallel with the native heart. In the current era there are two possible indications for this procedure:

1. if the pulmonary vascular resistance is high (pulmonary artery pressure >60 mmHg) and cannot be manipulated by the use of nitric oxide;
2. if the donor is considerably smaller than the recipient.

The results of heterotopic transplantation have been generally inferior to orthotopic heart transplantation. However, this technique is occasionally considered.[36]

Perioperative management

The principles of early management of the heart transplant patient are: (a) to maintain graft function, specifically to recognise and manage right ventricular impairment early; (b) to establish adequate immunosuppression; (c) to prevent and treat early infections; and (d) to allow recovery of other system functions, such as renal function.

Graft function

Most patients will have reduced myocardial function after heart transplantation and would require inotropic support, which is usually weaned over 24–48 hours. However, some patients develop either right or left ventricular dysfunction. Development of right ventricular dysfunction is multifactorial. Initially brain death-induced subendocardial ischaemia can occur in the donor. The right ventricle has a disposition to suffer the consequences of relative size mismatching of donor and recipient especially related to the level of PVR after transplantation, inadequate myocardial preservation and discrepancy in size of the donor (smaller donor). The management of this would consist of adjusting preload, aggressive Right Ventrical (RV) afterload reduction and inotropic support. If the PVR is elevated, nitric oxide is added and in extreme cases a right ventricular assist device is used. It is important that an anastomotic complication is excluded by measuring the pressure gradient across the pulmonary artery suture line; surgical revision is indicated if the systolic gradient is >10 mmHg.

Left ventricular dysfunction is again managed with adjustment in inotropic support; under rare circumstances the dysfunction can be life threatening and left ventricular assist may be required.

Immunosuppression

Triple therapy consisting of ciclosporin, azathioprine or mycophenolate mofetil and steroids is the standard regimen used by most centres. We employ induction therapy using equine antithymocyte globulin (ATG) in the paediatric group and in patients with severe renal impairment who are intolerant of ciclosporin. Tacrolimus has been used as a substitute for ciclosporin in patients with persistent rejection and in children with side-effects of ciclosporin, i.e. gingival hypertrophy and hirsutism. Newer immunosuppressive agents, such as rapamycin and interleukin-2 receptor antibodies (basiliximab and daclizumab), are being used in heart transplant rejection prophylaxis and treatment but have not yet found their place in routine immunosuppression protocols, despite a number of promising studies.

Monitoring of rejection following heart transplantation is crucial to short- and long-term survival of patients. The gold standard of rejection monitoring is endomyocardial biopsy. However, the histological findings of rejection are not uniformly present throughout the myocardium, and a high degree of clinical vigilance is needed in post-transplant follow-up of these patients. Approximately 7% of patients have an early rejection episode within 30 days of transplantation and at least 15% receive treatment for acute rejection within the first year.

Infection prophylaxis and treatment

Significant progress has been made in both treatment and prophylaxis for heart transplant recipients. In most centres infective episodes have been reduced to <15% during the first year after transplantation. Notably a dramatic reduction in cytomegalovirus (CMV), Pneumocystis Parinii (PCP) and toxoplasmosis infections have been achieved with strict prophylactic antimicrobial regimes. The most common type of infection is bacterial (50%) followed by viral (40%), fungal (5%) and protozoal (5%). The commonest single organism causing infection after heart transplantation is CMV.[37]

Survival

The registry of the International Society for Heart and Lung Transplantation (ISHLT) collects data on heart transplantation performed worldwide. The data have been collected since 1980; by 2005 more than 70 000 cardiac recipients had been reported.[1] The survival data show 5-year survival of approximately 65%, at 10 years approximately 50% and at 15 years 25–30%. The median survival time is now 10 years. For patients who are alive at 1 year, the median survival time is 13 years.

In our own centre, we have undertaken 554 adult heart transplantations between 1985 and 2003. Amongst these patients 147 have survived more than 10 years and 24 more than 15 years. During a similar period we have also undertaken 107 heart transplants in the paediatric group (<16 years). In this group 24 patients are alive at 10 years and 3 at 15 years.

Cause of death after heart transplantation

This information is available from the ISHLT registry but needs to be considered in the context of inherent difficulties associated with registry information, namely non-uniformity of definitions, validity of information and other difficulties of collecting information. The distribution of causes, modes and mechanisms of mortality after heart transplantation are time related. During the first 30 days after heart transplantation, graft failure accounts for 41% of the deaths, followed by non-CMV infection (14.2%) and multiorgan failure (13.9%). After 30 days and up to 1 year, non-CMV infection accounts for up to one-third of deaths (33%), followed by graft failure and acute rejection (together ≈20%). Beyond 5 years, cardiac allograft vasculopathy (CAV) and late graft failure (possibly due to CAV) are the predominant causes of deaths. Malignancies are increasingly common after 10 years (>30%) and account for approximately 20% and non-CMV infections 10% of late deaths. Renal failure and multiorgan impairment are significant contributors.[1]

Cardiac allograft vasculopathy

This is an unusually accelerated and diffuse form of obliterative coronary artery arteriosclerosis and is the commonest cause of graft failure in the long term after heart transplantation. Coronary arterial disease begins to develop relatively early after heart transplantation and nearly all patients after 1 year would show some histopathological evidence of this disease.[38] Two different types of lesions develop. Type A are discrete stenoses of proximal and middle thirds of epicardial arteries. Type B are present in the distal coronary arteries and consist of tubular constrictions. Small vessels of less than 100 μm in diameter less frequently involved. CAV is probably due to a complex interplay between immunological and non-immunological factors. Viral infections, immunosuppressive drugs, dyslipidaemias and oxidant stress may all play a part.[39] Risk factors within 5 years of heart transplantation include pretransplant coronary artery disease, PRA positivity of >20% and donor hypertension. Female donors yield a weakly protective effect.[1] Beyond 5 years, hospitalisation for rejection within 5 years of transplantation and donor mass index also become additional risk factors.

Early identification of CAV is necessary as patients generally remain asymptomatic due to cardiac denervation; moreover, early recognition can improve long-term prognosis. Surveillance for CAV can be undertaken by several techniques, including intravascular ultrasound (IVUS), determination of coronary flow reserve and dobutamine stress echocardiography. However, coronary angiography remains the commonest form of investigation. IVUS is a much more sensitive tool in the detection of early intimal thickening.[40] However, it suffers from theoretical shortcomings of a lack of universal grading system and absence of an initial estimate of intimal thickening of donor arteries. In addition the size of available catheters means that they can only be used in vessels exceeding 1.5 mm in diameter. Dobutamine stress echocardiography has the advantage of non-invasive monitoring of CAV but its reported sensitivity of 72%[41] does not make it a suitable tool for replacement of angiography. Angiography is generally undertaken on an annual basis but if new lesions are identified it can be repeated more frequently. We would also perform a baseline study if the presence of atherosclerosis is suspected in the donor heart at the time of transplant operation.

There is no conventional treatment for CAV and the emphasis has been on tertiary prevention of progression. The usual preventative measures of coronary artery disease have limited value in this setting. Calcium channel blockers, especially diltiazem and statins, have been shown to be effective.[42,43]

Newer immunosuppressive agents like rapamycin and everolimus, due to its proliferation inhibitor properties, also offer hope for the future.[44] In a prospective randomised study comparing everolimus with azathioprine, the incidence of CAV was significantly lower in the group receiving everolimus.

Malignancy

The incidence of malignancy after heart transplantation is three to four times higher than the general population. The three common cancers after heart transplantation in order of frequency are cutaneous malignancies, post-transplant lymphoproliferative disorder (PTLD) and lung cancers. The risk factors for development of malignancies are presence of pretransplant malignancy, pretransplant coronary artery disease and increasing age. Female gender, use of mycophenolate mofetil and tacrolimus seem to have protective effects.[1] The probability of dying from malignancy (other than lymphoma) after heart transplantation after 10 years is ≈18%.

The incidence of PTLD with ciclosporin-based immunosuppression is 2–4% and the incidence is highest ≈3–5 years after transplantation.[45] Epstein–Barr virus-negative PTLD has been described after heart transplantation but is considerably less common. There is a wide spectrum of presentation of PTLD from pulmonary, gastrointestinal, tonsillar to central nervous system involvement. Disseminated PTLD is associated with a poor prognosis. PTLD may respond to reduction of immunosuppression, and lymphomas showing CD20 expression can be successfully treated with CD20 antibody (rituximab)[46] (see Chapter 12).

Hypertension

Most patients will develop arterial hypertension after heart transplantation. It is of interest to note that in the pre-ciclosporin era the incidence of hypertension was 20%; now more than 60% will have elevated systemic blood pressure at 1 year and virtually all patients are expected to have it within 5 years following transplantation.[47] Hypertension needs to be aggressively treated in these patients to prevent CAV and renal dysfunction. The treatment consists of sodium restriction with addition of calcium channel blockers and angiotensin-converting enzyme (ACE) inhibitors.

Chronic renal dysfunction

Chronic renal failure is a well-recognised complication after heart transplantation. The incidence of this is 5–10% at 60 months after transplantation in most series.[48] Our incidence of dialysis is 10% at 10 years. The causation is multifactorial, with preoperative renal dysfunction, perioperative haemodynamic insult, hypertension and diabetes mellitus all contributing in some measure, but the principal cause is calcineurin inhibitor treatment. The major decline in renal function occurs during the first 12 months after transplantation; thereafter there is gradual working of dysfunction. Unless there is pre-existing intrinsic renal disease, early renal dysfunction following transplantation is not a predictor of chronic renal failure.

The management of chronic renal impairment consists of preventative measures with close monitoring of calcineurin inhibitors levels, treatment of hypertension and avoidance of other nephrotoxic agents. Calcium channel blockers and ACE inhibitors have been proposed as agents to reduce renal toxicity of ciclosporin, possibly by reduction of afferent arteriolar tone, but clinical trials have failed to substantiate this claim.[49] It is hoped that the use of newer immunosuppressive agents such as sirolimus may lead to less renal toxicity.

Heart and lung transplantation (HLT)

This form of pulmonary transplantation has had a transformation in its indications and the frequency with which it is done since it was first undertaken in 1982. At its inception it was the commonest form of pulmonary transplantation but with the success of isolated lung transplantation the indications are now mainly confined to pulmonary hypertension without congenital heart disease and pulmonary hypertension associated with Eisenmenger's syndrome and congenital heart disease. The annual activity in heart–lung transplantation has been more then halved since 1995.[50] A minority of centres have continued to use heart–lung transplantation in patients with cystic fibrosis (CF), utilising the healthy recipient heart as a domino procedure (healthy recipient heart is donated to a cardiac recipient).

Recipient selection criteria

The selection criteria for patients requiring HLT are similar to isolated lung transplantation. We have an upper limit of 50 years for acceptance to the transplant list. Specific guidelines for selection of patients in this group are as follows.

Pulmonary hypertension without congenital heart disease

These patients have either pulmonary hypertension as a result of thromboembolic disease, veno-occlusive disease or collagen vascular disease. Patients are considered for transplantation when they become symptomatic and in spite of medical or surgical treatment remain in NYHA grade III or IV. They should be resistant to vasodilator treatment with either prostacyclin or calcium channel blockers. Useful parameters for acceptance include cardiac index <2 L/min/m^2, right atrial pressure of >15 mmHg and a mean pulmonary artery pressure of >55 mmHg.[51]

Eisenmenger's syndrome

These patients behave differently from the above group in several ways. With a similar degree of pulmonary hypertension these patients have better cardiac function and better prognosis. The predictors of survival are also less reliable. The selection criteria are unclear but severe progressive symptoms with NYHA grade III or IV symptoms despite optimum medical treatment would constitute an indication for transplantation.

Heart–lung operation

The operation is undertaken via sternotomy. Cardiopulmonary bypass is similar to heart implantation. The heart is excised first as it improves visualisation for pneumonectomy. Both the phrenic and vagus nerves are carefully preserved while pneumonectomy is undertaken. The recurrent laryngeal nerve is particularly at risk while the left pulmonary artery is being divided as the pulmonary artery is markedly enlarged due to pulmonary hypertension. A cuff of pulmonary artery can be left around the ligamentum arteriosum region to add to protection. The trachea is divided two rings above the carina.

The implantation proceeds with tracheal anastomosis, which is similar to bronchial anastomosis as described with lung transplantation (see below). The left and right lungs are carefully placed in their respective cavities behind the phrenic nerves, taking care to avoid hilar torsion. The aortic and caval anastomoses are undertaken to complete the operation. Often the donor atrial appendage has been divided at the time of procurement of organs to decompress the heart and requires securing.

The postoperative care is similar to patients who have had heart transplantation.

Survival

The results of HLT are not dissimilar to pulmonary transplantation. Survival at 1 and 5 years is 61% and 40% respectively.[50] Recipients with Eisenmenger's syndrome have a better prognosis than patients with primary pulmonary hypertension.

Future direction in heart transplantation

It is gradually being recognised that due to donor limitations, the number of heart transplant procedures is unlikely to exceed the present numbers of <3000 in the USA and <300 in the UK. Unfortunately the demand for donor hearts continues to escalate.

Alternatives to heart transplantation are being explored worldwide, with research programmes exploring novel strategies including cell transplantation and regrowth of heart muscle, mechanical circulatory support and use of neurohumoral blockers to treat end-stage heart failure.

Xenotransplantation has continued to raise hopes for an unlimited supply of donor hearts. However, the feasibility of translating this technology into good long-term outcome in the foreseeable future looks rather remote. The median time of survival of transgenic pig hearts in baboon is only a few months.[52] According to a committee of the ISHLT, the current experimental results do not justify initiating a clinical trial.[52] The future of xenotransplantation for hearts is still indeterminate.[53]

Lung transplantation

The lung has historically been the most challenging of the human organs to be successfully transplanted in clinical practice. It took almost two decades since the first successful renal transplant before it was accepted that this procedure may bring significant improvements in quality of life to selected recipients. Since Hardy undertook the first single lung transplant in 1963,[54] the operation has been challenged by the frequent occurrence of bronchiolitis obliterans leading to progressive respiratory impairment in the longer term.

Essentially, an improvement of the quality of life is reported by most lung transplant receipients, but prolonged survival has been shown only for selected subgroups of patients with end-stage respiratory failure.

Demographically, the ISHLT registry indicates that 78% of recipients in Europe were between 35 and 65 years of age, with the majority receiving their transplant for chronic obstructive pulmonary disease, CF or pulmonary fibrotic disease (**Fig. 11.3**). Only 4.1% were re-transplant procedures.

It is possible to transplant lungs singly (SLT) or sequentially as a bilateral lung transplant (BSLT) depending on patient characteristics and the nature of the pathological lung condition present. In some situations combined transplantation of the heart and lungs en bloc is necessary, as previously described.

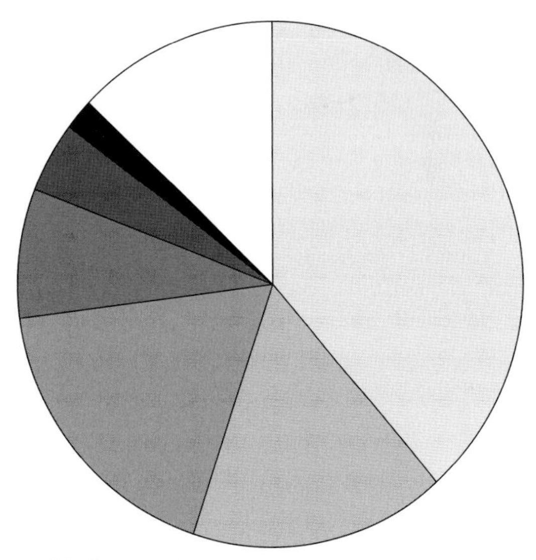

Figure 11.3 • Indication for lung transplantation.

Choice of lung transplant procedure

Bilateral sequential lung transplantation is performed where it is clinically necessary to remove all native lung tissue. In the context of chronic lung sepsis in CF or bronchiectasis, single lung transplantation would fail as infection may cross-contaminate from the native remaining lung into the graft. Similarly, extensive destruction of both lungs in emphysema may suggest the need for bilateral replacement to avoid air trapping in a remaining overly compliant native lung, resulting in mediastinal shift and compromise of the contralateral graft.

Single lung transplantation (SLT) is an attractive approach to the treatment of lung failure. The operation can often be performed without the potential for acute lung injury and other attendant risks associated with cardiopulmonary bypass. Most transplant centres experience an increasing disparity between the numbers of patients added onto the waiting list for lung transplantation and stagnating numbers of suitable donors providing allografts; therefore, there is an economy in the use of scarce donor organs, with two lung recipients benefiting from each donor and, in the event of acute or chronic injury to the graft, some viable native lung tissue will remain. Fibrotic lung conditions with normal pulmonary vasculature, a relatively immobile mediastinum and no significant native lung overinflation are most suited to this type of pulmonary transplant procedure. However, SLT is still used with varying enthusiasm between centres for selected patients with emphysema (with or without α_1-antitrypsin deficiency) and sarcoidosis.

A controversial area is lung transplantation for primary pulmonary hypertension. Here, in the absence of structural problems in the heart, SLT may suffice, although many centres still advocate bilateral lung transplantation or occasionally combined heart–lung transplantation with respect to potential cardiac involvement. Some centres are performing SLT alone in these circumstances.[55] Survival for all three modalities of treatment for primary pulmonary hypertension is similar.

Transplantation of both lungs en bloc with the heart for pulmonary pathology is now less frequent, although this was the early means of lung replacement. Few centres use this approach in circumstances where total lung replacement is required (the same indications as for bilateral lung transplantation). Although at first sight this may seem wasteful of scarce donor hearts where sequential lung transplantation will suffice, the structurally normal recipient heart is harvested and used for a heart transplant candidate (the 'domino' operation).[56]

Lung recipient assessment and selection

The general aspects of assessment for lung transplantation are identical to those for heart assessment. A detailed history of the patient's overall status and comorbidites as well as baseline evaluation of organ functions is essential. In view of the adverse effects of potent immunosuppressive regimens, a glomerular filtration rate (GFR) to determine renal function is performed routinely.

Again, after inpatient assessment, a recommendation is made to the patient by the surgeon after ensuring that all conventional treatments have been exhausted and, in terms of patient prognosis, that the timing for listing for transplantation is appropriate. Lung transplant assessment tests will typically also include specific sputum tests for otherwise difficult to eradicate pathogens including *Aspergillus* species and certain *Pseudomonas* species.

A 6-minute walk test is performed, which measures the distance a patient is able to walk in a given time and the degree of arterial desaturation that results during this exertion. Not only does this give a measure of symptomatic restriction, it also has prognostic value. Values of less than 300 m are seen in patients in end-stage pulmonary failure.[57]

Computed tomography (CT) is used to study the texture of lung parenchyma and identify areas of maximal lung destruction or bullous disease accurately. This, along with the results of ventilation-perfusion scanning, assists with the decision of whether to perform BSLT or SLT in those conditions where either might be considered and, if SLT is selected, which side should be transplanted. For SLT it is usual to replace the lung that has the poorer perfusion. It is always preferable to explant a lung if there is evidence of chronic sepsis within it or if there is a bulla that is likely to rupture. These investigations also help to identify those emphysematous patients who might be suitable for lung volume reduction surgery as an alternative to transplantation with the aim of improving their ventilatory mechanics with symptomatic relief.

Detailed microbiological screening is an essential part of the assessment, with attention also paid to cultures performed over the preceding months and years at the referring centre. This information identifies patients likely to be colonised with multiply resistant organisms (multi- or pan-resistant pseudomonads in the cystic population) and helps direct antibiotic prophylaxis in the perioperative period.

The particular importance of colonisation with *Burkholderia cepacia* has been recognised as an important predictor of post-transplant mortality,[58] which may influence acceptance or postoperative management.

In CF patients with chronic sepsis and malabsorption, a nutritional assessment is vital since wound healing is impaired and the loss of muscle bulk may result in insufficient respiratory effort to permit weaning from the ventilator in the postoperative period.

Cardiopulmonary bypass is used routinely for BSLT and in less then half of cases undergoing SLT. This depends mostly on patient oxygenation during single lung ventilation and haemodynamic stability after pulmonary artery clamping during the SLT procedure. Therefore, identifying coincidental cardiac disease is important. Older patients and those with relevant risk profiles should undergo cardiac catheterisation with coronary angiography. Right heart catheterisation is undertaken in patients considered for lung transplantation for pulmonary hypertension. This, occasionally can be supplemented by pulmonary angiography if pulmonary thromboendarterectomy might be considered as an alternative to transplantation.

The remainder of the assessment is designed primarily to identify contraindications to organ replacement. Unlike patients with cardiac failure most candidates for lung replacement have preserved renal function. The incidents of post-lung transplantation renal failure requiring dialysis is a major source of mobility and has been reported in up to 15% at 3 years post-transplantation. The pretransplant creatinine clearance (<15 ml/min per 1.73 m^2), as well as receipient height and the calcineurin inhibitor load after transplantation, correlate with higher risk of dialysis.[59]

Absolute contraindications to pulmonary transplantation include multiorgan failure and ongoing sepsis. Current malignancy, active peptic ulceration and inadequate conventional therapy are also important. Relative contraindications include peripheral and cerebral vascular disease, advanced diabetes, morbid obesity, osteoporosis and ischaemic heart disease.

On occasion, patients who have acute pulmonary failure – e.g. acute respiratory distress syndrome – and are ventilator dependent are referred to be considered for transplantation.[60] Pulmonary transplantation is seldom successful under these circumstances as sepsis and multiorgan failure are common and not usually reversible. However, if attempted, it is worth considering single lung replacement since the potential for recovery in the native lung is present in many cases of acute respiratory disease.

No universally accepted assist device technology has been established for patients with potential for 'bridge to lung transplantation'. Extracorporeal membrane oxygenation (ECMO) has been advocated by some centres but has not met expectations and utilises considerable resources in this setting. NOVA lung®, a pumpless arterio-venous CO_2 elimination device, has been used successfully by some centres for this purpose and may find more frequent application in the future despite the time limitation it imposes.[61]

Lung donor criteria and selection

The vast majority of lung allografts are from declared brain-dead donors according to straight criteria (deceased heart-beating donors, DBDs). Specific considerations for lung donation must include a demonstration of good gas exchange with no evidence of aspiration, embolism or pneumonia. Smokers and patients with a history of mild asthma may still be considered as potential lung donors. In practice, only a minority of multiorgan donors are suitable for lung donation as potential lung injury may arise in a number of ways. After chest trauma, a haemothorax and fractured ribs may indicate parenchymal damage, but this is not always so and the contralateral lung may be uninjured and available for use. Aspiration of gastric contents at the time of injury or cardiac arrest is not uncommon.

A recent consensus report issued by the ISHLT provides the acceptability criteria in the ideal scenario.[23]

Allografts from extended criteria donors (ECDs) may increase the primary graft dysfunction incidents. Bilateral transplantation from such donors may also be associated with higher early mortality but otherwise reasonable midterm outcomes.[62]

In exceptional circumstances donors that do not meet the strict brain death criteria may be declared non-heart-beating donors (donation after cardiac death, DCD). Such DCD lung allografts have recently been shown to provide very acceptable early and midterm lung function after transplantation.[63] The transfusion of blood and blood products in large volumes predisposes to lung injury (sometimes becoming apparent only after implantation). Fat embolus from long-bone fractures with catastrophic results has also been reported. In all these circumstances, infection in the donor is more likely and will compound the injury to the lung.

Examination of the chest radiograph is essential. Aspirates taken from the endotracheal tube should be examined microscopically and Gram stained at the donor hospital. Mixed Gram-negative and Gram-positive organisms and numerous polymorphs in the aspirate may indicate potentially unacceptable infection. Previous culture results should be requested and considered. Indiscriminate use of broad-spectrum antibiotics in the absence of specific organism sensitivities in the donor should also be treated with caution. Flexible bronchoscopy can be useful to facilitate full expansion of the lungs and obtain good specimens of pulmonary secretions.

Lung function is assessed by gas exchange. A useful standardised measure is the P_aO_2 with the donor ventilated on 100% oxygen and with 5 mmHg of positive end-expiratory pressure to optimise ventilation. A value of <35 kPa is an indicator of significant lung injury, and many centres will not use lungs where a value of <50 kPa is recorded. The aspiration of blood from individual upper and lower lobe pulmonary veins is often useful when evaluating single lungs, where 45 kPa is a reasonable level of acceptability.[64]

Final assessment of the lungs is performed by the donor surgeon, who can see bullae and traumatised lung and feel areas of consolidation. Oedematous lungs feel heavy and spongy and may lead the donor team to reject the organs for use.

Lung recipient–donor matching

As with heart donors, matching of donor and recipient for lung transplantation is a relatively crude process, focusing on blood group and dimensions with no prospective match for tissue typing due to time constraints.

Size can be assessed in a number of ways, including comparison of measurements taken from donor and recipient chest radiographs. Donor and recipient heights are a useful guide to matching, with a 10–15% mismatch permissible. However, it is now generally recognised that donors should be matched to the predicted lung size of the recipient rather than the pathological size, since thoracic capacity and chest wall mechanics will often normalise after transplantation.

CMV status of donor and recipient is an important consideration. CMV mismatch here has greater implications for a lung recipient in the event of seroconversion or reactivation in the grafted tissue.

Lung retrieval and preservation

As with heart retrieval, the median sternotomy is completed after initial mobilisation of the liver. Both pleurae are now opened widely and the lungs inspected. Any adhesions between visceral and parietal pleura are divided with electrocautery. The inferior pulmonary ligament on each side is divided up to the inferior pulmonary vein. The innominate vein is now ligated between ligatures and the pericardium is opened with the incision being continued up along the innominate artery, which is similarly divided. For this reason central venous access must be via the right internal jugular vein and arterial monitoring from the left radial artery. The pericardium is now opened and the aorta, superior vena cava (SVC) and inferior vena cava (IVC) are mobilised as before. Once the SVC has been mobilised the azygos vein can be identified, ligated and divided behind the SVC to facilitate the future removal of the bloc. It is now possible to mobilise the trachea above the aortic arch. It is important not to denude the trachea of its blood supply and a tape is simply passed around it. Perfusion cannulas are now inserted into the ascending aorta and into the main pulmonary artery.

When systemic heparinisation is administered and the perfusion apparatus has been set up for perfusion of the abdominal organs, removal of the heart and lung bloc can proceed. The SVC is divided between ligatures and the IVC is clamped above the diaphragm. The aorta is now cross-clamped and the heart is cardiopleged as described above for solitary heart transplant organ retrieval. However, under these circumstances the cardioplegia is vented from the heart by incision of the tip of the left atrial appendage, leaving the pulmonary veins intact. Once electromechanical arrest has been achieved, infusion of preservative into the lungs can proceed through the pulmonary artery catheter. Simultaneous topical cooling of heart and lungs with cold saline solution proceeds throughout this time.

Pulmonary preservative solutions exist in many forms. Traditionally Euro-Collins solution has been used, which essentially has an intracellular fluid electrolyte composition. Recently, other preservatives based on extracellular fluids have been used, such as low-potassium dextran, with more encouraging results and the potential to extend organ ischaemic times.[65]

Prostaglandins may help prevent leucocyte sequestration and also optimise perfusion of the pulmonary capillary bed, and are given before infusion of the pulmoplegic solution. Preservation is achieved by cooling with extracorporeal circulation in some centres but most units use a hypothermic cold flush perfusion technique. Ischaemic times of 6–8 hours can be safely achieved.

The anaesthetist is now asked to ventilate the lungs by hand with air to prevent alveolar collapse. Occasional cessation of ventilation will facilitate excision of the bloc, which is undertaken when cardioplegia and pulmonary preservative solutions have both been given. The heart is elevated and the pericardium incised posteriorly below the inferior pulmonary veins joining right and left pleural spaces. It is now possible to elevate the heart, the back of the left atrium and both hila, dividing the connective tissue between these structures and the descending aorta and oesophagus and vertebral column posteriorly. As the surgeon proceeds up the descending aorta the ligamentum arteriosum is encountered and divided. On the right-hand side, the divided azygos vein is seen as dissection proceeds in a cephalad direction. At this point the heart–lung bloc is placed back in the chest and attention turned to the aorta, which is divided below the cross-clamp. The anaesthetist is now asked to withdraw the endotracheal tube into the upper trachea whilst still ventilating. A clamp can now be placed across the trachea below the endotracheal tube and the trachea divided above. All that remains is to divide the connective tissue behind the ascending aorta and trachea and remove the entire heart–lung bloc. The trachea is stapled to allow removal of the clamp whilst leaving the lungs inflated for transfer.

If the lungs are to be sent to a different destination from the heart, it is now necessary to split the bloc. This is performed by incising the left atrium anterior to the hilum on each side to separate pulmonary veins from the left atrium. It is important to leave a small cuff of left atrium on the pulmonary veins to facilitate implantation in the lung recipient. The pulmonary artery is divided at its bifurcation leaving a good length of pulmonary artery attached to each lung. All that remains now is to separate the ascending aorta and heart from the pulmonary arteries on each side and from loose connective tissue connecting it to the carina posteriorly. Lungs and heart can now be transported separately to different destinations as required.

Single lung implantation

Anaesthesia is established with a double-lumen endotracheal tube to permit ventilation of the native lung whilst implantation proceeds. A pulmonary artery flotation catheter is often used to monitor pulmonary artery pressure during implantation and full arterial and venous monitoring is established. Facilities for cardiopulmonary bypass are made available but are used only if unacceptable desaturation during implantation occurs or if systemic hypotension or pulmonary hypertension develop.

A lateral thoracotomy is performed and the native lung is excised with ligation of inferior and superior pulmonary veins and pulmonary artery. The bronchus is divided and the native organ removed. Care is taken not to contaminate the pleural space with endobronchial secretions. Meticulous haemostasis at the hilum is established. The pericardium is now incised adjacent to the pulmonary veins and these are mobilised to develop a cuff of left atrium. The pulmonary artery is mobilised in a similar fashion. The donor lung is now prepared by trimming the left atrial cuff, cutting the pulmonary artery to length and excising the stapled end of the bronchus to deflate the lung.

Implantation commences with the bronchial anastomosis. The membranous part of the bronchus is anastomosed with a continuous 4/0 Prolene suture. A series of figure-of-eight interrupted sutures is now placed on the anterior cartilaginous part of the bronchus to complete the anastomosis. Loose connective tissue at the hilum of donor lung can now be used to cover the anastomosis. A long side-biting clamp is next placed across the native left atrial

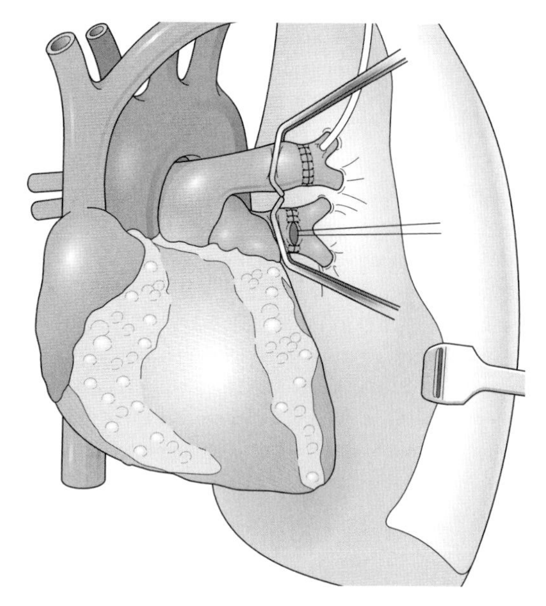

Figure 11.4 • The technique of vascular anastomosis in lung transplantation.

cuff, which encompasses the pulmonary veins; the cuff is then opened longitudinally. The left atrial anastomosis is now performed with a continuous 4/0 Prolene suture and the clamp is left applied. The pulmonary artery anastomosis is performed in the same fashion and the donor organ is now de-aired by partially releasing the clamp from the pulmonary artery and de-airing through the left atrial anastomosis. The left atrial clamp can now be removed. Ventilation of the new lung commences (**Fig. 11.4**).

Apical and basal chest drains are now inserted. The thoracotomy is closed and the patient is returned to the intensive care unit for further monitoring and care. It is usually possible to extubate the stable patient after the insertion of an epidural analgesic catheter. Typically the patient will return to the ward after approximately 24–48 hours.

Bilateral sequential lung implantation

Single-lumen intubation for anaesthesia is performed. At many centres, this procedure is undertaken as sequential single lung transplants to avoid the perceived increase in acute lung injury postoperatively that is said to accompany extracorporeal perfusion. At our institution and others, cardiopulmonary bypass is routinely used and seems to have little impact on lung function following surgery.[66]

The operation is approached through a submammary 'clam shell' incision that divides the sternum transversely. A median sternotomy is a less painful incision that can be used in some cases, although access can sometimes be more of a challenge.

The patient is fully heparinised, the pericardium is opened and the patient is placed on cardiopulmonary bypass with an ascending aortic inflow cannula and venous drainage from individual cannulation of the caval veins. Excision of each lung now proceeds with electrocautery division of adhesions. Care is taken to preserve the phrenic nerve while mobilising structures at each hilum, especially in patients with septic lung disease where large lymph nodes and dense hilar adhesions make excision of the lung difficult. Excision of each lung proceeds in turn, and in cases of pulmonary sepsis the pleural cavities are irrigated thoroughly with the antiseptic Taurolin. Implantation of the donor lungs is performed in exactly the fashion described for single lung transplantation, the right side being anastomosed first. The patient is usually cooled to 32 °C during implantation, the heart being allowed to continue to beat and eject in sinus rhythm. After implantation, de-airing and reperfusion are performed and ventilation is recommenced. At normothermia cardiopulmonary bypass can be weaned.

It is important that the pulmonary artery (PA) pressure at reperfusion is controlled. A number of experimental studies have shown reduced lung reperfusion injury when this is the case, and even a short period of controlled pressure reperfusion is beneficial.

We keep the mean PA pressure at less than 20 mmHg for at least 10 minutes while reperfusing a lung transplant.

Each thoracic cavity is drained with basal and apical drains and the wound is closed. The patient is then returned to the intensive care unit for further monitoring and can usually be extubated at 8–12 hours postoperatively. Epidural analgesia is essential following the 'clam shell' incision. Return to the ward is usually at about 24–48 hours.

Peri- and postoperative care for lung transplants

On notification of a possible donor the selected recipient is admitted and reassessed for deterioration or unexpected infection.

Infection and colonisation of the airway and lung is a major feature of lung transplant and a leading cause of morbidity and mortality in the post-operative period. Antibiotic prophylaxis is largely directed by the known flora of the recipient but flucloxacillin is used for Gram-positive cover and metronidazole for Gram-negative cover in the absence of other positive cultures. Colomycin is administered by nebuliser in the immediate postoperative period. Antibiotic therapy is modified in the first few days after transplant as the results of perioperative donor and recipient bronchoalveolar lavages become available. In the absence of infection, antibacterial agents are stopped after the first routine bronchoscopy and biopsy at 1 week provided airway anastomoses appear healthy. Aciclovir, antifungal agents and pneumocystis prophylaxis are used routinely as in cardiac transplantation.[67]

Immunosuppression, as in heart transplantation, commences preoperatively with the administration of azathioprine and ciclosporin A. In patients with normal GFR calcineruin inhibitor doses require adjustment to avoid postoperative renal impairment. Methylprednisolone is administered at reperfusion and continued intravenously for 24 hours in three divided doses, after which oral steroids may be commenced. However, since many pulmonary recipients have malabsorption and early ciclosporin levels may be erratic, antithymocyte globulin (ATG) remains an important modality to reduce acute rejection rates after lung transplantation and its use is reflected in the incidence of bronchiolitis obliterans syndrome.[68] ATG is administered routinely for the first 3 days as induction therapy, with dosage and timing being regulated by daily flow cytometry lymphocyte counts. With this exception, immunosuppression is managed in an identical fashion with the same dosage regimens as in cardiac transplantation.

In recent years, other immunosuppressants have been put forward for use in postoperative immunosuppressive regimens. In particular the use of induction therapy with ATG has been questioned due to concerns over increased rates of infection and of post-transplant lymphoma, though this remains controversial. Tacrolimus, mycophenolate mofetil and rapamycin have been investigated but no conclusive advantages have been demonstrated, although side-effect profiles may be subject to some improvements.[69]

If acute lung injury is present in the immediate postoperative period, ventilation can present great difficulties. Such reperfusion injury results from the sequestration of neutrophils in the lung parenchyma with release of injurious enzymes and oxygen free radicals. Lungs may be oedematous or infected, with poor gas exchange. Meticulous control of fluid balance, optimisation of ventilation and microbiological control are needed in this situation. ECMO has been shown to improve survival in patients severely affected by primary grafts dysfunction, but this requires early institution and has significant resource implications.[70]

Nitric oxide administration has many benefits in reperfusion injury and is distributed preferentially to ventilated areas of the lung. It improves ventilation–perfusion matching and lowers pulmonary artery pressures. It reduces the adhesion of neutrophils to the endothelium and so alleviates reperfusion injury.[71,72]

A number of other interventions (controlled pressure reperfusion, pentoxifylline, extracorporeal filters, adhesion molecule modulators) affecting neutrophil sequestration in the lung have been put forward to try to combat this problem postoperatively but have not been widely evaluated in clinical practice.[73]

In the case of the single lung transplant for emphysema, the residual overcompliant lung can inflate excessively with air-trapping and resultant mediastinal shift if the expiratory period of ventilation is insufficient. Modification of the ventilatory cycle can help but sometimes independent ventilation of each lung through a double-lumen endotracheal tube is needed. When the time comes to wean the recipient from the ventilator an epidural catheter to administer analgesics is essential. Epidural infusions can be continued for some days after extubation to assist expectoration of secretions.

Transbronchial biopsy and bronchoalveolar lavage with a flexible bronchoscope under sedation are performed at 1 week, 1 month and then every 3 months before reverting to annual biopsies to detect rejection and direct antimicrobial intervention. Additional biopsies are taken if rejection is suspected on the grounds of unexplained fever, symptomatic deterioration with arterial desaturation or a fall in pulmonary function tests including spirometry and transfer factor.

Rejection is graded according to a standard system adopted by the ISHLT. Treatment is by augmentation of steroid therapy – 3 days of intravenous methylprednisolone (500 mg) and subsequent augmentation of oral steroids. Treatment is required for grades 3 and 4 and for grade 2 if there is clinical concern.

Outcomes and complications of lung transplantation

Generally, the 30-day survival is approximately 85%, with 75% surviving to 1 year. At 5 years, 45% remain alive, and 25% after a decade. Survival curves for bilateral lung transplantation are a little better than for unilateral procedures. The early decline in survival mirrors that seen in heart transplantation and reflects operative mortality and donor organ dysfunction. The causes of perioperative mortality include unsuspected donor allograft injury (infection, oedema, embolic disease or poor preservation) and reperfusion injury with subsequent multiorgan failure complications.

Specific technical difficulties include anastomotic stenoses with pulmonary oligaemia (pulmonary arterial obstruction) or pulmonary oedema (venous stenosis)[74] and airway ischaemia and dehiscence with resultant mediastinitis and pleural sepsis.

The vascular supply of bronchial and tracheal anastomoses is compromised and early dehiscence with ischaemia is a life-threatening complication with prolonged air leak and mediastinitis. Attention to detail when the anastomosis is performed, with care not to denude the airway, minimises this risk. It is no longer thought necessary to wrap the anastomosis in a vascularised pedicle or omentum. Concurrent steroid therapy (once considered a contraindication to lung transplantation) may even reduce dehiscence as development of capillaries at the anastomosis is enhanced. Some early in-hospital deaths result from infection and acute rejection episodes.

Quality of life is significantly improved by transplantation for pulmonary failure. Studies in lung transplant patient groups consistently show improvements in functional status and the perception of symptoms, irrespective of the type of transplant performed or the primary pathology.[75]

Diagnosis of acute rejection is often made more difficult by concurrent infection, and the decision to treat can also be problematic because of the fear of resulting uncontrollable sepsis. Persistent or repeated episodes of rejection are managed with cytolytic therapy (ATG), monoclonal therapy or a change in immunosuppressive agent, perhaps to one of the newer agents (Box 11.4).

Box 11.4 • Grading of pulmonary allograft rejection

A	**Acute rejection**
0.	None
1.	Minimal – scattered mononuclear infiltrates
2.	Mild – frequent infiltrates of activated lymphocytes: 'endotheliatis'
3.	Moderate – vascular cuffing, alveolar macrophages, extension of infiltrate into perivascular and air spaces
4.	Severe – intra-alveolar necrosis, hyaline membrane, haemorrhage
B	**Airway inflammation**
0–4	According to severity
C	**Chronic airway rejection**
A.	Active
B.	Inactive
D	**Chronic vascular rejection**

Fungal and viral infections are seen commonly in the early postoperative period (*Aspergillus* and CMV) and carry a significant morbidity. *Pseudomonas* colonisation is common in the cystic population. If preoperative data suggest that multiresistant pseudomonads are present antibacterial therapy is kept to a minimum to allow growth of sensitive organisms.

CMV infection can be a major clinical problem. Diagnosis is by immunofluorescence at lavage, by transbronchial biopsy, estimation of antigenaemia and culture on bronchoalveolar lavage samples and quantitative CMV-PCR (polymerase chain reaction). Performed on a weekly basis to estimate viral load and identify those requiring treatment, this has provided improved treatment strategies with novel antiviral agents such as valgangcyclivir. Infection is common in all except the donor-negative/recipient-negative transplants (see Chapter 12).

Lymphoproliferative disease and other malignancies are seen as in cardiac transplantation, although lymphomas are more common (one series reports an incidence of 3.4% in heart recipients and 7.9% in lung recipients). Mortality is significantly higher in lymphomas appearing after the first year after transplantation. Reduction in immunosuppression can be highly effective in early disease but later conventional chemotherapy may be required.

Longer-term airway complications can arise with overgrowth of granulation tissue at the anastomosis, especially in the presence of chronic infection, or fibrotic stricture formation, which is usually ischaemic in origin. Treatment is local with dilatation or laser therapy, sometimes augmented with expandable stenting devices.

Chronic rejection in lung transplantation manifests itself as obliterative bronchiolitis or vascular atherosclerosis, the former being the greatest cause of long-term morbidity and mortality in recipients. Gradual deterioration in exercise tolerance and lung function arouse suspicion. Characteristic appearances are seen on the chest radiograph and CT scan, and the diagnosis is confirmed by transbronchial biopsy. Predisposing factors may include the occurrence of reperfusion injury and multiple episodes of acute rejection in the postoperative period, and CMV infection may be an aetiological factor. Treatment is directed towards decreased intensity of immunosuppression and total lymphoid irradiation may also help in some cases.

Recent advances and controversies

Over the last decade the outcomes in lung transplantation have significantly improved (**Fig. 11.5**).[1]

Recent changes in practice in lung transplantation are driven by a need to optimise the donor pool and reduce the impact of primary graft dysfunction in recipients. To this end, donation has been considered from both older and more marginal donors and in attempting to prolong the permissible ischaemic time. Encouraging results have been obtained and expansion of the donor pool through these methods has been achieved.[76] A further increase of donor organs available is now possible with the use of non-heart-beating donor lungs. In practice realistic only with controlled donors (category 3 of the Maastricht criteria for non-heart-beating donors), this has achieved very acceptable early outcomes. Both considerable logistic and ethical questions require local protocols.[63]

Living-related donation of lungs by blood relatives of patients needing pulmonary transplantation has been performed with reasonable results. This technique is most applicable to paediatric transplantation for CF or small recipients, where lobar donation by relatives alone may provide sufficient lung tissue to fill the recipient chest. Donor morbidity, reported even in experienced centres (up to 20% of lobectomies), can also be significant and this raises difficult ethical dilemmas.[77] Lung volume reduction surgery for symptomatic, non-prognostic emphysematous disease has diverted away from transplant lists some patients who might otherwise have been transplanted on the grounds of symptomatic restriction.

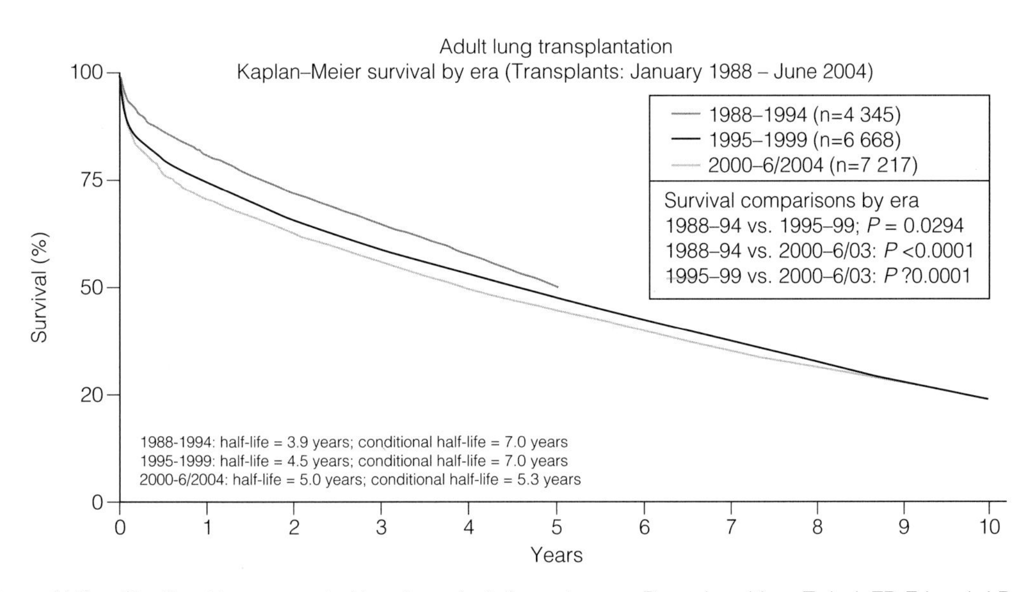

Figure 11.5 • Significant improvement of lung transplantation outcomes. Reproduced from Trulock EP, Edwards LB, Taylor DO et al. Registry of the International Society for Heart and Lung Transplantation: twenty-third official adult lung and heart–lung transplantation report – 2006. J Heart Lung Transplant 2006; 25:880–92. With permission from Elsevier.

Lung volume reduction surgery selectively excises the overexpanded, underventilated and underperfused parenchyma. The reduction in lung volume permits better ventilatory performance and ventilation–perfusion matching.

The specific subgroups of emphysema patients who are most likely to benefit from such lung volume reduction surgery have been established with the help of the NETT trial.[78,79]

There has been a recent focus both in the laboratory and in limited clinical practice to utilise lungs from ventilated non-heart-beating donors. Clearly the practice raises difficult ethical and logistic problems with regard to organ retrieval, but good lung function has been reported in large animal experimental models. The optimal criteria for retrieval and the preservation technique used are still under investigation but the procedure has some potential for further expanding the donor pool.[80]

Key points

- Heart transplantation improves survival and quality of life for selected patients with end-stage heart failure.
- Heart transplantation is most beneficial in patients with greatest risk of dying.
- More specific immunosuppression agents continue to decrease the impact of acute and chronic rejection and immunosuppression-related side-effects.
- Lung transplantation provides good survival and functional improvement in many forms of end-stage lung disease.
- Single or bilateral lung transplantation is possible depending on the underlying pathology.
- Colonisation with *Burkholderia cepacia* in cystic fibrosis patients is an important adverse prognostic marker.
- Reperfusion injury is a major challenge postoperatively but nitric oxide administration has the potential to limit this in clinical practice.

References

1. Taylor DO, Edwards LB, Boucek MM et al. Registry of the International Society for Heart and Lung Transplantation: twenty-fourth official adult heart transplant report – 2007. J Heart Lung Transplant 2007; 26(8):769–81.
2. McMurray JJ, Stewart S. Epidemiology, aetiology, and prognosis of heart failure. Heart 2000; 83(5):596–602.
3. Jessup M, Banner N, Brozena S et al. Optimal pharmacologic and non-pharmacologic management of cardiac transplant candidates: approaches to be considered prior to transplant evaluation: International Society for Heart and Lung Transplantation guidelines for the care of cardiac transplant candidates. J Heart Lung Transplant 2006; 25(9): 1003–23.
4. Redfield MM. Heart failure – an epidemic of uncertain proportions. N Engl J Med 2002; 347:1442–4.
5. McCullough PA, Philbin EF, Spertus JA et al. Confirmation of a heart failure epidemic: findings from the Resource Utilization Among Congestive Heart Failure. J Am Coll Cardiol 2002; 39(1):60–9.
6. Figulla HR, Rahlf G, Nieger M et al. Spontaneous hemodynamic improvement or stabilization and associated biopsy findings in patients with congestive cardiomyopathy. Circulation 1985; 71(6):1095–104.
7. Matitiau A, Perez-Atayde A, Sanders SP et al. Infantile dilated cardiomyopathy. Relation of outcome to left ventricular mechanics, hemodynamics, and histology at the time of presentation. Circulation 1994; 90(3):1310–18.
8. Mancini DM, Eisen H, Kussmaul W et al. Value of peak exercise oxygen consumption for optimal timing of cardiac transplantation in ambulatory patients with heart failure. Circulation 1991; 83(3):778–86.

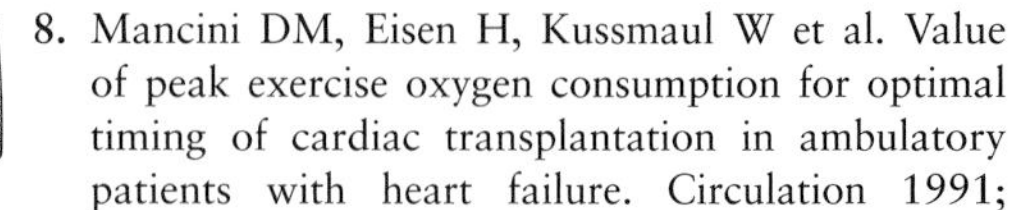

This clinical study nicely demonstrated the value of measurement of peak oxygen consumption during maximal exercise testing and has brought some objectivity in assessment of patients for heart transplantation.

9. Aaronson KD, Schwartz JS, Chery TM et al. Development and prospective validation of a clinical index to predict survival in ambulatory

patients referred for cardiac transplant evaluation. Circulation 1997; 95(12):2660–7.

10. Deng MC, De Meeste JM, Smits JM et al. Effect of receiving a heart transplant: analysis of a national cohort entering a waiting list, stratified by heart failure severity. Comparative Outcome Clinical Profiles in Transplantation (COCPIT) Study Group. BMJ 2000; 321(7260):540–5.
11. Khaghani A, Santini F, Dyke CM. Heterotopic cardiac transplantation in infants and children. J Thorac Cardiovasc Surg 1997; 113(6):1042–1048; discussion 1048–9.
12. Hermsen JL, Nath DS, del Rio AM et al. Combined heart–kidney transplantation: the University of Wisconsin experience. J Heart Lung Transplant 2007; 26(11):1119–26.
13. Olbrisch ME, Levenson JL. Psychosocial evaluation of heart transplant candidates: an international survey of process, criteria, and outcomes. J Heart Lung Transplant 1991; 10(6):948–55.
14. Mehra MR, Kobashigawa J, Starling R et al. Listing criteria for heart transplantation: International Society for Heart and Lung Transplantation guidelines for the care of cardiac transplant candidates. J Heart Lung Transplant 2006; 25(9):1024–42.
15. Alonso C, Leclercq C, Victory F et al. Electrocardiographic predictive factors of long-term clinical improvement with multisite biventricular pacing in advanced heart failure. Am J Cardiol 1999; 84(12):1417–21.
16. Higgins SL, Hummel JD, Niazi IK et al. Cardiac resynchronization therapy for the treatment of heart failure. J Am Coll Cardiol 2003; 42(8):1454–9.
17. Gras D, LeClercq C, Tang AS et al. Cardiac resynchronization therapy in advanced heart failure: the multicenter InSync clinical study. Eur J Heart Fail 2002; 4(3):311–20.
18. Olivari MT, Windle JR. Cardiac transplantation in patients with refractory ventricular arrhythmias. J Heart Lung Transplant 2000; 19(8 suppl): 538–42.
19. Lietz K, Miller LW. Improved survival of patients with end-stage heart failure listed for heart transplantation analysis of organ procurement and transplantation network/U.S. United Network of Organ Sharing Data, 1990–2005. J Am Coll Cardiol 2007; 50:1282–90.
20. Hunt SA, Baker DW, Chin MH et al. ACC/AHA guidelines for the evaluation and management of chronic heart failure in the adult. J Am Coll Cardiol 2001; 38(7):2101–13.
21. Rose EA, Moskowitz AJ, Packer M et al. The REMATCH trial: rationale, design, and end points. Ann Thorac Surg 1999; 67(3):723–30.
22. Rose EA, Gelijns AC, Moskowitz AJ, for the REMATCH Study Group. Long-term mechanical left ventricular assistance for end-stage heart failure. N Engl J Med 2001; 345:1435–43.
23. Rosengard BR, Feng S, Alfrey EJ et al. Report of the Crystal City meeting to maximize the use of organs recovered from the cadaver donor. Am J Transplant 2002; 2(8):701–11.
24. Wood KE, Becker BN, McCartney JG et al. Care of the potential organ donor. N Engl J Med 2004; 351:2730–9.
25. Marelli D, Laks H, Bresson S et al. Results after transplantation using donor hearts with preexisting coronary artery disease. J Thorac Cardiovasc Surg 2003; 126(3):821–5.
26. www.optn.org/AR2006/.
27. Potapov EV, Ivanitskaia EA, Loebe M et al. Value of cardiac troponin I and T for selection of heart donors and as predictors of early graft failure. Transplantation 2001; 71(10):1394–400.
28. Sethi GK, Lanause P, Rosado LJ et al. Clinical significance of weight difference between donor and recipient in heart transplantation. J Thorac Cardiovasc Surg 1993; 106(3):444–8.
29. West LJ, Pollock-Barziv SM, Dipchard AI et al. ABO-incompatible heart transplantation in infants. N Engl J Med 2001; 344(11):793–800.
30. Roche SL, Burch M, O'Sullivan J et al. Multicenter experience of ABO-incompatible pediatric cardiac transplantation. Am J Transplant 2008; 8(1):208–15.
31. Loh E, Bergin JD, Couper GS et al. Role of panel-reactive antibody cross-reactivity in predicting survival after orthotopic heart transplantation. J Heart Lung Transplant 1994; 13(2):194–201.
32. Wheeldon DR, Potter CD, Odano A et al. Transforming the 'unacceptable' donor: outcomes from the adoption of a standardized donor management technique. J Heart Lung Transplant 1995; 14(4):734–42.
33. Lower RR, Stofer RC, Shumway NE. Homovital transplantation of the heart. J Thorac Cardiovasc Surg 1961; 41:196–204.
34. Aziz TM, Burgess AI, Rahman A et al. Risk factors for tricuspid valve regurgitation after orthotopic heart transplantation. Ann Thorac Surg 1999; 68(4):1247–51.
35. Hasan A, Au J, Hamilton JR et al. Orthotopic heart transplantation for congenital heart disease. Technical considerations. Eur J Cardiothorac Surg 1993; 7(2):65–70.
36. Bleasdale RA, Bannen NR, Anyanwu AC et al. Determinants of outcome after heterotopic heart transplantation. J Heart Lung Transplant 2002; 21(8):867–73.
37. Miller LW, Naftal DC, Bourge RC et al. Infection after heart transplantation: a multi-institutional study. Cardiac Transplant Research Database Group. J Heart Lung Transplant 1994; 13(3):381–92; discussion 393.
38. Johnson DE, Gao SZ, Schroeden JS et al. The spectrum of coronary artery pathologic findings in

human cardiac allografts. J Heart Transplant 1989; 8(5):349–59.

39. Deng MC, Plenz G, Erren M et al. Transplant vasculopathy: a model for coronary artery disease? Herz 2000; 25(2):95–9.

40. St Goar FG, Pinto FJ, Alderman EL et al. Detection of coronary atherosclerosis in young adult hearts using intravascular ultrasound. Circulation 1992; 86(3):756–63.

41. Spes CH, Klauss V, Mudra H et al. Diagnostic and prognostic value of serial dobutamine stress echocardiography for noninvasive assessment of cardiac allograft vasculopathy: a comparison with coronary angiography and intravascular ultrasound. Circulation 1999; 100(5):509–15.

42. Wenke K, Meiser B, Thiery J et al. Simvastatin reduces graft vessel disease and mortality after heart transplantation: a four-year randomized trial. Circulation 1997; 96(5):1398–402.

43. Schroeder JS, Gao SZ, Alderman EL et al. A preliminary study of diltiazem in the prevention of coronary artery disease in heart-transplant recipients. N Engl J Med 1993; 328(3):164–70.

44. Eisen HJ, Tuzai EM, Dorent R et al. Everolimus for the prevention of allograft rejection and vasculopathy in cardiac-transplant recipients. N Engl J Med 2003; 349(9):847–58.

45. Penn I. Tumors after renal and cardiac transplantation. Hematol Oncol Clin North Am 1993; 7(2):431–45.

46. Zilz ND, Olson LJ, McGregor CG. Treatment of post-transplant lymphoproliferative disorder with monoclonal CD20 antibody (rituximab) after heart transplantation. J Heart Lung Transplant 2001; 20(7):770–2.

47. Corcos T, Tamburino C, Leger P et al. Early and late hemodynamic evaluation after cardiac transplantation: a study of 28 cases. J Am Coll Cardiol 1988; 11(2):264–9.

48. Ojo AO, Held PJ, Port FK et al. Chronic renal failure after transplantation of a nonrenal organ. N Engl J Med 2003; 349(10):931–40.

49. Brozena SC, Johnson MR, Ventura H . Effectiveness and safety of diltiazem or lisinopril in treatment of hypertension after heart transplantation. Results of a prospective, randomized multicenter trail. J Am Coll Cardiol 1996; 27(7):1707–12.

50. Trulock EP, Edwards LB, Taylor DO et al. The registry of the International Society for Heart and Lung Transplantation. Twentieth official adult lung and heart–lung transplant report – 2003. J Heart Lung Transplant 2003; 22(6):625–35.

51. D'Alonzo GE, Barst RJ, Ayres SM. Survival in patients with primary pulmonary hypertension. Results from a national prospective registry. Ann Intern Med 1991; 115(5):343–9.

52. Cooper DK. Clinical xenotransplantation – how close are we? Lancet 2003; 362(9383):557–9.

53. Cooper DK, Keogh AM, Brink J et al. Report of the Xenotransplantation Advisory Committee of the International Society for Heart and Lung Transplantation. The present status of xenotransplantation and its potential role in the treatment of end-stage cardiac and pulmonary diseases. J Heart Lung Transplant 2000; 19(12):1125–65.

54. Hardy JD, Alican F. Lung transplantation. Adv Surg 1966; 2:235–64.

55. Bando K, Armitage J, Paradis IL et al. Indications for and results of single, bilateral, and heart–lung transplantation for pulmonary hypertension. J Thorac Cardiovasc Surg 1994; 108(6):1056–65.

56. Anyanwu AC, Banner NR, Radley-Smith R et al. Long-term results of cardiac transplantation from live donors: the domino heart transplant. J Heart Lung Transplant 2002; 21(9):971–5.

57. Kadikar A, Maurer J, Kesten S. The six-minute walk test: a guide to assessment for lung transplantation. J Heart Lung Transplant 1997; 16(3):313–19.

58. Aris RM, Routh JC, Lipuma JJ et al. Lung transplantation for cystic fibrosis patients with *Burkholderia cepacia* complex. Survival linked to genomovar type. Am J Respir Crit Care Med 2001; 164(11):2102–6.

This is an important paper demonstrating the influence of colonisation with *B. cepacia*. The mortality rate was 33% in those infected compared to 12% in the control group. Moreover, genomovar III patients were at the highest risk of death.

59. Mason DP, Solovera-Rozas M, Feng J et al. Dialysis after lung transplantation: prevalence, risk factors and outcome. J Heart Lung Transplant 2007; 27:1155–62.

60. Flume PA, Egan TM, Westerman JH et al. Lung transplantation for mechanically ventilated patients. J Heart Lung Transplant 1994; 13(1):15–21; discussion 22–3.

61. Fischer S, Simon AR, Welte et al. Bridge to lung transplantation with the novel pumpless interventional lung assist device NovaLung. J Thorac Cardiovasc Surg 2006; 131(3):719–23.

62. Botha P, Trivedi D, Weir CJ et al. Extended donor criteria in lung transplantation; impact on organ allocation. J Thorac Cardiovasc Surg 2006; 131:1154–60.

63. Van Raemdonck DE, Rega FR, Neyrinck AP et al. Non heart beating donors. Semin Thorac Cardiovasc Surg 2004; 16:309–21.

64. Aziz TM, El-Gamel A, Saad RA et al. Pulmonary vein gas analysis for assessing donor lung function. Ann Thorac Surg 2002; 73(5):1599–604; discussion 1604–5.

65. Rabanal JM, Ibaanez AM, Mons R et al. Influence of preservation solution on early lung function (Euro-Collins vs Perfadex). Transplant Proc 2003; 35(5):1938–9.

66. Szeto WY, Kreisel D, Karakousis GC et al. Cardiopulmonary bypass for bilateral sequential lung transplantation in patients with chronic obstructive pulmonary disease without adverse

effect on lung function or clinical outcome. J Thorac Cardiovasc Surg 2002; 124(2):241–9.

67. Chan KM, Allen SA. Infectious pulmonary complications in lung transplant recipients. Semin Respir Infect 2002; 17(4):291–302.

68. Moffatt SD, Demers P, Robbins RC et al. Lung transplantation: a decade of experience. J Heart Lung Transplant 2005; 24:145–51.

69. Lama R, Santos F, Algar FJ et al. Lung transplants with tacrolimus and mycophenolate mofetil: a review. Transplant Proc 2003; 35(5):1968–73.

70. Wigfield CH, Lindsey JD, Steffens TG et al. Early institution of extracorporeal membrane oxygenation for primary graft dysfunction after lung transplantation improves outcome. J Heart Lung Transplant 2007; 26:331–8.

71. Adatia I, Lillemei C, Arnolds JH et al. Inhaled nitric oxide in the treatment of postoperative graft dysfunction after lung transplantation. Ann Thorac Surg 1994; 57(5):1311–18.

72. Bacha EA, Hervae P, Murakami S et al. Lasting beneficial effect of short-term inhaled nitric oxide on graft function after lung transplantation. Paris-Sud University Lung Transplantation Group. J Thorac Cardiovasc Surg 1996; 112(3):590–8.

73. Clark SC, Sudarshan CD, Dark JH et al. Controlled reperfusion and pentoxifylline modulate reperfusion injury after single lung transplantation. J Thorac Cardiovasc Surg 1998; 115(6):1335–41.

74. Clark SC, Levine AJ, Hasan A et al. Vascular complications of lung transplantation. Ann Thorac Surg 1996; 61(4):1079–82.

75. Charman SC, Sharples LD, McNeil AD et al. Assessment of survival benefit after lung transplantation by patient diagnosis. J Heart Lung Transplant 2002; 21(2):226–32.

This review of 653 patients undergoing lung transplantation used Cox regression analysis to demonstrate the survival advantages of postoperative patients irrespective of their primary pathology. There was no survival difference between patients having single or bilateral lung transplantation.

76. Meyer DM, Bennett LE, Novick RJ et al. Effect of donor age and ischemic time on intermediate survival and morbidity after lung transplantation. Chest 2000; 118(5):1255–62.

77. Bowdish ME, Barr ML, Schenkel FA et al. A decade of living lobar lung transplantation; peri-operative complications after 253 donor lobectomies. Am J Transplant 2004; 4:1283–8.

78. National Emphysema Treatment Research Group. A randomised trial comparing lung volume reduction surgery with medical therapy for severe emphysema. N Engl J Med 2003; 348:2059–73.

79. Tutic M, Lardinois D, Imfeld S et al. LVRS as an alternative or bridging procedure to lung transplantation. Ann Thorac Surg 2006; 82:208–13.

80. Corris PA. Non-heart beating lung donation: aspects for the future. Thorax 2002; 57:II53–6.

12

Cytomegalovirus, Epstein–Barr virus and BK virus infection following solid-organ transplantation

Imran A. Memon
Daniel C. Brennan

Cytomegalovirus

Cytomegalovirus (CMV) infection remains a significant contributor to post-transplant morbidity and mortality. Even asymptomatic CMV infection is associated with a 2.9 hazard ratio of death in renal transplant recipients by 4 years after transplant.[1]

Once active infection has been established, CMV replication is highly dynamic, with rapid increases in viral loads, development of CMV viraemia and an elevated risk of invasive disease. Exposure to the virus, as indicated by the presence of detectable IgG anti-CMV antibodies in the plasma, increases with age in the general population and is present in more than two-thirds of donors and recipients prior to transplantation. The virus can be transmitted from blood transfusion or by the transplanted kidney. The administration of immunosuppressive drugs, especially lymphocyte depleting agents, increases the risk of clinically relevant disease.[2]

CMV infection versus disease

There is an important distinction between CMV infection and disease. Infection is active if one or more of the following findings is noted: seroconversion with the appearance of anti-CMV IgM antibodies; a fourfold increase in pre-existing anti-CMV IgG titres; detection of CMV antigenaemia; detection of CMV DNAaemia by molecular techniques; or isolation of the virus by culture. CMV disease, in comparison, requires clinical signs and symptoms, such as fever, leukopenia or organ involvement.[3] In general terms, the total burden of CMV viral particles in the host correlates with clinical evidence of disease, disease severity or response to therapy.[4,5]

Quantitative CMV levels are lower in plasma compared to whole blood or buffy coat samples. Thus, it is difficult to interpret quantitative results among studies with different sample sources. The immune status of the host and timing and type of treatment also affect the impact of the viral load.

CMV disease

Symptomatic CMV infections typically occur 1–4 months after transplantation without prophylaxis or 1–4 months after discontinuation of prophylaxis, although cases may develop later. The onset usually follows a period of maximal immunosuppression for the prevention or treatment of acute rejection. The risk of reactivation is increased among those treated with lymphocyte-depleting agents.

The most common presentation of CMV disease is a mononucleosis-like syndrome with fever, malaise, myalgias and arthralgias, usually with leucopenia and mild (5–10%) atypical lymphocytosis. One of the cardinal features of CMV is leucopenia. This is more common in the absence of maintenance steroids, which are increasingly popular. A mild elevation in serum aminotransferase concentrations may also occur. With the use of mycophenolate mofetil (MMF), invasive CMV disease can occur in the absence of fever and leucopenia.

Interstitial pneumonitis and ulcerations in the oesophagus and colon cause major morbidity. For example, gastrointestinal bleeding among renal transplant recipients is commonly caused by erosions due to CMV. In this setting, invasive disease is usually confirmed with endoscopic biopsy and may be present in the absence of viraemia. By comparison, encephalopathy and chorioretinitis are unusual in renal transplant recipients but also may be present in the absence of viraemia (Box 12.1).

Diagnosis

Several diagnostic modalities are available, as shown in Table 12.1.[6]

Treatment of CMV disease

Therapy varies with the severity of the manifestations. CMV syndrome may resolve simply with discontinuation of azathioprine or MMF. More severe infections require discontinuation of the antimetabolite and administration of antiviral agents like intravenous ganciclovir or high-dose oral valganciclovir. Whether ciclosporin or tacrolimus should be discontinued remains controversial. Corticosteroids are generally lowered but continued to prevent adrenal insufficiency.

Patients with organ involvement may benefit from a 2- to 3-week course of intravenous ganciclovir.[7] Hyperimmune globulin (Cytogam) may also be added to the treatment regimen in those with organ involvement.[8]

The usual dose of ganciclovir is 5 mg/kg i.v. every 12 hours in patients with normal renal allograft function. Although dose reductions are recommended for renal insufficiency, lower doses increase the likelihood of recurrent CMV disease in renal transplant recipients.[7]

The excellent bioavailability and pharmacokinetics of valganciclovir, make it useful for treatment of CMV disease in transplant recipients.

The effectiveness of valganciclovir was best evaluated in the VICTOR study.[9] In this non-inferiority trial, 321 solid-organ recipients with CMV disease were randomly assigned to oral valganciclovir (900 mg twice daily for 21 days) or intravenous ganciclovir (5 mg/kg twice daily for 21 days), which was followed in both arms by valganciclovir (900 mg daily until day 49). Both agents were similarly effective in suppressing viraemia at 21 days.

Box 12.1 • Active CMV infection effects in immunocompromised hosts

Direct effects

CMV syndrome

Flu-like and mononucleosis-like syndrome

Tissue invasive disease

Nephritis
Hepatitis
Carditis
Pneumonitis
Pancreatitis
Colitis
Retinitis

Indirect effects

Cellular effects

Allograft injury (acute or chronic)
Allograft rejection (acute or chronic)
EBV-associated PTLD
Systemic immune response suppression
Opportunistic infections

CMV-induced renal disease

Whether the virus itself can cause allograft dysfunction is unclear.[10] Renal function may deteriorate in patients with CMV infection but factors such as decreased renal perfusion, acute tubular necrosis and transplant rejection may be more important than a direct viral effect on the kidney. CMV infection has been identified as an independent risk factor for the development of rejection.[11] Infection with CMV has also been implicated in the development of coronary artery narrowing.[12]

CMV has also been associated with thrombotic microangiopathy (TMA) in solid-organ and bone marrow transplant recipients, which may respond to immunoglobulin infusion.[13–15]

Prevention

Currently, there are two principal approaches to prevent CMV disease in patients undergoing solid-organ transplantation. A **prophylactic strategy** with the administration of antiviral agents to patients at

Table 12.1 • Diagnostic tests for CMV disease

Test	Method	Result	Important features
CMV antibody titre	Enzyme-linked immunosorbent assay (ELISA)	Fourfold increase in IgG or markedly positive IgM	Generally less useful than culture, or PCR
CMV culture	Using the shell vial technique which uses a fluorescence-tagged monoclonal antibody to detect CMV antigen early in viral replication	Positive or negative	Insensitive Requires immediate processing
pp65 antigenaemia assay	Antibody-based test using PMNs	Reported in terms of positive cells, i.e. ≥200 cells/ 20 000 PMNs	May discriminate between infection and disease. May not be helpful in patients with poor leucocyte function
pp67 mRNA detection	Qualitative test based on nucleic acid isolation and nucleic acid sequence determination based on amplification	Positive or negative	Predicts the onset of clinical disease and potential for pre-emptive therapy
Qualitative CMV PCR	Amplify an early antigen gene of CMV DNA	Positive or negative	May be positive in asymptomatic patients
Quantitative CMV PCR	Amplifies a 365-base-pair region of CMV polymerase gene	Detection range between 400 and 100 000 copies/mL	This test can be costly and results may not be available in a timely manner. Is more sensitive and helpful in screening
Hybrid-capture RNA–DNA hybridisation assay	Signal amplification method using an RNA probe that targets 17% of RNA genome. The target is detected by antibodies that bind to the RNA–DNA hybrid	Greater than 100 000 copies is suggestive of CMV disease	Sensitivity may allow its use in monitoring therapy

PMN, polymorphonuclear cell.

increased risk of developing CMV infection and a **pre-emptive strategy** with periodic monitoring for viraemia, principally using polymerase chain reaction (PCR), to permit treatment of asymptomatic early systemic infection. It is unclear which strategy is preferred. A number of meta-analyses of these and other strategies have been performed.[16–18]

A 2006 systematic review evaluated 10 trials involving 476 solid-organ transplant (SOT) recipients, of which six were pre-emptive therapy versus placebo or standard therapy (treatment once CMV disease occurred), three were pre-emptive versus prophylactic therapy, and one was oral versus intravenous pre-emptive therapy.[16] No difference was observed with pre-emptive versus prophylaxis therapy in the risks of CMV disease or mortality. Oral and intravenous pre-emptive therapy were also associated with similar risks of disease.

A 2006 meta-analysis of 17 trials (with the comparator being placebo or no treatment) involving 1980 kidney and liver transplant recipients found that CMV organ disease was significantly reduced with both prophylaxis and pre-emptive strategies.[17] Acute rejection was significantly reduced with both. Only prophylaxis resulted in decreased bacterial and fungal infections and mortality.

Prophylactic therapy

The incidence and risk of severe CMV disease as well as the approach to management are determined by the CMV status, defined serologically, of both the donor and the recipient. A number of different drugs and routes of administration have been used.

A meta-analysis of 19 trials in solid-organ transplant recipients found that antiviral prophylaxis significantly lowered the risk of CMV disease, CMV infection and mortality.[18] Ganciclovir was superior to acyclovir, while oral ganciclovir was similarly effective as valganciclovir and intravenous ganciclovir.

Ganciclovir and valganciclovir

Ganciclovir or valganciclovir has now supplanted either acyclovir or CMV hyperimmune globulin as the prophylactic therapy of choice for CMV infection or disease among transplant recipients. Although prophylactic **intravenous ganciclovir** can prevent CMV infection in transplant recipients it is unnecessary, expensive and associated with neutropenia, and oral ganciclovir or valganciclovir have taken over as prophylaxis. Although **oral ganciclovir** has poor bioavailability, one study, which evaluated 42 renal transplant recipients at risk for CMV infection in whom induction therapy was administered, found that prophylaxis with ganciclovir (1000 mg t.i.d.) was superior to deferred therapy with acyclovir (200 mg b.i.d.), both given for 12 weeks.[19]

Another prospective study of 101 high-risk renal transplant recipients, who were randomly assigned to prophylactic therapy with oral ganciclovir or acyclovir, found significantly fewer CMV infections after 3 months of therapy with ganciclovir.[20] A larger trial of liver transplant recipients (none of whom were seronegative recipients of seronegative donors) were randomised to either 1000 mg t.i.d. of oral ganciclovir or placebo. At 6 months, active therapy significantly reduced the incidence of CMV disease.[21]

Valganciclovir, a valyl-ester prodrug of oral ganciclovir, is rapidly metabolised to an active form (ganciclovir) in the intestinal wall and liver. It has a bioavailability of nearly 70% (compared to 7% for oral ganciclovir) and at doses of 450–900 mg produces serum ganciclovir levels that are similar to that measured with intravenous administration of ganciclovir administered at 2.5–5 mg/kg.[22]

In the phase III PV16000 international, double-dummy, double-blinded trial of oral valganciclovir versus oral ganciclovir for the prevention of CMV, valganciclovir and oral ganciclovir had similar efficacy and reaffirmed that a period of intravenous ganciclovir is unnecessary when an appropriate oral agent is used. During prophylaxis only 8.0% and 1.6% respectively developed symptomatic CMV infection; CMV disesase was common after prophylaxis, with disease occurring in 23% and 22% respectively by 6 months. These results suggest that a duration for CMV prophylaxis of 100 days may be insufficient for CMV D + R- transplant recipients[23] (**Fig. 12.1**).

Valacyclovir may prevent CMV disease among high-risk renal transplant patients. In a multicentre prospective study, use of valacyclovir was associated with a reduced incidence of disease among both seronegative and seropositive recipients.[24] Valacyclovir was also associated with a reduced incidence of acute rejection among seronegative patients (26% vs. 52% vs. placebo, $P = 0.001$). Since valacyclovir (even in high concentrations) has limited activity against CMV, the mechanism of action for these results remains unclear.

CMV hyperimmune globulin

CMV hyperimmune globulin is prepared from human serum that contains a high titre of anti-CMV antibodies. Initially developed in the 1980s, it was licensed for the prophylactic therapy of CMV disease, although it has been used for rescue therapy of tissue invasive disease. Infusions are typically commenced within 72 hours of engraftment, and are usually continued in the outpatient department for 4 months. The following protocol is recommended: 150 mg/kg within 72 hours; 100 mg/kg at 2, 4, 6 and 8 weeks; and 50 mg/kg at 12 and 16 weeks after transplantation.

A multicentre trial, performed before the availability of ganciclovir, showed that hyperimmune globulin decreased the incidence of CMV disease (20% vs. 60%), reduced the incidence of fungal or parasitic superinfection (0% vs. 20%) and diminished the incidence of marked leucopenia, a marker of severe infection (4% vs. 37%), but did not alter the incidence of viral isolation or seroconversion.[25]

Other antiviral agents

Several additional agents have been evaluated or are currently undergoing investigation as possible anti-CMV agents among solid-organ recipients.

Leflunomide

Leflunomide is a pyrimidine synthesis inhibitor and immunosuppressive agent currently utilised for the treatment of rheumatoid arthritis. Given its immunosuppressive properties, it has also been used in experimental and clinical transplantation. Leflunomide prevents CMV and herpes simplex virus (HSV) type I replication by interfering with virion assembly, and may have a role in the management of CMV in transplant recipients.[26–30]

Acyclovir

The efficacy of acyclovir as prophylaxis in CMV infection has been compared to other antiviral agents, particularly ganciclovir. A 3-month course of high-dose acyclovir (800 mg four times per day, with modification of dosage for impaired renal allograft function) after transplantation has also been utilised. In one randomised, placebo-controlled trial, symptomatic CMV disease

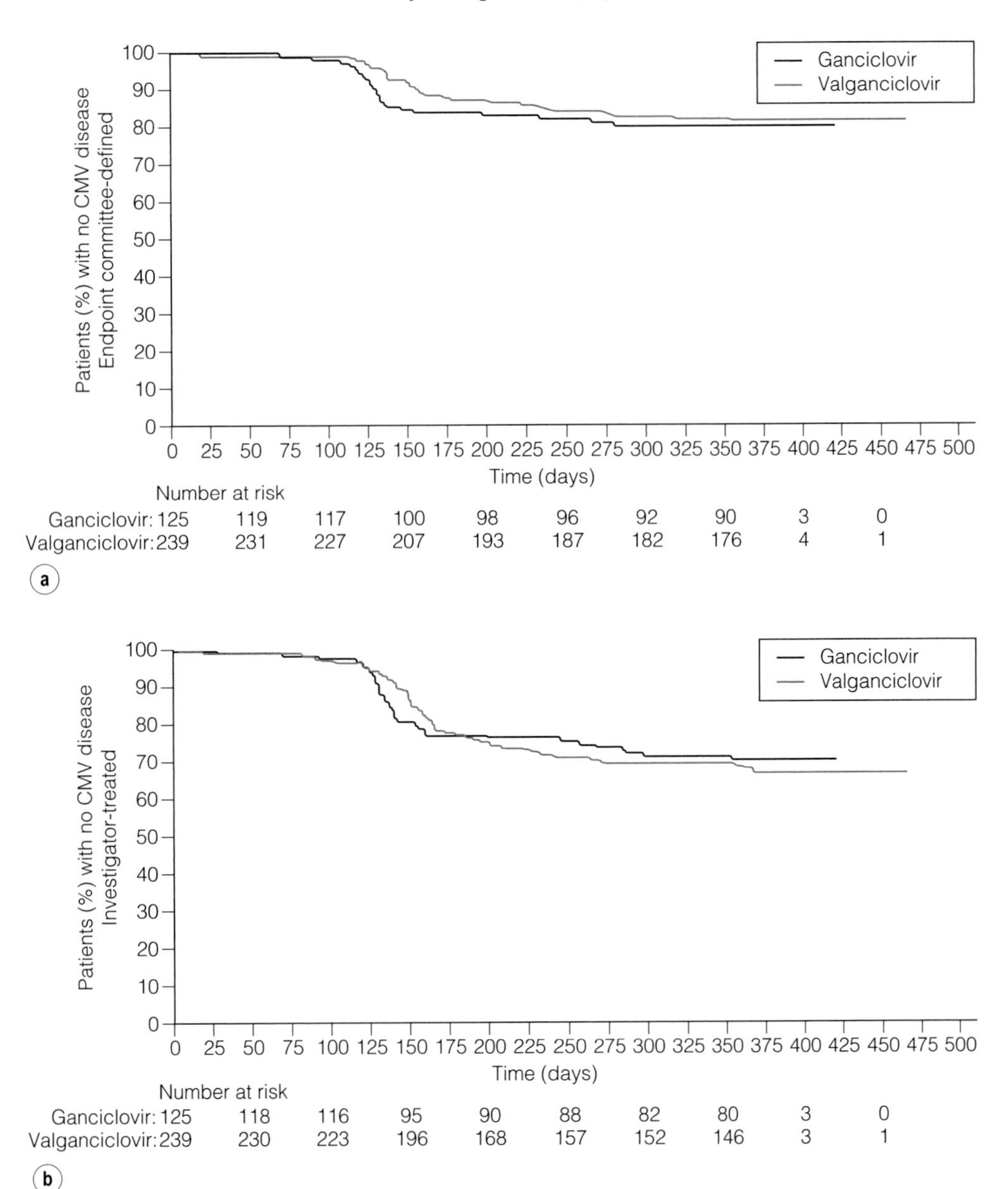

Figure 12.1 • Time (days) to CMV disease as assessed by the Endpoint Committee **(a)** and in investigator-treated CMV disease events **(b)** up to 12 months post-transplant in patients receiving antiviral prophylaxis with valganciclovir 900 mg once daily or oral ganciclovir 1000 mg three times daily for 100 days post-transplant (intent-to-treat (ITT) population). Reproduced from Paya C, Humar A, Dominguez E et al. Efficacy and safety of valganciclovir vs. oral ganciclovir for prevention of cytomegalovirus disease in solid organ transplant recipients. Am J Transplant 2004; 4(4):611–20. With permission from Blackwell Publishing.

developed in only one of six CMV-negative recipients who received a graft from a CMV-positive donor versus all seven patients treated with placebo.[31] More recent studies have shown that acyclovir has minimal utility for prophylaxis for CMV.[32]

Maribavir

Maribavir is a benzimidazole nucleoside that prevents viral DNA synthesis and capsid nuclear egress. It is therefore useful for CMV infections resistant to ganciclovir, cidofovir or foscarnet. In July 2004, a phase II trial was initiated.[33]

Drug-resistant CMV

Ganciclovir-resistant CMV isolates have emerged among seronegative recipients of seropositive organs. In one study of 240 recipients of liver, kidney or pancreas transplants, ganciclovir-resistant CMV disease developed in 7% of seronegative recipients of CMV-positive organs compared to none of 173 seropositive recipients.[34] In the PV16000 trial, three cases (2%) of those administered oral ganciclovir developed ganciclovir resistance compared to none of the valganciclovir patients.[23]

CMV status and prophylactic therapy

CMV-negative recipients of kidneys from CMV-negative donors have the lowest incidence of CMV infection. When infection does occur in these patients, it may be caused by false seronegativity in the donor or recipient, community exposure or perioperative transfusion of leucocyte-containing blood. The use of leucocyte-poor blood substantially decreases the risk of virus transmission. By comparison, there is a high risk of infection or disease when the donor or recipient, or both, is CMV positive.

CMV-positive donor, CMV-negative recipient (D^+R^-)

Historically, approximately 70–90% of CMV-positive donor, CMV-negative recipients developed 'primary' CMV infection when prophylaxis was not administered. Furthermore, 50–80% developed CMV disease and roughly 30% developed pneumonitis, a major contributor to morbidity and mortality and, in the absence of prophylactic therapy, the mortality rate was 15%. For unclear reasons, the rate of primary infection has decreased to 50% even in the absence of prophylaxis and despite the use of lymphocyte-depleting therapy.[35–37]

CMV-negative donor, CMV-positive recipient (D^-R^+)

Seropositive recipients of seronegative organs may have reactivation of latent CMV infection due to the administration of immunosuppressive drugs. CMV infection or disease may develop in up to 20% of cases, although progression to pneumonitis is rare. Some groups use prophylactic ganciclovir therapy in this population. This is especially true when lymphocyte-depleting agents are used for induction therapy or treatment of rejection.[36]

CMV-positive donor, CMV-positive recipient (D^+R^+)

Seropositive recipients of CMV-positive kidneys are at risk for reactivation of latent virus or superinfection with a new viral strain. The worst graft and patient survival at 3 years post-transplantation is observed among those in which the donor and recipient are both positive.[38,39] This unexplained increased risk of disease may be due to the presence of multiple CMV virotypes or reactivation of different viruses.

CMV-negative donor, CMV-negative recipient (D^-R^-)

The incidence of disease in the D^-R^- population is 5–10%. Given this low prevalence of disease, independent of type of immunosuppressive therapy, no treatment with anti-CMV prophylaxis is recommended.[36]

Duration of prophylaxis

The optimal duration of prophylaxis is unknown. Most institutions use approximately 3 months of prophylaxis.

At some transplant centres, a longer duration of prophylaxis has been used according to the CMV serostatus of the donor and recipient.[35,40,41] In one centre, prophylaxis with oral ganciclovir was prescribed for 90 days in D^-R^+ cases and for 180 days in D^+R^- and D^+R^+ cases.[35] There was a significant association of disease with donor CMV seropositivity when the recipient was CMV seronegative. By comparison, donor CMV serostatus was not significantly associated with CMV disease when the recipient was CMV seropositive. All CMV disease occurred after prophylaxis ended. Despite the relatively mild severity of disease, which was promptly detected and treated, the presence of disease was significantly associated with poorer graft survival.

In another study, the group in which recipients received a 24-week regimen of oral ganciclovir, compared with a historic control group administered a 12-week course, a markedly lower rate of CMV infection occurred (31% vs. 7% respectively).[40] A similar decrease in CMV was found with a 6-month course of valganciclovir.[41]

Pre-emptive strategy

A pre-emptive strategy is based on periodic PCR monitoring of whole blood for CMV DNAaemia to allow detection of infection. This permits prompt treatment, which may limit morbidity and mortality.

A recent systematic review showed that pre-emptive therapy, compared with placebo or standard therapy, significantly reduced the risk of CMV disease.[42]

In a single-centre study comparing pre-emptive and prophylactic strategies, 98 at-risk kidney transplant recipients (D^+R^-, D^+R^+, D^-R^+) were randomly assigned to prophylaxis (valganciclovir, 900 mg/day for 100 days) or pre-emptive therapy (valganciclovir, 900 mg twice per day for at least 21 days for CMV DNAaemia as defined by levels greater than 2000 copies/mL by PCR). All patients who developed CMV DNAaemia were treated with valganciclovir, which was continued until the PCR assay became negative.[37] A similar incidence of symptomatic infection (disease) was reported in both arms (overall 4%) and there were no deaths. CMV DNAaemia was significantly more likely in the pre-emptive group (59% vs. 29% for the prophylactic arm). However, the late onset of CMV DNAaemia (greater than 100 days) was significantly more likely with prophylaxis therapy (24% vs. 0%). Time of onset of disease occurred later than 100 days in all patients with symptomatic disease in the prophylactic arm, but at day 62 in one patient treated pre-emptively. Both approaches were associated with similar overall costs, minimal side-effects and low rates of adverse outcomes.

Recommendations

With a prophylactic strategy, oral ganciclovir or valganciclovir should be administered to all patients other than seronegative recipients of seronegative grafts for a total duration of 100 days.[3] The choice of agent depends upon the renal function of the patient and the relative affordability of the agents. In general, valganciclovir is preferred because of the reduced pill burden, increased likelihood of compliance with daily administration, and the potential for a decreased incidence of drug resistance. However, valganciclovir is contraindicated when the patient is on dialysis or has an estimated creatinine clearance less than 10 mL/min.

When lymphocyte-depleting therapy (ATGAM, thymoglobulin, OKT3, alemtuzumab) is administered in a quadruple immunosuppressive regimen and a pre-emptive approach is not used, prophylaxis should be extended to 6 months for D^+R^+ patients and to 6–12 months for D^+R^- patients.

With a pre-emptive strategy, whole blood quantitative CMV PCR should be monitored weekly for 12–16 weeks postsurgery. In patients in whom the CMV PCR becomes positive above 2000 copies/mL, the following approach could be considered, which varies in part upon the severity of the infection. In all cases of quantifiable CMV viraemia, the antimetabolite (azathioprine or mycophenolate) should be discontinued at least until viraemia clears. The severity of the CMV and the risk of rejection dictate whether to restart the antimetabolite. Weekly quantitative PCRs should be obtained during treatment to determine an adequate response. If the quantitative PCR level does not decrease by 50% in 2 weeks, viral resistance or recipient immunocompetence should be suspected. The dose of valganciclovir can then be increased or the patient switched to intravenous ganciclovir to overcome relative resistance. Consideration should also be given for treatment with hyperimmune CMV globulin for patients with suspected CMV immunocompetence. Treatment should continue for at least one week beyond the finding of a negative quantitative CMV PCR. Asymptomatic or mild disease is treated with valganciclovir for a minimum of 21 days or longer if necessary to clear viraemia. Invasive disease is initially treated with intravenous ganciclovir. Secondary prophylaxis for an additional 3 months should be considered after this initial treatment, particularly in those at high risk of recurrence (i.e. D^+R^- or two donor/recipient human leucocyte antigen (HLA)-mismatched patients).

Summary

That CMV should be prevented is clear. It is unclear whether a preemptive or prophylactic approach is preferred in patients at risk for CMV infection. The optimal strategy to control CMV is based upon the institution and available resources.

Epstein–Barr virus

Epstein–Barr virus (EBV) is a widely disseminated herpes virus spread by intimate contact between susceptible persons and asymptomatic EBV shedders. The majority of primary EBV infections throughout the world are subclinical. Approximately 90–95% of adults are EBV seropositive worldwide.[6]

Virology

EBV is a member of the gamma herpes virus family and the prototype for the lymphocryptovirus group of viruses that typically result in latent infection with persistence of the viral genome and expression of a restricted set of latent gene products. Host cells in humans are B lymphocytes, T lymphocytes, epithelial cells and myocytes. Unlike herpes simplex (HSV) or CMV, EBV is capable of transforming B cells and does not routinely display a cytopathic effect in cell culture.

Various components of EBV and the cells that the virus infects contribute to the pathogenesis of infection, including the viral receptor, penetration and uncoating, viral expression in latent infection, and cell transformation with the production of latent proteins.

Viral receptor

The EBV receptor on human cells is the B-cell-surface molecule CD21, which is the receptor for the C3d component of complement (also called CR2, complement receptor type 2).[43] Infection is initiated by binding of the major EBV outer envelope glycoprotein gp350/220 with CD21.

The virus is then transported to the nucleus and the EBV genome is replicated by cellular DNA polymerases during the cell cycle S phase.[44] It persists as multiple, extrachromosomal double-stranded EBV episomes, which are organised into nucleosomes similar to chromosomal DNA.[45]

Virus expression in latent infection

The hallmark of B-lymphocyte infection with EBV is the establishment of latency. The triggers for the shift from latency to lytic replication are not clearly defined. Latency is characterised by three distinct processes: viral persistence, restricted virus expression which alters cell growth and proliferation, and retained potential for reactivation to lytic replication.

EBV exploits normal pathways of B-cell differentiation to allow it to persist in a transcriptionaly quiescent state in memory B cells and thus minimise immune recognition.[46,47] Intracellular persistence of the entire viral genome is achieved through circularisation of the linear EBV genome, and maintenance of multiple copies of this covalently closed episomal DNA.[44]

Clinical manifestations

EBV is the primary agent of infectious mononucleosis and persists asymptomatically for life in nearly all adults. Reactivation is not prominent with EBV, in contrast to other common herpes viruses, but has been associated with an aggressive lymphoproliferative disorder in transplant recipients, B-cell lymphomas, T-cell lymphomas, Hodgkin lymphoma and nasopharyngeal carcinomas in other patients.

Primary infection

Congenital and perinatal infections

Intrauterine infection with EBV is rare because fewer than 5% of pregnant women are susceptible to primary infection.

Primary EBV infection in infants and children

Primary EBV infections in infants and young children are common and frequently asymptomatic.[48] When symptoms occur, a variety of manifestations have been observed, including otitis media, diarrhoea, abdominal complaints, upper respiratory infection and infectious mononucleosis. Children can have symptomatic primary EBV infection without the production of heterophile antibodies which are insensitive for diagnosis. Thus, EBV-specific serological studies are required to establish the diagnosis definitively. However, antiviral capsid antigen (anti-VCA) IgM was less frequently positive in infants; peak titres of VCA antibody were lower, and the development of antibodies to early antigen was less common in infants.

Acute infectious mononucleosis

Infectious mononucleosis is the best-known acute clinical manifestation of EBV. It often begins with malaise, headache, low-grade fever before development of the more specific signs of tonsillitis or pharyngitis, cervical lymph node enlargement and tenderness, and moderate to high fever.[48,49] Affected patients usually have peripheral blood lymphocytosis, composed of atypical lymphocytes. The lymphadenopathy is characteristically symmetric and involves the posterior cervical chain more than the anterior chain. Tonsilar exudates are a frequent component of the pharyngitis, and can be white, grey–green or necrotic in appearance. Splenomegaly occurs in up to 50%, but jaundice and hepatomegaly are uncommon. Acute symptoms resolve in 1–2 weeks, but fatigue often persists for months. The vast majority of individuals with primary EBV infection recover uneventfully and develop a high degree of durable immunity.

Other manifestations

Neurological syndromes include Guillain–Barré syndrome, facial nerve palsy, meningoencephalitis, aseptic meningitis, transverse myelitis, peripheral neuritis and optic neuritis.[50] Haematological abnormalities include haemolytic anaemia, thrombocytopenia, aplastic anaemia, thrombotic thrombocytopenic purpura/haemolytic–uraemic syndrome, and disseminated intravascular coagulation. EBV can affect virtually any organ system and has been associated with pneumonia, myocarditis, pancreatitis, mesenteric adenitis, myositis, glomerulonephritis and genital ulceration.[51]

Complications

EBV infection is associated with a number of acute complications, including a morbilliform rash following the administration of ampicillin and, to a lesser extent, penicillin. **Oral hairy leucoplakia** (OHL) is an unusual EBV-mediated mucocutaneous disease of the lingual squamous epithelium.[52] The OHL lesions appear to be relatively specific for human immunodeficiency virus (HIV) infection, since they are only rarely observed in patients with other immunodeficiencies.[52,53] The use of highly active antiretroviral therapy has reduced the incidence of OHL.[54] **Splenic rupture** is a rare but potentially life-threatening complication of infectious mononucleosis, estimated to occur in between one and two cases per 1000.[55] **Obstruction of the upper airway** due to massive lymphoid hyperplasia and mucosal oedema is an uncommon and potentially fatal complication of infectious mononucleosis.

Lymphoproliferative disorders

EBV infection is associated with a variety of lymphoproliferative disorders, including haemophagocytic lymphohistiocytosis, lymphomatoid granulomatosis, X-linked lymphoproliferative disease (also called Duncan syndrome) and EBV-associated post-transplant lymphoproliferative disease (PTLD).[56] EBV is a transforming virus and has also been causally linked to a variety of malignancies in addition to lymphomas in transplant recipients.[56] These include Burkitt's lymphoma (tumours in HIV-infected patients), Hodgkin lymphoma, nasopharyngeal and other head and neck carcinomas, and T-cell lymphoma.

Diagnosis

EBV infection is suspected when patients present with typical symptoms, and supportive evidence of infection is derived from the peripheral blood smear and antibody studies as above. A detailed discussion of the diagnosis of EBV-related PTLD is discussed below.

Treatment and prevention

Primary EBV infections rarely require more than supportive therapy. Even in clinical situations where an antiviral or immunomodulatory treatment might be considered, it is not clear that EBV responds. The use of corticosteroids in the treatment of EBV-induced infectious mononucleosis is controversial. Corticosteroids are warranted in individuals with impending airway obstruction, those suffering from severe overwhelming life-threatening infection or other severe complications such as aplastic anaemia. Antiviral agents such as acyclovir inhibit permissive EBV infection through inhibition of EBV DNA polymerase, but have no effect on latent infection. Short-term suppression of viral shedding has been demonstrated, but significant clinical benefit has been lacking.

In the majority of the EBV-associated malignancies, there is little evidence for permissive (lytic) infection. Since acyclovir is only effective in inhibiting replication of linear EBV DNA, there is little to be gained by its use in diseases associated with latent infection characterised by episomal circular DNA.

Anecdotal use of interleukin-2, interferon-α and intravenous immunoglobulins have been reported. No clear benefits of such modalities have been demonstrated, with the possible exceptions of lymphomatoid granulomatosis and PTLD.[57]

The large body of evidence implicating EBV in the aetiology of a variety of human neoplasms has made the prospect of developing a viral-based vaccine effective against human cancers appealing. The viral glycoprotein gp350/220 is the most abundant capsid protein present in lytically infected cell plasma membranes and the most abundant protein on the outer surface of the virus. It binds to the CD21 receptor on the B cell, and is responsible for the initiation of infection. Most of the human EBV-neutralising antibody response is directed against gp350/220.[58,59] Thus, gp350/220 is the major EBV lytic-cycle gene product being pursued in the development of a subunit vaccine. In animal studies, immunisation with partially purified gp350/220 antigen induced EBV-neutralising antibody and protected a portion of cotton-top tamarins against a normally lethal, lymphoma-producing challenge with EBV.[60]

Post-transplant lymphoproliferative disorder (PTLD)

Lymphoproliferative disorders occurring after transplantation have different characteristics from those that occur in the general population.[61] Non-Hodgkin's lymphoma (NHL) accounts for 65% of lymphomas in the general population compared to 93% in transplant recipients. These tumours are mostly large-cell lymphomas, the great majority of which are of the B-cell type.[61] Extranodal involvement is common, occurring in approximately 30–70% of cases.

Although uncommon, PTLD may also originate from T cells and more rarely from natural killer cells.[62–66] As of 2004, only 17 cases of T-cell PTLD had been reported among renal transplant recipients.[66]

Pathogenesis

The pathogenesis of post-transplant NHL in most patients is related to B-cell proliferation induced by infection with EBV in the setting of chronic immunosuppression.[56] However, EBV-negative disease can occur.[67]

Most PTLD cells in solid-organ allograft recipients are of host origin. The clinical manifestations, and course of PTLD, vary with the origin of the lymphoproliferative cells.[68–70] In a report of 12 renal transplant recipients, eight arose from the recipient and four from the donors.[69] Recipient-origin PTLD presented as multisystem disease at a mean of 76 months after transplantation; five of the eight patients with disease from the recipient died. In contrast, donor-origin PTLD was limited to the allograft, developed after a mean of 5 months after transplantation, and regressed after reduction of immunosuppression.

Clinical manifestations and epidemiology

Three types of EBV-related lymphoproliferative disease occur in transplant recipients.[71] Benign polyclonal lymphoproliferation is an infectious mononucleosis-type acute illness that develops 2–8 weeks after immunosuppressive therapy begins and accounts for 55% of cases.[71] It is characterised by polyclonal B-cell proliferation with normal cytogenetics and no evidence of immunoglobulin gene rearrangements to suggest malignant transformation. The second EBV-induced disorder is similar to the first in its clinical presentation, but is characterised by polyclonal B-cell proliferation with evidence of early malignant transformation, such as clonal cytogenetic abnormalities and immunoglobulin gene rearrangements and accounts for approximately 30% of cases.

The third type is characterised by monoclonal B-cell proliferation with malignant cytogenetic abnormalities and immunoglobulin gene rearrangements, and accounts for about 15% of cases. It is usually an extranodal condition presenting with localised solid tumours.[71] Involved organs include the gastrointestinal tract (stomach, intestine), lungs, skin, liver, central nervous system (CNS) and the allograft itself; 20–25% have CNS disease (which is rare in the general population) and a similar proportion have infiltrative lesions in the allograft.[72]

Non-EBV-associated PTLD differs clinically from EBV-related tumours.[67,73,74] Tumours not due to EBV present much later, suggesting that their incidence may increase with time, and are much more aggressive.

The overall incidence of PTLD is approximately 1%, 30–50 times higher than in the general population, with a recent trend towards increased frequency.[75] The risk of PTLD is greatest in patients with more marked degrees of immunosuppression, which explains part of the variability in the incidence of PTLD with different types of transplants.[76] The incidence is 1–2% in liver transplants, 1–3% in renal transplants, 2–6% in heart transplants, 2–9% in lung transplants and as high as 11–33% in intestinal or multiorgan transplants.[77–86]

Risk factors

The principal risk factors underlying the development of PTLD are the degree of overall immunosuppression and the EBV serostatus of the recipient. Additional risk factors include time post-transplant, recipient age and ethnicity.[75,87]

Degree of immunosuppression

Immunosuppression acts in part by impairing EBV-specific, T-cell-mediated immunity. EBV-infected B cells are thought to be kept quiescent by cytotoxic T cells, resulting in an equilibrium between cell division and death of EBV-infected B cells. This equilibrium is disturbed by impaired T-cell function, promoting the development of PTLD.

Solid-organ transplant patients at increased risk for PTLD are paediatric recipients and those treated with an increased degree of immunosuppression, particularly those exposed to certain types of lymphocyte-depleting therapy.[79,80,83,88–90]

The incidence ratio of PTLD, compared with the non-transplant population, was 21.5, 4.9, 29.0, 21.6 and 7.8 for those administered induction therapy with OKT3, antithymocyte globulin (ATG), ATGAM, thymoglobulin and interleukin-2 receptor antagonists respectively.[90]

For OKT3-treated patients, both the dose and the duration of therapy are important. The incidence was 11% in those treated with OKT3 for induction and 36% (5 of 14) in patients who received more than 75 mg of OKT3.[83] OKT3 is not currently used commonly for induction therapy.

The incidence of PTLD is highest in the first year and falls by about 80% thereafter. The incidence of PTLD is much greater in heart transplant recipients in whom a greater degree of immunosuppression is required because of the more serious consequences of transplant rejection. PTLD tends to be 'organotropic'. Renal transplant recipients are more likely to have renal lymphoma, while heart transplant recipients are more likely to develop lymphoma in the heart or lungs.[79] In a report of nine lung transplant recipients with PTLD, eight had isolated intrathoracic disease.[91] Local immune reaction against the graft may be one of the factors promoting malignant transformation.

The risk of PTLD is also significantly higher among those administered tacrolimus versus ciclosporin for maintenance therapy without induction therapy.[88,89] If induction therapy were given, there was a non-significant trend toward a higher risk with tacrolimus.[88] The use of MMF appears not be associated with an increased risk of PTLD.[92]

EBV serostatus

EBV serostatus is considered an important risk factor for development of PTLD. In the absence of the use of OKT3 and CMV seromismatch (i.e. a negative recipient and a positive donor), the incidence rate of PTLD for EBV-seronegative recipients was 24 times higher than that for EBV-seropositive recipients.[93] These patients, who had no preoperative immunity to EBV, usually acquire the infection post-transplant from the donor.[93] PTLD developed in 42% of patients who developed primary EBV infection after transplantation.[84,94]

Other risks

Additional risk factors for PTLD include a history of pre-transplant malignancy and fewer HLA matches.[95,96]

PTLD is also more common in children because more are EBV seronegative prior to transplantation. A prospective study of 50 paediatric heart transplant recipients found that the overall incidence of PTLD was 26%.[97] However, the incidence varied according to the initial and final EBV status, occurring in 12 of 19 (63%) who seroconverted after transplantation, 1 of 20 (5%) who were seropositive before transplantation, and 0 of 12 who were initially seronegative and remained so after transplantation.

The risk of PTLD also varies with time post-transplant. In a retrospective analysis of nearly 90 000 patients placed on the renal transplant waiting list over a 10-year period, 357 cases of lymphoma developed in transplant recipients.[87] The first year post-transplant was associated with the highest lymphoma rate. An increased risk was also noted in those under 25 years of age and Caucasian.

A combination of risk factors has been shown to markedly increase the overall risk of PTLD in non-renal transplant recipients.[93] Therapy with OKT3 for rejection and CMV sero-mismatch increases the risk four- to sixfold above that seen in EBV-seronegative recipients. The presence of all three risk factors increased the incidence rate of fatal and/or CNS PTLD by a factor of 654 compared to patients lacking all three factors.

Evaluation and diagnosis

An accurate diagnosis of PTLD requires a high index of suspicion, since the disorder may present subtly and/or extranodally.[98] Quantitative viral loads exceeding 1000 copies/mL in plasma or 5000 copies/mL are suggestive but not specific with a poor predictive value.[99] Radiological evidence of a mass or the presence of elevated serum markers (such as increased LDH levels) are suggestive of PTLD, with positive positron emission tomography (PET) scanning (possibly indicating metabolically active areas) also favouring the diagnosis.[100] PET/computed tomography (CT) is a useful tool for staging and therapy monitoring of PTLD after liver transplantation.[101]

The different forms of PTLD are diagnosed histologically and distinguished by a number of features. These include clonality, whether elements of a malignant process are present (such as the presence or absence of clonal cytogenetic abnormalities, immunoglobulin gene rearrangements, and disruption of underlying tissue architecture), donor versus recipient origin, and (increasingly) whether EBV can be detected within the tumour.[76] A number of classification schemes have been published.[102] The reliance upon histological descriptions of PTLD is limited because of the lack of uniformity concerning definitions of a polyclonal or monoclonal process, diagnostic use of non-histological features, such as DNA rearrangements, mutations and clonality, use of EBV positivity within the tumour or the pathological process, and donor or host origin of the tumour.

Neoplastic forms of EBV-positive PTLD should have disruption of underlying tissue architecture by a lymphoproliferative process, presence of mono- or oligoclonal populations as determined by cellular or viral markers and EBV infection of many cells.

If any two of these three features are present in combination with a lymphoid tumour, the diagnosis of a neoplastic PTLD is felt to be sufficiently established.

The diagnosis of CNS PTLD is difficult. The diagnosis is suspected in transplant recipients with mental status changes or new neurological findings. Diagnostic tests include gadolinium CT of the head, cerebral spinal fluid (CSF) analysis for EBV by PCR and cytology with cell markers by flow cytometry, and peripheral titration of circulating EBV load in plasma or white cells.

A CT scan with gadolinium enhancing lesions, a positive analysis of the spinal fluid for EBV by PCR and an increased EBV load in the peripheral blood are highly suggestive. However, the diagnosis should be confirmed either by the presence of malignant lymphocytes in the CSF or by direct biopsy of the lesion. Steroid use, commonly applied in the post-transplant setting, may confound testing, as these agents are lymphocytotoxic and may alter radiological and histopathological evaluation.

Confirming the presence of cardiac lymphoma in heart transplant recipients is similarly difficult. The diagnosis is frequently established on post-mortem examination.[103] It has also been made by endomyocardial biopsy in which the EBV genome was demonstrated in the lymphoid infiltrates.[104] These infiltrates consisted of a mixture of small lymphocytes, plasma cells, immunoblasts and atypical immunoblast-like cells. They are different from the infiltrates of cellular rejection seen on previous biopsies.

Prevention

Prevention depends on limiting patient exposure to excessive immunosuppression. Lower doses and the rapid tapering of tacrolimus may limit the development of PTLD. In a review of 82 children who received renal allografts with tacrolimus-based regimens, the incidence of PTLD fell from 17% to 4%.[105] This was attributed in part to a policy of aggressive tapering of tacrolimus and corticosteroids with chronic target trough concentration of 5–9 ng/mL.

Suppression of primary EBV infection or detection and early treatment may minimise the development of PTLD, especially in children.[106–108]

These approaches were evaluated in 40 children receiving a liver allograft.[106] Among 18 high-risk children, there were no cases of PTLD and one case of EBV infection (which resolved). Among 22 low-risk patients, two cases of PTLD occurred, both of which resolved after tacrolimus was stopped; there was one case of EBV increase, which also resolved, and a 50% reduction in the incidence of PTLD.

A retrospective multicentre case–control study of adult and paediatric renal transplant recipients found that prophylactic antiviral therapy with acyclovir or ganciclovir reduced the risk of PTLD.[109] The reduction was greatest with ganciclovir. Of 100 biopsy-confirmed cases and 375 matched controls, the risk of PTLD during the first year post-transplant decreased by 38% for every 30 days of treatment with ganciclovir (**Fig. 12.2**).

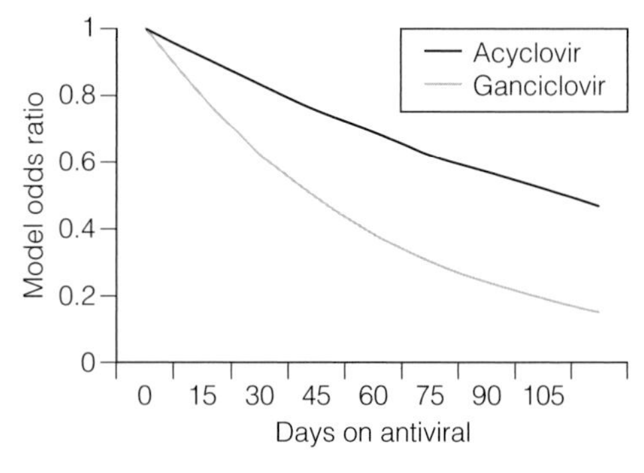

Figure 12.2 • Risk of post-transplant lymphoproliferative disorder with days on antiviral therapy during the first year following transplant. Reproduced from Funch DP, Walker AM, Schneider G et al. Ganciclovir and acyclovir reduce the risk of post-transplant lymphoproliferative disorder in renal transplant recipients. Am J Transplant 2005; 5(12):2894–900. With permission from Blackwell Publishing.

A randomised trial showed that the addition of immune globulin to ganciclovir did not provide additional benefit to ganciclovir alone among patients at increased risk for PTLD.[110] In one large retrospective database study, the use of prophylaxis with anti-CMV immunoglobulin during the first 4 months after kidney transplantation significantly reduced the incidence of PTLD during the first year post-transplant, but not in the subsequent 5 years.[111]

Treatment

Reduction in immunosuppression

Many polyclonal lymphoproliferative disease lesions resolve or improve significantly with reductions in immunosuppression.[112–116]

The response is best among patients with early-onset disease in whom immunosuppression is a major risk factor; in comparison, patients with late-onset or extensive disease are much less likely to benefit.[82,115,117] The optimal immunosuppression reduction regimen to ensure regression of disease is unknown. The regimen adopted is based upon the severity of the disease in combination with the health risk associated with the possible loss of the allograft.

Among the severely ill with extensive disease, one regimen is reduction of prednisone to a maintenance dose of 7.5–10 mg/day and stopping all other immunosuppressive agents.[102] If there is no response (as defined by a decrease in tumour mass by 10–20 days), additional therapeutic options can be entertained. Among those less severely ill with only limited disease, one regimen is the reduction by at least 50% of calcineurin inhibitor and prednisone, and the discontinuation of the antimetabolite. After 2 weeks, another 50% reduction of immunosuppression can be considered if necessary.

Reducing immunosuppressive therapy increases the risk of allograft rejection, which may result in death among those with heart, lung and liver allografts. However, a careful decrease in immunosuppressive therapy based upon the individual's clinical features and type of allograft can result in both a high response rate and a low rate of rejection and allograft loss.

One report evaluated 42 transplant recipients (which included kidney, heart, lung and liver allografts) with PTLD. All were treated with a reduction in immunosuppression and 12 also underwent surgical removal of all apparent disease.[115] Overall, 31 (74%) achieved complete remission. Among those treated with reduced immunosuppression alone, 63% had a complete or partial response with a median time to documentation of response of 3.6 weeks. Twelve developed acute rejection, but only one lost the graft because of rejection.[115] At a median follow-up of almost 3 years, 55% of patients were alive and 50% were in complete remission.

Antiviral therapy

Many early treatment algorithms for PTLD included antiviral therapy in an attempt to control EBV infection. Acyclovir inhibits viral DNA polymerase, and in clinical studies has been shown to decrease oropharyngeal shedding of EBV.[118] However, latent or transformed forms of EBV are not susceptible to antiviral agents in vitro[119] and there are no in vivo data to support therapeutic acyclovir for EBV-associated PTLD.[120] There is as yet no evidence of the efficacy of antiviral therapy and no consensus exists as to whether acyclovir or ganciclovir is preferred.[102] Thus, although immunological and antiviral therapy have been moderately effective for treating EBV-associated infections in the lytic phase, they have been less useful in the more common latent phase of the disease and for PTLD. The lack of viral thymidine kinase expression in EBV-positive tumour cells, due to viral latency, makes antiviral therapy alone ineffective as an antineoplastic therapy. However, given the low toxicity, most centres include either acyclovir or ganciclovir with decreased immunosuppression and other therapies in the treatment of PTLD.[121]

Monoclonal malignant lymphoma

The first step in the management of patients with malignant lymphoma is to reduce immunosuppression and consider antiviral therapy as described in the preceding section. Patients who do not respond have been treated with chemotherapy, immunotherapy and occasionally surgical resection.[115,117,122–124] Use of anti-B-cell antibodies is increasingly used as first-line therapy alone or in combination with other therapies.[125–127]

Anti-B-cell antibodies

The efficacy of anti-CD21 and anti-CD24 murine monoclonal antibodies was evaluated in a study of 58 transplant patients with severe PTLD.[125] Complete remission was observed in 36 of the 59 episodes of PTLD (61%), and the rate of relapse was only 8%.

The anti-CD20 monoclonal antibody rituximab has induced complete remissions in some patients with PTLD following either solid-organ transplantation or haematopoietic cell transplantation.[124,126,128–139]

Single-agent anti-CD20 rituximab (375 mg/m^2 per week for 4 weeks) was employed in a prospective trial involving 43 evaluable patients with previously untreated B-cell PTLD not responding to tapering of immunosuppression.[134] The overall response rate was 44%, while overall survival was 86% and 67% at 80 days and 1 year respectively. The only baseline factor predicting response at day 80 was a normal level of serum lactate dehydrogenase (LDH).

In a single-centre retrospective study, 19 patients with PTLD were treated with rituximab after reduction of immunosuppression, while three received rituximab after failure with chemotherapy.[124] A complete or partial remission was noted in 13 and two patients respectively, resulting in an overall response rate of 68%. Likelihood of response correlated with EBV positivity.

Five patients with PTLD following intestinal transplantation were treated with rituximab (375 mg/m^2 weekly until full remission was ascertained), reduction or interruption of immunosuppression, and antiviral therapy; no patient received chemotherapy.[139] All patients achieved full remission. Early treatment with rituximab, along with reduction of immunosuppression, appears to be the evolving standard of care for CD20$^+$ PTLD. Monitoring of the EBV viral load does not predict clinical response of the tumour to this agent, as even those with undetectable viral loads may progress.[128]

Chemotherapy and radiation therapy

Patients with monoclonal malignancies can be treated with chemotherapy, such as cyclophosphamide plus prednisone, CHOP (cyclophosphamide, doxorubicin, vincristine, prednisone), dose-adjusted ACVBP (doxorubicin, cyclophosphamide, vindesine, bleomycin, prednisone) and other novel therapies.[117,140–145] One report described prolonged complete remission in six of eight patients treated with an intensive ProMACE-CytaBOM regimen.[117] All immunosuppressive therapy was stopped and not restarted until the end of the final cycle of chemotherapy. Another study reported that CHOP resulted in an overall response rate of 70% among patients with refractory or relapsed disease after treatment with rituximab.[146]

For patients with localised disease and those with CNS involvement, field radiation therapy may be beneficial.[147–149]

Interferon-α

The largest series of interferon-α use described 16 solid-organ transplant recipients with systemic PTLD treated with interferon-α (3 million units per day).[123] Among the 14 patients who received at least 3 weeks of therapy, eight had total regression of their disease; therapy was continued for 6–9 months in the responders. None of these patients had a relapse with the same neoplastic clone but two developed a new neoplastic clone. Seven patients who failed interferon-α were treated with systemic chemotherapy. Only one patient died of uncontrolled PTLD and only three patients developed rejection, perhaps because other immunosuppression was not discontinued.

Other treatment strategies

Intravenous immunoglobulin (IVIG) has been successfully employed in a number of settings, usually in combination with other modalities.[127,150]

Immunotherapy with lymphokine-activated auto logous killer cells (adoptive immunity) was evaluated in one study of seven patients with PTLD (four patients were EBV positive and three were EBV negative).[151] Four of the seven tumours could be adequately assessed for clonality and were found to be monoclonal. All four EBV-positive patients had complete regression of tumour after one dose of interleukin-2-treated cells. Treatment precipitated allograft rejection in two of four patients, which was treated effectively with corticosteroids. Although tumour regression was not observed in the EBV-negative patients, the circumstances surrounding the clinical care of these three patients precluded any definitive conclusion regarding therapeutic efficacy.

Infusions of donor leucocytes or EBV-specific cytotoxic T-cell lines have been used in severe cases of polyclonal PTLD and in some cases of monoclonal B-cell lymphoma.[152–156] The rationale for donor leucocyte infusion is that the infection is usually acquired from the donor who presumably has cytotoxic T cells sensitised to EBV.

Extracorporeal photochemotherapy involves the administration of photosensitising agents, such as methoxalens, which are selectively concentrated in malignant lymphocytes. Patients then undergo photopheresis with extracorporeal ultraviolet irradiation, resulting in preferential destruction of malignant cells. The technique has been used in combination with reduction of the intensity of immunosuppression in lung transplant recipients and has resulted in significant remissions of PTLD.[157] However, no randomised trials have unequivocally demonstrated the efficacy of this technique.

Augmented antiviral strategies have been developed for the treatment of EBV-associated lymphomas/PTLD using pharmacological induction of the latent viral thymidine kinase gene and enzyme in the tumour cells, followed by treatment with ganciclovir.[158] Arginine butyrate selectively activates the EBV thymidine kinase gene in latently EBV-infected human lymphoid cells and tumour cells. In six patients with EBV-associated lymphomas or PTLD, all of which were resistant to conventional radiation and/or chemotherapy, the combination of arginine butyrate and ganciclovir combination produced complete clinical responses in four of six patients, with a partial response occurring in a fifth patient. Pathological examination in two of three patients demonstrated complete necrosis of the EBV lymphoma with no residual disease, following a single 3-week course of the combination therapy.

Rapamycin, an immunosuppressant with antiproliferative properties, can prevent proliferation of B-cell (but not T-cell) PTLD-derived tumour cell lines in vitro and in vivo and may be able to prevent or treat B-cell-derived PTLD.[159] However, PTLD has been reported in clinical trials of patients treated with rapamycin.

Prognosis

Published series suggest overall survival rates between 25% and 35%.[160] Mortality with monoclonal malignancies has been reported to be as high as 80%. T-cell lymphomas have an extremely poor prognosis.

In a multivariate analysis of 61 patients, a performance status graded as 2 and more than one site of involvement were associated with significantly worse outcomes.[161] At a median follow-up of 22 months, median survival for patients with one or two of these factors was 34 and 1 month(s) respectively.

In a study from the Israel Penn International Transplant Tumor Registry,[162] clinical features associated with survival were analysed among 402 patients with PTLD registered in this database from 1968 to 2000.[162] Increased mortality rates were associated with a diagnosis within 6 months versus after 6 months from surgery (64% vs. 54%), increasing age, multiple versus single sites (73% vs. 53%), absence of surgery (100% vs. 55%), and allograft plus other organ involvement versus allograft involvement alone (64% vs. 31%).

Re-transplantation

A paucity of data exists concerning solid-organ re-transplantation in patients with a history of PTLD.[163,164]

In the largest cohort study based upon the Organ Procurement Transplantation Network (OPTN)/United Network for Organ Sharing (UNOS) database, outcomes were reported among 69 transplant recipients who survived PTLD and underwent retransplantation.[164] The re-transplant surgeries included 27 kidney, 22 liver, nine lung, six heart, four intestine and one pancreas. The average time from PTLD to re-transplant, time from transplant to re-transplant, and time for patient survival after re-transplant were 945, 2081 and 784 days respectively. At follow-up, overall patient and allograft survival was 86% and 74% respectively.

One retrospective study reported the outcomes of six patients with kidney re-transplantation after cure of EBV-related monoclonal B-cell lymphoma.[163] Five of six patients had the PTLD confined to the graft, with all six undergoing allograft nephrectomy in addition to their other treatment. The subsequent time interval from PTLD to re-transplant was longer, ranging from 50 to 128 months. At 24–47 months, all patients had functioning grafts without recurrent PTLD.

Polyomavirus

Polyomavirus infection in kidney transplant recipients has emerged as one of the important causes of graft dysfunction. Although two human polyomaviruses, BK virus (BKV) and JC virus (JCV), were reported in 1971,[165,166] their influence and importance were limited. The emergence of polyomavirus nephropathies coincided with the use of new potent immunosuppressive medications.[167,168] It is usually associated with BKV, affects up to 8% of recipients and frequently results in allograft loss or permanent dysfunction.[169] It presents as an asymptomatic gradual rise in creatinine with a tubulointerstitial nephritis that mimics rejection, producing a treatment dilemma. The decrease in immunosuppression that is needed to treat infection is opposite to the increases that are needed to treat rejection.

Epidemiology

On the basis of serology, BKV is acquired during childhood, and seroprevalence stabilises or wanes with increasing age.[170,171] In contrast, JCV seroprevalence increases with age. The route of the primary infection may be faecal–oral, respiratory, transplacental or from donor tissue.[172–175] Presumably, during a viraemic phase, the virus infects target tissues, including the uroepithelium, lymphoid tissue and brain,[175,176] establishing a latent or permissively lytic infection. SV40, a simian virus, was introduced into the human population through contaminated polio and adenovirus vaccines.[177] It can be acquired through close contact with non-human primates and may spread at a low rate from person to person.[172,178,179] Although SV40 has been identified in kidney transplant biopsies and associated with native kidney diseases,[180–183] its importance in kidney transplantation is poorly defined.

Molecular biology of BK virus

Polyomaviruses, especially SV40, have been intensively studied to examine fundamental processes in eukaryotic molecular biology. Because of BKV's close resemblance to SV40, much of the current knowledge of the molecular biology of BKV derives from extrapolations from studies with SV40.

Structure of the BK virion

Biophysical characteristics

The polyomaviruses are small (30–45 nm), non-enveloped viruses with an icosahedral capsid and a core of circular double-stranded DNA in association with histones.[184] Virions consist of 88% proteins and 12% DNA. The capsid is composed of 72 capsomers, each consisting of three different structural proteins, VP1, VP2 and VP3.

The viral genome and proteins

The genome consists of a single copy of a circular double-stranded DNA of approximately 5 kilobases. The BKV genome shares 75% overall homology with JCV and a 70% overall homology with the SV40 genome. The genome is transcribed bidirectionally. The genome can be divided into three functional subregions: (a) the **early region** encodes the regulatory proteins large tumour antigen (TAg) and small tumour antigen (tAg); (b) the **late region** encompasses the genetic information for the capsid proteins and the agnoprotein; (c) the **NCCR (non-coding control region)** spans the origin of replication and the sequences involved in the transcriptional regulation of both the early and the late genes. The agnogene and its protein product help regulate the virus replication and disrupt host cell processes.[185–187]

Biology and pathogenesis

Primary infection usually occurs at an early age with the highest incidence between 1 and 6 years, and appears to be asymptomatic in most individuals. Mild respiratory tract disease, pyrexia and transient cystitis can occur. The route of infection has not been firmly defined.

After primary infection, BKV seems to be spread haematogenously. Lymphocytes possess BKV-specific receptors on their cell surface. BKV usually establishes a harmless latent infection, and viral proteins or nucleic acid sequences have been demonstrated in the kidney, bladder, tonsils, brain, liver, lymphocytes and bone tissue of normal individuals.

Reactivation of BKV in the urinary tract occurs under a wide variety of conditions, including pregnancy, kidney and bone marrow transplantations, heart transplantation, immunodeficiency diseases, systemic lupus erythematosus and immunosuppressive therapy. Non-haemorrhagic or haemorrhagic cystitis, ureteric stenosis, interstitial nephritis, encephalitis, transient hepatic dysfunction, nephrotic syndrome and retinitis may occur in association with BKV reactivation. BKV viraemia and BKV nephropathy, however, are rare outside of kidney transplantation. BKV viraemia occurs in 13% and BKV nephropathy in 8% of kidney transplant recipients.[169]

Pathogenesis

Polyomaviruses have a seroprevalence of 60–80%. Replication of BKV occurs during states of immune suppression. VP1 binds the sialic acid residues of its receptor onto permissive cells.[188] The gangliosides GD1b and GT1b and 2,3-linked sialic acids on N-linked glycoproteins can act as the receptor for BKV,[189,190] whereas 2,6-linked sialic acids and the serotonin receptor 5-HT_{2A} can act as the receptor for JCV.[191,192] After attachment, BKV is internalised via caveolae-mediated endocytosis, whereas JCV enters through a clathrin-dependent endocytosis.[193,194] Once inside the cell, the viruses traffic to the nucleus and establish a latent or lytic infection. Although JCV resides in the uroepithelium and commonly reactivates, it rarely causes nephropathy.[195–201]

In kidney transplant recipients, BKV reactivation early after transplant appears to come from the donor. Recipients who had BKV infection and received a kidney from the same donor have been shown to have identical BKV genotypes, supporting donor transmission.[174,202] Recipients whose donors had higher BKV antibody titres were more likely to develop BKV infection than those with lower titres, also supporting donor transmission.[174,203] Injury is also believed to contribute to reactivation. In a mouse polyomavirus model, mechanical or chemical injury allowed for initiation of acute infection and also reactivation of latent polyomavirus.[204] In humans, injury could come from ischaemia, stent placement or rejection.

Once the virus has reactivated, an ascending infection via cell-to-cell spread occurs.[205–207] Without appropriate immunological control, a progressive lytic infection ensues.[208] This results in large nuclear and perinuclear virus-containing inclusions in the tubule cells. Lysis of these infected cells results in viral seepage into the tubule lumen and urine, but also to the interstitium and propagation to surrounding cells. Subsequent tubular cell necrosis leads to cast formation and denudation of the basement membrane. Destruction of tubular capillary walls results in vascular spread of the virus. A heterogeneous interstitial infiltration of inflammatory cells as well as tubulitis may be absent, intermixed with the active infection, or noted in areas that lack cytopathic changes. Collateral damage with necrosis and apoptosis of non-infected tubule cells may occur. The resultant effect of continued intragraft inflammation, tubular injury and upregulation of profibrotic mediators is allograft dysfunction and loss.[209]

Early retrospective studies implicated tacrolimus and MMF as risk factors for BKV nephropathy.[168,208,210–212] More recent retrospective studies found BKV nephropathy associated with the combination of tacrolimus levels (8 ng/ml) and MMF dosages (1.5–2 g/day).[213,214] BKV nephropathy, however, has been reported with triple drug regimens that include a calcineurin inhibitor (tacrolimus or ciclosporin), an adjuvant agent (MMF, azathioprine or sirolimus) and prednisone,[209,210,214–217] calcineurin-free triple drug therapy,[218] double therapy with a calcineurin inhibitor and sirolimus,[209,219] tacrolimus monotherapy,[220] and with or without use of an induction agent.[217]

A prospective, randomised study showed that BK viruria and viraemia were not different among those who received tacrolimus compared with ciclosporin, azathioprine compared with MMF, and rabbit ATG induction compared with no induction.[221] The highest rates of viruria and viraemia were among those who received the combination of ciclosporin and azathioprine or tacrolimus and MMF. Taken together, these studies suggest that it is the net state of immunosuppression and not a specific drug that allows for development of progressive BKV infection **(Fig. 12.3)**.

Immunobiology

Although polyomavirus reactivation is common, clinically significant disease is unusual. This is because most recipients can control the viruses. Persistent viral infections, such as polyomaviruses, cannot be completely cleared and require continuous immune control.[222,223] BKV replication typically begins early after transplantation and after treatment of rejection when immunosuppression is greater and immune control is reduced. The contribution of the **humoral**, **cellular** and **innate immune** compartments to the control is not well known.

Although 60–80% of recipients are BKV seropositive before transplantation,[62–64,203,224] the presence of these BKV-specific antibodies has not been shown to prevent development of BKV infection. However, BKV-specific antibodies can inhibit BKV infectivity.[194,225] BKV seronegativity is also a risk factor for BKV viruria[224] and nephropathy[226] in children. In adults, Shah reported that seropositive donors and seronegative recipients (BKV D^+/R^-) developed a serologically defined BKV infection most frequently (43%).[227] However, it has been found that seropositive donor and recipients (BKV D^+/R^+) developed BKV viruria most frequently (50%).[174] In both studies, only 10% of seronegative donors and recipients developed BKV infection. Thus, BKV antibodies may play a role in the immune response, but they also may indicate a risk for reactivation. Reduction in immunosuppression results in a significant increase in BKV-specific IgG antibody titres,[228,229] emergence of BKV-specific cellular immunity,[228] clearance of viraemia and stabilisation of graft function.[221]

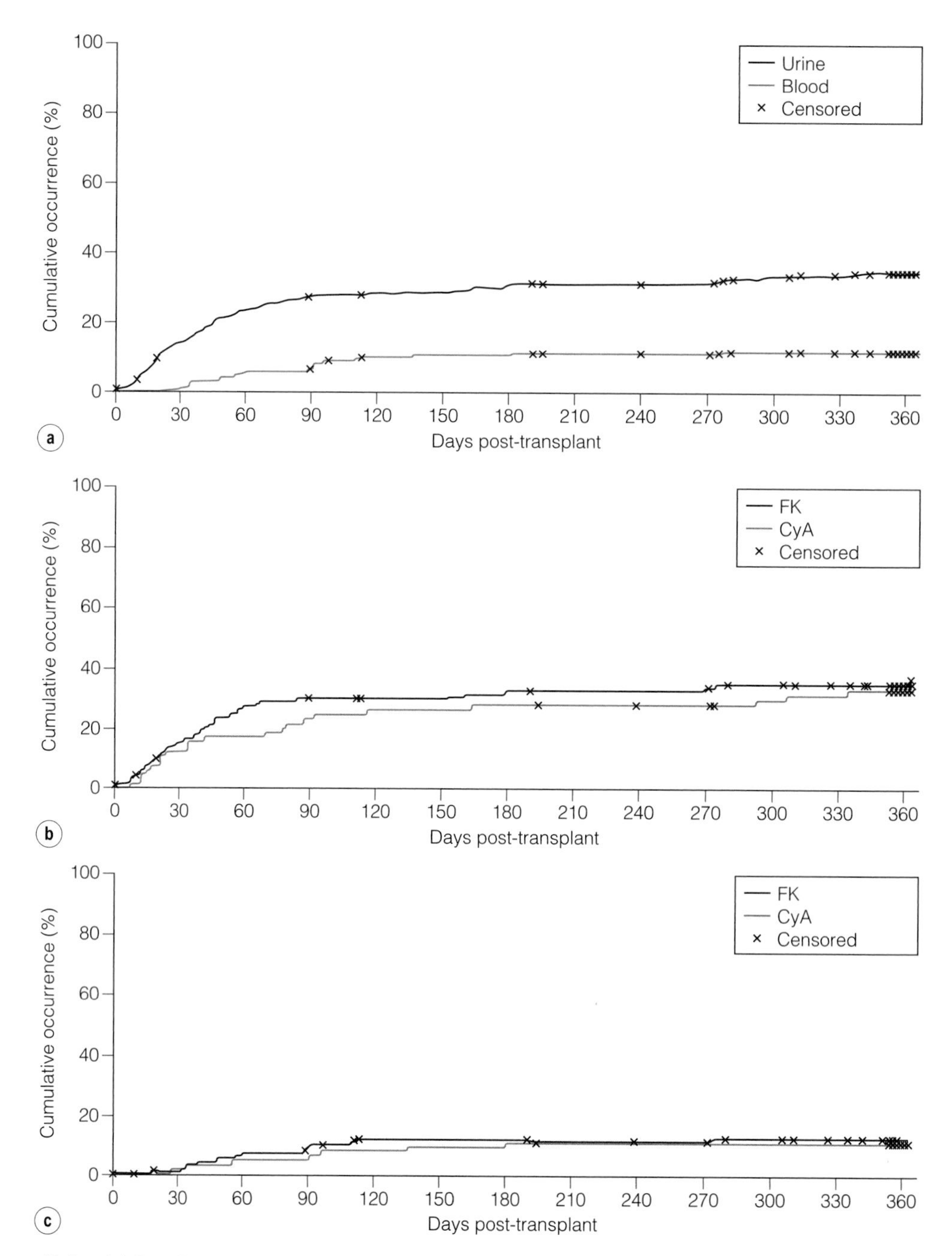

Figure 12.3 • (a) Overall cumulative incidence of BK viruria (yellow line) and viraemia (red line). **(b)** Cumulative incidence of BK viruria. **(c)** Cumulative incidence of BK viraemia. CyA, ciclosporin A; FK, FK506 (tacrolimus). Reproduced from Brennan DC, Agha I, Bohl DL et al. Incidence of BK with tacrolimus versus cyclosporine and impact of preemptive immunosuppression reduction. Am J Transplant 2005; 5(3):582–94. With permission from Blackwell Publishing.

Despite persistently elevated BKV antibody titres, recurrent BKV viraemia has been associated with a low frequency of interferon-producing cells,[228] high BKV antibody titres but weak cytotoxic T-lymphocyte responses.[230] However, in recipients with a strong cytotoxic T-lymphocyte response but low antibody titres, viraemia cleared and serum creatinine returned to the pre-BKV nephropathy baseline. The cellular immune response may also contribute to allograft dysfunction, with the CD8 functional response being more profibrotic than CD4 response.[209,231] The specificity of the cellular response may also be detrimental. Recipients with greater donor and recipient HLA mismatching had an increased incidence of BKV nephropathy,[232] possibly mediated by more episodes of rejection, intense immunosuppression and impaired cytotoxicity in an allogeneic environment, but less allograft loss.[233–236]

Diagnosis

Viral replication begins early after transplantation and progresses through detectable stages: viruria, then viraemia, then nephropathy.[169,221,237–239] Viruria can be detected by PCR for BKV DNA, reverse transcription–PCR for BKV RNA, cytology for BKV inclusion-bearing epithelial cells termed 'decoy cells', or electron microscopy for viral particles.[151,197,221,240] These tests are sensitive for detecting active BKV infections but lack specificity for nephropathy because the detected virus could originate anywhere along the urinary tract.

Detection of BKV DNA in the plasma or viraemia may be a better indicator of nephropathy. As the infection intensifies, the markers of viral replication increase. Threshold values have been suggested to predict BKV nephropathy, but considerable overlap of these values exists among recipients without BKV nephropathy, active BKV nephropathy and resolved BKV nephropathy.[241]

A transplant kidney biopsy remains the gold standard for diagnosing BKV nephropathy. Importantly, the interstitial nephritis and tubular cytopathic changes of BKV nephropathy can be focal or isolated to the medulla and missed on one-third of biopsies if only a single core is evaluated.[242] Therefore, at least two cores including medulla should be examined. If there are no cytopathic changes on routine histology but there is a high clinical suspicion, then adjunctive tests such as immunohistochemistry directed specifically against BKV or cross-reacting SV40 large T antigen should be performed because the histopathology of BKV infections may be misinterpreted.[210]

Histology

The characteristic findings on light microscopy are intranuclear basophilic and gelatinous-appearing viral inclusions in epithelial cells of the urothelium. These are found in the medulla or cortex and are multifocal with random distribution. Three histological patterns (A, B and C) have been described, as shown in Table 12.2.[242–244]

Table 12.2 • Histological pattern of BK nephropathy

Clinical disease	Pattern	Histology
Early disease	A	Cytopathic changes are present; little to no inflammation or tubular atrophy
Moderate disease	B	Viral cytopathic changes with varying degrees of inflammation, tubular atrophy and fibrosis
Late disease	C	Cytopathic changes often less apparent as a result of a background of tubular atrophy, interstitial fibrosis and chronic inflammatory infiltrate

The degree of damage corresponds to the degree of allograft dysfunction and allograft outcome.[242] The distinction of BKV nephropathy from acute tubular necrosis, interstitial nephritis and acute cellular rejection is difficult and aided by assessment of blood or urine PCR. Absence of definitive features of acute cellular rejection such as endotheliitis and absence of C4d deposits in peritubular capillaries are helpful. Other histopathological changes include glomerular crescents (10–20%),[211,245] ischaemic glomerulopathy (62%),[245] transplant glomerulopathy (62%),[245] abundant plasma cell infiltrates (up to 75%)[120,242,246] and tubular microcalcifications (25%).[243] Features of calcineurin inhibitor toxicity such as striped fibrosis thrombotic microangiopathy and tubular isometric vacuolisation may also be present.[210,211,242,243]

Treatment

The principle treatment for BKV nephropathy is reduction in immunosuppression. Various strategies include reduction or discontinuation of the calcineurin inhibitor and/or adjuvant agent, changing from MMF to azathioprine, sirolimus or leflunomide, or from tacrolimus to ciclosporin.[202,214,216,217,219,247–249] Importantly, BKV nephropathy seems to develop less frequently with maintenance protocols that involve steroid withdrawal.[237,250] When BKV nephropathy is diagnosed early within the first 6 months after transplantation and the creatinine is stable, survival is improved compared with when the diagnosis is made later and the creatinine is elevated.

Early or presumptive BKV nephropathy

Pre-emptive withdrawal of the antimetabolite upon detection of viraemia prevents BKV nephropathy without significantly increasing the risk for rejection.[221] In a study of 200 adult patients, viraemia cleared in 22 (96%) of 23 recipients with only one episode of acute rejection directly related to immunosuppression reduction. Of the 22 recipients whose viraemia resolved, 32% cleared before protocol decreases in immunosuppression, 32% cleared after withdrawal of the antimetabolite, 9% cleared after reduction in the calcineurin inhibitor and 27% required withdrawal of the antimetabolite followed by reduction in calcineurin inhibitor for persistent viraemia.

With prospective screening of paediatric recipients, viraemia cleared in 58% of recipients with presumptive nephropathy after a 50% reduction in the dosage of mycophenolate or sirolimus and targeting tacrolimus troughs of 3–5 g/dL.[251] With BKV nephropathy diagnosed on surveillance biopsy before an elevation in creatinine, the creatinine remained stable and the number of BKV-positive tubules on follow-up biopsy significantly decreased after a stepwise reduction in MMF plus reduction in tacrolimus or conversion to ciclosporin.[215] Conversion from tacrolimus to ciclosporin may lower MMF levels if dosages of MMF remain the same.[252]

Ciclosporin in vitro but not tacrolimus in vitro has been shown to inhibit BKV reactivation. Although complete cessation of MMF may be necessary if viraemia persists, MMF may limit proinflammatory and profibrotic cytokines.[253,254]

Late BKV nephropathy

The diagnosis of BKV nephropathy in the setting of allograft dysfunction often indicates more severe histological changes, and renal function may stabilise or continue to progress despite treatment.[255,256] Not treating or inadvertently treating with an antilymphocyte antibody often may lead to progression of disease.[257] With late disease, whether to reduce or discontinue one or more components of the maintenance regimen is not clear.[239,248,255] With a for-cause diagnosis, no difference was found in graft survival whether immunosuppression was reduced or continued or between reduction and discontinuation of tacrolimus or MMF among 67 recipients with BKV nephropathy.[217]

Another study reported no allograft loss and clearance of viruria in seven patients after discontinuation of MMF and reduction in calcineurin inhibitor dosage despite a mean creatinine of 3.2 mg/dL at diagnosis compared with persistent viruria and allograft loss in two recipients after only reduction in MMF and modification of calcineurin inhibitor.[214] Dose reduction of the adjuvant agent and calcineurin inhibitor in paediatric patients failed to clear viraemia with 50% graft loss.[251] Dose reduction of the adjuvant agent and calcineurin inhibitor in adults failed to improve tubular BKV burden with a rising mean creatinine.[215]

Adjuvant therapies

On the basis of in vitro activity against BKV, cidofovir, quinolones, leflunomide and intravenous Ig (IVIG) have been reported as treatment options for BKV nephropathy.

Cidofovir, a cytosine analogue and viral DNA polymerase inhibitor, inhibits BKV replication; the mechanism is unclear because BKV lacks a viral polymerase gene.[258–260] Rather than a direct effect on BKV replication, cidofovir may restore the function of p53 and pRB, targets of the large T antigen, and permit BKV-infected and non-infected cells to undergo apoptosis.[184,260] When used for treatment of BKV nephropathy, cidofovir has been given at dosages ranging from 0.25 to 1 mg/kg every 1–3 weeks with generally favourable results.[219,256,261–270] However, most studies were observational, and cidofovir was used in conjunction with immunosuppression reduction.[264] Cidofovir should be used with caution, frequent monitoring, and with informed consent because of the potential complications.[271,272]

Quinolones are DNA gyrase inhibitors and may interfere with the large T-antigen helicase activity.[273] They have in vitro and in vivo activity against BKV.[274–276] Two months after a 10-day course of gatifloxacin, seven of 10 recipients with active BKV replication had reduction in viraemia or urinary decoy cells. However, another study did not find improvement in viral clearance after a 10-day course of ciprofloxacin.[277]

IVIG has been used for treatment for BKV nephropathy because of its immunomodulatory activity as well as potential anti-BKV properties.[278] Although IVIG contains BKV-specific antibodies, seropositive recipients as well as recipients with active BKV infections may have high BKV-specific antibody titres, suggesting that antibody-mediated neutralisation does not contribute to viral control. Nevertheless, in combination with immunosuppression reduction, IVIG (2–3.5 g/kg over 2–7 days) treatment was used as initial treatment for BKV nephropathy and BKV nephropathy with concurrent acute rejection.[256,279] Because of the cost, potential adverse effects and unproven efficacy, IVIG use for BKV nephropathy should be limited until controlled studies suggest benefit.

Leflunomide use, along with discontinuation of MMF and dosage reduction of tacrolimus, resulted in 15% allograft loss in recipients who had BKV nephropathy, which was less than historical controls. However, requirement of high dosages (>40 mg/day), the unpredictable relationship between the drug dosage and drug level, absence of therapeutic drug monitoring and weak immunosuppressive potency have limited enthusiasm for its use.

Postinfection monitoring

Failure to clear BKV leads to worse graft function and outcomes. Because histological clearance of the virus[241] and disappearance of decoy cells[249] precede clearance from the blood, monitoring should be performed with quantitative assays, preferably BKV PCR, until the viral level is undetectable or at least falls below the threshold value that is associated with BKV nephropathy. On the basis of kinetic models[280] and prospective monitoring,[221,241,249,280] viraemia clears in 7–20 weeks, but the initial decrease may be delayed by 4–10 weeks after immunosuppression reduction. If viraemia persists, then further reduction of current maintenance therapy, conversion to sirolimus, or addition of leflunomide can be considered.[244]

Re-transplantation

BKV nephropathy shortens allograft survival. Damage from direct and indirect effects of the virus and rejection after immunosuppression reduction lead to early graft loss or chronic dysfunction.[281] In recipients who have advanced kidney disease or who have returned to dialysis from BKV nephropathy, re-transplantation has been successful.[282–286] In most cases, transplant nephrectomy and/or studies to confirm no active viral replication have been performed. However, in the setting of active viraemia, viral levels become undetectable within 14 days after pre-emptive re-transplantation with simultaneous allograft nephrectomy despite antibody induction. BKV viruria, viraemia, nephropathy and graft loss can recur.[285–288] Therefore, BKV nephropathy is not a contraindication for re-transplantation but has recurred in two (12%) of 17 reported recipients.

Evidence of BKV-specific immunity can be evaluated before re-transplantation but in the setting of a resolved infection can be inferred.[282] However, for pre-emptive transplantation in recipients of combined organ transplants, for whom reduction in immunosuppression is limited, evidence of no active viral replication or development of BKV-specific immunity should be determined.

BKV nephropathy and acute rejection

The treatment of recipients whose biopsy shows rejection with concurrent BKV nephropathy remains problematic. More than half of biopsies with BK may show tubulitis.[169,207,212,242] Reduction in immunosuppression can precipitate rejection in 10–30% of recipients.[289] The infiltrating mononuclear cells may represent a BKV-specific and/or allospecific response, and treatment is debatable.[290,291] Studies that compared BKV nephropathy with acute rejection have identified differences in the proportion and type of infiltrating cell,[242,246,247] protein expression[207,242] and proteomic profiles,[292] and gene expression profiles.[209] However, these differences have not been characterised serially after modification in immunosuppression. Clinically, reports have described improved, stable and worse graft function after steroid pulses.[169,210,215,255] Comparing recipients who initially received increased immunosuppression with those whose immunosuppression was decreased, no significant short- or long-term improvement in tubulitis or creatinine was found with brief steroid therapy.[255] On biopsies that were performed within the first 8 weeks, the histological viral load had improved significantly with initially decreased compared with increased immunosuppression, but was similar on later biopsies. The presence of atypical features such as strong peritubular capillary C4d staining, vasculitis, glomerulitis or interstitial haemorrhage would support rejection and require an individualised approach.[246,293] The delayed improvement in creatinine after reduction in immunosuppression likely reflects slow resolution of the cellular infiltrate. Once BKV nephropathy and viraemia have cleared, the benefit of augmenting immunosuppression to prevent chronic rejection or late acute rejection remains unknown.

Screening

It has been recommended that screening for BKV should be performed monthly for the first 6 months, at months 9 and 12, when allograft dysfunction occurs and when a transplant kidney biopsy is performed.[281]

Screening should be based on a urinary assay for decoy cells, BKV DNA or BKV RNA. A positive screening test should be confirmed within 4 weeks along with a quantitative assay. Recipients with persistently high viral levels for >3 weeks should undergo biopsy and intervention. Monitoring should continue every 2–4 weeks until the viral level falls below threshold values and preferably to undetectable levels. These recommendations are guidelines and should vary on the basis of assay availability and cost, recipient risk, and local incidence of BKV nephropathy. With a low incidence of BKV nephropathy, high false-positive testing rates, or high occurrence of acute rejection or chronic allograft dysfunction after reduction in immunosuppression, screening could produce greater cost and harm than not screening.[294] Screening protocols should be used in centres with higher incidences of BKV nephropathy, using triple drug therapy including tacrolimus and MMF, and with clinical trials to evaluate new therapeutic agents. At one institution, plasma screening was done monthly for the first 6 months and at months 9 and 12 after transplantation, at the time of a transplant kidney biopsy, and after augmentation in immunosuppression. BKV viraemia with stable allograft function triggered empiric immunosuppression reduction and continued monitoring.[221] This strategy has resulted in only one case of BK nephropathy in 5 years.[281]

Key points

- CMV remains an important pathogen after renal transplantation, as even asymptomatic CMV reactivation is associated with a marked increase in the likelihood of death.
- Increasing evidence is showing that prophylaxis is beneficial not only when the donor is seropositive and the recipient is seronegative, but whenever the donor or the recipient is seropositive.
- Valganciclovir has become the drug of choice for prophylaxis of CMV and has also been shown to be effective for treatment of CMV disease.
- EBV seronegativity increases the likelihood of developing PTLD, and more than half of all PTLDs are EBV positive.
- PTLD may progress even when EBV viral loads decrease, and the utility of monitoring EBV viral loads is unclear.
- In addition to immunosuppression reduction, treatment with the monoclonal antibody rituximab, which targets CD20, with or without chemotherapy, has emerged as the principal treatment regimen for $CD20^+$ PTLD.
- The polyomavirus BK has emerged as an important viral infection causing BK nephropathy in up to 10% of renal transplant recipients.
- BK nephropathy in solid-organ transplant recipients other than kidney recipients is extremely rare and supports clinically significant BK infection being usually donor derived.
- No specific immunosuppressive regimen or agent is clearly associated with BK infection. Rather, a net state of over-immunosuppression is associated with BK reactivation.
- There is no known antiviral agent effective for BK, and monitoring and pre-emptive reduction in immunosuppression are the most effective ways to prevent progression from BK viraemia to BK nephropathy.

References

1. Sagedal S, Hartmann A, Nordal KP et al. Impact of early cytomegalovirus infection and disease on long-term recipient and kidney graft survival. Kidney Int 2004; 66(1):329–37.
2. Brennan DC. Cytomegalovirus in renal transplantation. J Am Soc Nephrol 2001; 12(4):848–55.
3. Preiksaitis JK, Brennan DC, Fishman J et al. Canadian Society of Transplantation consensus workshop on cytomegalovirus management in solid organ transplantation final report. Am J Transplant 2005; 5(2):218–27.
4. Roberts TC, Brennan DC, Buller RS et al. Quantitative polymerase chain reaction to predict occurrence of symptomatic cytomegalovirus infection and assess response to ganciclovir therapy in renal transplant recipients. J Infect Dis 1998; 178(3):626–35.
5. Emery VC, Sabin CA, Cope AV et al. Application of viral-load kinetics to identify patients who develop cytomegalovirus disease after transplantation. Lancet 2000; 355(9220):2032–6.
6. Storch GA (ed.) Viral infections in immunocompromised patients. In: Diagnostic virology, 1st edn. New York: Harcourt Brace, 2000.
7. Snydman DR. Ganciclovir therapy for cytomegalovirus disease associated with renal transplants. Rev Infect Dis 1988; 10(Suppl 3):S554–62.
8. Waid TH, McKeown JW. Cytomegalovirus hyperimmune globulin for CMV disease refractory to ganciclovir in renal transplantation. Transplant Proc 1995; 27(5, Suppl 1):46.
9. Asberg A, Humar A, Rollag H et al. Oral valganciclovir is noninferior to intravenous ganciclovir for the treatment of cytomegalovirus disease in solid organ transplant recipients. Am J Transplant 2007; 7(9):2106–13.
10. Birk PE, Chavers BM. Does cytomegalovirus cause glomerular injury in renal allograft recipients? J Am Soc Nephrol 1997; 8(11):1801–8.
11. Pouteil-Noble C, Ecochard R, Landrivon G et al. Cytomegalovirus infection – an etiological factor for rejection? A prospective study in 242 renal transplant patients. Transplantation 1993; 55(4):851–7.
12. Pouria S, State OI, Wong W et al. CMV infection is associated with transplant renal artery stenosis. Q J Med 1998; 91(3):185–9.
13. Miller BW, Hmiel SP, Schnitzler MA et al. Cyclosporine as cause of thrombotic microangiopathy after renal transplantation. Am J Kidney Dis 1997; 29(5):813–14.
14. Jeejeebhoy FM, Zaltzman JS. Thrombotic microangiopathy in association with cytomegalovirus infection in a renal transplant patient: a new treatment strategy. Transplantation 1998; 65(12):1645–8.
15. Hochstetler LA, Flanigan MJ, Lager DJ. Transplant-associated thrombotic microangiopathy: the role of IgG administration as initial therapy. Am J Kidney Dis 1994; 23(3):444–50.
16. Strippoli GF, Hodson EM, Jones C et al. Preemptive treatment for cytomegalovirus viremia to prevent cytomegalovirus disease in solid organ transplant recipients. Transplantation 2006; 81(2):139–45.
17. Kalil AC, Levitsky J, Lyden E et al. Meta-analysis: the efficacy of strategies to prevent organ disease by cytomegalovirus in solid organ transplant recipients. Ann Intern Med 2005; 143(12):870–80.
18. Hodson EM, Jones CA, Webster AC et al. Antiviral medications to prevent cytomegalovirus disease and early death in recipients of solid-organ transplants: a systematic review of randomised controlled trials. Lancet 2005; 365(9477):2105–15.
19. Brennan DC, Garlock KA, Singer GG et al. Prophylactic oral ganciclovir compared with deferred therapy for control of cytomegalovirus in renal transplant recipients. Transplantation 1997; 64(12):1843–6.
20. Flechner SM, Avery RK, Fisher R et al. A randomized prospective controlled trial of oral acyclovir versus oral ganciclovir for cytomegalovirus prophylaxis in high-risk kidney transplant recipients. Transplantation 1998; 66(12):1682–8.
21. Gane E, Saliba F, Valdecasas GJ et al. Randomised trial of efficacy and safety of oral ganciclovir in the prevention of cytomegalovirus disease in liver-transplant recipients. The Oral Ganciclovir International Transplantation Study Group [corrected]. Lancet 1997; 350(9093):1729–33.
22. Pescovitz MD, Rabkin J, Merion RM et al. Valganciclovir results in improved oral absorption of ganciclovir in liver transplant recipients. Antimicrob Agents Chemother 2000; 44(10):2811–15.
23. Paya C, Humar A, Dominguez E et al. Efficacy and safety of valganciclovir vs. oral ganciclovir for prevention of cytomegalovirus disease in solid organ transplant recipients. Am J Transplant 2004; 4(4):611–20.
24. Lowance D, Neumayer HH, Legendre CM et al. Valacyclovir for the prevention of cytomegalovirus disease after renal transplantation. International Valacyclovir Cytomegalovirus Prophylaxis Transplantation Study Group. N Engl J Med 1999; 340(19):1462–70.
25. Snydman DR, Werner BG, Heinze-Lacey B et al. Use of cytomegalovirus immune globulin to prevent cytomegalovirus disease in renal-transplant recipients. N Engl J Med 1987; 317(17):1049–54.
26. Knight DA, Hejmanowski AQ, Dierksheide JE et al. Inhibition of herpes simplex virus type 1 by the experimental immunosuppressive agent leflunomide. Transplantation 2001; 71(1):170–4.
27. Waldman WJ, Knight DA, Blinder L et al. Inhibition of cytomegalovirus in vitro and in vivo

by the experimental immunosuppressive agent leflunomide. Intervirology 1999; 42(5–6):412–18.

28. Waldman WJ, Knight DA, Lurain NS et al. Novel mechanism of inhibition of cytomegalovirus by the experimental immunosuppressive agent leflunomide. Transplantation 1999; 68(6):814–25.
29. John GT, Manivannan J, Chandy S et al. Leflunomide therapy for cytomegalovirus disease in renal allograft recepients. Transplantation 2004; 77(9):1460–1.
30. Chong AS, Zeng H, Knight DA et al. Concurrent antiviral and immunosuppressive activities of leflunomide in vivo. Am J Transplant 2006; 6(1):69–75.
31. Balfour HH Jr, Chace BA, Stapleton JT et al. A randomized, placebo-controlled trial of oral acyclovir for the prevention of cytomegalovirus disease in recipients of renal allografts. N Engl J Med 1989; 320(21):1381–7.
32. Kletzmayr J, Kotzmann H, Popow-Kraupp T et al. Impact of high-dose oral acyclovir prophylaxis on cytomegalovirus (CMV) disease in CMV high-risk renal transplant recipients. J Am Soc Nephrol 1996; 7(2):325–30.
33. Lu H, Thomas S. Maribavir (ViroPharma). Curr Opin Invest Drugs 2004; 5(8):898–906.
34. Limaye AP, Corey L, Koelle DM et al. Emergence of ganciclovir-resistant cytomegalovirus disease among recipients of solid-organ transplants. Lancet 2000; 356(9230):645–9.
35. Schnitzler MA, Lowell JA, Hmiel SP et al. Cytomegalovirus disease after prophylaxis with oral ganciclovir in renal transplantation: the importance of HLA-DR matching. J Am Soc Nephrol 2003; 14(3):780–5.
36. Jassal SV, Roscoe JM, Zaltzman JS et al. Clinical practice guidelines: prevention of cytomegalovirus disease after renal transplantation. J Am Soc Nephrol 1998; 9(9):1697–708.
37. Khoury JA, Storch GA, Bohl DL et al. Prophylactic versus preemptive oral valganciclovir for the management of cytomegalovirus infection in adult renal transplant recipients. Am J Transplant 2006; 6(9):2134–43.
38. Schnitzler MA, Woodward RS, Brennan DC et al. The effects of cytomegalovirus serology on graft and recipient survival in cadaveric renal transplantation: implications for organ allocation. Am J Kidney Dis 1997; 29(3):428–34.
39. Schnitzler MA, Woodward RS, Brennan DC et al. Impact of cytomegalovirus serology on graft survival in living related kidney transplantation: implications for donor selection. Surgery 1997; 121(5):563–8.
40. Doyle AM, Warburton KM, Goral S et al. 24-week oral ganciclovir prophylaxis in kidney recipients is associated with reduced symptomatic cytomegalovirus disease compared to a 12-week course. Transplantation 2006; 81(8):1106–11.
41. Akalin E, Bromberg JS, Sehgal V et al. Decreased incidence of cytomegalovirus infection in thymoglobulin-treated transplant patients with 6 months of valganciclovir prophylaxis. Am J Transplant 2004; 4(1):148–9.
42. Jung C, Engelmann E, Borner K et al. Preemptive oral ganciclovir therapy versus prophylaxis to prevent symptomatic cytomegalovirus infection after kidney transplantation. Transplant Proc 2001; 33(7–8):3621–3.
43. Guinee D Jr, Jaffe E, Kingma D et al. Pulmonary lymphomatoid granulomatosis. Evidence for a proliferation of Epstein–Barr virus infected B-lymphocytes with a prominent T-cell component and vasculitis. Am J Surg Pathol 1994; 18(8):753–64.
44. Rooney CM, Gregory CD, Rowe M et al. Endemic Burkitt's lymphoma: phenotypic analysis of tumor biopsy cells and of derived tumor cell lines. J Natl Cancer Inst 1986; 77(3):681–7.
45. Gregory CD, Murray RJ, Edwards CF et al. Downregulation of cell adhesion molecules LFA-3 and ICAM-1 in Epstein–Barr virus-positive Burkitt's lymphoma underlies tumor cell escape from virus-specific T cell surveillance. J Exp Med 1988; 167(6):1811–24.
46. Billaud M, Rousset F, Calender A et al. Low expression of lymphocyte function-associated antigen (LFA)-1 and LFA-3 adhesion molecules is a common trait in Burkitt's lymphoma associated with and not associated with Epstein–Barr virus. Blood 1990; 75(9):1827–33.
47. Gregory CD, Rowe M, Rickinson AB. Different Epstein–Barr virus-B cell interactions in phenotypically distinct clones of a Burkitt's lymphoma cell line. J Gen Virol 1990; 71(Pt 7):1481–95.
48. Evans A, Niederman J. Epstein–Barr virus. In: Evans A (ed.) Viral infections of human epidemiology and control. New York: Plenum, 1989; p. 265.
49. Peter J, Ray CG. Infectious mononucleosis. Pediatr Rev 1998; 19(8):276–9.
50. Tselis A, Duman R, Storch GA et al. Epstein–Barr virus encephalomyelitis diagnosed by polymerase chain reaction: detection of the genome in the CSF. Neurology 1997; 48(5):1351–5.
51. Hudson LB, Perlman SE. Necrotizing genital ulcerations in a premenarcheal female with mononucleosis. Obstet Gynecol 1998; 92(4, Pt 2):642–4.
52. Triantos D, Porter SR, Scully C et al. Oral hairy leukoplakia: clinicopathologic features, pathogenesis, diagnosis, and clinical significance. Clin Infect Dis 1997; 25(6):1392–6.
53. Epstein JB, Sherlock CH, Greenspan JS. Hairy leukoplakia-like lesions following bone-marrow transplantation. AIDS 1991; 5(1):101–2.
54. Birnbaum W, Hodgson TA, Reichart PA et al. Prognostic significance of HIV-associated oral lesions and their relation to therapy. Oral Dis 2002; 8(Suppl 2):110–4.

55. Aldrete JS. Spontaneous rupture of the spleen in patients with infectious mononucleosis. Mayo Clin Proc 1992; 67(9):910–12.
56. Thorley-Lawson DA, Gross A. Persistence of the Epstein–Barr virus and the origins of associated lymphomas. N Engl J Med 2004; 350(13):1328–37.
57. Wilson WH, Kingma DW, Raffeld M et al. Association of lymphomatoid granulomatosis with Epstein–Barr viral infection of B lymphocytes and response to interferon-alpha 2b. Blood 1996; 87(11):4531–7.
58. Thorley-Lawson DA, Poodry CA. Identification and isolation of the main component (gp350-gp220) of Epstein–Barr virus responsible for generating neutralizing antibodies in vivo. J Virol 1982; 43(2):730–6.
59. North JR, Morgan AJ, Thompson JL et al. Purified Epstein–Barr virus Mr 340,000 glycoprotein induces potent virus-neutralizing antibodies when incorporated in liposomes. Proc Natl Acad Sci USA 1982; 79(23):7504–8.
60. Epstein MA, Morgan AJ, Finerty S et al. Protection of cottontop tamarins against Epstein–Barr virus-induced malignant lymphoma by a prototype subunit vaccine. Nature 1985; 318(6043):287–9.
61. Caillard S, Agodoa LY, Bohen EM et al. Myeloma, Hodgkin disease, and lymphoid leukemia after renal transplantation: characteristics, risk factors and prognosis. Transplantation 2006; 81(6):888–95.
62. Leblond V, Sutton L, Dorent R et al. Lymphoproliferative disorders after organ transplantation: a report of 24 cases observed in a single center. J Clin Oncol 1995; 13(4):961–8.
63. Hanson MN, Morrison VA, Peterson BA et al. Posttransplant T-cell lymphoproliferative disorders – an aggressive, late complication of solid-organ transplantation. Blood 1996; 88(9):3626–33.
64. Draoua HY, Tsao L, Mancini DM et al. T-cell post-transplantation lymphoproliferative disorders after cardiac transplantation: a single institutional experience. Br J Haematol 2004; 127(4):429–32.
65. Ravat FE, Spittle MF, Russell-Jones R. Primary cutaneous T-cell lymphoma occurring after organ transplantation. J Am Acad Dermatol 2006; 54(4):668–75.
66. Rajakariar R, Bhattacharyya M, Norton A et al. Post transplant T-cell lymphoma: a case series of four patients from a single unit and review of the literature. Am J Transplant 2004; 4(9):1534–8.
67. Leblond V, Davi F, Charlotte F et al. Posttransplant lymphoproliferative disorders not associated with Epstein–Barr virus: a distinct entity? J Clin Oncol 1998; 16(6):2052–9.
68. Weissmann DJ, Ferry JA, Harris NL et al. Posttransplantation lymphoproliferative disorders in solid organ recipients are predominantly aggressive tumors of host origin. Am J Clin Pathol 1995; 103(6):748–55.
69. Petit B, Le Meur Y, Jaccard A et al. Influence of host-recipient origin on clinical aspects of posttransplantation lymphoproliferative disorders in kidney transplantation. Transplantation 2002; 73(2):265–71.
70. Hjelle B, Evans-Holm M, Yen TS et al. A poorly differentiated lymphoma of donor origin in a renal allograft recipient. Transplantation 1989; 47(6):945–8.
71. Nalesnik MA, Jaffe R, Starzl TE et al. The pathology of posttransplant lymphoproliferative disorders occurring in the setting of cyclosporine A-prednisone immunosuppression. Am J Pathol 1988; 133(1):173–92.
72. Penn I, Porat G. Central nervous system lymphomas in organ allograft recipients. Transplantation 1995; 59(2):240–4.
73. Dotti G, Fiocchi R, Motta T et al. Epstein–Barr virus-negative lymphoproliferate disorders in long-term survivors after heart, kidney, and liver transplant. Transplantation 2000; 69(5):827–33.
74. Dotti G, Fiocchi R, Motta T et al. Lymphomas occurring late after solid-organ transplantation: influence of treatment on the clinical outcome. Transplantation 2002; 74(8):1095–102.
75. Caillard S, Lelong C, Pessione F et al. Post-transplant lymphoproliferative disorders occurring after renal transplantation in adults: report of 230 cases from the French Registry. Am J Transplant 2006; 6(11):2735–42.
76. Green M, Avery RK, Preiksaitis JK. Guidelines for the prevention and management of infectious complications of solid organ transplantation. Am J Transplant 2004; 4(Suppl 10):51.
77. Walker RC, Paya CV, Marshall WF et al. Pretransplantation seronegative Epstein–Barr virus status is the primary risk factor for posttransplantation lymphoproliferative disorder in adult heart, lung, and other solid organ transplantations. J Heart Lung Transplant 1995; 14(2):214–21.
78. Penn I. Posttransplantation de novo tumors in liver allograft recipients. Liver Transplant Surg 1996; 2(1):52–9.
79. Opelz G, Henderson R. Incidence of non-Hodgkin lymphoma in kidney and heart transplant recipients. Lancet 1993; 342(8886–8887):1514–16.
80. Cockfield SM, Preiksaitis JK, Jewell LD et al. Post-transplant lymphoproliferative disorder in renal allograft recipients. Clinical experience and risk factor analysis in a single center. Transplantation 1993; 56(1):88–96.
81. Faull RJ, Hollett P, McDonald SP. Lymphoproliferative disease after renal transplantation in Australia and New Zealand. Transplantation 2005; 80(2):193–7.
82. Armitage JM, Kormos RL, Stuart RS et al. Posttransplant lymphoproliferative disease in thoracic organ transplant patients: ten years of cyclosporine-

based immunosuppression. J Heart Lung Transplant 1991; 10(6):877–886; discussion 886–7.

83. Swinnen LJ, Costanzo-Nordin MR, Fisher SG et al. Increased incidence of lymphoproliferative disorder after immunosuppression with the monoclonal antibody OKT3 in cardiac-transplant recipients. N Engl J Med 1990; 323(25):1723–8.

84. Aris RM, Maia DM, Neuringer IP et al. Post-transplantation lymphoproliferative disorder in the Epstein–Barr virus-naive lung transplant recipient. Am J Respir Crit Care Med 1996; 154(6, Pt 1):1712–17.

85. Levine SM, Angel L, Anzueto A et al. A low incidence of posttransplant lymphoproliferative disorder in 109 lung transplant recipients. Chest 1999; 116(5):1273–7.

86. Cockfield SM. Identifying the patient at risk for post-transplant lymphoproliferative disorder. Transplant Infect Dis 2001; 3(2):70–8.

87. Smith JM, Rudser K, Gillen D et al. Risk of lymphoma after renal transplantation varies with time: an analysis of the United States Renal Data System. Transplantation 2006; 81(2):175–80.

88. Bustami RT, Ojo AO, Wolfe RA et al. Immunosuppression and the risk of post-transplant malignancy among cadaveric first kidney transplant recipients. Am J Transplant 2004; 4(1):87–93.

89. Opelz G, Dohler B. Lymphomas after solid organ transplantation: a collaborative transplant study report. Am J Transplant 2004; 4(2):222–30.

90. Opelz G, Naujokat C, Daniel V et al. Disassociation between risk of graft loss and risk of non-Hodgkin lymphoma with induction agents in renal transplant recipients. Transplantation 2006; 81(9):1227–33.

91. Rappaport DC, Chamberlain DW, Shepherd FA et al. Lymphoproliferative disorders after lung transplantation: imaging features. Radiology 1998; 206(2):519–24.

92. Funch DP, Ko HH, Travasso J et al. Posttransplant lymphoproliferative disorder among renal transplant patients in relation to the use of mycophenolate mofetil. Transplantation 2005; 80(9):1174–80.

93. Walker RC, Marshall WF, Strickler JG et al. Pretransplantation assessment of the risk of lymphoproliferative disorder. Clin Infect Dis 1995; 20(5):1346–53.

94. Shahinian VB, Muirhead N, Jevnikar AM et al. Epstein–Barr virus seronegativity is a risk factor for late-onset posttransplant lymphoproliferative disorder in adult renal allograft recipients. Transplantation 2003; 75(6):851–6.

95. Caillard S, Dharnidharka V, Agodoa L et al. Posttransplant lymphoproliferative disorders after renal transplantation in the United States in era of modern immunosuppression. Transplantation 2005; 80(9):1233–43.

96. Bakker NA, van Imhoff GW, Verschuuren EA et al. HLA antigens and post renal transplant lymphoproliferative disease: HLA-B matching is critical. Transplantation 2005; 80(5):595–9.

97. Zangwill SD, Hsu DT, Kichuk MR et al. Incidence and outcome of primary Epstein–Barr virus infection and lymphoproliferative disease in pediatric heart transplant recipients. J Heart Lung Transplant 1998; 17(12):1161–6.

98. Kasiske BL, Vazquez MA, Harmon WE et al. Recommendations for the outpatient surveillance of renal transplant recipients. American Society of Transplantation. J Am Soc Nephrol 2000; 11(Suppl 15):S1–86.

99. van Esser JW, van der Holt B, Meijer E et al. Epstein–Barr virus (EBV) reactivation is a frequent event after allogeneic stem cell transplantation (SCT) and quantitatively predicts EBV–lymphoproliferative disease following T-cell-depleted SCT. Blood 2001; 98(4):972–8.

100. Bakker NA, Pruim J, de Graaf W et al. PTLD visualization by FDG-PET: improved detection of extranodal localizations. Am J Transplant 2006; 6(8):1984–5.

101. McCormack L, Hany TI, Hubner M et al. How useful is PET/CT imaging in the management of post-transplant lymphoproliferative disease after liver transplantation? Am J Transplant 2006; 6(7):1731–6.

102. Paya CV, Fung JJ, Nalesnik MA et al. Epstein–Barr virus-induced posttransplant lymphoproliferative disorders. ASTS/ASTP EBV-PTLD Task Force and The Mayo Clinic Organized International Consensus Development Meeting. Transplantation 1999; 68(10):1517–25.

103. Sibley RK, Olivari MT, Ring WS et al. Endomyocardial biopsy in the cardiac allograft recipient. A review of 570 biopsies. Ann Surg 1986; 203(2):177–87.

104. Montone KT, Budgeon LR, Brigati DJ. Detection of Epstein–Barr virus genomes by in situ DNA hybridization with a terminally biotin-labeled synthetic oligonucleotide probe from the EBV Not I and Pst I tandem repeat regions. Mod Pathol 1990; 3(1):89–96.

105. Shapiro R, Scantlebury VP, Jordan ML et al. FK506 in pediatric kidney transplantation – primary and rescue experience. Pediatr Nephrol 1995; 9(Suppl):S43–8.

106. McDiarmid SV, Jordan S, Kim GS et al. Prevention and preemptive therapy of postransplant lymphoproliferative disease in pediatric liver recipients. Transplantation 1998; 66(12):1604–11.

107. Ellis D, Jaffe R, Green M et al. Epstein–Barr virus-related disorders in children undergoing renal transplantation with tacrolimus-based immunosuppression. Transplantation 1999; 68(7): 997–1003.

108. Holmes RD, Sokol RJ. Epstein–Barr virus and post-transplant lymphoproliferative disease. Pediatr Transplant 2002; 6(6):456–64.

109. Funch DP, Walker AM, Schneider G et al. Ganciclovir and acyclovir reduce the risk of post-transplant lymphoproliferative disorder in renal transplant recipients. Am J Transplant 2005; 5(12):2894–900.
110. Humar A, Hebert D, Davies HD et al. A randomized trial of ganciclovir versus ganciclovir plus immune globulin for prophylaxis against Epstein–Barr virus related posttransplant lymphoproliferative disorder. Transplantation 2006; 81(6):856–61.
111. Opelz G, Daniel V, Naujokat C et al. Effect of cytomegalovirus prophylaxis with immunoglobulin or with antiviral drugs on post-transplant non-Hodgkin lymphoma: a multicentre retrospective analysis. Lancet Oncol 2007; 8(3):212–18.
112. Allen U, Hebert D, Moore D et al. Epstein–Barr virus-related post-transplant lymphoproliferative disease in solid organ transplant recipients, 1988–97: a Canadian multi-centre experience. Pediatr Transplant 2001; 5(3):198–203.
113. Flinner R. Neoplasms occurring in solid organ transplant recipients. In: Hammond E (ed.) Solid organ transplantation pathology. Philadelphia: Saunders, 1994; pp. 262–73.
114. Starzl TE, Nalesnik MA, Porter KA et al. Reversibility of lymphomas and lymphoproliferative lesions developing under cyclosporin–steroid therapy. Lancet 1984; 1(8377):583–7.
115. Tsai DE, Hardy CL, Tomaszewski JE et al. Reduction in immunosuppression as initial therapy for posttransplant lymphoproliferative disorder: analysis of prognostic variables and long-term follow-up of 42 adult patients. Transplantation 2001; 71(8):1076–88.
116. Rees L, Thomas A, Amlot PL. Disappearance of an Epstein–Barr virus-positive post-transplant plasmacytoma with reduction of immunosuppression. Lancet 1998; 352(9130):789.
117. Swinnen LJ, Mullen GM, Carr TJ et al. Aggressive treatment for postcardiac transplant lymphoproliferation. Blood 1995; 86(9):3333–40.
118. Datta AK, Colby BM, Shaw JE et al. Acyclovir inhibition of Epstein–Barr virus replication. Proc Natl Acad Sci USA 1980; 77(9):5163–6.
119. Sixbey JW, Pagano JS. Epstein–Barr virus transformation of human B lymphocytes despite inhibition of viral polymerase. J Virol 1985; 53(1):299–301.
120. Zutter MM, Martin PJ, Sale GE et al. Epstein–Barr virus lymphoproliferation after bone marrow transplantation. Blood 1988; 72(2):520–9.
121. MacMahon EM. More on virostatic therapy for advanced lymphoproliferation associated with Epstein–Barr virus in an HIV-infected patient. N Engl J Med 2000; 343(1):71–2.
122. DeMario MD, Liebowitz DN. Lymphomas in the immunocompromised patient. Semin Oncol 1998; 25(4):492–502.
123. Davis CL, Wood BL, Sabath DE et al. Interferon-alpha treatment of posttransplant lymphoproliferative disorder in recipients of solid organ transplants. Transplantation 1998; 66(12):1770–9.
124. Elstrom RL, Andreadis C, Aqui NA et al. Treatment of PTLD with rituximab or chemotherapy. Am J Transplant 2006; 6(3):569–76.
125. Benkerrou M, Jais JP, Leblond V et al. Anti-B-cell monoclonal antibody treatment of severe posttransplant B-lymphoproliferative disorder: prognostic factors and long-term outcome. Blood 1998; 92(9):3137–47.
126. Cook RC, Connors JM, Gascoyne RD et al. Treatment of post-transplant lymphoproliferative disease with rituximab monoclonal antibody after lung transplantation. Lancet 1999; 354(9191):1698–9.
127. Schaar CG, van der Pijl JW, van Hoek B et al. Successful outcome with a "quintuple approach" of posttransplant lymphoproliferative disorder. Transplantation 2001; 71(1):47–52.
128. Yang J, Tao Q, Flinn IW et al. Characterization of Epstein–Barr virus-infected B cells in patients with posttransplantation lymphoproliferative disease: disappearance after rituximab therapy does not predict clinical response. Blood 2000; 96(13):4055–63.
129. Zilz ND, Olson LJ, McGregor CG. Treatment of post-transplant lymphoproliferative disorder with monoclonal CD20 antibody (rituximab) after heart transplantation. J Heart Lung Transplant 2001; 20(7):770–2.
130. Blaes AH, Peterson BA, Bartlett N et al. Rituximab therapy is effective for posttransplant lymphoproliferative disorders after solid organ transplantation: results of a phase II trial. Cancer 2005; 104(8):1661–7.
131. Faye A, Quartier P, Reguerre Y et al. Chimaeric anti-CD20 monoclonal antibody (rituximab) in post-transplant B-lymphoproliferative disorder following stem cell transplantation in children. Br J Haematol 2001; 115(1):112–18.
132. Verschuuren EA, Stevens SJ, van Imhoff GW et al. Treatment of posttransplant lymphoproliferative disease with rituximab: the remission, the relapse, and the complication. Transplantation 2002; 73(1):100–4.
133. Milpied N, Vasseur B, Parquet N et al. Humanized anti-CD20 monoclonal antibody (Rituximab) in post transplant B-lymphoproliferative disorder: a retrospective analysis on 32 patients. Ann Oncol 2000; 11(Suppl 1):113–16.
134. Choquet S, Leblond V, Herbrecht R et al. Efficacy and safety of rituximab in B-cell post-transplantation lymphoproliferative disorders: results of a prospective multicenter phase 2 study. Blood 2006; 107(8):3053–7.
135. Garnier JL, Stevenson G, Blanc-Brunat N et al. Treatment of post-transplant lymphomas with

anti-B-cell monoclonal antibodies. Recent Results Cancer Res 2002; 159:113–22.

136. Oertel SH, Verschuuren E, Reinke P et al. Effect of anti-CD 20 antibody rituximab in patients with post-transplant lymphoproliferative disorder (PTLD). Am J Transplant 2005; 5(12):2901–6.

137. Jain AB, Marcos A, Pokharna R et al. Rituximab (chimeric anti-CD20 antibody) for posttransplant lymphoproliferative disorder after solid organ transplantation in adults: long-term experience from a single center. Transplantation 2005; 80(12):1692–8.

138. Gong JZ, Stenzel TT, Bennett ER et al. Burkitt lymphoma arising in organ transplant recipients: a clinicopathologic study of five cases. Am J Surg Pathol 2003; 27(6):818–27.

139. Berney T, Delis S, Kato T et al. Successful treatment of posttransplant lymphoproliferative disease with prolonged rituximab treatment in intestinal transplant recipients. Transplantation 2002; 74(7):1000–6.

140. Smets F, Vajro P, Cornu G et al. Indications and results of chemotherapy in children with posttransplant lymphoproliferative disease after liver transplantation. Transplantation 2000; 69(5):982–4.

141. Mamzer-Bruneel MF, Lome C, Morelon E et al. Durable remission after aggressive chemotherapy for very late post-kidney transplant lymphoproliferation: A report of 16 cases observed in a single center. J Clin Oncol 2000; 18(21):3622–32.

142. Oertel SH, Papp-Vary M, Anagnostopoulos I et al. Salvage chemotherapy for refractory or relapsed post-transplant lymphoproliferative disorder in patients after solid organ transplantation with a combination of carboplatin and etoposide. Br J Haematol 2003; 123(5):830–5.

143. Gross TG, Bucuvalas JC, Park JR et al. Low-dose chemotherapy for Epstein–Barr virus-positive post-transplantation lymphoproliferative disease in children after solid organ transplantation. J Clin Oncol 2005; 23(27):6481–8.

144. Fohrer C, Caillard S, Koumarianou A et al. Long-term survival in post-transplant lymphoproliferative disorders with a dose-adjusted ACVBP regimen. Br J Haematol 2006; 134(6):602–612.

145. Taylor AL, Bowles KM, Callaghan CJ et al. Anthracycline-based chemotherapy as first-line treatment in adults with malignant posttransplant lymphoproliferative disorder after solid organ transplantation. Transplantation 2006; 82(3): 375–81.

146. Trappe R, Riess H, Babel N et al. Salvage chemotherapy for refractory and relapsed posttransplant lymphoproliferative disorders (PTLD) after treatment with single-agent rituximab. Transplantation 2007; 83(7):912–18.

147. Koffman BH, Kennedy AS, Heyman M et al. Use of radiation therapy in posttransplant lymphoproliferative disorder (PTLD) after liver transplantation. Int J Cancer 2000; 90(2):104–9.

148. Hsi ED, Singleton TP, Swinnen L et al. Mucosa-associated lymphoid tissue-type lymphomas occurring in post-transplantation patients. Am J Surg Pathol 2000; 24(1):100–6.

149. Snanoudj R, Durrbach A, Leblond V et al. Primary brain lymphomas after kidney transplantation: presentation and outcome. Transplantation 2003; 76(6):930–7.

150. Cantarovich M, Barkun JS, Forbes RD et al. Successful treatment of post-transplant lymphoproliferative disorder with interferon-alpha and intravenous immunoglobulin. Clin Transplant 1998; 12(2):109–15.

151. Nalesnik MA, Rao AS, Furukawa H et al. Autologous lymphokine-activated killer cell therapy of Epstein–Barr virus-positive and -negative lymphoproliferative disorders arising in organ transplant recipients. Transplantation 1997; 63(9):1200–5.

152. Haque T, Taylor C, Wilkie GM et al. Complete regression of posttransplant lymphoproliferative disease using partially HLA-matched Epstein Barr virus-specific cytotoxic T cells. Transplantation 2001; 72(8):1399–402.

153. Haque T, Wilkie GM, Taylor C et al. Treatment of Epstein–Barr-virus-positive post-transplantation lymphoproliferative disease with partly HLA-matched allogeneic cytotoxic T cells. Lancet 2002; 360(9331):436–42.

154. Sun Q, Burton R, Reddy V et al. Safety of allogeneic Epstein–Barr virus (EBV)-specific cytotoxic T lymphocytes for patients with refractory EBV-related lymphoma. Br J Haematol 2002; 118(3):799–808.

155. Comoli P, Maccario R, Locatelli F et al. Treatment of EBV-related post-renal transplant lymphoproliferative disease with a tailored regimen including EBV-specific T cells. Am J Transplant 2005; 5(6):1415–22.

156. Savoldo B, Goss JA, Hammer MM et al. Treatment of solid organ transplant recipients with autologous Epstein Barr virus-specific cytotoxic T lymphocytes (CTLs). Blood 2006; 108(9):2942–9.

157. Schoch OD, Boehler A, Speich R et al. Extracorporeal photochemotherapy for Epstein–Barr virus-associated lymphoma after lung transplantation. Transplantation 1999; 68(7):1056–8.

158. Mentzer SJ, Perrine SP, Faller DV. Epstein–Barr virus post-transplant lymphoproliferative disease and virus-specific therapy: pharmacological re-activation of viral target genes with arginine butyrate. Transplant Infect Dis 2001; 3(3):177–85.

159. Majewski M, Korecka M, Kossev P et al. The immunosuppressive macrolide RAD inhibits growth of human Epstein–Barr virus-transformed B lymphocytes in vitro and in vivo: a potential approach to prevention and treatment of posttransplant lymphoproliferative disorders. Proc Natl Acad Sci USA 2000; 97(8):4285–90.

160. Savage P, Waxman J. Post-transplantation lymphoproliferative disease. Q J Med 1997; 90(8): 497–503.
161. Leblond V, Dhedin N, Mamzer Bruneel MF et al. Identification of prognostic factors in 61 patients with posttransplantation lymphoproliferative disorders. J Clin Oncol 2001; 19(3):772–8.
162. Trofe J, Buell JF, Beebe TM et al. Analysis of factors that influence survival with post-transplant lymphoproliferative disorder in renal transplant recipients: the Israel Penn International Transplant Tumor Registry experience. Am J Transplant 2005; 5(4, Pt 1):775–80.
163. Karras A, Thervet E, Le Meur Y et al. Successful renal retransplantation after post-transplant lymphoproliferative disease. Am J Transplant 2004; 4(11):1904–9.
164. Johnson SR, Cherikh WS, Kauffman HM et al. Retransplantation after post-transplant lymphoproliferative disorders: an OPTN/UNOS database analysis. Am J Transplant 2006; 6(11): 2743–9.
165. Gardner SD, Field AM, Coleman DV et al. New human papovavirus (B.K.) isolated from urine after renal transplantation. Lancet 1971; 1(7712):1253–7.
166. Padgett BL, Walker DL, ZuRhein GM et al. Cultivation of papova-like virus from human brain with progressive multifocal leucoencephalopathy. Lancet 1971; 1(7712):1257–60.
167. Purighalla R, Shapiro R, McCauley J et al. BK virus infection in a kidney allograft diagnosed by needle biopsy. Am J Kidney Dis 1995; 26(4):671–3.
168. Binet I, Nickeleit V, Hirsch HH et al. Polyomavirus disease under new immunosuppressive drugs: a cause of renal graft dysfunction and graft loss. Transplantation 1999; 67(6):918–22.
169. Hirsch HH, Knowles W, Dickenmann M et al. Prospective study of polyomavirus type BK replication and nephropathy in renal-transplant recipients. N Engl J Med 2002; 347(7):488–96.
170. Knowles WA, Pipkin P, Andrews N et al. Population-based study of antibody to the human polyomaviruses BKV and JCV and the simian polyomavirus SV40. J Med Virol 2003; 71(1):115–23.
171. Stolt A, Sasnauskas K, Koskela P et al. Seroepidemiology of the human polyomaviruses. J Gen Virol 2003; 84(Pt 6):1499–504.
172. Vanchiere JA, Nicome RK, Greer JM et al. Frequent detection of polyomaviruses in stool samples from hospitalized children. J Infect Dis 2005; 192(4):658–64.
173. Bofill-Mas S, Formiga-Cruz M, Clemente-Casares P et al. Potential transmission of human polyomaviruses through the gastrointestinal tract after exposure to virions or viral DNA. J Virol 2001; 75(21):10290–9.
174. Bohl DL, Storch GA, Ryschkewitsch C et al. Donor origin of BK virus in renal transplantation and role of HLA C7 in susceptibility to sustained BK viremia. Am J Transplant 2005; 5(9):2213–21.
175. Reploeg MD, Storch GA, Clifford DB. Bk virus: a clinical review. Clin Infect Dis 2001; 33(2):191–202.
176. Eash S, Tavares R, Stopa EG et al. Differential distribution of the JC virus receptor-type sialic acid in normal human tissues. Am J Pathol 2004; 164(2):419–28.
177. Shah KV. Simian virus 40 and human disease. J Infect Dis 2004; 190(12):2061–4.
178. Paracchini V, Garte S, Pedotti P et al. Molecular identification of simian virus 40 infection in healthy Italian subjects by birth cohort. Molec Med 2005; 11(1–12):48–51.
179. Engels EA, Switzer WM, Heneine W et al. Serologic evidence for exposure to simian virus 40 in North American zoo workers. J Infect Dis 2004; 190(12):2065–9.
180. Li RM, Branton MH, Tanawattanacharoen S et al. Molecular identification of SV40 infection in human subjects and possible association with kidney disease. J Am Soc Nephrol 2002; 13(9):2320–30.
181. Milstone A, Vilchez RA, Geiger X et al. Polyomavirus simian virus 40 infection associated with nephropathy in a lung-transplant recipient. Transplantation 2004; 77(7):1019–24.
182. Li RM, Mannon RB, Kleiner D et al. BK virus and SV40 co-infection in polyomavirus nephropathy. Transplantation 2002; 74(11):1497–504.
183. Butel JS, Arrington AS, Wong C et al. Molecular evidence of simian virus 40 infections in children. J Infect Dis 1999; 180(3):884–7.
184. Imperiale M. The human polyomaviruses: an overview. In: Khalili K, Stoner G (eds) Human polyomaviruses: molecular and clinical perspectives. New York: Wiley-Liss, 2001; pp. 53–71.
185. Akan I, Sariyer IK, Biffi R et al. Human polyomavirus JCV late leader peptide region contains important regulatory elements. Virology 2006; 349(1):66–78.
186. Khalili K, White MK, Sawa H et al. The agnoprotein of polyomaviruses: a multifunctional auxiliary protein. J Cell Physiol 2005; 204(1):1–7.
187. Suzuki T, Okada Y, Semba S et al. Identification of FEZ1 as a protein that interacts with JC virus agnoprotein and microtubules: role of agnoprotein-induced dissociation of FEZ1 from microtubules in viral propagation. J Biol Chem 2005; 280(26):24948–56.
188. Gee GV, Tsomaia N, Mierke DF et al. Modeling a sialic acid binding pocket in the external loops of JC virus VP1. J Biol Chem 2004; 279(47):49172–6.
189. Low JA, Magnuson B, Tsai B et al. Identification of gangliosides GD1b and GT1b as receptors for BK virus. J Virol 2006; 80(3):1361–6.

190. Dugan AS, Eash S, Atwood WJ. An N-linked glycoprotein with alpha(2,3)-linked sialic acid is a receptor for BK virus. J Virol 2005; 79(22):14442–5.

191. Elphick GF, Querbes W, Jordan JA et al. The human polyomavirus, JCV, uses serotonin receptors to infect cells. Science 2004; 306(5700):1380–3.

192. Komagome R, Sawa H, Suzuki T et al. Oligosaccharides as receptors for JC virus. J Virol 2002; 76(24):12992–3000.

193. Pho MT, Ashok A, Atwood WJ. JC virus enters human glial cells by clathrin-dependent receptor-mediated endocytosis. J Virol 2000; 74(5):2288–92.

194. Eash S, Querbes W, Atwood WJ. Infection of vero cells by BK virus is dependent on caveolae. J Virol 2004; 78(21):11583–90.

195. Boldorini R, Veggiani C, Barco D et al. Kidney and urinary tract polyomavirus infection and distribution: molecular biology investigation of 10 consecutive autopsies. Arch Pathol Lab Med 2005; 129(1):69–73.

196. Hogan TF, Borden EC, McBain JA et al. Human polyomavirus infections with JC virus and BK virus in renal transplant patients. Ann Intern Med 1980; 92(3):373–8.

197. Gardner SD, MacKenzie EF, Smith C et al. Prospective study of the human polyomaviruses BK and JC and cytomegalovirus in renal transplant recipients. J Clin Pathol 1984; 37(5):578–86.

198. Muller A, Beck B, Theilemann K et al. Detection of polyomavirus BK and JC in children with kidney diseases and renal transplant recipients. Pediatr Infect Dis J 2005; 24(9):778–81.

199. Randhawa P, Uhrmacher J, Pasculle W et al. A comparative study of BK and JC virus infections in organ transplant recipients. J Med Virol 2005; 77(2):238–43.

200. Wen MC, Wang CL, Wang M et al. Association of JC virus with tubulointerstitial nephritis in a renal allograft recipient. J Med Virol 2004; 72(4):675–8.

201. Kazory A, Ducloux D, Chalopin JM et al. The first case of JC virus allograft nephropathy. Transplantation 2003; 76(11):1653–5.

202. Vera-Sempere FJ, Rubio L, Felipe-Ponce V et al. Renal donor implication in the origin of BK infection: analysis of genomic viral subtypes. Transplant Proc 2006; 38(8):2378–81.

203. Andrews CA, Shah KV, Daniel RW et al. A serological investigation of BK virus and JC virus infections in recipients of renal allografts. J Infect Dis 1988; 158(1):176–81.

204. Atencio IA, Shadan FF, Zhou XJ et al. Adult mouse kidneys become permissive to acute polyomavirus infection and reactivate persistent infections in response to cellular damage and regeneration. J Virol 1993; 67(3):1424–32.

205. Meehan SM, Kraus MD, Kadambi PV et al. Nephron segment localization of polyoma virus large T antigen in renal allografts. Hum Pathol 2006; 37(11):1400–6.

206. Drachenberg CB, Papadimitriou JC, Wali R et al. BK polyoma virus allograft nephropathy: ultrastructural features from viral cell entry to lysis. Am J Transplant 2003; 3(11):1383–92.

207. Nickeleit V, Hirsch HH, Zeiler M et al. BK-virus nephropathy in renal transplants – tubular necrosis, MHC-class II expression and rejection in a puzzling game. Nephrol Dial Transplant 2000; 15(3):324–32.

208. Low J, Humes HD, Szczypka M et al. BKV and SV40 infection of human kidney tubular epithelial cells in vitro. Virology 2004; 323(2):182–8.

209. Mannon RB, Hoffmann SC, Kampen RL et al. Molecular evaluation of BK polyomavirus nephropathy. Am J Transplant 2005; 5(12):2883–93.

210. Howell DN, Smith SR, Butterly DW et al. Diagnosis and management of BK polyomavirus interstitial nephritis in renal transplant recipients. Transplantation 1999; 68(9):1279–88.

211. Nickeleit V, Hirsch HH, Binet IF et al. Polyomavirus infection of renal allograft recipients: from latent infection to manifest disease. J Am Soc Nephrol 1999; 10(5):1080–9.

212. Randhawa PS, Finkelstein S, Scantlebury V et al. Human polyoma virus-associated interstitial nephritis in the allograft kidney. Transplantation 1999; 67(1):103–9.

213. Mengel M, Marwedel M, Radermacher J et al. Incidence of polyomavirus-nephropathy in renal allografts: influence of modern immunosuppressive drugs. Nephrol Dial Transplant 2003; 18(6):1190–6.

214. Rocha PN, Plumb TJ, Miller SE et al. Risk factors for BK polyomavirus nephritis in renal allograft recipients. Clin Transplant 2004; 18(4):456–62.

215. Buehrig CK, Lager DJ, Stegall MD et al. Influence of surveillance renal allograft biopsy on diagnosis and prognosis of polyomavirus-associated nephropathy. Kidney Int 2003; 64(2):665–73.

216. Hirsch HH, Mohaupt M, Klimkait T. Prospective monitoring of BK virus load after discontinuing sirolimus treatment in a renal transplant patient with BK virus nephropathy. J Infect Dis 2001; 184(11):1494–1495; author reply 1495–6.

217. Ramos E, Drachenberg CB, Papadimitriou JC et al. Clinical course of polyoma virus nephropathy in 67 renal transplant patients. J Am Soc Nephrol 2002; 13(8):2145–51.

218. Lipshutz GS, Flechner SM, Govani MV et al. BK nephropathy in kidney transplant recipients treated with a calcineurin inhibitor-free immunosuppression regimen. Am J Transplant 2004; 4(12):2132–4.

219. Josephson MA, Gillen D, Javaid B et al. Treatment of renal allograft polyoma BK virus infection with leflunomide. Transplantation 2006; 81(5):704–10.
220. Randhawa PS, Gupta G, Vats A et al. Immunoglobulin G, A, and M responses to BK virus in renal transplantation. Clin Vaccine Immunol 2006; 13(9):1057–63.
221. Brennan DC, Agha I, Bohl DL et al. Incidence of BK with tacrolimus versus cyclosporine and impact of preemptive immunosuppression reduction. Am J Transplant 2005; 5(3):582–94.
222. Dorries K. Molecular biology and pathogenesis of human polyomavirus infections. Dev Biol Stand 1998; 94:71–9.
223. Kemball CC, Lee ED, Vezys V et al. Late priming and variability of epitope-specific CD8+ T cell responses during a persistent virus infection. J Immunol 2005; 174(12):7950–60.
224. Ginevri F, De Santis R, Comoli P et al. Polyomavirus BK infection in pediatric kidney-allograft recipients: a single-center analysis of incidence, risk factors, and novel therapeutic approaches. Transplantation 2003; 75(8):1266–70.
225. Flaegstad T, Traavik T, Christie KE et al. Neutralization test for BK virus: plaque reduction detected by immunoperoxidase staining. J Med Virol 1986; 19(3):287–96.
226. Smith JM, McDonald RA, Finn LS et al. Polyomavirus nephropathy in pediatric kidney transplant recipients. Am J Transplant 2004; 4(12):2109–17.
227. Shah KV. Human polyomavirus BKV and renal disease. Nephrol Dial Transplant 2000; 15(6):754–5.
228. Comoli P, Azzi A, Maccario R et al. Polyomavirus BK-specific immunity after kidney transplantation. Transplantation 2004; 78(8):1229–32.
229. Hariharan S, Cohen EP, Vasudev B et al. BK virus-specific antibodies and BKV DNA in renal transplant recipients with BKV nephritis. Am J Transplant 2005; 5(11):2719–24.
230. Chen Y, Trofe J, Gordon J et al. Interplay of cellular and humoral immune responses against BK virus in kidney transplant recipients with polyomavirus nephropathy. J Virol 2006; 80(7):3495–505.
231. Hammer MH, Brestrich G, Andree H et al. HLA type-independent method to monitor polyoma BK virus-specific CD4 and CD8 T-cell immunity. Am J Transplant 2006; 6(3):625–31.
232. Awadalla Y, Randhawa P, Ruppert K et al. HLA mismatching increases the risk of BK virus nephropathy in renal transplant recipients. Am J Transplant 2004; 4(10):1691–6.
233. Drachenberg CB, Papadimitriou JC, Mann D et al. Negative impact of human leukocyte antigen matching in the outcome of polyomavirus nephropathy. Transplantation 2005; 80(2):276–8.
234. Sharma MC, Zhou W, Martinez J et al. Cross-reactive CTL recognizing two HLA-A*02-restricted epitopes within the BK virus and JC virus VP1 polypeptides are frequent in immunocompetent individuals. Virology 2006; 350(1):128–36.
235. Krymskaya L, Sharma MC, Martinez J et al. Cross-reactivity of T lymphocytes recognizing a human cytotoxic T-lymphocyte epitope within BK and JC virus VP1 polypeptides. J Virol 2005; 79(17):11170–8.
236. Li J, Melenhorst J, Hensel N et al. T-cell responses to peptide fragments of the BK virus T antigen: implications for cross-reactivity of immune response to JC virus. J Gen Virol 2006; 87(Pt 10):2951–60.
237. Bressollette-Bodin C, Coste-Burel M, Hourmant M et al. A prospective longitudinal study of BK virus infection in 104 renal transplant recipients. Am J Transplant 2005; 5(8):1926–33.
238. Limaye AP, Jerome KR, Kuhr CS et al. Quantitation of BK virus load in serum for the diagnosis of BK virus-associated nephropathy in renal transplant recipients. J Infect Dis 2001; 183(11):1669–72.
239. Nickeleit V, Klimkait T, Binet IF et al. Testing for polyomavirus type BK DNA in plasma to identify renal-allograft recipients with viral nephropathy. N Engl J Med 2000; 342(18):1309–15.
240. Ding R, Medeiros M, Dadhania D et al. Noninvasive diagnosis of BK virus nephritis by measurement of messenger RNA for BK virus VP1 in urine. Transplantation 2002; 74(7):987–94.
241. Randhawa P, Ho A, Shapiro R et al. Correlates of quantitative measurement of BK polyomavirus (BKV) DNA with clinical course of BKV infection in renal transplant patients. J Clin Microbiol 2004; 42(3):1176–80.
242. Drachenberg CB, Papadimitriou JC, Hirsch HH et al. Histological patterns of polyomavirus nephropathy: correlation with graft outcome and viral load. Am J Transplant 2004; 4(12):2082–92.
243. Drachenberg RC, Drachenberg CB, Papadimitriou JC et al. Morphological spectrum of polyoma virus disease in renal allografts: diagnostic accuracy of urine cytology. Am J Transplant 2001; 1(4):373–81.
244. Hirsch HH, Brennan DC, Drachenberg CB et al. Polyomavirus-associated nephropathy in renal transplantation: interdisciplinary analyses and recommendations. Transplantation 2005; 79(10):1277–86.
245. Celik B, Randhawa PS. Glomerular changes in BK virus nephropathy. Hum Pathol 2004; 35(3):367–70.
246. Meehan SM, Kadambi PV, Manaligod JR et al. Polyoma virus infection of renal allografts: relationships of the distribution of viral infection, tubulointerstitial inflammation, and fibrosis suggesting viral interstitial nephritis in untreated disease. Hum Pathol 2005; 36(12):1256–64.

247. Ahuja M, Cohen EP, Dayer AM et al. Polyoma virus infection after renal transplantation. Use of immunostaining as a guide to diagnosis. Transplantation 2001; 71(7):896–9.

248. Vasudev B, Hariharan S, Hussain SA et al. BK virus nephritis: risk factors, timing, and outcome in renal transplant recipients. Kidney Int 2005; 68(4):1834–9.

249. Wali RK, Drachenberg C, Hirsch HH et al. BK virus-associated nephropathy in renal allograft recipients: rescue therapy by sirolimus-based immunosuppression. Transplantation 2004; 78(7):1069–73.

250. Matas AJ, Kandaswamy R, Humar A et al. Long-term immunosuppression, without maintenance prednisone, after kidney transplantation. Ann Surg 2004; 240(3):510–6; discussion 516–7.

251. Hymes LC, Warshaw BL. Polyomavirus (BK) in pediatric renal transplants: evaluation of viremic patients with and without BK associated nephritis. Pediatr Transplant 2006; 10(8):920–2.

252. Filler G, Zimmering M, Mai I. Pharmacokinetics of mycophenolate mofetil are influenced by concomitant immunosuppression. Pediatr Nephrol 2000; 14(2):100–4.

253. Morath C, Schwenger V, Beimler J et al. Antifibrotic actions of mycophenolic acid. Clin Transplant 2006; 20(Suppl 17):25–9.

254. Allison AC, Eugui EM. Mechanisms of action of mycophenolate mofetil in preventing acute and chronic allograft rejection. Transplantation 2005; 80(2, Suppl):S181–90.

255. Celik B, Shapiro R, Vats A et al. Polyomavirus allograft nephropathy: sequential assessment of histologic viral load, tubulitis, and graft function following changes in immunosuppression. Am J Transplant 2003; 3(11):1378–82.

256. Wadei HM, Rule AD, Lewin M et al. Kidney transplant function and histological clearance of virus following diagnosis of polyomavirus-associated nephropathy (PVAN). Am J Transplant 2006; 6(5, Pt 1):1025–32.

257. Hussain S, Bresnahan BA, Cohen EP et al. Rapid kidney allograft failure in patients with polyoma virus nephritis with prior treatment with antilymphocyte agents. Clin Transplant 2002; 16(1):43–7.

258. Andrei G, Snoeck R, Vandeputte M et al. Activities of various compounds against murine and primate polyomaviruses. Antimicrob Agents Chemother 1997; 41(3):587–93.

259. Farasati NA, Shapiro R, Vats A et al. Effect of leflunomide and cidofovir on replication of BK virus in an in vitro culture system. Transplantation 2005; 79(1):116–18.

260. De Clercq E. Clinical potential of the acyclic nucleoside phosphonates cidofovir, adefovir, and tenofovir in treatment of DNA virus and retrovirus infections. Clin Microbiol Rev 2003; 16(4):569–96.

261. Papadopoulos EB, Ladanyi M, Emanuel D et al. Infusions of donor leukocytes to treat Epstein–Barr virus-associated lymphoproliferative disorders after allogeneic bone marrow transplantation. N Engl J Med 1994; 330(17):1185–91.

262. Tong CY, Hilton R, MacMahon EM et al. Monitoring the progress of BK virus associated nephropathy in renal transplant recipients. Nephrol Dial Transplant 2004; 19(10):2598–605.

263. Araya CE, Lew JF, Fennell RS 3rd et al. Intermediate-dose cidofovir without probenecid in the treatment of BK virus allograft nephropathy. Pediatr Transplant 2006; 10(1):32–7.

264. Kuypers DR, Vandooren AK, Lerut E et al. Adjuvant low-dose cidofovir therapy for BK polyomavirus interstitial nephritis in renal transplant recipients. Am J Transplant 2005; 5(8):1997–2004.

265. Lim WH, Mathew TH, Cooper JE et al. Use of cidofovir in polyomavirus BK viral nephropathy in two renal allograft recipients. Nephrology (Carlton) 2003; 8(6):318–23.

266. Keller LS, Peh CA, Nolan J et al. BK transplant nephropathy successfully treated with cidofovir. Nephrol Dial Transplant 2003; 18(5):1013–14.

267. Kadambi PV, Josephson MA, Williams J et al. Treatment of refractory BK virus-associated nephropathy with cidofovir. Am J Transplant 2003; 3(2):186–91.

268. Vats A, Shapiro R, Singh Randhawa P et al. Quantitative viral load monitoring and cidofovir therapy for the management of BK virus-associated nephropathy in children and adults. Transplantation 2003; 75(1):105–12.

269. Bjorang O, Tveitan H, Midtvedt K et al. Treatment of polyomavirus infection with cidofovir in a renal-transplant recipient. Nephrol Dial Transplant 2002; 17(11):2023–5.

270. Burgos D, Lopez V, Cabello M et al. Polyomavirus BK nephropathy: the effect of an early diagnosis on renal function or graft loss. Transplant Proc 2006; 38(8):2409–11.

271. Lopez V, Sola E, Gutierrez C et al. Anterior uveitis associated with treatment with intravenous cidofovir in kidney transplant patients with BK virus nephropathy. Transplant Proc 2006; 38(8):2412–13.

272. Lalezari JP, Stagg RJ, Kuppermann BD et al. Intravenous cidofovir for peripheral cytomegalovirus retinitis in patients with AIDS. A randomized, controlled trial. Ann Intern Med 1997; 126(4):257–63.

273. Stenlund A. Initiation of DNA replication: lessons from viral initiator proteins. Nat Rev Molec Cell Biol 2003; 4(10):777–85.

274. Leung AY, Chan MT, Yuen KY et al. Ciprofloxacin decreased polyoma BK virus load in patients who

underwent allogeneic hematopoietic stem cell transplantation. Clin Infect Dis 2005; 40(4):528–37.

275. Randhawa PS. Anti-BK virus activity of ciprofloxacin and related antibiotics. Clin Infect Dis 2005; 41(9):1366–1367; author reply 1367.

276. Portolani M, Pietrosemoli P, Cermelli C et al. Suppression of BK virus replication and cytopathic effect by inhibitors of prokaryotic DNA gyrase. Antiviral Res 1988; 9(3):205–18.

277. Thamboo TP, Jeffery KJ, Friend PJ et al. Urine cytology screening for polyoma virus infection following renal transplantation: the Oxford experience. J Clin Pathol 2007; 60(8):927–30.

278. Kazatchkine MD, Kaveri SV. Immunomodulation of autoimmune and inflammatory diseases with intravenous immune globulin. N Engl J Med 2001; 345(10):747–55.

279. Sener A, House AA, Jevnikar AM et al. Intravenous immunoglobulin as a treatment for BK virus associated nephropathy: one-year follow-up of renal allograft recipients. Transplantation 2006; 81(1):117–20.

280. Funk GA, Steiger J, Hirsch HH. Rapid dynamics of polyomavirus type BK in renal transplant recipients. J Infect Dis 2006; 193(1):80–7.

281. Bohl DL, Brennan DC. BK virus nephropathy and kidney transplantation. Clin J Am Soc Nephrol 2007; 2(Suppl 1):S36–46.

282. Ginevri F, Pastorino N, de Santis R et al. Retransplantation after kidney graft loss due to polyoma BK virus nephropathy: successful outcome without original allograft nephrectomy. Am J Kidney Dis 2003; 42(4):821–5.

283. Lipshutz GS, Mahanty H, Feng S et al. BKV in simultaneous pancreas–kidney transplant recipients: a leading cause of renal graft loss in first 2 years post-transplant. Am J Transplant 2005; 5(2):366–73.

284. Poduval RD, Meehan SM, Woodle ES et al. Successful retransplantation after renal allograft loss to polyoma virus interstitial nephritis. Transplantation 2002; 73(7):1166–9.

285. Ramos E, Vincenti F, Lu WX et al. Retransplantation in patients with graft loss caused by polyoma virus nephropathy. Transplantation 2004; 77(1):131–3.

286. Womer KL, Meier-Kriesche HU, Patton PR et al. Preemptive retransplantation for BK virus nephropathy: successful outcome despite active viremia. Am J Transplant 2006; 6(1):209–13.

287. Azzi A, De Santis R, Salotti V et al. BK virus regulatory region sequence deletions in a case of human polyomavirus associated nephropathy (PVAN) after kidney transplantation. J Clin Virol 2006; 35(1):106–8.

288. Boucek P, Voska L, Saudek F. Successful retransplantation after renal allograft loss to polyoma virus interstitial nephritis. Transplantation 2002; 74(10):1478.

289. Trofe J, Roy-Chaudhury P, Gordon J et al. Outcomes of patients with rejection post-polyomavirus nephropathy. Transplant Proc 2005; 37(2):942–4.

290. Nickeleit V, Mihatsch MJ. Polyomavirus allograft nephropathy and concurrent acute rejection: a diagnostic and therapeutic challenge. Am J Transplant 2004; 4(5):838–9.

291. Randhawa P, Shapiro R. Conceptual problems in the diagnosis and therapy of acute rejection in patients with polyomavirus nephropathy. Am J Transplant 2004; 4(5):840.

292. Jahnukainen T, Malehorn D, Sun M et al. Proteomic analysis of urine in kidney transplant patients with BK virus nephropathy. J Am Soc Nephrol 2006; 17(11):3248–56.

293. Drachenberg CB, Papadimitriou JC. Polyomavirus-associated nephropathy: update in diagnosis. Transplant Infect Dis 2006; 8(2):68–75.

294. Kiberd BA. Screening to prevent polyoma virus nephropathy: a medical decision analysis. Am J Transplant 2005; 5(10):2410–6.

13

Chronic transplant dysfunction

Lorna P. Marson

Introduction

The success of organ transplantation is marred by the fact that some initially successful transplants have suboptimal function or deteriorate over time. Despite advances in the control of acute rejection, the problem of chronic transplant dysfunction (CTD) remains significant.

CTD is characterised by a progressive deterioration in graft function over the months and years following transplantation, and is associated with characteristic histological features of fibrosis and graft arteriosclerosis. By the time of diagnosis, the changes that have taken place are usually irreversible and lead inexorably to graft loss. For the majority of solid organs transplanted, CTD is the leading cause of graft loss after the first year. At 5 years post-transplantation, 30–50% of kidney, heart, lung and pancreas, and 5–20% of liver allografts demonstrate typical morphological changes.

These data must be considered in conjunction with the consistent improvement in 1-year graft survival rates that have been achieved, due to more effective immunosuppressive therapy and recipient/donor care. The 1-year survival rate is 80–95% for kidney, liver, heart and lung transplantation, and so it is now necessary to turn our attention to the causes of late graft loss.

The corresponding chapter in the previous edition of this book provided an excellent overview of CTD.[1] Although a brief description of organ-specific CTD is provided, this chapter will primarily adopt renal transplantation as the paradigm for study of CTD. It will explore developments in our understanding of the condition, and expand upon immunological factors that influence this pathological entity. The adoption of protocol biopsies in some centres across the world has allowed the investigation of early changes in renal histology that lead to CTD, and so this concept will be discussed.

Organ-specific findings

After excluding organ-specific disease (such as hepatitis C or recurrent glomerulonephritis) characteristic but non-specific patterns of pathology can be described in different organs.

Heart

CTD in transplanted hearts remains the most common cause of graft loss after the first post-transplant year. It is manifested as accelerated cardiac allograft arteriosclerosis or cardiac allograft vasculopathy (CAV). The diagnosis of CAV relies on clinical and radiographic features of arteriosclerosis. The pathognomonic lesion of CAV in heart allografts is concentric intimal thickening affecting the coronary vessels, particularly the smaller distal intramyocardial vessels.

Diagnostic criteria for CAV were presented in the consensus document of 1993 from the fourth Alexis Carrel Conference.[2]

The document proposed the following guidelines:

1. Histological diagnosis should be based on the presence of tubular myointimal hyperplasia with an intact elastic lamina containing few breaks. There should be little or no calcification.
2. The histopathological diagnosis should be made on explanted transplants or at autopsy, recognising that endomyocardial biopsies rarely contain arteries.
3. The clinical diagnosis should be made on angiography and reported according to categories described by the Stanford group:[3]
   a. Type A lesions – discrete stenoses in the proximal, middle or distal segment branches.
   b. Type B lesions – diffuse concentric narrowing arising in the middle to distal arteries.
   c. Type C lesions – narrowed irregular vessels with occluded side branches.

Type A changes in the proximal arteries are likely to represent prior coronary artery disease or new-onset atherosclerosis typical of that found in native hearts.

Liver

Chronic liver rejection is again one of the most common causes of late graft loss, but it occurs far less frequently than in other solid-organ transplants. The overall incidence of chronic transplant dysfunction is between 5% and 20% of all liver transplants.[3]

The liver is immunologically privileged, with all forms of rejection being less common. Specifically, CTD is becoming less common,[4] with graft loss more likely from recurrent disease, late acute rejection or non-compliance with immunosuppression, but it is still a major indication for re-transplantation.[5,6]

Histological diagnosis is also essential, with liver cell dropout, vanishing bile duct syndrome and obliterative vasculopathy being the cardinal features.[7] 'Vanishing bile duct syndrome' when diagnosed on biopsy is a good indicator for severe chronic allograft damage. It is a uniform loss of small bile ducts (<75 μm) throughout the liver[8] without replacement by fibrosis. Ductule proliferation does not occur as in other cholestatic conditions, such as primary biliary cirrhosis. When seen in biopsy specimens, large and medium-sized arteries have a foam cell and macrophage-laden intima as is found in other allografted organs.

Lung

Bronchiolitis obliterans syndrome (BOS) is the clinical manifestation of CTD in the lung, and the typical histopathological features are referred to as obliterative bronchiolitis. BOS accounts for over one-third of late graft losses and the prevalence of BOS in patients greater than 3 months post-transplantation may be as high as 68%.[9] The mortality rate is 50% once the diagnosis of BOS has been made, with few patients coming for re-transplantation. In common with the disease process in other organs, BOS is progressive and has no effective treatment at present. Clinically it presents with slowly progressing dyspnoea on exertion. Worsening airflow obstruction follows as a consequence of deteriorating graft function. Physiologically there is a mixed picture of airflow obstruction and restrictive pulmonary disease.[10]

Histologically there is inflammation and fibrosis of the cartilaginous airways and particularly within the smaller airways. Bronchioles usually demonstrate areas of inflammation and fibrosis in the lamina propria and luminal surfaces whilst the larger bronchi show peribronchial fibrosis and bronchiectasis. Airway narrowing follows, accounting for the decline in spirometry readings.[11] Surrounding alveoli and interstitium are often, but not always, normal. As a pathological entity it is no different from obliterative bronchiolitis caused by non-transplant-related aetiological factors, such as toxic fume inhalation, drug side-effects and connective tissue disorders.

In 1990 an ad hoc committee under the auspices of the International Society for Heart and Lung Transplantation proposed a working formulation for the standardisation of nomenclature and for clinical staging of chronic dysfunction in lung allografts.[12]

The formulation described the following staging system:

1. Stage 0 – no significant abnormality; forced expiratory volume in 1 second (FEV_1) 80% or more of baseline value.
2. Stage 1 – mild bronchiolitis obliterans syndrome; FEV_1 66–88%.

3. Stage 2 – moderate bronchiolitis obliterans syndrome; FEV_1 51–65%.
4. Stage 3 – severe bronchiolitis obliterans syndrome; FEV_1 50% or less.

Within each stage, 'a' and 'b' categories exist: 'a' denotes no histological evidence of bronchiolitis obliterans or an absence of biopsy material and 'b' denotes a positive diagnosis obtained on biopsy material.

Pancreas

CTD accounts for over half of all pancreas graft losses 5 years post-transplantation.[13] Deterioration in graft function is not easy to detect and is usually manifested by hyperglycaemic episodes with a low C-peptide level following a glucose challenge.[14] There are no serial markers of graft deterioration in pancreas transplantation and, in the absence of concurrent acute rejection, serum amylase is often normal.

Graft biopsy is required to confirm the diagnosis. The typical histological features seen are septal fibrosis with acinar loss and fibrointimal vascular proliferation. Vessels are seldom seen on graft biopsies and therefore vascular changes cannot be relied on for diagnosis.

Thus it is observed that different organs deteriorate with different patterns, and these echo to some extent the patterns of ageing: the atrophic and fibrotic processes in kidney transplants resemble the much slower shutdown of nephrons in normal ageing, the 'Achilles heel' of the transplanted and native heart is coronary artery deterioration, and lung transplants develop increased small airway resistance, just as normal lungs do with age and environmental factors, such as smoking.

Renal transplantation: the paradigm for chronic allograft injury

Chronic renal allograft dysfunction remains one of the most common causes of graft loss beyond the first year, and has been the subject of study over recent years. The adoption of serial protocol biopsies in some centres has led to an improvement in our understanding of the process. The pattern of injury is similar to that seen in other organs, such as the heart and lungs, and so much of the work is relevant to all organs.

Protocol biopsies

Renal allograft biopsy is the most accurate method for identifying pathophysiological events that have a bearing on short- and long-term graft outcomes. Allograft biopsies are traditionally performed in the setting of acute or chronic deterioration in graft function. Recent years have seen increasing recognition that a measurable decrease in renal function does not always accompany acute rejection or chronic changes such as tubular atrophy or interstitial fibrosis, and this has given rise to the practice of performing biopsies on stable allografts at predefined post-transplant periods, known as protocol biopsies.

Protocol biopsies and subclinical rejection

Protocol biopsies allow the early detection of rejection, at a time before this is manifested clinically, known as subclinical rejection (SCR). SCR is defined as histological evidence of acute rejection in patients with stable renal function (less than 25% change in serum creatinine), or borderline changes, a term that is used when no intimal arteritis is present, but there are foci of tubulitis that do not meet criteria for rejection diagnosis, also in patients with stable renal function. Thus, SCR may be important as a precursor to clinically relevant rejection episodes, allowing early intervention. Moreover, persistent, unrecognised inflammatory injury, secondary to a low-grade rejection process, may result in long-term graft fibrosis and chronic dysfunction without overt clinical rejection, and this could be identified on protocol biopsies.[15]

In order to justify a protocol biopsy with its attendant risks, a clear link must be defined between SCR and subsequent development of chronic graft dysfunction. In one study, SCR diagnosed in early biopsies (1–3 months) was treated with corticosteroids, and this was associated with both a reduction in late acute rejection and also in chronic damage compared with controls.[16] This has been supported in other studies, suggesting that the use of protocol biopsies for the early detection and treatment of subclinical rejection may be a major factor in preserving long-term graft function.[17,18]

Protocol biopsies and chronic allograft changes

In addition, protocol biopsies may allow early detection of specific chronic allograft changes, such as tubular atrophy and interstitial fibrosis (TA/IF). Such chronic changes are a frequent finding in protocol biopsies, and are seen to progress rapidly over the first year and more slowly thereafter.[19,20] Helantera et al. undertook a study to determine the optimal timing of protocol biopsies and used the chronic allograft damage index (CADI), defined at the Banff 1997 meeting, to define chronic injury.[21] Follow-up was for 18 months, and this demonstrated that CADI at 6 and 12 months was associated with long-term graft survival, but not that at 3 months.[22] If the diagnosis of early changes of fibrosis and atrophy are detected on protocol biopsy, the real challenge is to identify effective therapeutic strategies leading to improved outcome. The diagnosis alone is unlikely to have a positive impact on the patient.

Protocol biopsies and non-immunological contributors to chronic graft dysfunction

Non-immunological factors also play a significant role in the subsequent development of chronic allograft dysfunction and these, having been identified early in protocol biopsies, may be amenable to specific treatment. One of the most common factors is calcineurin inhibitor (CNI) toxicity, which can be reversed by modifying immunosuppressive therapy. A small study showed that the histological change characteristic of CNI toxicity (hyaline arteriolar sclerosis), when seen in early protocol biopsies (<1 year), was an independent risk factor for subsequent development of chronic allograft nephropathy.[23] This supports earlier work by Nankivell et al., who undertook a longitudinal study of biopsies from 120 kidney pancreas patients, and described an increasing presence of arteriolar hyalinosis over time.[19] Such changes have been detected in up to 42% of protocol biopsies,[24] identifying a potentially important contributor to chronic transplant dysfunction, which is amenable to therapeutic intervention.

Protocol biopsies may also be a useful tool to detect renal diseases such as BK nephropathy. Activation of BK virus is increasingly common in renal transplant recipients and can lead to a rapid decline in transplant function. Effective and early treatment of BK nephritis by reducing immunosuppressive therapy significantly improves outcome.[25]

Another potential benefit of protocol biopsies is their implementation in clinical trials designed to prevent CTD. Any such clinical trials are challenging to establish, requiring large numbers of patients to be followed up over a long period. The findings that early diagnosis of TA/IF in early protocol biopsies is an independent predictor of long-term survival, and that such changes are superior to other predictors of outcome, such as acute rejection or serum creatinine, make their use for clinical trials attractive.[26,27]

Risks of protocol biopsies

When making a decision about the routine implementation of protocol biopsies after renal transplantation, the risks of the procedure must be taken into account. These have been addressed by Schwarz et al., in a published experience of 1171 protocol biopsies. The complications were gross haematuria (3.1%), perirenal haematoma (3.3%), vasovagal reaction (0.3%) and A-V fistula with spontaneous resolution (9%). There was no associated mortality, and all complications were apparent within the 4-hour post-biopsy surveillance period.[28]

Protocol biopsies: should they be adopted more widely?

Good evidence supporting the general implementation of protocol biopsies is lacking as large, prospective trials have not been performed. In those studies that have been published, there are many variables, making comparison between studies difficult: the timing of biopsies, immunosuppressive regimens, as well as different histological parameters used to define SCR. In addition, the impact of treatment based on the findings from protocol biopsies can only be fully determined in the long term, and the question as to whether treatment should be instigated on the basis of a single biopsy or sequential biopsies remains unclear.[29] Consistent, effective therapeutic strategies for CTD are not available, and this limits the clinical purpose of protocol biopsies in this setting.

An alternative to the histological analysis of protocol biopsies is to adopt molecular technology to seek other strategies for early identification of allografts at risk of rejection.[30] For example, high-density array analysis of protocol biopsies demonstrated upregulation of genes associated with inflammation and matrix remodelling, which correlated with histological evidence of TA/IF.[31] Previous studies implicated growth factors such as transforming growth factor-β (TGF-β), as an important molecule in the development of fibrosis in allografts. Additional non-invasive methods under development seek to examine gene and protein expression profiles in peripheral blood and urine.

The natural history of chronic transplant dysfunction

The chronic changes observed following renal transplantation occur as histological sequelae of a series of pathological insults to the kidney from the time of organ retrieval, which leads to incremental and cumulative damage to nephrons, and ultimate loss of graft function.[32] Factors affecting long-term graft survival can be defined by era: peritransplant factors include **donor factors**, such as age, donor serum creatinine and comorbid conditions such as hypertension; **brainstem death** with the accompanying catecholamine storm; **preservation and implantation injury**, e.g. ischaemic damage and reperfusion injury, perhaps leading to delayed graft function. Factors affecting early transplant function include acute rejection and early infection. Over time, recipient factors such as hypertension, hyperlipidaemia and viral infections, e.g. cytomegalovirus, play a more significant role in mediating renal tubular injury. Calcineurin inhibitor toxicity is becoming increasingly recognised as a significant contributor to late injury, giving rise to the paradox that reduction in acute rejection rates has not led to improvements in long-term graft survival.

Until recently it has not been possible to study the progressive changes that culminate in the loss of graft function through tubular atrophy and interstitial fibrosis.

A recent longitudinal study of 961 protocol biopsies performed on 120 renal transplant patients over a period of 10 years has provided information about the natural history of the process.[19]

Two distinct phases of injury were evident:

- **Early phase** of tubulointerstitial injury, with rapidly increasing TA/IF. Early changes of chronic damage were seen at 3 months, and were associated with ischaemic injury, prior severe rejection and SCR.
- **Late phase** of injury, occurring beyond 1 year, was associated with microvascular and glomerular changes. Progressive high-grade hyalinosis with luminal narrowing, increasing glomerulosclerosis and additional tubulointerstitial damage was accompanied by the use of calcineurin inhibitors. By 10 years, severe evidence of chronic damage was seen in 58.4% of patients, with sclerosis in 37.3% of glomeruli.

Chronic allograft nephropathy: an obsolete term

Our increased understanding of the processes which lead ultimately to chronic allograft failure led to a review of the nomenclature at the eighth Banff conference on allograft pathology held in July 2005.[33] The proposed changes to the diagnostic categories are shown in Box 13.1.

The term chronic allograft nephropathy (CAN) was coined in 1991 as a more generic alternative to the

Box 13.1 • Banff 07 diagnostic criteria for renal allograft biopsies

1. Normal
2. Antibody-mediated rejection (AMR)
   a. Acute AMR
      i. ATN-like C4d$^+$, minimal inflammation
      ii. Capillary margination and/or thromboses, C4d$^+$
      iii. Arterial-v3, C4d$^+$
   b. Chronic active AMR
      Glomerular double contours and/or peritubular capillary basement membrane multilayering and/or interstitial fibrosis/tubular atrophy and/or fibrous intimal thickening in arteries, C4d$^+$
3. Borderline changes: suspicious for T-cell-mediated rejection, no intimal arteritis, foci of tubulitis
4. T-cell-mediated rejection
   a. Acute T-cell-mediated rejection
   b. Chronic active T-cell-mediated rejection
      Chronic allograft arteriopathy
5. **Tubular atrophy/interstitial fibrosis (TA/IF), no evidence of specific aetiology**
   Grade:
      I. Mild TA/IF (<25% cortical area)
      II. Moderate TA/IF (26–50%)
      III. Severe (>50%)
6. Other: changes not considered to be due to rejection

Chronic categories are shown in bold. Reproduced from Solez K, Colvin RB, Racusen LC et al. Banff 07 classification of renal allograft pathology: updates and future directions. Am J Transplant 2008; 8(4):753–60. With permission from Blackwell Publishing.

Table 13.1 • Morphology of specific chronic diseases resulting in TA/IF (non-immune mediated)

Aetiology	Morphology
Chronic hypertension	Arterial/fibrointimal thickening with reduplication of elastic, usually with small artery and arteriolar changes
CNI toxicity	Arterial hyalinosis
	Tubular cell injury with vacuolisation
Chronic obstruction	Marker tubular dilatation
	Large protein casts with extravasation into interstitium and/or lymphatics
Bacterial pyelonephritis	Intra- and peritubular neutrophils, lymphoid follicle formation
Viral infection	Viral inclusions on histology, immunohistochemistry and/or electron microscopy

misleading term 'chronic rejection', which led to the misconception that all late scarring was due to alloimmune injury.[21] However, over the years, the term CAN has become used to describe a specific disease entity rather than a descriptive term for non-specific scarring and has led to an acceptance of the inevitability of the process, rather than seeking specific causes. Thus, chronic changes within the kidney should be defined according to aetiology where possible, such as chronic active T-cell-mediated rejection or chronic antibody-mediated rejection. In addition, evidence for CNI toxicity, with arteriolar hyalinosis, should be sought and defined. Other causes of fibrosis and atrophy include recurrent disease, hypertension resulting in glomerulosclerosis, with duplication of internal elastica and arterial fibrointimal thickening, and chronic polyomavirus. Changes detected in specific chronic disease states are shown in Table 13.1. If there is no known aetiology, the term 'interstitial fibrosis and tubular atrophy, no specific aetiology' should be adopted.

Aetiological factors

The definition of two phases of injury resulting in chronic changes of tubular atrophy is in keeping with the model of injury in CTD, described in a recent review,[34] in which late allograft failure was described as a composite phenotype reflecting the total burden of stressful injury, mediated by five groups of risk factors:[35]

- donor age;
- brainstem death;
- ischaemia–reperfusion injury;
- immune-mediated injury;
- post-transplant stressors.

Donor age

Donor age has a powerful impact on graft survival. Organs removed from donors at the extremes of age are associated with poorer long-term graft survival compared with young to middle-aged donors.[36] This may be related to the reduction in the transplanted nephron mass. Long-standing donor stresses such as hypertension and diabetes also affect long-term outcome.

Other donor risk factors include death from an acute cerebral haemorrhage, female gender (again perhaps due to nephron mass) and Afro-Caribbean ethnic origin.[1]

Brainstem death

Donation after brainstem death (rather than live donation) is significantly associated with poor long-term graft survival and an increased incidence of delayed graft function and acute rejection.[37] Although the condition of brainstem death is well defined legally and neurologically, the systemic sequelae on the potential donor are very poorly understood. The effects of an irreversible and catastrophic injury to the brainstem include labile blood pressure, alterations in thermal regulation, endocrine and biochemical derangements plus pulmonary changes. Surges in catecholamine release are experienced, with resultant physical and structural changes affecting the vital organs. It may also be the trigger for the systemic release of a range of cytokines, causing a greater exposure of antigenic surface molecules, possibly as a result of endothelial cell retraction. Thereby the transplanted organ may prove more immunogenic to the recipient as a result.

Ischaemia–reperfusion injury

Removal of the graft from the donor includes a variable but short period of warm ischaemia, a more prolonged period of cold ischaemia, further warm ischaemia and subsequent reperfusion. All events have been shown to produce organ damage, and the degree of ischaemia correlates with the subsequent development of CTD. There is a considerable body of evidence supporting a strong association between cold ischaemia time and delayed graft function (DGF).[38] In turn, DGF is associated with increased rates of acute rejection and is predictive of higher rates of graft loss at 5 years.

Immune injury

Acute rejection episodes

In kidney, heart, lung and liver transplant recipients, acute rejection episodes correlate strongly with the subsequent development of CTD. In one series, recipients of cadaveric renal allografts who had never experienced an acute rejection episode were found to have a 5-year graft survival rate of 92%. In contrast, recipients experiencing one or more acute rejection episodes had an overall graft survival at 5 years of 45%.[39]

Further studies of both living-donor and cadaveric renal allografts showed that it was not simply the presence or absence of an acute rejection episode that predicted the likelihood of subsequent TA/IF but rather the nature of the rejection episode.[40–43] Early acute rejection episodes that are completely reversed do not appear to confer any greater degree of risk for the later development of TA/IF.[42]

The greatest risk factors for the development of chronic allograft injury following an acute rejection episode are:[44]

1. Rejection episodes occurring after 3 months.
2. Recurrent rejection episodes.
3. Rejection episodes where the predominant histological changes are vascular rather than interstitial.
4. Incompletely reversed rejection episodes.

Combinations of these risk factors may prove to be more than additive when determining the relative risk for the development of CTD.

There are several theories to explain the late injury caused by these early immunological events, although there is little scientific evidence to favour one explanation over another. In kidneys, the damage caused by the rejection episode may cause a reduction in the functioning nephron mass, subjecting the remaining nephrons to hyperfiltration, thereby producing fibrotic changes. Alternatively, the rejection episode may persist in a subclinical form, despite apparent treatment, allowing ongoing immunologically mediated cell damage.[45]

At a cellular level, tubulitis is the most common manifestation of acute T-cell-mediated rejection, and this is the most logical candidate for the ensuing loss of nephrons. It is likely that tubulitis damages the epithelium beyond the normal capacity for repair, triggering loss of nephrons, with TA/IF.

Despite newer improved immunosuppressive agents resulting in less acute rejection, there has been little apparent improvement in long-term graft survival from the first year onward. The main reason for this is that studies used to assess acute rejection rates were never powered to look at long-term graft outcomes. However, it may be that a greater number of marginal recipients are being transplanted with expanded criteria donor organs and therefore improvements caused by reduced acute rejection have been masked. Alternatively, acute rejection may be subclinical in its presentation and therefore not diagnosed and treated.

Despite the current lack of evidence for improved long-term graft survival with a reduction in acute rejection episodes, the association between acute and chronic rejection is so strong that, empirically, it seems justified to continue to aim to minimise the number and severity of acute rejection episodes, and evidence is emerging to support this strategy.[46]

Antibody-mediated rejection

The extremely low incidence of CTD in grafts from human leucocyte antigen (HLA)-identical living-related donors supports the role for immunological factors in developing TA/IF,[47] and the predictive value of anti-donor antibody production on graft outcome has long been emphasised.[48] The precise mechanisms by which the immune system seeks to destroy the graft are still somewhat unclear, but there is evidence for the role of anti-HLA antibodies in the process.

Initial studies assessed the presence of antibodies against a panel of HLA antigens (PRA), but more recently the detection of C4d permits the definitive diagnosis of antibody-mediated rejection to be made. C4d is a fragment of C4b, an activation product of the classic complement pathway.

In a 5-year longitudinal study of 493 sera from 54 kidney transplant patients, only 3 of 22 patients without antibodies rejected a graft, in contrast to 17 of 32 patients with antibody ($P = 0.003$). Antibody expression preceded a rise in serum creatinine, and the authors suggest a role for antibody expression as a marker of late failure.[49]

The relative involvement of major histocompatibility complex (MHC) class I versus class II antibodies is unclear, although one study has supported the role of MHC II. However, the percentage of recipients expressing anti-HLA class I antibodies was low in this study, perhaps because these were associated more commonly with early graft loss.[50] Clearly not all patients with anti-HLA antibodies go on to develop CTD and so this may not be sufficient in itself. For further details of the link between HLA antibodies and CTD, please see the section on transplant glomerulopathy below.

The case for non-HLA antibodies contributing to chronic graft injury is also growing. These include anti-glomerular Basement Merrbrane (GBM) antibodies and antibodies directed against the vascular endothelial cells. In addition detection of antibody to MHC class I-associated (MICA) proteins is uncommon but associated with worse graft outcome.[51]

Post-transplant stressors

Hypertension

Hypertension is a risk factor for the development of CTD and is particularly important in relation to heart and kidney transplantation, and the majority of research has focused on these two areas. It is very common in kidney transplant recipients, with as many as 76% of recipients with a systolic blood pressure >130 mmHg in a study of European Transplant Centres. Worsening degrees of post-transplant hypertension[52] have been shown to correlate with both the development of CTD and its rate of progression over time. There are currently no clinical trials to show that treating hypertension has any impact on either of these facts, although experimental animal models suggest that this may be the case.

Hyperlipidaemia

The post-transplant lipid profile is characterised by an increase in total cholesterol concentration with a predominance of cholesterol carried in low-density lipoprotein (LDL). There is also a significant increase in the percentage of cholesterol and triglycerides carried in very-low-density lipoprotein (VLDL).[53] The effects of renal impairment and the propensity for hyperlipidaemia because of immunosuppressive agents (steroids, ciclosporin and sirolimus in particular) bring about these abnormalities.

Elevated levels of cholesterol and triglyceride are independent risk factors for CTD.[53] They are associated with adverse morphological changes on the Banff criteria, reduced glomerular filtration rate and proteinuria. There is also a strong correlation between hyperlipidaemia and the other main cause of late graft loss, death with a functioning graft.[54]

Lipid-lowering agents may prove beneficial not only in the prevention of CTD but also in its treatment. Statins have been studied in experimental models and their action in reducing CTD appears to be related to inhibition of smooth muscle cell migration and proliferation in the endothelium, and their use has been supported in clinical studies,[55] but not all because of improved renal function. These agents are probably more important in lowering the cardiovascular risk to the transplant patient.[56]

Viral infections

Cytomegalovirus

Cytomegalovirus (CMV) replication can be detected in over 50% of allograft recipients and CMV remains one of the most clinically significant organisms in solid-organ transplantation.[57] Apart from its direct role in causing clinical manifestations of active disease it may also be implicated in the pathogenesis of CTD, a theory proposed as far back as 1971.[58] Experimental evidence supports this theory in animal models of cardiac, renal and lung transplantation.[57] The true relevance of CMV and its involvement in the progression of CAN is still a topic of much deliberation. One large study in renal transplant recipients[59] suggested that CMV infection alone is not a risk factor for CAN. However, if acute rejection and CMV infection were present together, then the progression of changes was far more rapid than if there was acute rejection alone. This result is contradicted by several studies in heart transplant recipients, where transplant-associated coronary artery disease was much more common in those patients who received grafts from CMV-seropositive donors or who were themselves CMV seropositive either pretransplant or who subsequently acquired CMV seropositivity. Everett et al.[60] found in their cohort of 129 heart transplant recipients that neither primary CMV infection nor previous exposure to CMV antigens were risk factors for the development of graft vasculopathy. However, the presence of ongoing CMV viraemia and its duration was significant in the progression of vascular changes.

Experimentally, treatment of CMV disease and CMV viraemia with ganciclovir reduces the incidence and severity of CTD in experimental allografts, but there are still no long-term data in the clinical setting to support this. A more in-depth discussion on CMV in transplantation is available in Chapter 12 of this book.

Polyoma virus

Interest is now turning to the role of other viral infections (particularly of the herpes class) and the role that they play in the development of chronic allograft disease.

Polyoma viruses (also called BK and JC virus) are an increasing problem and are posing a threat to improving renal transplant graft survival. Polyoma virus causes no symptoms in healthy individuals, but in the presence of immunosuppression it causes a tubulointerstitial nephritis in kidney allografts and usually presents in a way that mimics acute rejection.[61] Occurring as an early event post-transplantation, usually within the first year, patients usually remain asymptomatic until they experience renal insufficiency. The diagnosis is made by detection of BK virus DNA in urine and plasma and is confirmed by transplant biopsy. Risk factors for the development of BK virus include increasing HLA mismatches, a history of acute rejection and immunosuppressive load.

BK viral infections have been detected with increasing frequency since the beginning of the tacrolimus and mycophenolate mofetil (MMF) era, but these seem to be related more to a heavy burden of immunosuppression, rather than to specific agents. Until recently, BK viral nephritis resulted in irreversible graft failure in 30–60% of cases. This was due to lack of awareness, late diagnosis and use of over-immunosuppression to treat presumptive acute rejection. This has improved over the last decade, with 1-year graft survival of 95%, and 57.6% at 5 years for patients with BK viral nephritis described in a single centre.[62] Treatment involves alteration in immunosuppressive therapy with or without antiviral therapies, and such improvements have been observed due to increasing awareness of the impact of the infection, and early diagnosis and treatment. However, the long-term impact of BK nephritis and its treatment on the development of chronic rejection in solid-organ transplants is still not clear[63] (again, see Chapter 12).

Immunosuppression

Ciclosporin and tacrolimus have had a dramatic impact on reducing the number of acute rejection episodes in the first year post-transplantation but, as yet, have not demonstrated long-term benefit in avoidance of CTD. Similarly, sirolimus and MMF have been used, mostly in combination with CNI, to reduce acute rejection rates still further, but their use is now being pushed more directly towards the management of CTD.

There is an increasing wealth of evidence for CNI toxicity as a major contributor to the late phase of injury, leading ultimately to graft loss through TA/IF. As discussed previously, these agents are effective at reducing acute rejection rates, thus having a significant and positive impact on the early phase of chronic injury due to acute rejection, but paradoxically it is the same agents that may cause graft damage beyond 1 year after transplantation. This damage appears to occur independently of drug levels or period of exposure to CNIs,[64] although data in this quoted study were short term and based only on findings in the first year post-transplant. Histologically protocol biopsies performed beyond 1 year demonstrate a pattern of persistent arteriolar hyalinosis characteristic of CNI toxicity and the prevalence of these findings increases annually, such that it is almost universally present at 10 years.[19]

Pathophysiology

Thus, the aetiological factors promoting graft injury are both immune (antigen dependent) and non-immune (alloantigen independent), but regardless of the cause of the initial injury, the result is endothelial cell activation and inflammation. Endothelial cell activation occurs as an early event after transplant, as these cells form the interface between the

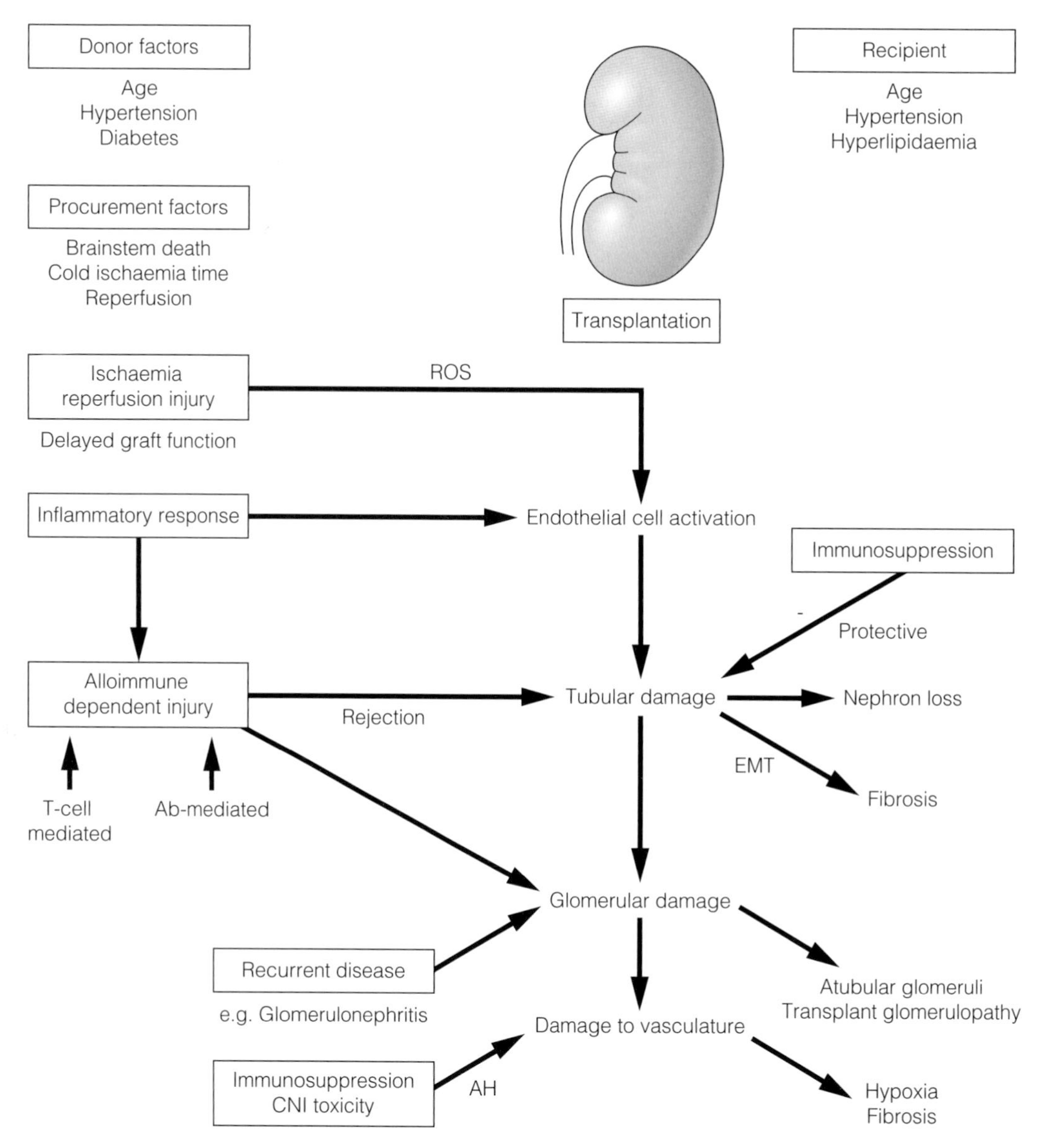

Figure 13.1 • Summary of the pathophysiological events that culminate in interstitial fibrosi (IF) and tubular atrophy (TA), with resultant loss of graft function. AH, arteriolar hyalinosis; EMT, epithelial-to-mesenchymal transition; ROS, reactive oxygen species.

donor and recipient circulations. A summary of the pathophysiological events that culminate in the development of TA/IF is shown in **Fig. 13.1**.

Endothelial cell injury

The maintenance of endothelial integrity and function is clearly important to normal vascular performance, which in turn sustains normal organ structure and function. Maintenance of integrity requires a balance between endothelial losses and regeneration. Since the vascular endothelium forms the interface of the donated graft with the recipient circulation, it is particularly vulnerable to injury from recipient circulating cells, humoral factors, drugs and hyperlipidaemia. As a result of activation, the endothelial cells are seen to retract, causing breaks in their functional integrity and exposure of adhesion molecules such as intercellular adhesion molecule 1 (ICAM-1) and vascular cell adhesion molecule 1 (VCAM-1).[52] Antigenic molecules are also exposed, and this fact, in combination with exposure of adhesion molecules, causes chemotaxis of T lymphocytes and monocytes. There is also marked release of platelet-derived growth factor (PDGF), tumour necrosis factor-α (TNF-α), epidermal growth factor (EGF) and thromboxane A (TXA).[36,52] Damaged areas of endothelium cause platelet aggregation, which in turn releases growth factors such as PDGF and eicosanoids such as leucotriene B_4 (LTB_4) and thromboxane A_2 (TXA_2).[65]

Breaches in the endothelium allow inward migration of monocytes, macrophages and lymphocytes from the luminal surface. Smooth muscle cells migrate in the opposite direction from the media to the intimal surface to begin the process of vascular remodelling. This migration of cells is a key aspect of the early alloimmune response and this involves a complex cascade of interaction between soluble cytokines, chemokines, adhesion molecules and their receptors. For example, the chemokine CCR1 is a receptor for various proinflammatory cytokines, and its blockade has been shown to reduce renal injury in experimental models.[66,67]

Over the lifetime of the graft, dynamic changes occur within the microvasculature of the transplanted kidney. There is an early angiogenic response following transplantation with upregulation of angiogenic growth factors, such as vascular endothelial growth factor (VEGF). Over time, there is evidence of microvascular rarefaction, which will result in tissue hypoxia, and this has been demonstrated in grafts that failed with histological features of TA/IF and despite evidence of increased endothelial cell proliferation.[68,69] Therapy targeted at preservation of the microvasculature may play a role in improving long-term survival of renal allografts, although proangiogenic therapy is likely to lead to a proinflammatory state within the graft and may thus promote further alloimmune injury.[70]

Migration of monocytes and macrophages into the graft results from endothelial cell activation and injury, and stimulates the initial response of the innate immune system.

Innate immune-mediated allograft damage

Innate immunity is a highly conserved, rapid first line of defence against invading pathogens. It is now well established that the activation of adaptive responses requires direction through the innate immune system, and the signals for such activation are largely provided by dendritic cells. Many cell types are involved in the innate immune response, specifically monocytes, macrophages and dendritic cells.

Macrophages in CTD

Macrophages arise from the mononuclear phagocytic system, which consists of macrophages, blood monocytes and their precursors. They are expressed in low numbers in normal tissue and are significantly upregulated in response to local inflammation. Macrophages play a role in both innate and adaptive immunity, and so have the potential to shape both these responses. During the innate response, macrophages promote inflammation through release of cytokines such as TNF-α and interleukin-1 (IL-1), and act as effector cells partly through the production of reactive oxygen species. Macrophages contribute to adaptive immunity by presenting antigen to primed T cells and by effecting cell death under the direction of T cells. They also play a role in immunomodulation and resolution of inflammation, depending on their state of activation.

Macrophages are present throughout the lifetime of the allograft, initially in high numbers, as they accumulate in response to ischaemia–reperfusion injury within the first 24 hours. With resolution of the ischaemic insult and in the absence of rejection, the number of macrophages decreases, but they are still present in low numbers. In acute rejection, their numbers increase significantly as they comprise between 38% and 60% of cellular infiltrate. In CTD, macrophage accumulation predates and accompanies the chronic changes of interstitial fibrosis and tubular atrophy.[40,71]

There are experimental data to suggest that macrophages promote the development of CTD. In a miniature swine model, an initial episode of acute rejection is followed by chronic changes (TA/IF) in 50% of animals and spontaneous recovery in the other 50%. Persistence of macrophage infiltrate following resolution of acute rejection was predictive of chronic damage.[41] Administration of a macrophage inhibitor abrogated the development of CTD in a rat model.[42]

Human biopsy studies support these experimental findings, with association between early macrophage infiltration and subsequent development of CAN.[71] This was independent of severity of acute rejection and renal function at the time of biopsy.

The mechanisms by which macrophages contribute to chronic transplant dysfunction are not understood, but are likely to involve smooth muscle proliferation (PDGF release), cytokine-mediated inflammation (TNF-α, IL-1), cell destruction (generation of reactive oxygen species), promotion of fibrosis (TGF-β) and potentially through destruction of the microvasculature.[68]

The interface between innate and adaptive immune responses

Cells involved in innate immunity, such as macrophages and dendritic cells, are equipped with pattern recognition receptors, such as Toll-like receptors (TLRs). These receptors recognise specific structures of microorganisms or injury-induced altered host-derived structures. After recognition, TLRs initiate intracellular signal transduction pathways that result in the activation of inflammatory cytokines and the maturation of immature dendritic cells (DCs). It is the maturation of DCs that initiates the adaptive immune response.[43]

Reactive oxygen species

There is accumulating evidence that allograft injury mediated by reactive oxygen species (ROS) occurs during donor brain death. The subsequent ischaemia–reperfusion injury, leading to TLR-mediated maturation of DCs, both donor- and recipient-derived, stimulates the adaptive immune response.

Oxidative stress is an interesting candidate in mediating injury in CTD. It denotes damage to cells, tissues and organs caused by the production of ROS, such as superoxide anion (O_2^-) and hydrogen peroxide. Peroxynitrite is another potent oxidising agent generated from nitric oxide.

A possible mechanism by which ROS may contribute to CTD is through upregulation of TGF-β and resultant epithelial to mesenchymal transition of tubular epithelial cells.

There is evidence for increased generation of ROS in kidney allografts with TA/IF compared with controls, in part due to increased macrophage production of inducible nitric oxide synthase (iNOS) and other oxidase enzymes, e.g. NADPH. Other potential sources of ROS in allografts are immunosuppressive drugs such as ciclosporin, hypoxic injury and comorbid condition such as hypertension, diabetes and proteinuria.[72]

Epithelial-to-mesenchymal transition

Tubular epithelial cells comprise approximately 75% of renal parenchymal cells and are susceptible to early injury. Such injury is associated with fibrosis with the development of fibroblasts within the interstitium, and it is likely that tubular epithelial cells undergo a phenotypic change in the form of epithelial-to-mesenchymal transition (EMT). Such a mechanism occurs in embryogenesis and in tissue repair, and is a multistep process. It commences with loss of epithelial adhesiveness and cytoskeletal reorganisation, followed by disruption of the basement membrane and migration of tubular epithelial cells into the interstitium. A recent study has demonstrated that EMT is associated with SCR, with high donor serum creatinine and increasing cold ischaemia time. These findings suggest that EMT occurs as a consequence of early injury, which may be immunological or ischaemic,[73] and it may be a marker of subsequent fibrosis.[74] Manipulation of the process is an attractive therapeutic goal. However, there are many factors that regulate EMT; TGF-β is one of the most potent, capable of initiating and completing the entire process, but manipulation of TGF-β must be considered with caution as this pleomorphic cytokine plays an important role in regulatory T-cell function.[75]

Adaptive immunity in chronic transplant dysfunction

Undoubtedly alloimmune injury plays a significant role in the development of CTD, and recent developments have led to an increased understanding in the contribution of antibody-mediated rejection in this process.

Antibody-mediated rejection in the graft is defined histologically by the presence of C4d within more than 50% of peritubular capillaries (**Fig. 13.2**), with characteristic morphological changes, such as transplant glomerulopathy, defined by duplication of the glomerular capillary basement membrane (**Fig. 13.3**), and peritubular capillary basement membrane multilayering (PTCBMML), the severity of which is classified according to the number of layers. The diagnosis is confirmed by the presence of circulating donor-specific antibodies,[33] and this

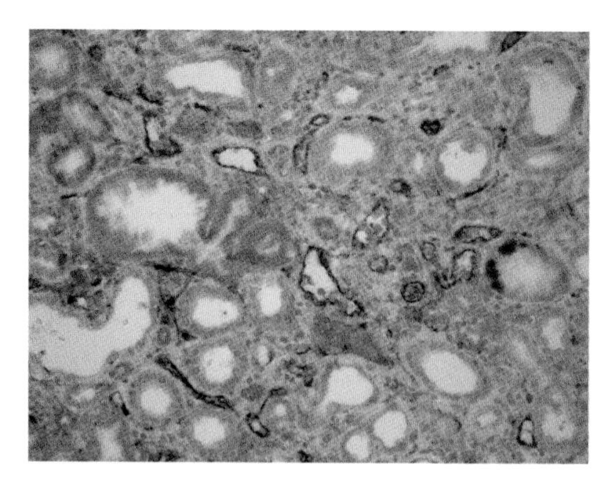

Figure 13.2 • C4d staining of peritubular capillaries (PTCs), characteristic of antibody-mediated rejection, when >50% of PTCs stain C4d positive (PTCs highlighted in brown).

combination has been described as the ABCD tetrad for diagnosis of late antibody-mediated rejection.[76] The diagnostic criteria are shown in Box 13.2.

C4d staining

The routine use of biopsy staining for C4d has led to an increased acceptance of the role of antibody-mediated damage in CTD. The published prevalence of C4d deposition in peritubular capillaries reflects whether biopsies were performed for cause or in protocol settings; in the latter, C4d positivity is seen relatively infrequently.

Box 13.2 • ABCD tetrad for diagnosis of chronic antibody-mediated rejection

1. Duplication of glomerular basement membrane (TG)
2. Multilaminated peritubular capillary basement membrane
3. Evidence for antibody action/deposition e.g. C4d in peritubular capillaries
4. Serological evidence of anti-HLA or other anti-donor antibody

Detection of C4d-positive staining in >50% of peritubular capillaries in the first 6 months following transplantation is associated with a 50% reduction in long-term survival from an average of 8 to 4 years,[77] and this was more significant in this study than the presence of circulating antibody.

This association with long-term outcome has been confirmed in other studies.[78,79]

Transplant glomerulopathy

Another entity that is interesting in the context of CTD is transplant glomerulopathy,[80] defined by characteristic duplication of glomerular basement membrane and associated with poor long-term outcome.[81]

Transplant glomerulopathy (TG) is relatively rarely encountered on biopsies in the first year after transplantation, but it is likely that the clinical presentation lags behind histological changes, with mounting evidence to support this from protocol biopsies. This suggests that the incidence of TG stands at approximately 4% at 1 year, increasing progressively and reaching 20% at 5 years.[82] Clinically, early TG presents with proteinuria, hypertension and loss of kidney function.

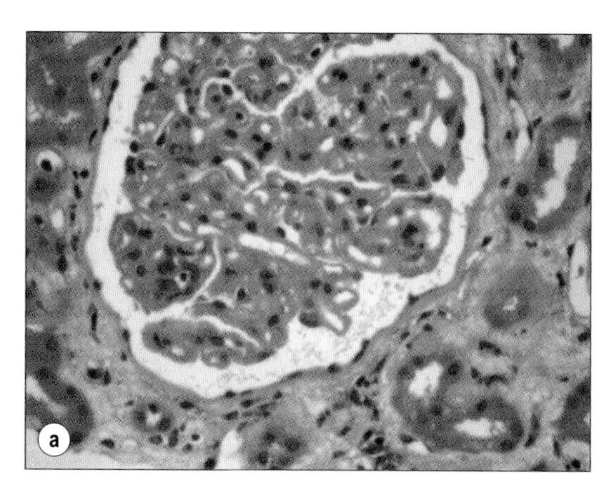

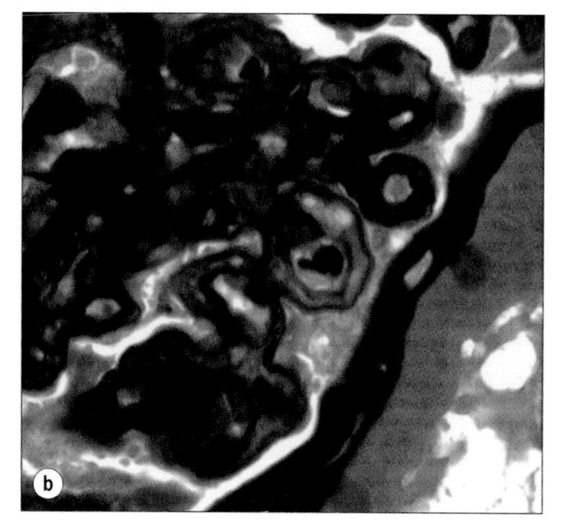

Figure 13.3 • Transplant glomerulopathy with characteristic multilayering of the glomerular basement membrane.

There is early injury to the endothelial cells, with acute cell swelling, followed by multilayering of the basement membrane. These changes occur independently of the TA/IF characterising 'chronic allograft injury, with no distinct aetiology'. The development of TG is associated with glomerular inflammatory infiltrate.

TG appears to be associated with antibody-mediated injury:

- it is significantly more common in patients with anti-HLA antibodies;[76,83]
- in a study of protocol biopsies performed at 1 year, TG was seen in 22% of patients with donor-specific HLA antibodies detected by positive crossmatch, compared with 8% with negative crossmatch;[82]
- the risk of TG is increased in patients with a history of antibody-mediated rejection.

Specifically, there is a particularly strong association between TG and anti-HLA class II antibodies. This combination has been associated with particularly poor survival. In one study, 26% of patients with pretransplant anti-HLA II antibodies developed TG, compared to 8% without these antibodies.[82]

A confounding factor in these observations is that the association between TG and C4d positivity is weaker than expected, perhaps suggesting a limitation of C4d staining. In one study, at the time of diagnosis of TG, C4d was positive in fewer than half of the cases.[83] This may be due to the existence of two distinct types of TG, C4d positive and negative, or it may be that not all antibody-mediated rejection episodes are associated with C4d positivity, and the latter finding has been supported in other studies.[82]

Thus, there is a body of evidence to support the role of chronic antibody production, both HLA and non-HLA directed, as an important contributor to chronic transplant dysfunction; however, the management of this disorder remains poorly defined. Although there are anecdotal reports of management with the anti-B-cell antibody, rituximab (anti-CD20) and plasmapheresis, the evidence supporting this strategy is not available[54] and long-term reduction of anti-donor antibodies is not feasible.

Over recent years, there have been significant changes in the understanding of chronic allograft damage, its multifactorial aetiology and the final common pathway of damage, TA/IF and graft arteriosclerosis. But whether these changes in understanding will lead to improvement in outcome from TA/IF remains to be seen.

Therapeutic strategies

Principles of management of chronic graft injury

The multiple pathophysiological causes of injury outlined in this chapter suggest that there will be no single therapeutic intervention in the treatment of chronic graft injury, but instead it is more likely that several approaches will be required to abrogate specific aetiological insults. Such therapy will need to be tailored to meet the needs of individual patients.

In general, the development of late graft failure is a variant of chronic kidney disease, and so the same standard management principles apply. These include adequate blood pressure control, adequate diabetic control and treatment of other factors such as hyperlipidaemia and anaemia. Indeed, both hypertension and diabetes are significant risk factors for the development of coronary artery disease, which is a contributor to death with a functioning graft.

The time-dependent manner in which CTD develops also leads to the observation that therapy should be ideally initiated prior to or during periods of active injury and that some treatments will be detrimental if used too late. The most obvious example of this is treatment with CNIs, the use of which should be maximal during the early post-transplant period when alloimmune injury plays the major role in long-term outcome, and then reduced over time beyond 1 year, in order to avoid toxicity.

In this chapter, the strategies that are known to alter progression of CTD will be described, followed by the novel treatment strategy aimed at abrogating the matrix deposition that leads to graft fibrosis.

General treatment strategies

Hypertension

Due to the wealth of evidence to support a role for hypertension in the development of chronic graft injury, tight control of blood pressure <130/80 mmHg is recommended. A variety of agents have been studied in transplant recipients, but no single class of agent has been demonstrated to have superior effects compared with others.

Hyperlipidaemia

Lipid-lowering agents include statins and fenofibrates, which control LDL and triglycerides. Statins have been implicated as beneficial agents because of their potential to regulate fibrogenic mechanisms and their impact on endothelial dysfunction. There is little evidence to demonstrate that treatment with statins has a direct effect on improving graft survival, but it does reduce the risk of coronary events, and treatment should be aimed at improving overall patient health and survival.

Altering immunosuppression

As discussed previously, there is little doubt that CNIs contribute to late graft injury leading to TA/IF. Thus much work has been performed focusing on CNI minimisation or withdrawal. Strategies to minimise CNI use have been studied, including CNI minimisation in the presence of chronic changes, reduction during stable renal function and de novo strategies to reduce or delay the introduction of CNIs. Other strategies have aimed to avoid CNI use, and to replace them with non-nephrotoxic agents such as MMF or sirolimus.[84,85]

Flechner et al. undertook a randomised controlled trial of sirolimus-based versus ciclosporin-based immunosuppression and demonstrated similar rates of acute rejection and morbidity. Over the 5-year follow-up period, there was longer graft survival and fewer episodes of chronic allograft dysfunction in the CNI-free group.[84]

For further discussion of immunosuppression, see Chapter 5.

A recently published review demonstrated that persistent damage was observed in biopsies as long as CNIs were continued. CNI withdrawal may be the best option for management by delivering CNIs during the early period of immunological graft injury and then converting them to less nephrotoxic agents before significant renal damage occurs.[86]

Abrogating matrix deposition: novel therapies for CTD

Irrespective of the cumulative causes, kidneys with interstitial fibrosis have an imbalance between matrix deposition and degradation. The major regulators of extracellular matrix turnover are matrix metalloproteinases (MMPs) and growth factors. MMPs are a large family of matrix-degrading enzymes and are inhibited by inhibitors of MMPs. Thus, targeting therapy to block matrix deposition is an attractive prospect, and such treatments are undergoing investigation in experimental model systems and in clinical trials. In experimental model systems, early inhibition of MMPs resulted in reduction of chronic injury, although later inhibition led to increased injury compared with controls, emphasising the necessity of an appropriate era of treatment.[87] Another example of targeting the matrix is the blockade of prolyl-4-hydroxylase, a rate-limiting step in collagen synthesis; this resulted in reduction of fibrosis and graft inflammation, and improved graft function in a murine model.[88] Retinoids have been shown to reduce rejection severity[89] and to limit fibrosis in rat models of chronic allograft damage.[58] Thus, there are a number of agents which target the matrix as a mediator of fibrotic injury following transplantation, and these are examples of exciting possible therapeutic targets for chronic transplant dysfunction. It is likely that no single treatment will be effective in overcoming this challenge, and other interesting strategies include blocking chemokine receptors, as mentioned earlier in the chapter. This would thus alter leucocyte trafficking to and from the graft, and modulate the early alloimmune response.[90,91]

Key points

- CTD remains a significant cause of late graft loss following solid-organ transplantation.
- Characteristic but non-specific histological changes are seen in different organs.
- Protocol biopsies have improved our understanding of the role of subclinical rejection in CTD and its natural history.
- The natural history of CTD in the kidney can be described in two phases:
  - early tubulointerstitial injury;
  - late phase of injury, involving glomeruli and microvasculature.
- The multifactorial aetiology of CTD reflects ongoing time-dependent changes that occur throughout the lifespan of the graft.
- There is increasing evidence for the role of antibody-mediated rejection in CTD.
- The pathophysiology of CTD involves:
  - endothelial cell activation;
  - the inflammatory response to the graft;
  - T-cell- and antibody-mediated activation of adaptive immune response;
  - non-immunological factors such as hypertension, hyperlipidaemia and CNI toxicity.
- Treatment strategies include:
  - optimal management of risk factors such as hypertension and hyperlipidaemia;
  - reduction or withdrawal of CNI drugs;
  - novel strategies such as inhibitors of matrix deposition are under trial.

References

1. Graetz KP, Rigg KM. Chronic transplant dysfunction. In: Forsythe JLR (ed.) Transplantation, 3rd edn. London: Elsevier Saunders, 2005; pp. 277–303.

2. Paul LC, Hayry P, Foegh M et al. Diagnostic criteria for chronic rejection/accelerated graft atherosclerosis in heart and kidney transplants: joint proposal from the Fourth Alexis Carrel Conference on Chronic Rejection and Accelerated Arteriosclerosis in Transplanted Organs. Transplant Proc 1993; 25(2):2022–3.

   Consensus view of diagnostic criteria for chronic rejection in heart and kidney transplants, arising from a multidisciplinary conference.

3. Freese DK, Snover DC, Sharp HL et al. Chronic rejection after liver transplantation: a study of clinical, histopathological and immunological features. Hepatology 1991; 13(5):882–91.
4. Pirsch JD, Kalayoglu M, Hafez GR et al. Evidence that the vanishing bile duct syndrome is vanishing. Transplantation 1990; 49(5):1015–18.
5. Knechtle SJ. Rejection of the liver transplant. Semin Gastrointest Dis 1998; 9(3):126–35.
6. Junge G, Tullius SG, Klitzing V et al. The influence of late acute rejection episodes on long-term graft outcome after liver transplantation. Transplant Proc 2005; 37(4):1716–17.
7. Demetris AJ, Seaberg EC, Batts KP et al. Chronic liver allograft rejection: a National Institute of Diabetes and Digestive and Kidney Diseases interinstitutional study analyzing the reliability of current criteria and proposal of an expanded definition. National Institute of Diabetes and Digestive and Kidney Diseases Liver Transplantation Database. Am J Surg Pathol 1998; 22(1):28–39.
8. Oguma S, Belle S, Starzl TE et al. A histometric analysis of chronically rejected human liver allografts: insights into the mechanisms of bile duct loss: direct immunologic and ischemic factors. Hepatology 1989; 9(2):204–9.
9. Reichenspurner H, Girgis RE, Robbins RC et al. Stanford experience with obliterative bronchiolitis after lung and heart–lung transplantation. Ann Thorac Surg 1996; 62(5):1467–1472; discussion 1472–3.
10. Neuringer IP, Chalermskulrat W, Aris R. Obliterative bronchiolitis or chronic lung allograft rejection: a basic science review. J Heart Lung Transplant 2005; 24(1):3–19.
11. Tazelaar HD, Yousem SA. The pathology of combined heart–lung transplantation: an autopsy study. Hum Pathol 1988; 19(12):1403–16.
12. Cooper JD, Billingham M, Egan T et al. A working formulation for the standardization of nomenclature and for clinical staging of chronic dysfunction in lung allografts. International Society for Heart and Lung Transplantation. J Heart Lung Transplant 1993; 12(5):713–16.

    These are the published working formulations for staging and standardisation of chronic dysfunction in lung transplants.

13. Sutherland DE, Gruessner A. Long-term function (> 5 years) of pancreas grafts from the International Pancreas Transplant Registry database. Transplant Proc 1995; 27(6):2977–80.

14. Allen RDM. Pancreas transplantation. In: Forsythe JLR (ed.) Transplantation surgery, 1st edn. London: WB Saunders, 1997; pp. 167–201.

15. Cosio FG, Grande JP, Wadei H et al. Predicting subsequent decline in kidney allograft function from early surveillance biopsies. Am J Transplant 2005; 5(10):2464–72.

16. Rush DN, Nickerson P, Jeffery JR et al. Protocol biopsies in renal transplantation: research tool or clinically useful? Curr Opin Nephrol Hypertens 1998; 7(6):691–4.

17. Shishido S, Asanuma H, Nakai H et al. The impact of repeated subclinical acute rejection on the progression of chronic allograft nephropathy. J Am Soc Nephrol 2003; 14(4):1046–52.

18. Nankivell BJ, Borrows RJ, Fung CL et al. Natural history, risk factors, and impact of subclinical rejection in kidney transplantation. Transplantation 2004; 78(2):242–9.

19. Nankivell BJ, Borrows RJ, Fung CL et al. The natural history of chronic allograft nephropathy. N Engl J Med 2003; 349(24):2326–33.

This study of 961 renal protocol biopsies performed on 120 kidney–pancreas recipients over a period of 10 years yields important and interesting data on the natural history of chronic allograft dysfunction, and highlights two phases of injury, with different aetiological factors contributing to them.

20. Seron D, Moreso F. Protocol biopsies and risk factors associated with chronic allograft nephropathy. Transplant Proc 2002; 34(1):331–2.

21. Racusen LC, Solez K, Colvin RB et al. The Banff 97 working classification of renal allograft pathology. Kidney Int 1999; 55(2):713–23.

22. Helantera I, Ortiz F, Helin H et al. Timing and value of protocol biopsies in well-matched kidney transplant recipients – a clinical and histopathologic analysis. Transplant Int 2007; 20(11):982–90.

23. Yehia M, Matheson PJ, Merrilees MJ et al. Predictors of chronic allograft nephropathy from protocol biopsies using histological and immunohistochemical techniques. Nephrology (Carlton) 2006; 11(3):261–6.

24. Takeda A, Morozumi K, Yoshida A et al. Studies of cyclosporine-associated arteriolopathy in renal transplantation: does the long-term outcome of renal allografts depend on chronic cyclosporine nephrotoxicity? Transplant Proc 1994; 26(2):925–8.

25. Hirsch HH, Brennan DC, Drachenberg CB et al. Polyomavirus-associated nephropathy in renal transplantation: interdisciplinary analyses and recommendations. Transplantation 2005; 79(10):1277–86.

26. Seron D, Moreso F, Bover J et al. Early protocol renal allograft biopsies and graft outcome. Kidney Int 1997; 51(1):310–16.

27. Dimeny E, Wahlberg J, Larsson E et al. Can histopathological findings in early renal allograft biopsies identify patients at risk for chronic vascular rejection? Clin Transplant 1995; 9(2):79–84.

28. Schwarz A, Gwinner W, Hiss M et al. Safety and adequacy of renal transplant protocol biopsies. Am J Transplant 2005; 5(8):1992–6.

29. Wilkinson A. Protocol transplant biopsies: are they really needed? Clin J Am Soc Nephrol 2006; 1(1):130–7.

30. Solez K, Colvin RB, Racusen LC et al. Banff 07 classification of renal allograft pathology: updates and future directions. Am J Transplant 2008; 8(4):753–60.

31. Hoffmann SC, Hale DA, Kleiner DE et al. Functionally significant renal allograft rejection is defined by transcriptional criteria. Am J Transplant 2005; 5(3):573–81.

32. Nankivell BJ, Chapman JR. Chronic allograft nephropathy: current concepts and future directions. Transplantation 2006; 81(5):643–54.

33. Solez K, Colvin RB, Racusen LC et al. Banff 05 Meeting Report: differential diagnosis of chronic allograft injury and elimination of chronic allograft nephropathy ('CAN'). Am J Transplant 2007; 7(3):518–26.

This presents changes in diagnostic categories for chronic graft dysfunction and in the accepted nomenclature for describing interstitial fibrosis and tubular atrophy following kidney transplantation, reflecting an increasing understanding of the complex aetiology contributing to CTD.

34. Halloran PF. Call for revolution: a new approach to describing allograft deterioration. Am J Transplant 2002; 2(3):195–200.

35. Gourishankar S, Halloran PF. Late deterioration of organ transplants: a problem in injury and homeostasis. Curr Opin Immunol 2002; 14:576–83.

36. Alexander JW, Bennett LE, Breen TJ. Effect of donor age on outcome of kidney transplantation: a two-year analysis of transplants reported to the United Network of Organs Sharing Registry. Transplantation 1994; 57:871–6.

37. Pratschke J, Wilhelm M, Kusaka M et al. Brain death and its influence on donor organ quality and outcome after transplantation. Transplantation 1999; 67:343–8.

38. Ojo AO, Wolfe RA, Held PJ et al. Delayed graft function: risk factors and implications for renal allograft survival. Transplantation 1997; 63(7):968–74.

39. Kahan BD. Toward a rational design of clinical trials of immunosuppressive agents in transplantation. Immunol Rev 1993; 136:29–49.

40. Croker BP, Clapp WL, Abu Shamat AR et al. Macrophages and chronic renal allograft nephropathy. Kidney Int 1996; 57(Suppl):S42–9.
41. Shimizu A, Yamada K, Sachs DH et al. Mechanisms of chronic renal allograft rejection. II. Progressive allograft glomerulopathy in miniature swine. Lab Invest 2002; 82(6):673–86.
42. Azuma H, Nadeau KC, Ishibashi M et al. Prevention of functional, structural, and molecular changes of chronic rejection of rat renal allografts by a specific macrophage inhibitor. Transplantation 1995; 60(12):1577–82.
43. Land WG. The role of postischemic reperfusion injury and other nonantigen-dependent inflammatory pathways in transplantation. Transplantation 2005; 79(5):505–14.
44. Leggat JE Jr, Ojo AO, Leichtman AB et al. Long-term renal allograft survival: prognostic implication of the timing of acute rejection episodes. Transplantation 1997; 63(9):1268–72.

This study of 31 600 transplant recipients investigated the impact of acute rejection episodes on subsequent loss of graft function, and found a correlation between severity of rejection episodes and late graft loss, but more significant was the effect of timing of acute rejection episodes on outcome.

45. Rush DN, Karpinski ME, Nickerson P et al. Does subclinical rejection contribute to chronic rejection in renal transplant patients? Clin Transplant 1999; 13(6):441–6.
46. Kasiske BL, Gaston RS, Gourishankar S et al. Long-term deterioration of kidney allograft function. Am J Transplant 2005; 5(6):1405–14.
47. Opelz G, Wujciak T, Dohler B et al. HLA compatibility and organ transplant survival. Collaborative Transplant Study. Rev Immunogenet 1999; 1(3):334–42.
48. McKenna RM, Takemoto SK, Terasaki PI. Anti-HLA antibodies after solid organ transplantation. Transplantation 2000; 69(3):319–26.
49. Mao Q, Terasaki PI, Cai J et al. Extremely high association between appearance of HLA antibodies and failure of kidney grafts in a five-year longitudinal study. Am J Transplant 2007; 7(4):864–71.

This important and recent study emphasises the importance of antibody expression in graft failure.

50. Campos EF, Tedesco-Silva H, Machado PG et al. Post-transplant anti-HLA class II antibodies as risk factor for late kidney allograft failure. Am J Transplant 2006; 6(10):2316–20.
51. Zou Y, Stastny P, Susal C et al. Antibodies against MICA antigens and kidney-transplant rejection. N Engl J Med 2007; 357(13):1293–300.
52. Cheigh JS, Haschemeyer RH, Wang JC et al. Hypertension in kidney transplant recipients. Effect on long-term renal allograft survival. Am J Hypertens 1989; 2(5, Pt 1):341–8.
53. Arnadottir M, Berg AL. Treatment of hyperlipidemia in renal transplant recipients. Transplantation 1997; 63(3):339–45.
54. Jevnikar AM, Mannon RB. Late kidney allograft loss: what we know about it, and what we can do about it. Clin J Am Soc Nephrol 2008; 3(Suppl 2):S56–67.
55. Lisik W, Schoenberg L, Lasky RE et al. Statins benefit outcomes of renal transplant recipients on a sirolimus–cyclosporine regimen. Transplant Proc 2007; 39(10):3086–92.
56. Fellstrom B, Holdaas H, Jardine AG et al. Effect of fluvastatin on renal end points in the Assessment of Lescol in Renal Transplant (ALERT) trial. Kidney Int 2004; 66(4):1549–55.
57. Tolkoff-Rubin N, Rubin RH. The impact of cytomegalovirus infection on graft function and patient outcome. Graft 1999; 2(2):S101–3.
58. Adams J, Kiss E, Arroyo AB et al. 13-Cis-retinoic acid inhibits development and progression of chronic allograft nephropathy. Am J Pathol 2005; 167(1):285–98.
59. Humar A, Gillingham KJ, Payne WD et al. Association between cytomegalovirus disease and chronic rejection in kidney transplant recipients. Transplantation 1999; 68(12):1879–83.
60. Everett JP, Hershberger RE, Norman DJ et al. Prolonged cytomegalovirus infection with viremia is associated with development of cardiac allograft vasculopathy. J Heart Lung Transplant 1992; 11(3, Pt 2):S133–7.
61. Randhawa PS, Finkelstein S, Scantlebury V et al. Human polyoma virus-associated interstitial nephritis in the allograft kidney. Transplantation 1999; 67(1):103–9.
62. Medipalli R, Vasudev BZ, Saad E et al. Improved outcomes of BKVN: impact on BK virus surveillance protocol (abstract). Am J Transplant 2007; 7(Suppl 2):150.
63. Dall A, Hariharan S. BK virus nephritis after renal transplantation. Clin J Am Soc Nephrol 2008; 3(Suppl 2):S68–75.
64. Naesens M, Lerut E, Damme BV et al. Tacrolimus exposure and evolution of renal allograft histology in the first year after transplantation. Am J Transplant 2007; 7(9):2114–23.
65. Tullius SG, Tilney NL. Both alloantigen-dependent and -independent factors influence chronic allograft rejection. Transplantation 1995; 59(3):313–18.
66. Gao W, Topham PS, King JA et al. Targeting of the chemokine receptor CCR1 suppresses development of acute and chronic cardiac allograft rejection. J Clin Invest 2000; 105(1):35–44.
67. Anders HJ, Vielhauer V, Frink M et al. A chemokine receptor CCR-1 antagonist reduces renal fibrosis after unilateral ureter ligation. J Clin Invest 2002; 109(2):251–9.
68. Adair A, Mitchell DR, Kipari T et al. Peritubular capillary rarefaction and lymphangiogenesis in

chronic allograft failure. Transplantation 2007; 83(12):1542–50.

69. Ishii Y, Sawada T, Kubota K et al. Injury and progressive loss of peritubular capillaries in the development of chronic allograft nephropathy. Kidney Int 2005; 67(1):321–32.

70. Reinders ME, Rabelink TJ, Briscoe DM. Angiogenesis and endothelial cell repair in renal disease and allograft rejection. J Am Soc Nephrol 2006; 17(4):932–42.

71. Pilmore HL, Painter DM, Bishop GA et al. Early up-regulation of macrophages and myofibroblasts: a new marker for development of chronic renal allograft rejection. Transplantation 2000; 69(12):2658–62.

72. Djamali A. Oxidative stress as a common pathway to chronic tubulointerstitial injury in kidney allografts. Am J Physiol Renal Physiol 2007; 293(2):F445–55.

The potential contribution of oxidative stress to allograft injury and development of interstitial fibrosis and tubular atrophy is addressed in this paper.

73. Hertig A, Verine J, Mougenot B et al. Risk factors for early epithelial to mesenchymal transition in renal grafts. Am J Transplant 2006; 6(12):2937–2946.

74. Hertig A, Anglicheau D, Verine J et al. Early epithelial phenotypic changes predict graft fibrosis. J Am Soc Nephrol 2008 (Epub ahead of print).

75. Csencsits K, Wood SC, Lu G et al. Transforming growth factor-beta1 gene transfer is associated with the development of regulatory cells. Am J Transplant 2005; 5(10):2378–2384.

76. Sis B, Campbell PM, Mueller T et al. Transplant glomerulopathy, late antibody-mediated rejection and the ABCD tetrad in kidney allograft biopsies for cause. Am J Transplant 2007; 7(7):1743–52.

77. Lederer SR, Kluth-Pepper B, Schneeberger H et al. Impact of humoral alloreactivity early after transplantation on the long-term survival of renal allografts. Kidney Int 2001; 59(1):334–41.

In this study of 218 kidney transplants, the presence of C4d staining was associated with circulating antibody levels and, when present in the first 6 months after transplantation, was found to be a highly significant predictor of subsequent graft loss.

78. David-Neto E, Prado E, Beutel A et al. C4d-positive chronic rejection: a frequent entity with a poor outcome. Transplantation 2007; 84(11):1391–8.

79. Ranjan P, Nada R, Jha V et al. The role of C4d immunostaining in the evaluation of the causes of renal allograft dysfunction. Nephrol Dial Transplant 2008; 23(5):1735–41.

80. Zollinger HU, Moppert J, Thiel G et al. Morphology and pathogenesis of glomerulopathy in cadaver kidney allografts treated with antilymphocyte globulin. Curr Top Pathol 1973; 57:1–48.

81. Cosio FG, Gloor JM, Sethi S et al. Transplant glomerulopathy. Am J Transplant 2008; 8(3):492–6.

82. Gloor JM, Sethi S, Stegall MD et al. Transplant glomerulopathy: subclinical incidence and association with alloantibody. Am J Transplant 2007; 7(9):2124–32.

This study of 582 renal transplant recipients investigated the impact of subclinical and clinical transplant glomerulopathy on outcome; it demonstrated that subclinical TG is as likely to be associated with poor outcome as clinical TG, and that it is a progressive condition, associated with poor outcome.

83. Colvin RB. Antibody-mediated renal allograft rejection: diagnosis and pathogenesis. J Am Soc Nephrol 2007; 18(4):1046–56.

84. Flechner SM, Goldfarb D, Solez K et al. Kidney transplantation with sirolimus and mycophenolate mofetil-based immunosuppression: 5-year results of a randomized prospective trial compared to calcineurin inhibitor drugs. Transplantation 2007; 83(7):883–92.

This prospective randomised controlled trial supports the use of a CNI-free immunosuppressive regime, with comparable rates of acute rejection in the sirolimus group, but fewer episodes of TA/IF.

85. Mulay AV, Hussain N, Fergusson D et al. Calcineurin inhibitor withdrawal from sirolimus-based therapy in kidney transplantation: a systematic review of randomized trials. Am J Transplant 2005; 5(7):1748–56.

86. Flechner SM, Kobashigawa J, Klintmalm G. Calcineurin inhibitor-sparing regimens in solid organ transplantation: focus on improving renal function and nephrotoxicity. Clin Transplant 2008; 22(1):1–15.

A recent review of the literature has demonstrated the superiority of CNI withdrawal rather than minimisation.

87. Lutz J, Yao Y, Song E et al. Inhibition of matrix metalloproteinases during chronic allograft nephropathy in rats. Transplantation 2005; 79(6):655–61.

88. Franceschini N, Cheng O, Zhang X et al. Inhibition of prolyl-4-hydroxylase ameliorates chronic rejection of mouse kidney allografts. Am J Transplant 2003; 3(4):396–402.

89. Kiss E, Adams J, Grone HJ et al. Isotretinoin ameliorates renal damage in experimental acute renal allograft rejection. Transplantation 2003; 76(3):480–9.

90. Bedke J, Kiss E, Schaefer L et al. Beneficial effects of CCR1 blockade on the progression of chronic renal allograft damage. Am J Transplant 2007; 7(3):527–37.

91. Mayer V, Hudkins KL, Heller F et al. Expression of the chemokine receptor CCR1 in human renal allografts. Nephrol Dial Transplant 2007; 22(6):1720–9.

Index

Note: Page numbers in *italics* refer to figures and page numbers in **bold** refer to tables.

A

B

C

E

F

G

H

J

K

L

M

N

O

P

Q

R

S

V

W

X

Z